Special Publications 68

35 Seasons of U.S. Antarctic Meteorites (1976–2010)

A Pictorial Guide to the Collection

Kevin Righter
Catherine M. Corrigan
Timothy J. McCoy
Ralph P. Harvey
Editors

This work is a copublication between the American Geophysical Union and John Wiley & Sons, Inc.

WILEY

This work is a copublication between the American Geophysical Union and John Wiley & Sons, Inc.

Published under the aegis of the AGU Publications Committee.

Brooks Hanson, Director of Publications
Robert van der Hilst, Chair, Publications Committee
Richard Blakely, Vice Chair, Publications Committee

Published by John Wiley & Sons, Inc., Hoboken, New Jersey
Published simultaneously in Canada

For general information on our other products and services or for technical support, please contact our Customer Care Department within the United States at (800) 762-2974, outside the United States at (317) 572-3993 or fax (317) 572-4002.

Wiley also publishes its books in a variety of electronic formats. Some content that appears in print may not be available in electronic formats. For more information about Wiley products, visit our web site at www.wiley.com.

Library of Congress Cataloging-in-Publication Data is available.

ISBN: 978-1-118-79832-4

Cover image: A meteorite on the blue ice of the Miller Range, Antarctica, from the 2011-2012 field season. (Antarctic Search for Meteorites Program / Anne Peslier, NASA Johnson Space Center)

Printed in Singapore

10 9 8 7 6 5 4 3 2 1

CONTENTS

PREFACE

This book was inspired by a great many mysteries and many great people.

All four of us (the editors) are scientists who use meteorites as tools to explore the history of our solar system. At the same time, our roles in the U.S. Antarctic Meteorite program mean we share the goal of preserving and providing these samples to the planetary sciences community so that OTHER scientists can make great discoveries. We are called on to consistently step aside and let others lead the way when (in the field or in the curatorial lab) something fundamentally new or unique comes into view. Fortunately, our understanding of the importance and altruistic nature of this program helps us serve the common good rather than ourselves.

The inherent altruism of the U.S. Antarctic Meteorite program is one of several important legacies of William (Bill) Cassidy, to whom this book is dedicated. In the very earliest days of the program's history, Bill showed astounding foresight, making some counter-intuitive decisions that still seem astonishing today. Immediately following Bill's first successes in the field, dozens of museums and institutions began readying themselves to aggressively seek their own share of the Antarctic meteorite bonanza. Uncompromising competition between these institutions for the extremely limited resources of the U.S. Antarctic Program would almost certainly have led to a relatively modest Antarctic meteorite collection divided among many institutions, available only with difficulty and collected and curated under a wide variety of conditions and protocols.

With his single season of fieldwork giving him only modest advantages, Bill saw what was coming. He could have chosen to join the race and fought to preserve his singular place in U.S. meteoritics. Or he could have simply gathered his specimens, taken them home to the University of Pittsburgh, and worked on them for years. But recognizing that the scientific impact of a unified Antarctic meteorite collection could be exponentially larger than what one, or even dozens of institutions, could do on their own, Bill found a way to create an entirely new science support mechanism. He allied with the two most powerful planetary materials institutions of the time (NASA and the Smithsonian), giving up his privileged position in the program if they would do the same. At the Smithsonian, Brian Mason took up the challenge

of classifying meteorites on an unprecedented scale, undertaking nearly 10,000 classifications and continuing his efforts well into retirement. Roy Clarke Jr. also played a key role in administering the program and in detailed characterization of the small but important subset of iron meteorites. At NASA's Johnson Space Center, Don Bogard (and later Marilyn Lindstrom) served as curators of the newly arrived samples, ensuring adequate facilities for storage and processing of the meteorites and applying lessons learned from Apollo to optimize access to the samples by the scientific community. Together they created the U.S. Antarctic meteorite collection, a continuous sample return mission still serving science today.

And here we are, more than 20,000 meteorites and almost forty field seasons later. Our predecessors created a program so valuable and so stable that it has become inter-generational; we (the editors) represent a later generation of curators and field team leaders brought up from within the U.S. Antarctic meteorite program. Yet that sense of altruism so firmly established at the beginning remains. Curators and field party members still have no more rights to the samples than anyone else; field teams are made up primarily of volunteers wishing to serve their science, and samples are allocated by a panel of peers. Most importantly, we are still committed to encouraging and expediting the scientific utilization of the Antarctic meteorite samples.

That is our broad goal for this book: to encourage further use of these extraordinary samples. Countless times over the years, particularly during sample allocations, we've recognized that a specific sample or a tidbit of curatorial insight promised potential rewards for someone's research, if only they knew. The collection is now so large that key samples (both old and new) can be easily overlooked, lost among the hundreds of new sample descriptions published yearly. Following the example set by Cassidy and others, we conceived of this book as a way to add context to the U.S. Antarctic meteorite program and illuminate key collected samples, helping the collection serve the planetary science community to its fullest. Ursula Marvin's chapter covers the early history of the program, while chapters by Harvey et al. and Righter et al. describe current field and curatorial practices. Chapters by Weisberg and Righter and Mittlefehldt and McCoy explore the nebular and planetesimal history of our solar system through key specimens, while chapters by Korotev and Zeigler and McSween et al. explore the significance of key samples from larger bodies (the Moon and Mars). A chapter by McCoy catalogs specific samples we feel hold unrealized promise for research, while chapters by Herzog et al. and Corrigan et al. look at the collection more broadly across

statistical, geographical, and temporal scales. Finally, we'd like to acknowledge the beneficial reviews provided by W. Cassidy, D. Sears, D. Bogard, N. Chabot, D. Mittlefehldt, R. Score, L. Folco, C. Alexander, J.-A. Barrat, B. Cohen, P. Warren, T. Usui, A. Peslier, M. Weisberg, R. Wieler, C. Smith, G. Benedix, and several anonymous reviewers.

Kevin Righter
National Aeronautics and Space Administration

Cari Corrigan
Smithsonian Museum of Natural History

Tim McCoy
Smithsonian Museum of Natural History

Ralph Harvey
Case Western Reserve University

CONTRIBUTORS

Marc W. Caffee
Department of Physics
Purdue University
West Lafayette, IN

Catherine M. Corrigan
Department of Mineral Sciences
National Museum of Natural History
Smithsonian Institution
Washington, DC

Ralph P. Harvey
Department of Earth, Environmental,
and Planetary Sciences
Case Western University
Cleveland, OH

Gregory F. Herzog
Department of Chemistry and Chemical
Biology
Rutgers University
Piscataway, NJ

A. J. Tim Jull
Arizona Accelerator Mass Spectrometry
Laboratory
Department of Physics
University of Arizona
Tucson, AZ

Jim Karner
Department of Earth, Environmental,
and Planetary Sciences
Case Western University
Cleveland, OH

Randy L. Korotev
Department of Earth and Planetary
Sciences
McDonnell Center for the Space Sciences
Washington University
Saint Louis, MO

Ursula B. Marvin
Harvard–Smithsonian Center
for Astrophysics
HCO/SAO
Cambridge, MA

Kathleen M. McBride
Jacobs Technology
NASA Johnson Space Center
Houston, TX

Timothy J. McCoy
Department of Mineral Sciences
National Museum of Natural History
Smithsonian Institution
Washington, DC

Harry Y. McSween Jr.
Department of Earth and Planetary Sciences
University of Tennessee
Knoxville, TN

David W. Mittlefehldt
Astromaterials Research Office
NASA Johnson Space Center
Mailcode KR
Houston, TX

Kevin Righter
Mailcode KT
NASA Johnson Space Center
Houston, TX

Cecilia E. Satterwhite
Jacobs Technology
NASA Johnson Space Center
Houston, TX

John Schutt
Ferndale, WA

Michael K. Weisberg
Department of Physical Sciences
Kingsborough Community College
City University of New York
Brooklyn, NY
and
Department of Earth and Environmental
Sciences
The Graduate Center
City University of New York
365 Fifth Avenue
New York, NY
and

Department of Earth Planetary Sciences
American Museum of Natural History
New York, NY

Linda C. Welzenbach
Department of Mineral Sciences
National Museum of Natural History
Smithsonian Institution
Washington, DC

Ryan A. Zeigler
Astromaterials Research and Exploration Science
Directorate, Acquisition and Curation
NASA Johnson Space Center
Mailcode KT
Houston, TX

1

The Origin and Early History of the U.S. Antarctic Search for Meteorites Program (ANSMET)

Ursula B. Marvin

The information that would first lead U.S. teams to search for meteorites in Antarctica was presented at an evening session of the Meteoritical Society on 27 August 1973 in Davos, Switzerland. On that occasion, Dr. Makoto Shima of the Institute of Physical and Chemical Research of Japan described four meteorite fragments with differing mineralogical and chemical compositions that had been collected in 1969 from a downhill sloping patch of bare ice in the Yamato Mountains of eastern Antarctica.

In the audience sat William A. Cassidy, of the University of Pittsburgh. Bill Cassidy wrote later that, on hearing that report a comic-strip lightbulb appeared in his mind with a message reading: "Meteorites are *concentrated* on the ice!" To him, this was a new and electrifying idea. Cassidy expected the whole room to be excited, but looking around he found the audience looking as comatose and glassy-eyed as audiences sometimes do. I was chairing the session that evening, but I was much too preoccupied with keeping the speakers more or less on schedule to be having any eureka experiences.

After the session, Cassidy talked with Dr. Shima and his wife, Dr. Masako Shima, both of whom are chemists who were then visiting the Max-Planck-Institut für Chemie in Mainz. Dr. Shima explained to Cassidy that the team of glaciologists in the Yamato Mountains had collected five more meteorites from the same patch of ice. Of the nine meteorites, only the four they had reported on had been analyzed for their chemical compositions and rare gas contents. These had been identified as (a) an enstatite chondrite, (b) a Ca-poor achondrite, (c) a probable carbonaceous chondrite, and (d) an olivine-bronzite chondrite. The remaining five also clearly were meteorites of differing types. Earlier that summer the

Harvard–Smithsonian Center for Astrophysics, Cambridge, MA

Shimas had coauthored an article about the four analyzed meteorites with Dr. Heinrich Hintenberger of Mainz, in *Earth and Planetary Science Letters* [*Shima et al.*, 1973], and the Shimas also had published a brief summary of their chemical results in the abstract volume of the meeting at Davos [*Shima and Shima*, 1973]. But Cassidy had not seen the article and had skimmed too quickly through the abstracts.

At the meeting, Cassidy was captivated by the evidence that meteorites from different falls sometimes are concentrated by the dynamics of ice motion. Within the hour, he began planning a proposal to the National Science Foundation's Division of Polar Programs to lead an expedition to search for meteorite concentrations on patches of ice in Antarctica. He assumed that the concentration in the Yamato Mountains could not be unique in a huge continent making up 9% of the Earth's land surface, so he would propose to work out of McMurdo Station, the U.S. base that lies near the opposite edge of Antarctica from the Yamato Mountains (Figure 1.1).

1.1. HISTORY OF METEORITE FINDS IN ANTARCTICA

Cassidy was well aware of the historical record of random meteorite finds in Antarctica, in which only four meteorites had been encountered since 1912. In that year, Douglas Mawson led the Australian-Antarctic Expedition on a five-year study of the Adelie Land coast. Mawson's three-man party discovered a stony meteorite lying on hard snow on their fourth day after breaking camp. That stone, Adelie Land, was the only known Antarctic meteorite for the next 50 years. Then, soon after the International Geophysical Year (July 1957–December 1958) had generated a widespread interest in Antarctica,

35 Seasons of U.S. Antarctic Meteorites (1976–2010): A Pictorial Guide to the Collection, Special Publication 68,
First Edition. Edited by Kevin Righter, Catherine M. Corrigan, Timothy J. McCoy and Ralph P. Harvey.

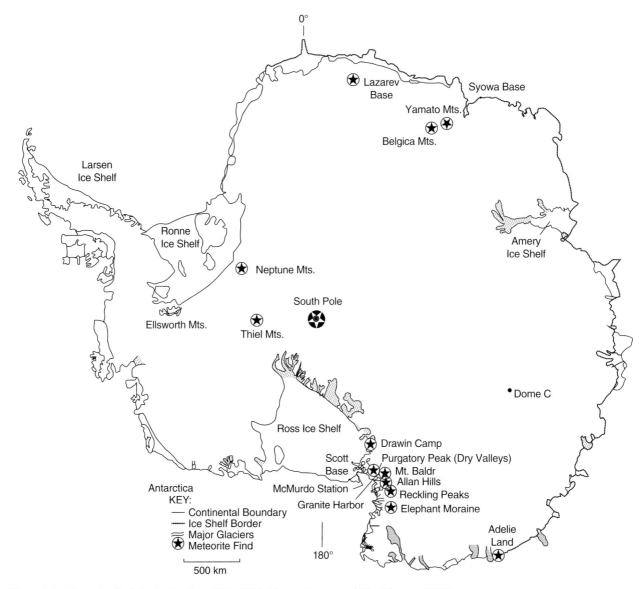

Figure 1.1. Meteorite finds in Antarctica: 1911–2012. From *Marvin and MacPherson* [1989].

three more meteorites had been found during geological surveys. In 1961 Russian geologists discovered two fragments of Lazarev, an iron meteorite lying on a rocky spur of the Humboldt Mountains. Later in that same year, geologists of the U.S. Geological Survey picked up two pieces of Thiel Mountains, a pallasite, lying on hard irregularly surfaced glacier ice, where it was associated with debris of a moraine. Three years later, in 1964, another team of U.S. Survey geologists found Neptune Mountains, an iron on a mountainside in the Pensacola Range. Of these three finds, only Thiel Mountains lay on ice. Cassidy realized that this sparse record of four meteorite finds in the past 63 years could not be read as particularly encouraging, but to him it emphasized the importance of the new discovery reported by the Shimas

of nine different kinds of meteorites on a patch of ice in the Yamato Mountains. He took it as positive proof that a concentration mechanism, unknown on bare ground, can be effective on ice. And he never doubted that additional concentrations of meteorites would be discovered in Antarctica once serious searches for them began.

1.2. CASSIDY'S FIRST PROPOSAL TO THE NSF FOR METEORITE SEARCHES

Before the deadline of 1 June 1974, Cassidy submitted his first proposal to the National Science Foundation, which supported all projects of U.S. scientists in Antarctica. He even included in it the possibility of finding lunar meteorites there [*Cassidy*, 1998, personal communication]:

Many scientists agree that lunar material can be transferred to Earth as a result of large impacts on the Moon. Such material has not been recognized yet but would be concentrated in the same manner as, and together with, concentrations of meteorites. Lunar material, therefore, might also be recovered as a result of this proposed work.

He had high hopes that the referees would share his excitement. But in due time, his proposal was rejected. The reasons for proposal denials are always kept secret to protect the referees. However, we can hazard a guess or two. First, most referees would be experienced members of Antarctic teams familiar with the difficulties of living and working, between storms, in the snow and ice. To them, the idea of looking for meteorites there might appear ludicrous; in fact, the word *ludicrous* surfaced at one time, rightly or wrongly, having been leaked out by a referee. Next, the question might be raised as to why we should collect meteorites at all. By training and experience, most geologists are not interested in loose rocks, which cannot yield direct clues to the formations from which they are derived. Meteorites are the ultimate loose rocks which, it then seemed, never could be matched to their original sources. So meteorites might be of interest to collectors and dealers but not to earth scientists. I was made acutely aware of this attitude as late as 1978 when I was preparing for my first trip to Antarctica on Cassidy's team. More than once during visits to universities, I was asked by professors and students: "Why do you want to collect meteorites?" and "What would you want to do a thing like that for?" Faced with questions like these, I wondered how so many people could seem to be so unaware that we were living in the Space Age.

The Space Age had dawned two decades earlier on 4 October 1957, when the Russians launched Sputnik I and sent it beeping around the world. In the following year, President Eisenhower announced the establishment of NASA on 29 July 1958. And on 25 May 1961, President John F. Kennedy declared that Americans would fly to the Moon and return safely back to Earth within that decade. So, by the time Cassidy submitted his proposal to the NSF in 1973, six *Apollo* missions already had returned from the Moon with samples of its crust and soils. There was great public interest in space flight and the Moon but not in meteorites, which may have seemed more than ever like orphaned rocks from space.

Perhaps we may stop here to ask why Bill Cassidy was so interested in meteorites. Cassidy had attended the Institute for Meteoritics in Albuquerque, where he served as a research assistant to the director, Dr. Lincoln LaPaz, who took all his students to see Meteor Crater and its irons. Cassidy won the first Fulbright Fellowship in meteoritics and spent a year investigating tektites, meteorites, and craters in Australia and Thailand. On his way home, he arranged through Theodore Monod, Director of l'Institut Francais d'Afrique Noire in Dakar, to visit the

newly discovered Aouelloul crater in Mauritania. Later on, he examined the Campo del Cielo craters in the Argentine chaco and a crater in the deserts of Chile. For a time, in the 1950s, Cassidy must have held a world's record for the number of meteorite impact craters he had seen. His PhD thesis, at the Pennsylvania State University, dealt with the high-temperature chemistry of meteorite and tektite systems.

1.2.1. A Link with Japan

On arriving home to Pittsburgh from Davos in the fall of 1973, Cassidy met with Professor Takesi Nagata, who was paying one of his regular visits to the University of Pittsburgh. He had been a visiting professor there since 1961, and he typically came to the department once or twice a year to collaborate on research with two members of the regular faculty. In his own book, *Cassidy* [2003, p. 19] describes the high honors accorded to Dr. Nagata internationally. He was one of the few non-U.S. members of the National Academy of Sciences, and in Japan the emperor had designated him as a National Living Intellectual Treasure. Furthermore, a few weeks later, on 29 September 1973, the National Institute of Polar Research was opening in Tokyo with Takesi Nagata as director general.

Cassidy assumed Dr. Nagata would know all about the meteorite concentrations reported by Dr. Shima at Davos. But Dr. Nagata pushed his chair away from his desk, seeming to be quite taken aback to learn, in this way, of meteorite concentrations that Japanese scientists had collected four years earlier! A short description of them had been published by the team leader, Dr. Masao Yoshida, in 1971, and as we noted above, the Shimas had published their results with Dr. Heinrich Hintenberger at Mainz early in the summer of 1973. But neither Nagata nor Cassidy had read these papers. Nagata immediately sent telegrams to Japan asking for details, and he encouraged the current field party, the 14th Japanese Antarctic Research Expedition (JARE-14), in the Yamato Mountains, to collect more meteorites. During the remainder of that season, 1973–1974, the Japanese team of glaciologists collected 12 more meteorites from the same patch of ice in the Yamato Mountains. Cassidy added that news to his proposal and resubmitted it to the NSF, which responded with the same negative decision as before.

1.3. JAPANESE INTEREST IN ANTARCTIC METEORITES

In Japan, an interest in Antarctic meteorites had arisen rather casually, almost as a joke. The story begins with a short letter written in 1970 by Professor Masao Gorai, of the Tokyo University of Education, who kept a meteorite collection at the University. His letter appeared under the title "Meteorite Museum" in *Magma,* a small newsletter on petrology. Here he used the term *meteorite museum* to refer to a naturally occurring collection of differing species of meteorites.

Dr. Gorai wrote that Dr. Yoshida of Hokkaido University had greeted him in his laboratory one day in the fall of 1968 just before Yoshida's departure for the 10th Japanese expedition to Antarctica:

> I sent him off saying that the rock samples around the Syowa base do not continue to interest me because I have plenty of them. This time please get me a meteorite, or some *getemonos* (uncommonly odd rocks) as a souvenir.

Yoshida spent a year and a half in Antarctica (because the trip on an icebreaker from Japan to Syowa was too long for a return after a single summer), and it was May or June of 1970 when he visited Dr. Gorai again. At that time, Gorai was away, so Yoshida left a bagful of samples with Mr. Sugiyama of the laboratory. Several days later, Sugiyama brought the bag to Gorai, who had completely forgotten about his own request for meteorites or *getemonos* as a souvenir. So he decided to open the bag sometime later on.

On his next trip to Tokyo, Yoshida visited Gorai and reminded him that he had brought him the *getemonos* as requested. At last, Gorai remembered that he had asked Yoshida for meteorites or some *getemonos*. He opened the bag, not really expecting anything special, and found something very odd inside. In *Magma,* Gorai wrote:

> The colors are dark gray and gray-green and the shapes are rounded and covered with a thin skin; they looked something like meteorites but it seemed incomprehensible that such various kinds of meteorites are found in the very limited area near the Yamato Mts. I thought they were some sort of ordinary moraine rocks and that they were weathered by the special environment of Antarctica. So, half-believing and half-doubting, I was more or less convinced, at 95% of my confidence level, that they were not meteorites. But, just in case, I took pictures of them and weighed them and had thin sections made of them. I was surprised at the thin sections; they were all meteorites: 8 of them are chondrites and 1 is an achondrite. Literally, I was astonished and sent a wire with this news to Mr. Yoshida. Under the circumstance, the details of the meteorite collection in Antarctica will be presented soon by Yoshida et al. in 1971. I have decided to ask them to investigate from the standpoint of glaciology how the "Museum of Meteorites" had formed. I also want to proceed in mineralogical, petrological, and cosmochemical research of the samples, consulting appropriate people.

It is interesting to note that in Yoshida's article of 1971 he reports that members of the field party said it seemed easy to recognize the meteorites because they were black rocks on the bare white ice. However, these black rocks were unrecognizable to Gorai when they came out of the bag in which they had been packed and carried for so long. But when he saw the thin sections he sent a telegram to Yoshida, saying: "All were found to be meteorites!" Then he wrote Yoshida a letter listing the differing species of the meteorites. It was Gorai's list that the Shimas reported on in Davos. Gorai added that his earlier letter requesting

meteorites or *getemonos* had been meant as a half joke. But he now was searching for possible explanations of the formation of meteorite concentrations on the ice.

Searches for meteorite concentrations were then formally incorporated into the work schedules of the glaciology group of JARE-14 in the polar summer of 1975–1976. In that season, when the field team was deliberately searching for meteorites rather than just picking them up in the course of other work, the team returned from the same large ice patch in the Yamato Mountains with a spectacular collection of 663 more specimens!

When this news reached Cassidy, he called Mortimer Turner, the program manager in the Division of Polar Programs at the NSF, and reported this figure. Turner reconsidered the situation. He told Cassidy that the panel had just declined his proposal again, but he advised him to add this new information and resubmit his proposal immediately; he thought it might pass this time. And indeed it did. Cassidy was approved to lead a team of two members to search for meteorites out of McMurdo Station in December and January of 1976–1977.

When Dr. Nagata arrived for his next visit to Pittsburgh, Cassidy hastened to inform him that his proposal had been accepted, and he planned to search for meteorites out of McMurdo Station in the polar summer of 1976–1977. Nagata cordially congratulated him, and then he delivered a profound blow: he told Cassidy that he was planning to send a man to McMurdo to search for meteorites in that same season! Bill was thunderstruck. There had been no suggestion of possibly working together, and the very thought of a competitive search out of McMurdo left him speechless. But Bill soon learned that JARE had had a cooperative arrangement with the U.S. program at McMurdo for a number of years. So he realized that Nagata could send a meteorite hunter there whenever he chose. Meanwhile, Bill had so much to do that he soon stopped worrying about how things might turn out with Nagata.

1.4. THE U.S. ANTARCTIC SEARCH FOR METEORITES GETS ORGANIZED

One of Cassidy's first problems was to decide whom he should invite to accompany him to Antarctica. At just about that time, he received a letter from Edward Olsen, the curator of minerals at the Field Museum in Chicago. Ed had read the abstracts of the meeting at Davos and decided that further searches for meteorite concentrations should be made in Antarctica. He contacted his friend Carleton B. Moore, director of the Center for Meteorite Studies at Arizona State University, and proposed that they submit a joint proposal to search for meteorites in Antarctica. Carleton had been a reviewer of Cassidy's third proposal, so he told Ed he was too late. Then, after proper hesitancy, Carleton broke the rules and told Ed whose proposal had been accepted. So Olsen wrote to Cassidy and they arranged to work together in Antarctica.

In 1976, the NSF shipped its scientists and other personnel from Port Hueneme in California to Christchurch, New Zealand, in the Military Air Transport System, which consisted of C-141 cargo planes. Bill and Ed climbed aboard the one assigned to them and found much of the floor space filled with chairs bolted to the floor. The plane had virtually no insulation against sounds, so the cabin not only was crowded and noisy, but it could be smoky, whenever the nature of the cargo would allow for smoking. The flight was about 22 hours long, but passengers had opportunities to emerge and walk around a bit at two nighttime refueling stops: Honolulu and Pago Pago. Identical box lunches with hearty slices of steak, French fries, a salad vegetable, and a cookie or an apple, were distributed three times during the trip.

1.4.1. Christchurch, New Zealand

Early one morning in late November, they arrived at Christchurch after losing a day from the calendar by crossing the international date line. Christchurch impressed Cassidy with its elegant British-style homes surrounded by gardens, which were just then bursting into bloom. He and Ed had a few days to rest and relax before leaving for McMurdo. So Cassidy took advantage of this opportunity to visit Oxford Terrace and contemplate the statue of Robert Falcon Scott, standing in a heroic pose gazing southward. Cassidy remarked that Scott occupies a mystical niche in the British psyche, but in no place more so than in Christchurch, which lies close to the harbor at Lyttleton, from which Scott's last expedition set out. Scott's final letter, written as death approached, made him a hero wherever English is spoken. However, in subsequent years, excerpts from Scott's own diary and those of some of his team members have become available, indicating that he made serious mistakes and alienated some members of his expedition. Cassidy concluded that Scott may have overreached his abilities and brought death upon all five members of the team that had arrived at the Pole. Cassidy recommended that interested readers look into the currently available literature before forming opinions on Scott.

The NSF maintains a huge clothing distribution center (CDC) at the Christchurch airport, stocked with Antarctic clothing of all types and sizes. Travelers to McMurdo are expected to try on pieces of inner- and outerwear they feel they will need. More specifically, each person is required to carry at least three types of well-fitted boots and shoes for working on the ice. And each must have a long, hooded Antarctic parka. Although beautiful parkas were available at the CDC, I was among those who bought my own, to keep as a souvenir. When Bill and Ed were scheduled to leave for McMurdo, they dressed in their Antarctic clothing at the CDC and packed their street clothing into suitcases that they checked there for the season. They then were presented with huge, orange waterproofed bags for carrying their spare Antarctic clothing. They boarded a C-141 Starlifter, a wheeled jet with four engines, that could reach McMurdo in 5½ hours, provided it did not have to turn back for bad weather or any other problem.

1.5. ANSMET SEASON I: 1976–1977

At McMurdo, Bill and Ed checked into the NSF headquarters at the administration's chalet and were assigned living quarters for the time they would spend there. They also were informed that Nagata and two other Japanese scientists had already arrived and had gone on a helicopter search for meteorites. They had found none, but in any case, the chief administrator, Duwayne Anderson, was unhappy about the prospect of being asked to support two meteorite-hunting groups when none at all had been scheduled in previous years. He proposed that the two field parties should work together. Nagata agreed to that, so Anderson asked Cassidy to write an agreement for them both to sign. Cassidy consulted with Nagata, thought about it overnight, and then wrote out the following memo with four main provisions:

1.5.1. The U.S-Japan Agreement

December 9, 1976

1. Logistics and base facilities of the USARP program at McMurdo will be used by a joint U.S.-Japan team to search for meteorites in the Dry Valleys and adjacent parts of the ice cap during the 1976–1977 field season.

2. Any meteorite specimens recovered will be distributed in the following way:

   a. Specimens larger than 300 g will be cut in two approximately equal pieces at the Thiel Earth Science Laboratory (in McMurdo). One piece will be utilized by the U.S. group and the other by the Japan group.

   b. Specimens 300 g or smaller will be distributed in equal numbers between the groups on an alternate-choice basis.

3. As observations from helo pilots and other groups come in, we may find it desirable to visit other field areas. The arrangements described above will apply to any meteorites recovered as a result of such change of plans.

4. Even though specimens will be distributed between our two groups, we will remain in contact about our current research programs on them, in order to avoid duplication of effort and in order to plan better how they may be utilized. We feel it would be appropriate to acknowledge the efforts of the joint U.S.-Japan team in any subsequent publication of research results.

Cassidy and Nagata both signed the document and it went into the official records of the Office of Polar Programs of the National Science Foundation. Cassidy then gave the U.S. effort a name: ANSMET, the Antarctic Search for Meteorites.

This project was a radical departure from anything previously done at McMurdo, and it generated considerable hilarity among the old hands. "What program are you with?" asked a young British petrologist who fell into step with Ed Olsen on the way to the galley.

"I'm collecting meteorites."

"Ah, yes, we saw that meteorite project on the list the other night. We had rather a good laugh over that one."

Of the three Japanese men who had taken that first helo ride, only one was planning to search for meteorites. Nagata himself could not spend time on the high ice plateau because of a heart condition, and Katsu Kaminuma was in Antarctica to conduct other research activities. So Keizo Yanai joined Bill and Ed for the 1976–1977 season. (In reviewing this situation, one is tempted to wonder if Nagata had not planned it this way all along. It seems inconceivable that he had brought Yanai there to hunt for meteorites by himself. So, by not sharing his plans with Cassidy, he might well have been playing a bit of a joke on him, since Cassidy had caught him out on the information from the Shimas at Davos.) Keizo, who had led the Japanese search that acquired 663 meteorites in the Yamato Mountains was of much help to Bill and Ed. They soon got used to Keizo's efforts in English, and he adopted a phonetic approach to writing their lingo.

In preparation for this effort, Cassidy had made a detailed study of satellite and aerial photographs in the library of the U.S. Geological Survey at Reston, Virginia. Searching for streams of bare ice, he had found a very promising tongue that descended from the plateau and penetrated about 2 km into one of the Dry Valleys before sublimating or melting and dropping its cargo of meteorites, if any, into its terminal moraine. Cassidy planned to camp on the bare rock below the moraine. For a second camp he chose a small ice patch at an elevation of 1 km on nearby Mt. Baldr. When their helo pilot had deposited them and their camping gear below the moraine, he found he still had time enough for some reconnaissance. So Cassidy asked to be taken to their second camping spot. That required flying 10 km horizontally while the surface below them rose 1 km vertically. Up there, they found a small patch of ice measuring about 3 x 3 km and landed on it. They got out of the helo and Keizo noticed a black rock nearby. It was a meteorite! While Bill and Ed were admiring it and photographing it, Keizo gazed through his binoculars and spotted another black rock. He started running. Bill and Ed followed him. Then the pilot jumped into the helo and followed, 20 meters behind them at an elevation of about 2 meters. Bill glanced back at the

helo and thought this must be what an insect feels like being chased by a praying mantis. That stone was another meteorite! Keizo Yanai had found two meteorites in the first 20 minutes of the field season! Bill felt he could declare the season a success regardless of what more might happen.

Nothing more did happen for the next six weeks, in which they camped and searched on numerous patches of ice. Finally, it was time to pack up their camping equipment and return to McMurdo. Back at McMurdo, Cassidy requested a reconnaissance flight to the Allan Hills. He was told that the Allan Hills lay beyond the permitted range for helicopter flights. However, that afternoon, one of the pilots told Cassidy he actually had put a field party into the Allan Hills in the previous season. With that information, Cassidy gained permission for a reconnaissance flight to the Allan Hills on 18 January 1977.

On this reconnaissance flight, the pilot set them down near a rock. It was a meteorite. They then found three more before it was time to leave. In a few days they were allowed one more flight to Allan Hills. This time their helo pilot found four meteorites for them. But his flight time was ending, and he said there was nothing up ahead except a scattered moraine. Bill thought it was an odd place for a moraine, scattered or not, so he asked the pilot to let them check it out. Bill, Ed, and Keizo got out of the helo and each of them quickly found a meteorite. In fact, the "scattered moraine" consisted entirely of yellowish-brown meteorite fragments with no terrestrial rocks mixed in. The pilot moved the helo close to the largest of the meteorite fragments from which all the others had broken off. With great difficulty, Bill, Ed, and Keizo hoisted the large mass onto the floor of the helo cabin. Then they collected the remaining 33 fragments of it. Figure 1.2 shows Bill at the site reaching for one of the final fragments. All together, the pieces added to the main mass would have made up a meteorite weighing about 407 kg.

During that season, they had collected the 2 meteorites at Mt. Baldr, and 6 more of them on two helicopter stops, plus 34 pieces of the same one on the final day. The 2 Mt. Baldr stones have since been paired as 1, and the 34 Allan Hills meteorites count as 1, so that leaves a seasonal take of 42 fragments of up to 5 different meteorites.

Cassidy had proved that meteorite concentrations do occur within reach of McMurdo Station. And after the successes of the first day and the last day of his initial field season, Cassidy's proposal was certain to be renewed. Indeed, ANSMET is still collecting meteorites 38 years later, under the guidance of Dr. Ralph Harvey of Case Western Reserve University, whom Cassidy recommended to the NSF as his replacement beginning in the austral summer of 1994–1995.

Figure 1.2. Bill Cassidy reaching for a fragment of the large meteorite at the Allan Hills.

1.5.2. Processing and Distribution of Samples

At McMurdo, Bill, Ed, and Keizo moved the specimens to the Thiel Earth Science Laboratory and followed the procedures described in the joint memo: first, they cut the large specimen in half using the available rock saw. Then they divided all their smaller specimens into two groups and Keizo packed the Japanese share for shipment to Japan. Bill and Ed Olsen sawed the U.S. portion of the large stone in half and sorted out their shares of the smaller stones. Each of them shipped his samples home, as field geologists generally do in order to carry out their research on them, even though actual ownership of the rocks traditionally remains with their funding agency.

Cassidy was appalled at the sight of meteorites that had been collected from the most sterile environment on the Earth being cut on a rock saw in everyday use for terrestrial rocks. He now was giving serious thought to the expected worldwide interest and demand for samples of Antarctic meteorites. No procedures had been discussed for receiving and curating these meteorites, so he decided to write a proposal for a curation center using clean facilities at the University of Pittsburgh, from which samples for research would be distributed around the world.

For planning purposes, he sent a questionnaire to every member of the Meteoritical Society, and to every other interested person he could think of, asking for opinions on how Antarctic meteorites should be collected, stored for transport, processed, and distributed for research. He received answers from nearly 90 meteoriticists in 15 countries. Cassidy summarized their responses in a 22-page survey with four appendixes and circulated it for discussion at the meeting of the Meteoritical Society in Cambridge, England, on 24–29 July 1977. In it he listed the four major concerns expressed by his correspondents:

1. An oblique and/or stereo photos should be taken of each meteorite *in situ*. Detailed information should be recorded on its field occurrence, its degree of weathering, completeness of fusion crust, and, if possible, its magnetic orientation as found.

2. Meteorites should be altered as little as possible during collection, storage, and transport. A specimen must not be touched by hands or gloves; it must be collected in cleaned containers, such as glass jars, teflon bags, polyethylene bags, or aluminum cans or foil. They must be shipped to the U.S. to avoid having them x-rayed at airport security. For carbonaceous chondrites, exposing them to Pb-bearing helo exhaust must be avoided (assuming that they can be identified in the field.)

3. Meteorites, and subdivided parts of them, must be maintained in a chemically non-reactive, non-contaminated environment at sub-zero temperatures in a dry environment, or in dry nitrogen. Rock saws lubricated by water or organics must be avoided; specimens should be broken by stainless steel implements or cut with dry wire saws.

4. Complete laboratory documentation should be carried out while meteorites are being subdivided. A need was expressed for eventual archival storage of a part of each meteorite, but there was no general agreement on how much of each meteorite should be saved, or where it should be stored.

In summary, the international scientific community clearly favored clean handling of Antarctic meteorites, based on procedures used for the lunar samples, with a few extra-special precautions to minimize terrestrial contamination.

Cassidy wrote his proposal for processing ANSMET meteorites at Pittsburgh, although he could hear what he described as the presence of leviathans lumbering about half-seen at the edge of the clearing. He was referring to the new and competitive interests being shown in the care and handling of Antarctic meteorites by two huge organizations: NASA and the Smithsonian Institution.

NASA had had the unique experience of processing and distributing the lunar samples, and its Building 31 at the Johnson Space Center was well-equipped with cleanroom facilities. NASA also had a highly trained and dedicated staff. By 1976, however, NASA was constructing a new building for the curation of lunar samples and was beginning to look into possible uses of Building 31 for meteorite research. Early in 1977, one NASA staff member, John O. Annexstad, suggested that Building 31 might be put into good use for processing meteorites such as those Cassidy had begun collecting in Antarctica. Some NASA managers reportedly resisted that idea at first, but upon learning about it in more detail they lent it their full support. Thus, NASA played an active role in the meeting held at the NSF on 11 November 1977.

The Smithsonian Institution (SI) also took an early interest in the Antarctic program, mainly through the efforts of Brian Mason. Mason saw its importance to meteoritics the moment he heard from Cassidy about his plans for going to Antarctica. Mason volunteered to have thin sections made of each meteorite collected in Antarctica and to publish descriptions of them. He also hoped to join the field team some year soon. Meanwhile, the Smithsonian management pursued its century-old tradition of claiming ownership of specimens or artifacts found on federal lands or collected by projects on federal funds. At the meeting on November 11, it yielded to NASA the processing of Antarctic meteorites and their primary distribution for research. But it successfully pressed its claims to serve as the final, archival curator of samples of each ANSMET meteorite.

1.6. A MEETING OF MINDS AT THE NSF, 11 NOVEMBER 1977

On Armistice Day in 1977, shortly before the start of ANSMET's second field season, the NSF convened an ad hoc group of meteorite specialists from NASA, the Smithsonian, and various universities to formulate procedures for collecting, processing, and distributing Antarctic meteorites. I was invited in a surprise call from Mort Turner. He said he was calling people who were recommended by other people. (Perhaps I was on somebody's list because I had served as president of the Meteoritical Society in 1975 and 1976.) Mort said he would like me to come, but he added: "We have no funds to pay for your travel." I knew the Smithsonian would pay my way, so I agreed to attend the meeting, at which I obtained my first intimate knowledge of the Antarctic meteorite project and of big government bureaucracy.

At the meeting, Cassidy willingly relinquished his proposal to process meteorites at Pittsburgh. And the NSF, NASA, and SI worked out a three-agency agreement that was unique in the U.S. government: the NSF would continue to fund and provide field support for the expeditions from the University of Pittsburgh, but the NSF stipulated its technical ownership of all Antarctic meteorites collected with NSF funding; NASA agreed to serve as the processor and distributor of meteorite samples for research, and the U.S. National Museum (Smithsonian Institution) would become the final archival curator of ANSMET meteorites. The Smithsonian also would publish reports on each season's activities and describe its collection of meteorites.

During the meeting, the NSF set up two advisory groups: the Meteorite Working Group (MWG), and the Meteorite Steering Group (MSG). The MWG consists of about 10 people with a rotating membership from the three agencies and the wider meteoritical commuity. Its main responsibility is to review requests for Antarctic meteorite samples for research and to prepare an allocation plan for approval by the MSG, which consists of three members: one each from the NSF, NASA, and SI. In an effort to inform the worldwide scientific community of the number, character, and availability of each new batch of specimens, the MWG proposed the issuing of what became the *Antarctic Meteorite Newsletter*, which is composed and distributed by the curatorial staff at NASA's Johnson Space Center (JSC). It includes descriptions of each available sample and its thin section, and is accompanied by a sheet for submitting requests for research material to be reviewed by the MWG.

These new procedures were to be put into practice in the upcoming 1977–1978 season. Meanwhile, the field team lost one member and gained two new ones. Ed Olsen did not take part again because he could not add his finds to the collection at the Field Museum. In fact, he would be collecting for the benefit of a competitor: the Smithsonian Institution. Bill found an enthusiastic partner in Billy P. Glass, a professor of geology at the University of Delaware. Billy was well known for having discovered microtektites in deep sea cores and greatly extending the sizes of certain tektite strewnfields. Keizo Yanai and Minoru Funaki made up the Japanese contingent. Incidentally, from the beginning of these searches, there was a clear understanding among participants that this was to be a group effort. No counts would be kept of the numbers of specimens found by each person. Only the final totals would be recorded. Everyone approved of this policy, which served to keep the team members friendly.

1.7. ANSMET SEASON II: 1977–1978

For this season, Cassidy decided to return to the Allan Hills and do some more searching. This proved to be an excellent choice. They soon discovered what we now call the Allan Hills Main Icefield. It is a large exposure of blue ice bearing a rich concentration of meteorites with no terrestrial rocks among them. The team erected its tents at the edge of the ice for an extended stay. During that season, they recovered about 350 specimens, each of which was collected according to the new protocols.

On being discovered, each specimen was described in a short note and photographed in situ beside a measuring device with a 6-cm scale. Figure 1.3a illustrates a chondrite, from a rock formation about 4.5 billion years old, that fell so recently that it broke into two pieces when it struck the ice. Figure 1.3b shows an achondrite, a polymict breccia likely from the surface of asteroid 4 Vesta (Plate 57), that has been carried within the moving ice for perhaps several hundred thousand years before appearing at the surface.

Figure 1.3a. A chondrite, about 4.5 billion years old, that fell so recently that it broke into two pieces when it struck the ice.

Figure 1.3b. An achondrite, ALH A81006, a polymict breccia likely from the surface of asteroid 4 Vesta (Plate 57), that has been carried within the moving ice for perhaps several hundred thousand years before appearing at the surface.

NASA's curatorial facility at JSC had supplied the field crews with all the newly cleaned equipment they needed for collecting specimens without touching them. Sometimes nicknamed "*Apollo* surplus," this included teflon bags, stainless steel tongs, and teflon tape designed for use at sub-zero temperatures. Each specimen was placed in a Teflon bag and then that bag was dropped into a second bag carrying a numbered aluminum tag. The second bag was then sealed shut. In order to maintain the specimens at sub-zero temperatures while they were being stored and shipped, NASA provided the teams with burglar-proof padded steel boxes measuring about $60 \times 60 \times 90$ cm. One of these boxes was brought to the camp site at the Allan Hills.

The harvest of meteorites in that season was acquired under especially trying circumstances: the team had no

snowmobiles, so they were obliged to take turns trudging to each rock, carrying the camera and collecting equipment, and then carrying each wrapped-up stone back to the camp to be stored in the shipping box.

1.7.1. Early Procedures at Johnson Space Center

Many of the meteorites had ice or snow on them when they were collected, so at JSC each new specimen was put into a glove box in a stream of dry nitrogen to be thawed and dried and then sawed with a clean blade, or chipped apart depending on its size. It was photographed at each step of its processing. Three chips for thin sections were taken from each specimen that weighed more than 100 g. The sections were cut and polished at the Smithsonian in Washington for distribution to Japan, and to libraries at the NASA Johnson Space Center in Houston and the Smithsonian's Natural History Museum in Washington. The mineralogy of each thin section was described by Brian Mason, who, as noted above, had volunteered for the job. At the earliest meeting of the MWG in Houston, Klaus Keil raised the question of who would describe all these thin sections. He didn't want to do it himself, and he didn't want his students to spend time that way, either. But Brian, knowing that this question would arise, had given me a copy of one of his succinct descriptions of a thin section to read out to the members. Keil relaxed, seeming fully satisfied. Brian, during his long career of field and laboratory investigations, and his writing of books on geochemistry and meteorites, had developed a quick and effective technique of identifying meteorites in thin sections. He said later that he had had a wonderful time going through the several thousand new thin sections of Antarctic meteorites. And he was pleased that this was recognized as the essential service it was.

Each specimen unpacked at JSC was assigned a unique label. The labels had been worked out by the MWG in discussions with the Committee for Nomenclature of the Meteoritical Society. Each label would begin with three letters identifying its location, followed by the letter A, followed by two digits indicating the year of its discovery, and three more digits indicating the sequence in which the specimen was opened at JSC (not when it was found in the field). For example, ALH A78362 designated the following: Allan Hills, Expedition A, 1978, the 362nd to be opened at JSC. "Expedition A" was adopted at the insistence of Paul Pellas, the member from France, who insisted that many countries might begin sending collecting expeditions each year and would need different expedition letters. The MWG agreed to assign new letters whenever additional expeditions were fielded in a given year, but this never came to pass, so the letter A, which had been part of every label since 1975, was dropped in 1982.

Figure 1.4a. Iron ALH A77283.

Figure 1.4b. A polished and etched slice of the iron. Minute grains of diamond and lonsdaleite occur within inclusions such as the dark one at lower right.

In this, its second season, ANSMET discovered a unique iron, ALH A77283, which measured about 16 × 16 × 12 cm and weighed 10.5 kg (Figure 1.4a). It proved to be a carbon-rich octahedrite similar in composition to Canyon Diablo. That seemed simple enough until its inclusions of troilite-carbon-schreibersite-cohenite were found to contain carbonado-like material rich in minute diamonds and lonsdaleite (Figure 1.4b)! Had the iron smashed into the ground with enough force to form tiny diamonds? No, because the meteorite had a heat-altered ablation zone along one side of it, indicating that it had had an uninterrupted passage through the atmosphere

and a nonexplosive impact on the Earth. This identifies it as the only known iron meteorite showing evidence of a diamond-forming impact in space [*Clarke Jr.*, 1982, p. 51].

Unlikely as it seems, this iron gave rise to some second thoughts about Canyon Diablo: Could its diamonds be preterrestrial? Was Meteor Crater volcanic after all? No, it was not. That yawning crater, nearly a mile wide and surrounded by large fragments of an iron meteorite, bears testimony to a powerful diamond-producing impact on the Earth. ALH A77283 shows us that an impact in space also can create diamonds in irons.

1.7.2. Repercussions from Japan

At the end of the 1977–1978 field season, all the meteorites were shipped to NASA/JSC for processing. But nobody had thought to inform Takesi Nagata of this new arrangement. So when Keizo Yanai arrived home without Japan's half of the meteorites, Nagata raised a storm. Although it seemed to others that Nagata surely would approve of this effort to keep the meteorites uncontaminated, nationalism evidently outweighed science when he realized that he had not been consulted. It took a face-to-face meeting in Washington with Edward Todd, the Director of NSF's Division of Polar Programs, to win his agreement, as long as (1) the United States promised not to circulate any preliminary descriptions of meteorites before Japan received its share, and (2) a Japanese member was present when the collection was opened. That promise was easily fulfilled by inviting Keizo Yanai to be present at JSC during the opening and distribution of that season's meteorites.

1.8. BECOMING A MEMBER OF ANSMET

By then, I had developed a yearning to go to Antarctica. At the NSF meeting on 11 November 1977 there had been talk of maybe sending several field parties each season. So I asked Mort Turner about submitting proposals, and he loaded me with maps, forms, and booklets. However, when I checked back with him, he said, "Realistically, we won't be sending more than one team a year; call up Bill Cassidy and ask to join his team."

I had known Bill for years. In fact, he recently had written a favorable review of a paper I had submitted for publication. However, it certainly did no harm to my cause to call Bill and be able to say, "Mort Turner told me to call you and ask to join your team in Antarctica." After we talked awhile, Bill agreed to take me the next season: 1978–1979.

At about that time, Cassidy was formulating his policy on choosing team members. First, he always would take a crevasse expert. They are the most essential members of the teams. Next, he would look for senior people actively involved in meteorite research, including Europeans.

Next came graduate students in meteoritics or allied fields, and if there still was any room he would consider specialists who could contribute to the well-being of the field teams: first aid, communications, snowmobile maintenance, and such. When I learned about his list, I was glad to note that I would have fitted into his Category 2, even without a recommendation from Mort Turner.

In 1978, I had been invited to serve as a visiting professor for the fall semester at Arizona State University in Tempe. Both my husband and I were Arizonans by preference, if not by birth, so we found a comfortable motel near the campus and Tom led me through a fitness program for going to Antarctica. It began by running together several times around the university track early each morning and then performing exercises for improving my balance. To our delight, we soon found that I was sharing the office of Dr. Robert S. Dietz, in his absence, with Dr. James F. Hays of Harvard, who often had gone running with Tom and had led both of us on birding expeditions. Hays is a master birder, so we sometimes had late afternoon forays into the desert followed by dinners with Jim and his wife, Diane, at the French cafeteria Le Café Cazino in Phoenix. (I wrote to the manager from Antarctica urging him to open a branch in Harvard Square. He responded with thanks, but without a word about expanding his restaurant chain.)

In Tempe, I took the opportunity to examine meteorites in the new Center for Meteorite Studies. And I searched through the archives of the Meteoritical Society that were stored in file cabinets in the basement. I found interesting old letters discussing the need for a society devoted to meteorite research.

1.9. ANSMET SEASON III: 1978–1979

When the semester ended, Tom and I drove to San Diego and took a birding cruise to San Clemente Island. Then, on to Port Hueneme to catch my cargo plane for the flight to Christchurch. Dean Clauter, one of Cassidy's students, boarded the plane too, as did three graduate students from Tempe, bound for a different project. I found the flight to be much more comfortable than the one Cassidy had described. There was a padded pallet of cargo at center front of the cabin, with seats filling the rest of the floor. The five of us rushed for the middle of the front row of seats where we could rest our feet on the pallet. Shortly before takeoff, it was announced that due to the nature of the cargo there would be no smoking on the flight. That was wonderful news to us! It would be a long trip with the same box lunches and two refueling stops that Cassidy reported: at Honolulu and Pago-Pago. I had been to Honolulu before, and have been there since, but that was the only stroll of my lifetime under the tropical trees at Pago-Pago.

In Christchurch a young marine biologist named Susan Patla and I quickly discovered our common interest in birding, so we explored the incredibly beautiful botanical garden. Then we hired a taxi to take us birding in the countryside and along the shore. The driver, who owned her vehicle, was so pleased by such a mission that she charged us very little.

When the time came to leave, we assembled our Antarctic clothing at the CDC and took off for McMurdo. Shortly after we arrived there, Bill Cassidy came in from the Darwin Camp and arranged for Dean Clauter and me to replace him there while he set off in a helicopter for the Allan Hills. Counting Cassidy, there were seven active participants in three major ANSMET projects during that season. The projects were (1) the design and erection of a geodetic network across the meteorite-rich portion of the Allan Hills Main Icefield, (2) meteorite searches from the Darwin camp near the head of the Darwin Glacier, and (3) searches for more meteorite concentrations in the Allan Hills.

1.9.1. The Geodetic Network at the Allan Hills

John Annexstad of the Johnson Space Center in Houston, who had wintered over in Antarctica during the International Geophysical Year of 1957–1958, took the responsibility for laying out a geodetic network across the meteorite concentration on the Allan Hills Main Icefield. He was joined by Minoru Funaki and Fumihiko Nishio, both of the National Institute for Polar Research in Tokyo. They set up their camp at the Allan Hills on 7 December and stayed there for 26 days, through one 4-day blizzard and several other storms that kept them tent bound. Their net consisted of 20 stations stretching westward across the icefield for 15 kilometers. They anchored Stations 1 and 2 to the bedrock of the Allan Hills. For the rest, they bored auger holes 50 to 100 cm deep into the ice. They filled the holes with bamboo or aluminum flag poles and established the position of each station by means of a Wild-2 theodolite (Figure 1.5). Their plan was to come back in future years and remeasure the station locations to determine the direction and rates of ice motion and ablation. While they were constructing the network, the three of them picked up 103 meteorites, which they added to the season's collections.

1.9.2. The Darwin Camp

At the beginning of the season, Cassidy and Kazuyuki Shiraishi of the National Institute of Polar Research in Japan went to a temporary camp the NSF had erected at the head of the Darwin Glacier for use by several projects. The camp consisted of Jamesway huts linked together for sleeping, dining, working space, laundry,

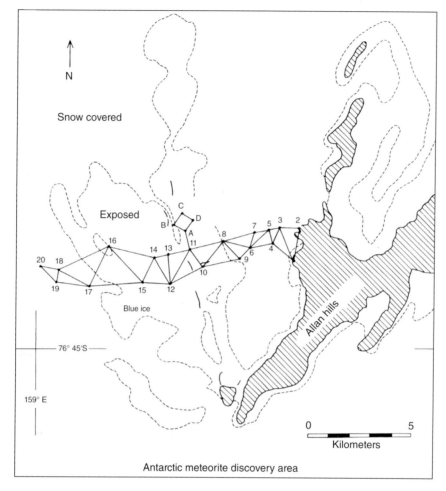

Figure 1.5. The geodetic network across the Main Icefield. Note Stations 1 and 2 on bedrock and the rest in auger holes bored in the ice. The small strain net with Stations A, B, C, and D was added three seasons after the original was completed. The meteorite distribution under the net is not shown.

plumbing, and an electric generator. There was a pad outside with three helos, one of which always was available for emergencies. A cook rustled up three meals a day for everybody.

On the high plateaus within helicopter range of the Darwin camp, Cassidy and Shiraishi found large patches of ice with no meteorite concentrations on them. They collected only eight stones up there. However, a major attraction at that site proved to be the scattered fragments of an iron meteorite that geologists from New Zealand had discovered and from which they had already recovered six fragments on the rocky, ice-free slopes of Derrick Peak. They radioed Bill to come and take a look, so searching jointly with them, Cassidy and Shiraishi found six more irons, which the Kiwis urged them to keep. All of these irons were coarsest octahedrites that appeared to lie in a strewnfield on the mountainside. One of these pieces was the enormous 160-kg iron that Bill found and photographed (Figure 1.6a; see also Plate 78).

After this run of remarkably good luck, Cassidy departed from the Darwin Camp, leaving Shiraishi to act as a guide for Dean and me when we would arrive there. On 20 December after a couple of days of being socked in at McMurdo, we took off from the ice at Williams field, next to Scott Base, in a Hercules LC-130, and flew to the Darwin Camp. There, I was happily surprised to discover that the chief scientist at the Darwin Camp was Dr. George Denton, who had been my best student back in 1960 when I was teaching mineralogy at Tufts College.

On 21 December, Shiraishi led us to a moraine near the head of the Darwin Glacier where there seemed to be nothing in view except a broad, closely packed assemblage of terrestrial boulders. I thought it would be helpful to look closely at a few of the boulders to get an idea of what sort of rocks were there. I reached out with my pick and pulled up a big stone at random, took it in my hands (which I should not have done), and saw that it was a meteorite! My first meteorite from Darwin was already

Figure 1.6a. The enormous 160-kg iron found and photographed by Cassidy on Derrick Peak.

Figure 1.6b. The 139-kg iron found by Shiraishi on Derrick Peak on 24 December. The iron is inside the helo cabin, with Shiraishi at the right of it and Clauter behind it.

contaminated! So I carried it with me when I went to join the others. By then, they had climbed down a steep hillside where Clauter had found a chunk of a fusion crust but no meteorites. When we got back to where we would meet our helo, we searched for more meteorites like the one I had found. Then, at the last minute, as the helo was descending, Shiraishi discovered a big stone similar to mine.

On the next afternoon, we found an ice patch on the Darwin glacier with 23 stones on it (21 of which appeared to be pieces of the same meteorite). We had to spend the next day, December 23, in camp because the helo time was all signed up for. So our next search took place on December 24, on the lower slopes of Derrick Peak. There, Shiraishi found one more rather small but nicely shaped iron. Then, just as the helo was landing, he (as was typical) spotted one more iron, but this one weighed 139 kg and was so heavy that both of the helo pilots helped Shiraishi and Clauter to lift it into the helo cabin (Figure 1.6b; see also Plate 78).

Christmas was upon us and all the field parties had come in and pitched their tents at the camp for the holiday. On Christmas Eve, a gale wind arose, but it subsided before morning, making way for a heat wave. At Darwin, the temperature climbed to −5° C and by December 31 it had broken all records at McMurdo: 9.4° C. Meanwhile, at 5:00 p.m. on Christmas Day, the cook served up a most royal feast: trays of cold oysters, shrimp, and lobster tails, three huge roasted stuffed turkeys with gravy and cranberry sauce, baked ham with fruit sauce, sweet potatoes, white potatoes, succotash, onions, fresh rolls, cornbread, and four kinds of pies. Wines were available if ordered in advance, but for those of us who had not known about putting in orders, there were plenty of wine bottles being passed around, and everyone was encouraged to sample any wines that appealed to them. It was a high-spirited celebration with much laughter and singing.

We spent the next day in camp with high winds and light snow and nothing flying. Bill called us by radio via the South Pole station to wish us happy holidays and ask how we were doing. He soon was leaving for the Allan Hills and hoped we would catch up with him there. We told him we planned to leave for McMurdo, carrying 26 stony meteorites and 8 irons on the earliest Hercules we could catch. He congratulated us and signed off. Then we palletized our samples to carry them off on a moment's notice.

I then learned that George Denton had arranged for me to fly over the huge Byrd Glacier on the 27th. He told me the Byrd is one of the few geological features on the Earth that can be seen from Mars. I think he was serious, but much as I would have liked to see the Byrd glacier, I could not risk being separated from my party when we were so close to leaving for McMurdo.

On December 30, we were awakened at 4:30 a.m. to get ready to leave the Darwin Camp on a Hercules. We had come on a Hercules, so we had no expectations of excitement. We dressed and had some breakfast before the plane landed close by with a great roar. It never shut off its motors while the captain handed over the mail he was carrying (he brought me two letters from Tom) and unloaded the cargo, including a much-needed fuel tank for use at Darwin. As soon as they loaded the new cargo, the three of us climbed aboard and settled into canvas seats held in place by netting. The Herc started forward and then, suddenly, we felt a frightful bumping and thumping and things falling to the floor. We stopped abruptly and the fight master ordered everybody back aft. Four Navy men and three of us USARP members squeezed past the cargo to the tail, where we lay on the tilted flap and were tied down with ribbons. Then we started up again with the same thumping and bumping. On the third try, we gained a little altitude and then kept climbing. Once we were securely aloft, we were directed to return to our seats. Forty minutes later we landed on the ice at Willy Field. Later that day at McMurdo, I described our takeoff to a woman pilot, who told me we had had a jet-assisted takeoff. Even so, it was no fun.

1.9.3. The Allan Hills

The Allan Hills camp was close to the edge of the Main Icefield, which made it easy to find and collect meteorites. I had a yellow Scott tent to myself, Dean and Bill shared one, and Shiraishi had his own tent. I had no stove for cooking because I was expected to have my meals with the others. This was a pleasure, considering Shiraishi's store of frozen foods from Japan. He particularly enjoyed munching on frozen Brussels sprouts as though they were candy. I had my own snowmobile and fully enjoyed it without realizing how lucky I was. This was the first season that the NSF had furnished snowmobiles to ANSMET members. Some people at McMurdo seemed to regard the new meteorite project as rather a bother, taking money and equipment away from more traditional pursuits. So, for the first two seasons, the ANSMET teams had to await helicopters to move long distances, and had to hike between close ones. This made no sense at all to our Japanese companions, who finally arranged for snowmobiles to be shipped directly to them from Japan. Suddenly, the NSF managers saw that they could not send ANSMET members out on foot, or as hitchhikers, alongside snowmobiles bearing the huge letters JARE. Thus we received our own snowmobiles for exploring the scene and carrying meteorites from the field to camp or pulling Nansen sleds for setting up temporary camps.

We made daily searches for meteorites on the Main Icefield, in a moraine on the western edge of it, and sometimes on the polar plateau beyond it. I found that the wind scoops around boulders almost invariably held small, cherry-sized stony meteorites, which, no doubt, had skittered across the ice in the wind and fallen into the scoops. A study of 145 of these cherry-sized stones, weighing less than 150 grams each, was published by *McKinley and Keil* [1984], who showed that although all of these stones were chondrites, they included some new and rare types. They found 19 specimens of a chondrite containing previously unknown intergrowths of graphite and magnetite in their matrixes. They also found one example of a new type of carbonaceous chondrite and two specimens of a rare enstatite chondrite.

Having 24 hours of daylight and a bright yellow-orange tent to live in suited me perfectly. I quickly found that being cold was no real problem, given padded clothing and a cozy sleeping bag. One night, I awoke about midnight and heard tremendous winds coursing down northward past my tent. This was something unusual so I snatched up my camera and burst out of the tent. There, I saw lines of snow and ice crystals sweeping over the low-lying end of the range and all along the ice sheet. The midnight Sun was shining over the highest peak of the Allan Hills so brightly that my light meter indicated I could not get a picture. I closed down the meter as far as it would go and snapped, expecting the worst. But in

Figure 1.7. The midnight sun on the Allan Hills at 12:15 a.m. 6 January 1979. A corner of my yellow tent is in the wind scoop at lower right. South is at the top of the picture.

Figure 1.7 we have the midnight Sun in full glory shining on the driven snow.

As noted above, Annexstad, Nishio, and Funaki, who had already left for home, collected 103 meteorites while setting up their geodetic network across the Allan Hills Main Icefield. At the head of the Darwin Glacier, including Derrick Peak, our party of four, Cassidy, Shiraishi, Clauter, and Marvin collected 8 irons from a single shower and 32 stones. At the Allan Hills camp, we collected 67 stones, making our total for the season of 204 stones and 8 fragments of a single iron.

1.10. CATALOGING THE HISTORY OF ANSMET

As was agreed at the November 11th meeting at the NSF, the Smithsonian members would report on each season's fieldwork and its harvest of meteorites. Brian Mason volunteered to write these reports, and then he persuaded me to join him. We styled them to be published in the well-established series *Smithsonian Contributions to the Earth Sciences*. However, from the first, we departed from the format of a standard catalog by including descriptive articles on field occurrences, collection and curation procedures, measurements of ice motion, terrestrial ages, and overviews of selected meteorite species. We dropped the name *catalog* altogether from issue Number 26 that described the 1981–1982 season.

In summary, Brian and I published the following issues of the *Contributions:* Number 23, on season 1977–1978; Number 24, on seasons 1978–1979 and 1979–1980; and Number 26, on seasons 1980–1981 and 1981–1982. Then Brian retired from this effort and was succeeded by Glenn MacPherson of the Smithsonian Natural History Museum, who contributed to Number 28, which reported on seasons 1982–1983 and 1983–1984. By then, we both found this to be an overwhelming assignment on top of our other responsibilities. We also realized that the *Antarctic Meteorite Newsletter*, which was issued and distributed by the curatorial staff at the Johnson Space Center, was fulfilling the needs of scientists in a much more timely fashion than we ever could. Thus ended the ANSMET series of *Smithsonian Contributions to the Earth Sciences*.

1.11. ANSMET: WELL ESTABLISHED BY 1980

By the end of its third season, ANSMET had become highly successful at carrying out this new mode of scientific inquiry. It would continue to do so for going on four decades, adding to the list of meteorite concentrations in Antarctica and, needless to say, adding thousands of meteorites, some of which were new to the science but would now become available to the international community.

Several changes to the procedure lay directly ahead: 1978–1979 was the final season in which Japanese members would take part in the fieldwork and share meteorites with ANSMET. The NSF soon would abandon the use of military cargo planes and negotiate government discounts to send its scientists to New Zealand via scheduled airlines. Cassidy remarked that the main advantage to this change was that passengers would arrive in Christchurch with smiles on their faces. In 1980–1981, Cassidy added to the team Ludolf Schultz of the Max-Planck-Institut für Chemie at Mainz, the first of numerous team members from Europe. He also acquired a highly skilled crevasse expert, John Schutt, who quickly became adept at recognizing meteorites and sometimes authored or coauthored seasonal reports for the *Catalogs*. Schutt remained with ANSMET for more than three decades, and his services were recognized in 2008 by the awarding of a well-deserved honorary Doctor of Science degree by Case Western Reserve University.

But the most exciting and significant event in the early history of ANSMET would occur on 18 January 1982, when John Schutt guided a visiting glaciologist, Ian Whillans, to the Middle Western Icefield and they found a lunar meteorite! Inasmuch as we have seen that Cassidy had included the possibility of finding lunar meteorites in his original proposal of 1974, we will extend this early history long enough to include that event.

1.12. ANSMET SEASON IV: 1979–1980

In that season, Bill Cassidy; John Annexstad; Lou Rancitelli of Batelle Memorial Institute at Columbus, Ohio; and Lee Benda, a crevasse expert from the University of Washington, conducted an expedition northward from the Allan Hills to Reckling Peak to check out a stretch of bare ice 100 km long by 3–5 km wide that extends westward from Reckling Peak. A year earlier, Philip Kyle of Ohio State University had visited this area and found five meteorites near Reckling Peak. The Allan Hills lie at the limit of helicopter range from McMurdo, so Cassidy's party had to make this trip by towing Nansen sledges, packed with camping equipment, behind their snowmobiles. They left the Allan Hills on 5 January and drove 24 km northward, carefully skirting crevasses before setting up Windy Camp, where they spent two days due to high winds and low visibility. On 8 June they broke camp and drove 32 km farther north and up a rather steep slope to the vicinity of Reckling Peak. There, crevasses became so large and numerous that the party stopped and erected Crevasse Camp for rest and relaxation. The next day they turned westward and carefully made their way downslope until they reached bare ice. There, they used snowmobiles in tandem to lower each Nansen sledge: one snowmobile in front, pulling and steadying, and one in the rear, braking. At the bottom they were on a part of the 80-km strip of bare ice where they found traces of Kyle's trail markers. Nearby, they found a clear, snow-covered spot adjacent to a moraine and set up Moraine Camp, about 12 km west of Reckling Peak. During a day and a half at Moraine Camp they recovered 13 meteorites.

Traveling westward along the northern edge of the ice patch, they soon were off the bare ice and thus learned nothing about whether the entire ice strip has meteorites on it. However, at a point about 65 km west of Reckling Peak they found a second large moraine, which they called Elephant Moraine, for its rounded form and long protuberance reminiscent, from the air, of an elephant's trunk. There, they built a camp and found 12 more meteorites. In the future, Elephant Moraine would prove to harbor a rich concentration of meteorites, but on that first visit, they turned southeastward toward a site they called Carapace Camp. They found no meteorites there, but it was not far from the Allan Hills, so they were picked up by a helicopter and flown back to the Allan Hills camp.

On their Reckling Peak traverse they collected a total of 26 specimens representing no more than 15 falls. These included two rare types: one iron meteorite and three achondrites, one of which was a shergottite (EET A79001; Plate 70), which we now classify as a rock from Mars. They found no large concentrations but concluded that the Reckling Peak and Elephant Moraine areas would be well worth another visit in the future.

1.13. ANSMET SEASON V: 1980–1981

Cassidy began his report of this season by pointing out that the team already knew of the major concentration of meteorites on the Allan Hills Main Icefield, and of lesser concentrations on the Near Western Icefield, Reckling Moraine, and Elephant Moraine. Now they were eager to investigate new sites in hopes of locating additional concentrations to be exploited in future seasons. Accordingly, in 1980–1981 they emphasized reconnaissance while devoting as little time as possible to collecting enough meteorites to assure a successful season.

That season, the participants were Bill Cassidy, John Annexstad, John Schutt, Lou Rancitelli, and Ludolf Schultz, plus two new members: Harry "Hap" McSween, a professor of geology at the University of Tennessee, and Joanne Danielson, a student at the University of Pittsburgh.

They collected 32 specimens at the Allan Hills Main Icefield (which had begun to serve as a reliable bank account), and then traversed back to Reckling Moraine, where they had to repeat the tandem lowering of their supply-laden Nansen sledges down over the steep slope to the 80-km strip of ice. Cassidy concluded that this ice strip is associated with a bedrock barrier, parallel to its long dimension, over which ice spills down the east-facing slope, just as it does down the "monocline" that bounds the Allan Hills meteorite concentration on the west. This return to Reckling Peak yielded 32 more fragments: 30 chondrites, one iron, and one achondrite, making a seasonal total of 62 specimens.

At Reckling Moraine they received an air-drop of eight drums of snowmobile fuel, which enabled them to travel further northward to new sites that were seen from the air to include ice patches. These were Outpost Nunatak, Griffin Nunatak, Brimstone Peak, Tent Rock, and Sheppard Rocks. To their disappointment, none of these sites had any meteorite concentrations, and the party collected only one isolated meteorite specimen near Outpost Nunatak. Cassidy wrote that the net result of this exploration was to give them a greater understanding of conditions under which concentrations do and do not form.

1.14. ANSMET SEASON VI: 1981–1982

This season is remembered as being, by any measure, a highly successful one. The activities were planned by Cassidy, but after he got things well started, he left to spend Christmas at home for the first time in five years. The team included Annexstad, Schultz, Schutt, myself, and two new members: Ghislaine Crozaz of Washington University at St. Louis, and Robert Fudali of the Smithsonian's National Museum of Natural History in Washington. For three of us, Crozaz, Fudali, and me, the season began on 29 November 1981 at the airport in Los Angeles, where we found a gate serving travelers associated with the U.S. Antarctic Program (USARP) and boarded a scheduled flight to Auckland, New Zealand. From there we flew to Christchurch (Figure 1.8) and visited the botanical garden, among other places. As the time approached to leave for McMurdo, we selected our Antarctic clothing (although Ghislaine and I already had our own parkas) and packed it all into the huge orange waterproof cloth bags that had proven to be very useful when camping on the ice (Figure 1.8). At McMurdo, we moved into the small dormitory and dined at the mess hall until it was time to leave for our assignments.

1.14.1. A Second Remeasurement of the Geodetic Net at the Allan Hills

Two members of this season's party, Ludolf Schultz and John Annexstad, had arrived at the Allan Hills on 15 November and set up their camp in the triangular space between stations 6, 7, and 8 of the network. They began remeasuring, for the second time, the horizontal displacement and degree of ablation at each of the 18 stations on the ice of the network. They planned to extend the network

Figure 1.8. Marvin (left) and Crozaz (right) in the C-141 Starlifter en route from Christchurch to McMurdo.

Figure 1.9. Measuring the 5-cm ablation of ice at a flagpole of the geodetic net.

a short distance westward. However, a succession of storms that barely allowed them to complete their measurements prevented them from making an addition at that time. They left the field on 13 December.

Schultz and Annexstad reported the following results: the ablation rate of the ice at the Allan Hills Main Icefield averages about 5 cm per year (Figure 1.9). This serves well to expose new meteorites that have been frozen within the ice. Their measurements of horizontal ice displacement since 1978 indicated that ice moves rather rapidly eastward from the western plateau toward the Allan Hills, but it slows down and becomes stagnant at the foot of a steep slope that leads to a narrow flat-bottomed valley about 4 km wide. On its icy surface, it holds the meteorite concentration. Concentrations like this one

develop most readily on ice that has no outlet but loses volume due to ablation of its exposed surface.

1.14.2. The North Victoria Land Camp

In 1978–1979, the NSF had built a season-long field camp, similar to the Darwin Camp, to serve several projects in North Victoria Land (NVL), 600 km north of McMurdo Station. John Schutt and Bob Fudali arrived at the NVL Camp on 24 November to carry out reconnaissance of ice patches that were visible on air photos. They also planned to visit a small circular feature at Littel Rocks to determine whether it could be an impact crater. They made numerous helo flights over ice patches that were strewn with rocks, but frequent stops revealed that the rocks were terrestrial, with no noticeable meteorites among them. When they left the area, they reported that they could not declare there were *no* meteorites in the rocky debris at NVL, but using their best efforts they did not find any.

The crater at Littel Rocks proved to be a former lake that had been overridden by ice more than once. They found no sign of impact features associated with it. So Schutt and Fudali returned to McMurdo.

1.14.3. A Visit to Granite House

Before the season began, I had persuaded Bill Cassidy by telephone to request a helicopter flight to Granite House, a historic edifice that was built in 1911 by four members of Scott's expedition, led by Griffith Taylor, who were mapping the coast of Victoria Land. When they exhausted their kerosene, they fashioned this structure out of natural ledges and boulders to spare a tent from being used for cooking with an oily, smelly, frequently boiling over blubber stove. Finally, they mounted one of their sledges across the top as a roof-tree and covered it with seal skins. They named it Granite House after the edifice in Jules Verne's book, *The Secret of the Island*, which they were carrying with them. Granite House never was a comfortable place for relaxing, but it served its purpose as a kitchen well enough.

Taylor and his group were expecting to be picked up by the ship *Terra Nova* about 15 January 1912, but the sea ice never broke out that season, so by mid-January they decided they must abandon their camp and make their way back at least as far as Cape Roberts, 14.5 km away at the entrance to Granite Harbor. They reclaimed the roof-tree sledge and stacked it with about 260 kg of specimens, including some unique Gondwana fossils, plus extra clothing, books, and tins of food to be recovered later. They then used their better sledge for carrying their needs with them. At Cape Roberts, they found the sea ice stretching toward the horizon, so they lightened their load once again by building a cache with their remaining possessions and tins of food they thought they could do without. Fortunately, their food cache saved the life of Frank Browning, a member of a five-man group that had been forced to winter over for two successive years on short rations. Taylor's group then proceeded southward along the shoreline toward their headquarters at Cape Evans, 160 km away. They were in luck: eight days later, on 14 February 1912, they were sighted from the *Terra Nova,* which sent a boat to their rescue. Within the following year, two colleagues revisited Granite House and brought back their specimens and books and other personal possessions [*Marvin*, 1983].

Granite House had no known visitors for the next 46 years. Then, in November of 1946, Professor Robert L. Nichols of Tufts College and three of his students stopped there in their man-hauling expedition from McMurdo. They took numerous photographs and made an inventory of all the items they found at Granite House. The pictures proved to be of special interest to me because one of them showed a stone that looked remarkably like a meteorite on the ledge behind Granite House. It was about 25 cm across, subangular, and mostly black except along the edges where its light interior was exposed. I had an enlargement made of the meteorite picture and mounted it among the posters at the March 1981 Lunar and Planetary Science Conference in Houston. Beside it, I posted a blank sheet asking for comments. The picture elicited much interest. A few viewers wrote that it surely was a meteorite, or probably was a meteorite. A few words of caution also were expressed, but a strong majority wrote that it must be examined in the field. (Unfortunately, my picture, which I carried to McMurdo, was irretrievably lost in transit.)

At about the same time that Cassidy requested a helicopter trip to Granite House, two young historians from New Zealand's Scott Base also requested one. So on 19 December 1981, a helicopter carried me, Ghislaine Crozaz, Bob Fudali, and two Antarctic historians, Jack Fry and Jerry Turner, on a brief afternoon visit to Granite House. As we circled over Granite House, the stone still looked like a meteorite, but on the ground, that vision faded quickly. We found a light-colored granite boulder with a dark surface over most of it. This was, of course, a disappointment to me. But Granite House with or without a meteorite was well worth a visit, and since I recently had been told it had become a tourist attraction subject to vandalism, I was pleased to learn that it was being properly documented by New Zealand historians who were devoted to the preservation of historic Antarctic huts.

1.14.4. The Allan Hills, Continued

On 13 December after remeasuring the location of each station of the geodetic network across the Allan Hills Main Icefield, Schultz and Annexstad left for home. About one week later, on 22 December, Schutt and Crozaz moved into the camp. Fudali and I joined them in the next available helicopter, which arrived late in the morning of 24 December when we all changed partners in the tents. That afternoon, John Schutt guided us to nearby Man Haul Bay, where exposures of Permian coal and shale beds contain petrified wood and *Glossopteris* seeds and leaves. Man Haul Bay is the open space formed by the Y-shaped arms of the Allan Hills. It is icy enough to allow easy access to the fossils. Afterward, with the Sun still shining, we all celebrated Christmas Eve with nips of Scotch whiskey, an excellent dinner, much singing, and Fudali on the harmonica.

On Christmas Day, we started measuring the force of gravity at each marker in the geodetic network, which by then consisted of 24 stations. This procedure required readings, at each station, of a gravimeter, an altimeter, and a thermometer. We completed these three measurements at 19 stations, leaving 5 more for the next day. When we finished the stations on the net, we measured 8 more stations on bedrock along the Allan Hills, including one on the top of Peak 2330, which stands like a beacon at the southern end of the range. The whole party climbed the peak. We then moved on to nearby Carapace Nunatak, which has its own small icefield. Carapace is a stunning sight with vertical crenulated cliffs of pillow lavas and paragonite lenses. Geodes are plentiful at its base, but we found no meteorites there.

The results of the gravity survey, reported by *Fudali and Schutt* [1984, p. 26], show the bedrock just above ice level at Stations 1 and 2 sloping gently down westward under the ice for about 110 km from both stations. But the two profiles are not identical. The profile beginning at Station 2 shows a slight dip in the bedrock under the valley holding the meteorite concentration. It loses 500 meters in its first 3.5 km, then the bedrock rises 250 m in the next 2.5 kilometers, and finally it slopes steadily downward to a depth of 750 meters under Station 20. The dip is not repeated in the second profile, from Stations 1 to 19, where the bedrock slopes gently westward with only a slight flattening after the first 2.5 km.

1.14.5. Systematic Sweeping of the Allan Hills Icefields

A major advance in field surveys, begun in this season, was the systematic sweeping of icefields. In a systematic sweep, three or more steel-cleated snowmobiles line up abreast, several meters apart, and drive straight ahead in a given direction, with each driver examining the ice on both sides. At the end of the first traverse, all snowmobiles reverse their direction. The driver who occupied one edge of the group (often the right-hand edge) simply pivots in place and follows his or her own track back to the starting line. The others line up at appropriate distances beyond

that one and drive to the starting line. Sweeps continue to be made until the area of interest has been completely searched. This technique is a very efficient way to spot meteorites down to a few millimeters in size.

In the area to the west and southwest of the Allan Hills Main Icefield lie three icefields called the Near Western, Middle Western, and Far Western Icefields. In addition, there is the Battlements Nunatak Icefield, which lies north of the Main Icefield (See Figure 1.10).

The Main Icefield is about 22 km long and encompasses some 75 km² of blue ice. The Near Western Icefield consists of five separate ovoid patches lying about 18 km NNW of Peak 2330 and includes more than 14 km² of bare ice. The Middle Western Icefield lies 31 km WSW of the Peak and consists of 30 km² of ice. The Far Western Icefield is larger than the Main Icefield, being more than 40 km long and 2 to 8 km wide with an area of more than 100 km². A planned reconnaissance of that large field was cancelled for this season because of poor weather and insufficient time. It seems very possible that all of these icefields are the currently exposed portions of one gigantic field separated by variable zones of snow cover. Changes in wind patterns and drifting snow, together

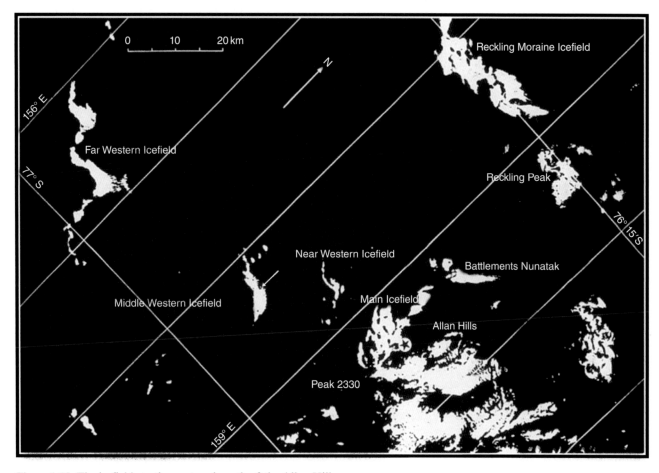

Figure 1.10. The icefields to the west and north of the Allan Hills.

with ablation, may ultimately reveal fresh exposures of ice dotted with meteorites covering this entire area.

The Battlements Nunatak Icefield lies about 10 km north of the Main Icefield. It is composed of two small ice patches a short distance west of the Nunatak and one long, narrow ice patch that extends northeastward for 12 km. The terrain south of the Nunatak is so badly crevassed that no visit to it had been made until this season, when team members Schutt and Fudali found a way through Battlements Nunatak and discovered the ice sheet beyond it to be strewn with terrestrial rocks with no readily visible meteorites mixed in with them. It resembled the patches they had seen in North Victoria Land. Although they found no sign of a meteorite concentration, they recommended that it should be visited again and studied with more care.

During this season, the flagging and mapping of meteorite finds was applied for the first time to create location maps that have since proven to be invaluable in solving pairing problems. In addition, oriented blocks of ice were collected for petrographic studies by Ian Whillans at Ohio State University. Ian spent the last 10 days of the season at the camp.

Schutt, Fudali, Crozaz, and I spent two and a half days sweeping the Near Western Icefield, where we collected 78 specimens (Figure 1.11). When paired, these represented at most 24 meteorites. Fifty-two of the specimens were weathered and appeared to come from one individual. These strongly resembled 30 fragments collected at the same site in previous years. Schutt and Fudali spent one day searching the Middle Western Icefield, where they

collected 14 specimens, 11 of which were paired. They found and flagged several more.

On 5 January, Bill Cassidy arrived back at McMurdo after enjoying Christmas at home. He came to the Allan Hills camp on 7 January, when we had about two more weeks to go. The first of these weeks was a very busy one spent collecting, flagging, and searching new areas. 14 January was an exceptionally fine day when we were looking forward to the arrival on the next day of Ian Whillans, hoping he would tell us all about the ice regime governing our icefields. At the end of the day when we arrived back at camp, I got off my snowmobile and started to toss its cover over it. The wind caught the cover and I lunged for it, not knowing that my left foot was wedged into the strut of the ski on my snowmobile. As I was twisted and thrown to the ground, I heard a crack and immediately was in pain. That ended my second sojourn at the Allan Hills. The following day, the helo that brought Ian Whillans to camp carried me to the infirmary at McMurdo, where x-rays showed that I had a spiral fracture of my left leg bone (fibula). Such a fracture is not serious, and the pain had ended long since, so with a proper walking cast and a cane for balance, I was told I could walk around McMurdo. I was walking around home via Los Angeles within 12 days.

1.14.6. "ANSMET Finds a Lunar Sample"

This is the heading that *Bill Cassidy* uses in his book [2003, p. 147] to introduce this topic. Bill led Ian Whillans through the Main Icefield, where they did some coring

Figure 1.11. Schutt takes notes while Marvin examines a small meteorite on the Near Western Icefield. Crozaz snapped the picture.

and mapping of find sites. Then, on 18 January, Ian wanted to see the Near and Middle Western Icefields, so the whole group rode to the Near Western and proceeded to search systematically and collect meteorites. Snow began coming down about noon so Cassidy, Crozaz, and Fudali rode back to camp while John Schutt guided Ian Whillans to the Middle Western Icefield. There, at the edge of the ice, they noted a small stone about the size of a golf ball (See the arrow pointing to the edge of the Middle Western Icefield in Figure 1.10). Schutt, who was keeping the notes, described the stone:

> #1422—Strange meteorite. Thin, tan-green fusion crust, ~50%, with possible ablation features. Interior is dark grey with numerous white to grey breccia (?) fragments. Somewhat equidimensional at ~3 cm.

They went on to collect 10 more meteorites, but that first one made history. It would prove to be the world's first recognized meteorite from the Moon. Some reports honor John Schutt as the finder and some pass the baton to Ian Whillans. In fact, they discovered it together. When the stone arrived in Houston, months later, it was seen by some of the world's leading experts on lunar rocks. They assigned to it the number ALH A81005 (Figure 1.12; see also Plate 64). When Brian Mason saw the thin section, he put in writing what many had been thinking: "Some of the clasts resemble the anorthositic clasts described from lunar rocks." There it was, all 31.4 g of it, almost certainly a meteorite from the Moon, but specific measurements would be required to prove it.

Cassidy, who was chairing the Meteorite Working Group, realized that numerous members of that committee would submit requests for a research sample, which meant there would be a series of absences while their requests were discussed. So he decided to form an

ad hoc committee of scientists who were familiar with meteorites and lunar rocks but did not intend to submit sample requests. On 4 December 1982, Cassidy held a special meeting in Houston to assess which lines of research would be the most definitive of lunar origin. Of the requests for samples submitted to the committee, those that were granted proposed to investigate one of the following six lines of research: oxygen isotope ratios, noble gas measurements, cosmogenic nuclides, nuclear particle tracks in feldspar grains, neutron activation analysis, and rare-earth element analyses. The committee itself solicited two more studies: measurement of the magnetic properties of several small chips, and passive counting of a major piece of the specimen for aluminum-26, which indicates how long a meteorite has been lying on the Earth's surface. It sounds like a lot of material, but the total weight of the eight samples allocated summed up to 2.6 g.

At the annual Lunar and Planetary Science Conference, held in Houston in March of 1983, 16 scientists presented evidence that this was a lunar meteorite. In his book [2003] on page 158, *Cassidy* reports that at the end of a long session of papers supporting a lunar origin, Randy Korotev of Washington University at St. Louis announced that he would like to present some evidence for why we believe ALH A81005 is *not* from the Moon. A hush fell over the room. Then Randy added: "Unfortunately, we were unable to find any such evidence, so I will have to talk about something else."

The very existence of even one meteorite from the Moon forced new lines of thought in planetary science. No longer could it be argued that lunar craters could not be due to meteorite impacts. Nor could it be argued that inasmuch as we have no meteorites from the nearby Moon, we certainly cannot have any from Mars. And it weakened the argument that the force of an impact required to send a martian meteorite into an earth-crossing orbit would totally destroy the rock. In fact, rather quickly after the verification of ALH A81005, studies redoubled of the strangely youthful, 180 million to 1.3 billion (instead of 4.5 billion) -year-old meteorites called shergottites, nakhlites, and chassignites until it was demonstrated that bubbles in the glass of EET A79001 (Plate 70) contain the martian atmosphere.

Some meteoriticists saw this Antarctic expedition as being of equal importance to an *Apollo* Mission, two of which had been cancelled within the previous year.

> With the *Apollo* 18 mission cancelled,
> the ANSMET expedition of 1981–1982
> to the Allan Hills of Antarctica
> collected lunar meteorite
> ALH A81005
> and changed the history
> of planetary science.

Figure 1.12. The lunar meteorite ALH A81005 after one chip has been taken off at Houston. (NASA photo)

REFERENCES

Cassidy, W. A. (2003), *Meteorites, Ice, and Antarctica. A Personal Account.* Cambridge University Press, Cambridge, UK.

Clarke Jr., R. S. (1982), Descriptions of iron meteorites. In *Catalog of Meteorites from Victoria Land, Antarctica, 1978–1980*, U. B. Marvin and B. Mason, eds. (pp. 50–51).

Fudali, R. F., and J. F. Schutt (1984), The field season in Victoria Land, 1981–1982, in *Field and Laboratory Investigations of Meteorites from Victoria Land, Antarctica*, edited by U. B. Marvin and B. Mason, pp. 9–16, Smithsonian Inst. Press, Washington, D. C.

Gorai, M. (1970), Meteorite museum. In *Magma, 23*, p. 8 (In Japanese).

Marvin, U. B. (1983), Granite House: 1911–1981. *Antarctic J., XVIII*, 15–16.

Marvin, U. B., and G. J. MacPherson (1989), Field and laboratory investigations of meteorites from Victoria Land and the Thiel Mountains Region, Antarctica, 1982–1983 and 1983–1984, *Smithsonian Contributions to the Earth Sciences, 28.*

McKinley, S. B., and K. Keil (1984), Petrology and classification of 145 small meteorites from the 1977 Allan Hills collection, in *Field and Laboratory Investigations of Meteorites from Victoria Land, Antarctica*, edited by U. B. Marvin and B. Mason, pp. 55–71, Smithsonian Institute Press, Washington, D. C.

Shima, M., M. Shima, and H. Hintenberger (1973), Chemical composition and rare gas content of four new detected Antarctic meteorites. *Earth and Planetary Science Letters, 19,* 246–249.

Shima, M., and M. Shima (1973), Mineralogical and chemical composition of new Antarctic meteorites, *Meteoritics, 8,* 439–440.

2

Fieldwork Methods of the U.S. Antarctic Search for Meteorites Program

Ralph P. Harvey, John Schutt, and Jim Karner

2.1. INTRODUCTION

The U.S. Antarctic Search for Meteorites (ANSMET) program has recovered more than 20,000 meteorite specimens since fieldwork began in 1976. The methods employed during fieldwork have evolved considerably over that interval in response to demand, logistical support, and an improved understanding of the links between Antarctic meteorite concentrations and their geographical and glaciological setting. This chapter describes how ANSMET fieldwork has evolved over the years to produce the current meteorite recovery methods and discusses how they relate to the complex phenomena of Antarctic meteorite concentrations, both in theory and in practice.

2.2. ANSMET FIELD SEASONS YESTERDAY AND TODAY

2.2.1. ANSMET's Place Among Modern Antarctic Meteorite Recovery Efforts

Meteorites have played a role in Antarctic science since the earliest years of the twentieth century. The first meteorite recovered from Antarctica, about 10 cm across and fully fusion crusted, was found by one of Douglas Mawson's field parties in 1912, lying on hard snow on the Adelie Coast [*Mawson*, 1915]. F. L. Stillwell, a geologist in the field party, immediately recognized the rock as a meteorite and studied it in detail after the expedition returned to Australia [*Bayly and Stillwell*, 1923]. Four decades later cooperative international scientific exploration of the Antarctic continent commenced with the

Department of Earth, Environmental, and Planetary Science, Case Western Reserve University, Cleveland, OH, 44106-7216

1957 International Geophysical Year. The global and developmental nature of that effort led to the high level of scientific activity in Antarctica that continues today.

During those early years, three meteorites were discovered during geological surveys: Lazarev, an iron recovered in two fragments from the Humboldt Mountains in January of 1961; Thiel Mountains, a pallasite recovered in two fragments in December of the same year; and Neptune Mountains, a single iron recovered from the Pensacola Range in February of 1964. Both Lazarev and Neptune Mountains were discovered on mountain slopes during geological surveys and were not associated with any obvious glacial processes [*Tolstikov*, 1961; *Turner*, 1962; *Ravich and Revnov*, 1963; *Duke*, 1965]. Thiel Mountains, on the other hand, was a harbinger of the future; the two fragments were found on "hard, irregularly cupped glacier ice" to the northeast of Mount Wrather, associated with morainal debris, as described by *Ford and Tabor* [1971]. These authors also noted that the association of the specimens with morainal debris implied that the specimens had been transported from their original fall site, and that the weathering state of the specimens implied that abrasion in the cold, dry katabatic winds of the polar plateau was extremely effective as a local mechanism of erosion. Their observations proved prescient; the Thiel Mountains pallasite deserves consideration as the first meteorite recovered from an Antarctic meteorite concentration surface (as later recoveries from the region would confirm). Unfortunately, it was the only meteorite located at that time, and thus the concentration at Thiel Mountains would not be recognized until 1982 [*Schutt*, 1989].

There is little ambiguity as to the event that revealed the existence of Antarctic meteorite concentrations. On 21 December 1969, Renji Naruse of the tenth Japanese

35 Seasons of U.S. Antarctic Meteorites (1976–2010): A Pictorial Guide to the Collection, Special Publication 68,
First Edition. Edited by Kevin Righter, Catherine M. Corrigan, Timothy J. McCoy and Ralph P. Harvey.

Antarctic Research Expedition (JARE-10) was one of several glaciologists establishing a network of survey stations in the East Antarctic ice sheet to allow the study of glacial movement. As they extended their survey across a blue icefield uphill from the Yamato (Queen Fabiola) Mountains, they found a total of nine meteorite specimens [*Yoshida et al.*, 1971; *Yoshida*, 2010). Within a few years, mounting numbers of meteorite recoveries by the Japanese eventually convinced the United States Antarctic Program (USAP) to begin supporting active searches, as described in chapter 1 [*Marvin*, 2014 (this volume)].

As of this writing, ANSMET has recovered more than 20,000 meteorites, but these numbers account for only part of the program's success. Consistent initial characterization of recovered specimens, curation at the highest level, and rapid, cost-free availability have given ANSMET meteorites unique value within the planetary materials research community, as described in chapter 3 [*Righter et al.*, 2014 (this volume)].

The early successes of the U.S. Antarctic meteorite program quickly led to increasing demand for new specimens and to the well-supported, institutionalized programs of recovery in place today. With a strong backbone of aerial logistics, U.S. expeditions have ranged widely across Antarctica, predominantly in East Antarctica along the Transantarctic Mountains. ANSMET has been one of the most active governmentally-supported meteorite recovery programs, with 45 independent field parties deployed to more than 75 different sites during 36 seasons of fieldwork (Figure 2.1 a through e). Continued demand for specimens recovered by ANSMET has been the primary driver for annual field parties, which are enabled by improvements in remote sensing of polar regions, an increased understanding of meteorite concentrations, and better access to remote locations. The fieldwork has evolved with these changes, resulting in a field program that is highly adapted to available logistics and the needs of the planetary materials community. This chapter documents the field practices that have helped ANSMET support research through the past four decades.

2.2.2. Preseason Planning and Site Selection

U.S. activities in the Antarctic are carried out within the U.S. Antarctic Program (USAP), funded and managed by the Office of Polar Programs (OPP) of the National Science Foundation. During its history, ANSMET has been supported through USAP, both directly (through OPP grants) and indirectly (through logistical support funded by NASA). As a result, planning for any given season may begin as many as seven years before deployment (at the time a grant is funded). Grants supporting ANSMET have been competitively selected,

with durations ranging from as many as six seasons to as few as one. Grant proposals may request support for field seasons dedicated to systematic meteorite recovery from known sites (hereafter called a systematic activity), reconnaissance efforts dedicated to improving our understanding of poorly known or previously unvisited sites (hereafter called a reconnaissance or recon activity), or some combination of these. A typical proposal will therefore include a list of sites prioritized from among potential targets based on our understanding of each site's potential. The highest-priority fieldwork targets are, not surprisingly, those we think will yield the most meteorite specimens. However, this is typically based on prior experience at the site, which is always incomplete during early visits. More practical issues such as logistical availability can trump potential meteorite yield. For example, when a remote helicopter camp allows us to reach otherwise inaccessible locations, those targets become a higher priority; and when aircraft support is predicted to be limited, we may choose targets demanding fewer flight hours.

Recon and systematic targets are both mixed into the long-term plan; the former typically result in fewer meteorite recoveries but are essential to ensure a continuous supply of new specimens. On a few occasions we have also adjusted the recon/systematic activity mix to reduce stress on the curatorial system, favoring recon activities when a characterization backlog is growing. The desire for geographical separation between ANSMET field parties (to minimize the effects of individual weather systems) is also considered. Every ANSMET proposal also includes alternate targets for either style of activity, allowing the project to adjust to rapid changes in USAP logistics and programmatic issues.

Meteorite concentration sites tend to occur on exposed blue ice in a variety of specific geographical and glaciological settings; the characteristics of these icefields and meteorite concentration mechanisms are discussed in detail in *Harvey* [2003]. Identification of such sites through examination of maps and imagery has been a natural first step in ANSMET's work since the program began. USAP-produced topographic maps and aerial photography documenting much of the Transantarctic Mountains became available throughout the 1960s and 1970s, and during ANSMET's early years these served as a primary means for identification of meteorite concentration sites. These maps, however, were primarily meant to serve navigational and geological needs, and do not document blue ice. Similarly, while aerial photography coverage was excellent, many blue ice areas were visible only in low-angle oblique images that mask their full extent. The most powerful "remote sensing" tool employed by ANSMET in this era was opportunistic reconnaissance flights, during which field personnel would sit glued to the windows of an aircraft, annotating maps and photographing key

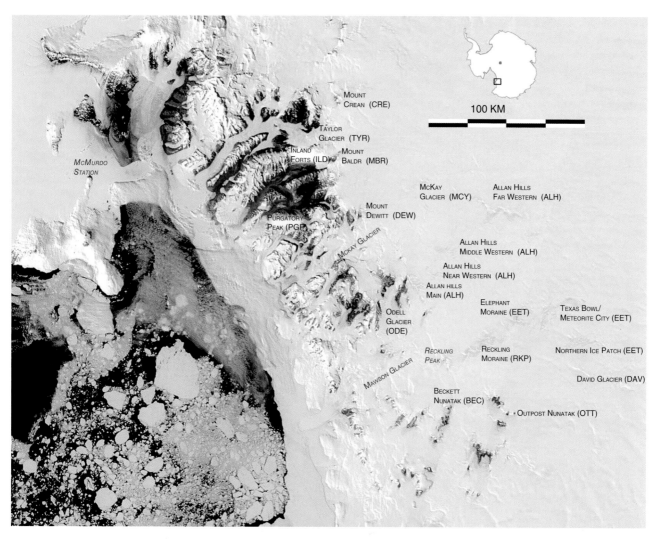

Figure 2.1(a through e). Meteorite concentration localities explored by ANSMET to date. The localities shown represent targets of ANSMET field seasons, typically icefields or groups of icefields within a target region. All location names should be considered informal, and where meteorites have been recovered the appropriate three-letter location code assigned to those specimens (e.g., ALH) is shown. In many cases a single code is used for several icefields, particularly where smaller geographical features were unnamed. The outline of Antarctica above the scale bar shows the approximate context of the figure within the Transantarctic Mountains. For additional context, a few geographical features are also shown in blue. A mosaic of MODIS Rapid Response Terra images (250-m resolution) is used as a base for all sections of the figure.

Figure 2.1a. ANSMET meteorite localities in the McMurdo Sound region, including many of the sites explored in the earliest period of ANSMET activity.

features of promising sites. Such flights were a common feature of early ANSMET seasons and remain a part of our reconnaissance tool kit, given their ability to reveal current surface features and conditions (rather than those in maps or images that can be decades old). On several occasions, "low and slow" flights led to the identification of meteorites from the air in areas where terrestrial rock was known to be absent; but in general, such discoveries have been very rare due to the small average size of meteorite specimens, the vibration of the aerial platform, and the limits of human visual acuity.

Satellite imagery became publicly accessible at about the same time ANSMET was formed, and with each technological advance it has played an increasing role in the project. ANSMET first used Landsat satellite imagery for reconnaissance purposes in the late 1970s, with significant help from the U.S. Geological Survey (USGS) [e.g., *Lucchitta et al.*, 1987]. Although initially restricted to latitudes north of 80° S and with limited surface resolution (80 m initially, 15 m later), the "bird's eye view" and geolocation afforded by this imagery dramatically improved ANSMET's identification of targets for

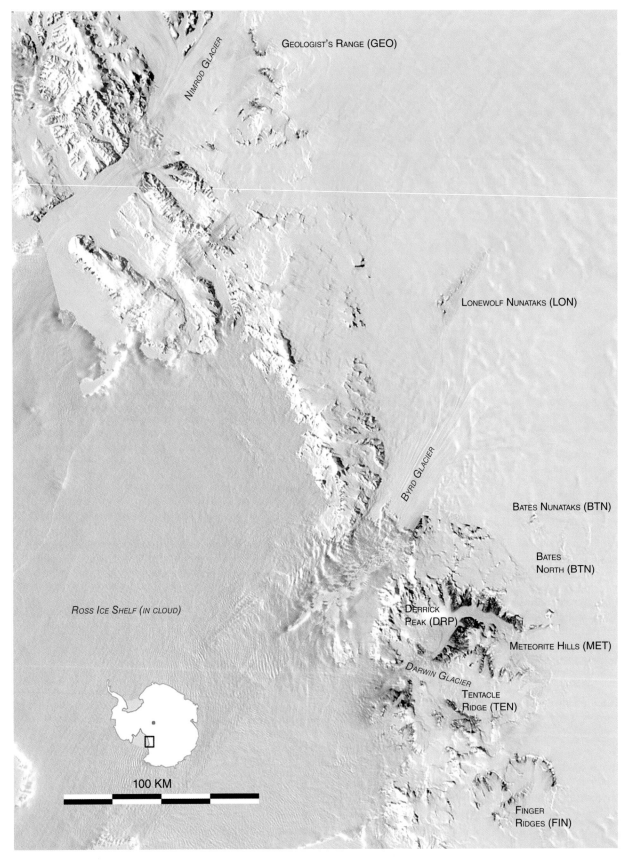

Figure 2.1b. Icefields further south and east along the Transantarctic Mountains between the Darwin Glacier region to the north and the Nimrod Glacier to the south.

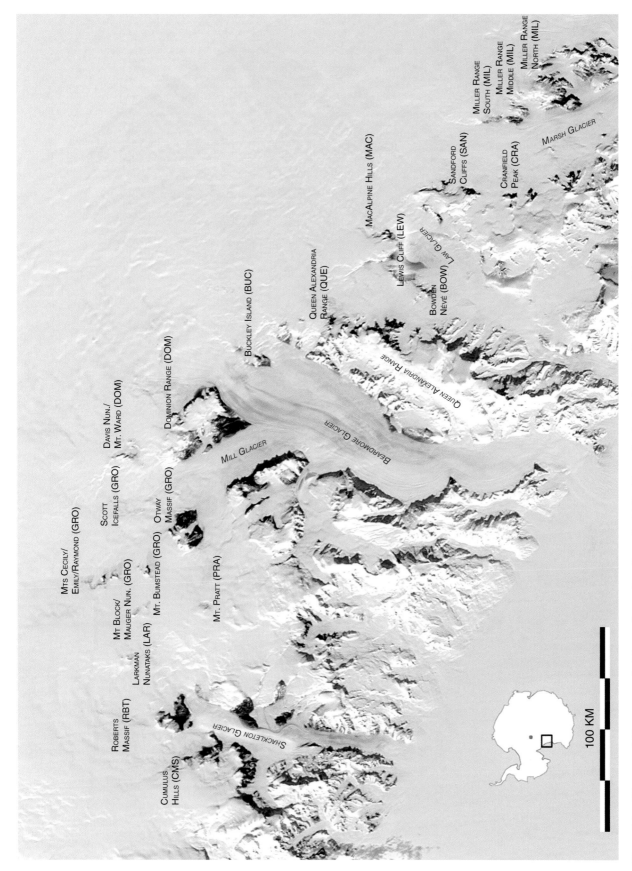

Figure 2.1c. The central Transantarctic Mountains region, from the Miller Range in the northwest to Roberts Massif in the southeast.

100 KM

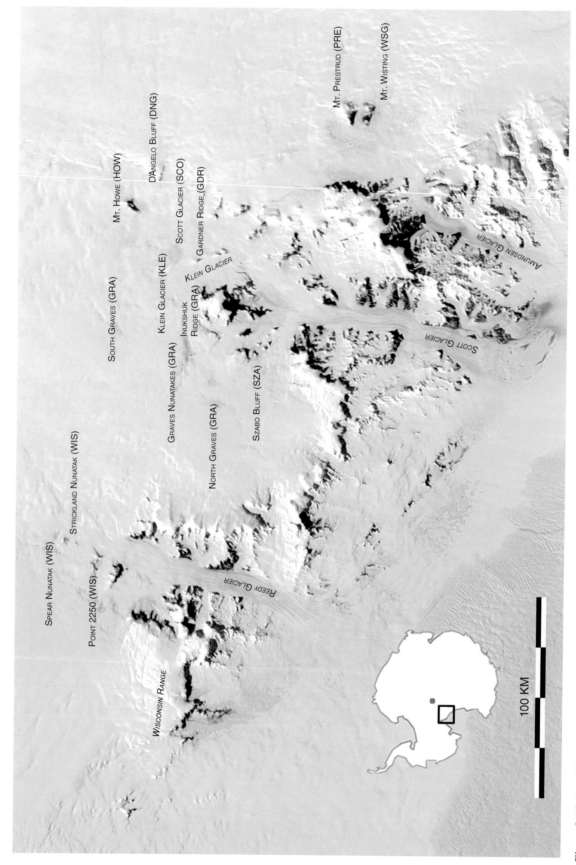

Figure 2.1d. Localities in the southernmost Transantarctic Mountains between the Amundsen glacier to the northwest and the Wisconsin Range to the east.

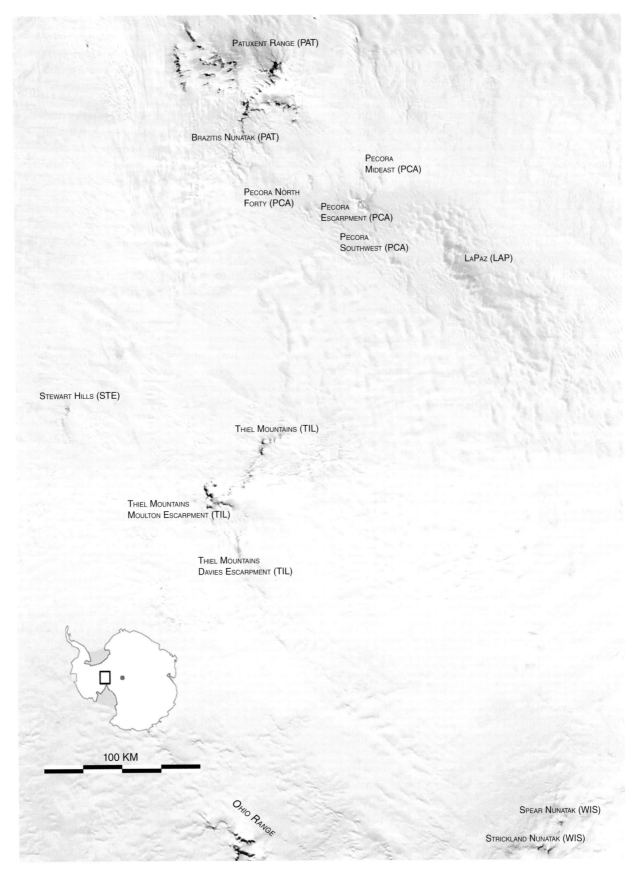

Figure 2.1e. Localities in the easternmost part of the Transantarctic Mountains, in the Weddell Seas sector of Antarctica, ranging from the Wisconsin Range to the west (bottom of figure) to the Patuxent Range in the east (top of figure).

exploration. A continent-wide mosaic of Advanced Very High Resolution Radiometer (AVHRR) satellite imagery prepared by the USGS in 1991 and revised in 1996 afforded a similar leap forward [*Ferrigno et al.*, 1996]. Although significantly poorer in resolution than Landsat (about 1 km per pixel), the AVHRR satellite map removed the latitude restriction and documented the East Antarctic ice sheet in its entirety. This in turn led to the discovery of several key icefields distant from the Transantarctic Mountains, notably the informally named LaPaz icefields. In the same time frame, Radarsat imagery also proved useful to ANSMET. It too was relatively limited in resolution (around 125 m per pixel) but had no latitude restrictions and when properly processed distinguished dense, bubble-free blue ice from surface snow. Comparisons of Radarsat and AVHRR datasets proved valuable in the identification of many of the icefields studied by ANSMET, particularly those in the most southerly Transantarctic Mountains.

Today ANSMET relies on two current-generation products for its remote sensing needs, both provided through the USAP-supported Polar Geospatial Center (PGC) at the University of Minnesota. First is imagery from the MODIS (Moderate Resolution Imaging Spectroradiometer) Rapid Response System [*Justice et al.*, 1998]. MODIS instruments aboard both the *Terra* and *Aqua* satellites image the entire Earth's surface every one to two days at ~250 m/pixel, acquiring data in 36 different spectral bands. The Rapid Response System provides daily images of Antarctica in true color. These images have proven exceptionally useful for direct confirmation of the presence of blue ice, and their daily recurrence allows selection for cloud-free views with limited snow cover and maximum Sun angles, helping us identify many smaller icefields throughout the Transantarctics and reducing our dependence on reconnaissance overflights.

When localized detail is needed, such as to serve as a base map for meteorite searches, the PGC also provides ANSMET with high-resolution satellite imagery. This imagery, licensed by the PGC from GeoEye, Digital Globe, Ikonos, and other sources, can have resolutions as high as 1 m/pixel. The main value of such images is in the tremendous geographical control they provide when used in concert with GPS-derived meteorite and base station locations.

Planning for individual seasons typically begins about eight months before any given austral summer, with the preparation and submission of a detailed support request to USAP. Called the Support Information Package (SIP), this document summarizes and formalizes ANSMET's needs across a broad spectrum of categories, including the specific targets for field work; a schedule of support events; participant lists; permitting needs (to comply with Antarctic treaties and federal regulations); shipping and cargo handling; potential environmental impacts; staging and storage needs; field equipment requests; laboratory, computing, and communication needs; food and fuel requirements; and myriad other details. This document, often 70+ pages for each field party, becomes the basis for negotiations between the project and USAP contractors, eventually leading to concurrence before the start of fieldwork in November.

Field party selection typically takes place in the same time frame as the support request SIP preparation. ANSMET field teams typically consist of a science leader (ANSMET project personnel), a mountain guide who also serves as camp manager, and a mixture of ANSMET veterans and first-time volunteers, with a targeted ratio of experienced to first-time field party members around 50:50. ANSMET is relatively unique among Antarctic field projects in that we welcome the involvement of volunteers from the research community, including international participants. Preference is given to those whose research involves Antarctic meteorites, but individuals with related research and/or significant related experience are also considered. After selection by the ANSMET principal investigator, all applicants must pass a set of qualifying medical and dental screening exams required by USAP due to the limited emergency treatment available in Antarctica. Inclusion of volunteers from the planetary research community does pose challenges, given that some have limited field experience in isolated or cold weather environments. However, the payback has been significant, since inclusion of volunteers reinforces the altruistic nature of the Antarctic meteorite program as a whole, encourages continuous and conservative field safety training, promotes highly efficient and robust field practices, lowers costs, and perhaps most importantly, injects new energy into the fieldwork each season. As of this writing, more than 170 scientists have participated in ANSMET fieldwork, and our record of safe field operations and continuous recoveries validates the inclusion of newcomers to the program.

2.2.3. *Field Season Structure and Logistics*

The basic unit of ANSMET activity is the field season: the annual period when one or more teams are deployed to target icefields for as long as six weeks. Fieldwork typically begins in early December, when the Sun is at its highest and the lack of diurnal atmospheric cooling helps to minimize katabatic winds on the East Antarctic Plateau. The season often continues until late January, when logistical support in the McMurdo region shifts away from science and toward preparations for the coming Antarctic winter.

McMurdo Station, the largest U.S. base in Antarctica, serves as USAP's hub for operations in the Transantarctic

Mountains and as the starting point for ANSMET expeditions. During a typical season, a few experienced ANSMET personnel will arrive in McMurdo in mid-November to begin assembling and preparing expedition gear. The remainder of the team typically arrives in McMurdo in late November and immediately engages in an intense 7–10 day preparation period. In addition to assembling and testing the remaining field gear and entering it into the cargo stream, the team spends several days training both to meet the challenges of the Antarctic environment and to introduce ANSMET procedures and protocols.

The remote nature of ANSMET field sites requires that the material needs of the field team be minimized, allowing the team and all its gear to be efficiently moved by aircraft in as few loads as possible. Aircraft use varies considerably between field seasons, but most seasons require the team to first move from McMurdo to an intermediate site suitable for landings by large, ski-equipped aircraft (usually the iconic LC-130 Hercules). From there the team will move to the target site, either by smaller aircraft (Twin Otter or helicopter) or by overland traverse using snowmobiles and sleds to move our gear. Travel to and from target sites can consume a significant proportion of a field season and logistical resources. Systematic field parties are typically larger and less mobile than reconnaissance parties, with six to eight people and one or two main targets for a given season. In contrast, reconnaissance teams are smaller (two to four people) and may move many times during a season, with stays at target icefields as short as a few hours or as long as a few weeks. Living in tents and conducting most searches from aboard snowmobiles, the field team will typically deploy with enough fuel, food, and other expendables to cover a significant portion of an entire six-week season; one or two resupply visits by light aircraft make up the difference and provide the opportunity for swapping out waste and damaged gear. When distances between target icefields are low and aircraft are available, ANSMET sometimes conducts a "flying traverse," with bulk cargo moving from one site to the other via airplane while the field team transports itself and survival gear overland by snowmobile.

ANSMET fieldwork is supported with rugged and functional equipment that serves both survival and scientific needs. USAP provides each participant with a basic wardrobe of extreme cold weather (ECW) clothing, including the infamous big red parka. Many ANSMET participants supplement this clothing with more personal or specialized gear for improved function and mobility (notably eyewear, gloves, and underwear). USAP also provides the four-sided, double-walled pyramidal Scott tents that serve as shelter, each occupied by two field party members and containing a propane stove for warmth and cooking. Plywood and thick insulating pads on the floors help keep the tents warm, while thick sleeping bags provide overnight comfort. In recent years, each tent has also been equipped with a 65W solar panel/field power station that makes modest electrical power available for electronic devices such as computers, GPS, cameras, and satellite phones. Expendables for the field camp primarily include food and fuel (propane, gasoline, and aviation fuel), with all solid waste recovered for recycling and/or removal from the continent.

Snowmobiles are a key part of the ANSMET tool kit. Not only do they serve as an individual mode of transportation, they also dramatically increase the range over which an ANSMET field party can conduct searches. They also allow independent mobility and constant team restructuring, serve as mobile storage and measurement stations, and dramatically reduce the fatigue associated with human-powered transport in the Antarctic. Unquestionably, the logistical costs associated with snowmobile use are high, due to the need for aerial transport not only of the vehicles but also of up to 700 kg of fuel and spare parts for each in a typical season, as well as off-season maintenance and storage. However, the gains in terms of mobility and search efficiency are equally large. Snowmobiles are to ANSMET field party members what horses are to cowboys, serving a multitude of needs and dramatically increasing the effectiveness of each individual. ANSMET personnel have tested the human-powered model, and while such efforts are good for soul and body, they can dramatically reduce the effectiveness of meteorite searches [*Haack et al.*, 2008].

2.3. SEARCHING FOR METEORITES IN ANTARCTICA

The goal of ANSMET fieldwork is to recover a complete and representative sample of the extraterrestrial materials falling to Earth so that it can be made available for research. While to some ANSMET's main task may seem simple (only slightly elevated above an Easter egg hunt), meeting these goals efficiently and with high standards requires planning and a particularly methodological approach. ANSMET has developed procedures and protocols to systematically recover meteorite specimens, ensuring few missed specimens, avoiding preferential sorting by type or by size, and maximizing scientific returns through contamination control and detailed record keeping. These procedures and protocols are a critical component of ANSMET success, helping ensure that the U.S. collection is both representative of the materials coming to Earth from space and contains the maximum number of samples of the rarest types of lithologies.

2.3.1. Reconnaissance Procedures

The goal of ANSMET reconnaissance work is to determine whether a blue ice area harbors a meteorite concentration, and if so, to understand its full extent. While individual rocks can be seen from the air (even satellite-based imagery can now pick out rocks in the 10's of cm range), determining the nature of these rocks (extraterrestrial or otherwise) still requires a personal visit. Reconnaissance visits typically take a variety of forms in a hierarchy of effort levels, from first visits to full-scale expeditions (Figure 2.2 a and b).

2.3.1.1. Early visits. Even a single meteorite find can prove the value of the site to ANSMET and serve as the impetus for future larger-scale recoveries. As a result (and given time constraints) the goal for most early visits is to examine as much high-priority blue ice as time and equipment allow. Often the first visit to a site will occur as part of a long day trip, where icefields within reach of an existing ANSMET camp are visited by snowmobile and examined for a few hours. Day trips are fairly common during the early sessions of systematic searching at a given icefield, since they are easily supported by the larger field team and fuel supply associated with such a camp. These efforts can also be very effective; day trips taken during the early years of systematic searching at the Lewis Cliff Ice Tongue led to the discovery of several additional meteorite concentrations in the Walcott Névé region, including the Foggy Bottom/Goodwin Nunatak and MacAlpine Hills icefields, home of the QUE and MAC meteorites, respectively. Thus a few reconnaissance day trips more than tripled the number of specimens recovered in the Walcott Névé region. Aerial support has also been effective in spite of obvious limits on the number of people and amount of equipment that can be transported. The first meteorite concentrations discovered by ANSMET in the Allan Hills region were all found during helicopter-supported day trips, as at the Lewis Cliff Ice Tongue and the Miller Range icefields about a decade later. Similar day trips, supported by helicopter or Twin Otter, take place whenever aerial support and potential targets coincide.

When a target icefield is simply too distant to be visited without an overnight stay, the need for increased survival gear scales up the complexity of a reconnaissance visit. For such visits ANSMET will typically send a team of two equipped with snowmobiles and survival gear sufficient for several days. Many major meteorite concentrations were first explored through two-person visits; notable examples include the LaPaz icefields, first visited in 1991, and more recently the Buckley Island icefields. The additional time and mobility available during such visits dramatically increases the area of ice that can be examined and allows the visit to do more than establish the presence or absence of meteorites; the full scope of the concentration can often be gauged. These extended first visits therefore often lead to some important decision making for ANSMET. Sometimes a two-person field party will find too few meteorites to support full-scale systematic recovery, and under those conditions the team may try to complete systematic recovery themselves. Examples of this kind of action from recent ANSMET history include the icefields near Bates Nunatak, Lonewolf Nunatak, and in the Geologist's Range. Similarly, an icefield may be considered "uneconomical" for other reasons, such as a very low concentration of meteorites, or an overwhelming abundance of terrestrial rock that compares poorly with other known icefields where recoveries are easier. The best situation is when the team quickly encounters large numbers of meteorites, immediately proving the need for large-scale systematic recovery. Having met the primary goal of reconnaissance, that team will typically move on to new targets as soon as possible. Nature, of course, has provided us with icefields exhibiting every level of concentration between these end members, and as the availability of logistical support and the demand for meteorites change, some icefields thought uneconomical by earlier explorers may be targets of systematic work in the future.

2.3.1.2. Large-scale reconnaissance. ANSMET has periodically dedicated whole field teams and seasons to reconnaissance efforts, usually when we have identified a broad region that contains numerous potential target icefields. Such a season is designed to send a lightly-equipped four-person team to a number of icefields, either on a long overland traverse or with extensive aerial support. Such a season may include opportunistic first visits to icefields, as well as visits to sites where meteorites have been previously recovered but the extent of the concentration (if any) remains unknown.

The amount of time planned for each icefield is estimated from previous visits or its geographical extent, with visits lasting from a few days to a week or more. With the many unknowns including rapidly changing weather conditions and challenging logistical schedules, preseason reconnaissance plans rarely survive first contact with the target icefield. We have learned that a dedicated reconnaissance season requires a very flexible timeline that allows for dramatic shifts in priorities as the season progresses. As noted previously, even a single meteorite find can lead to many days of searching a new icefield; and on many occasions long days of searching can lead to few or no finds at all. High numbers of meteorite recoveries are not typically expected during large-scale reconnaissance: the increased number of days dedicated to travel between icefields alone limits searching time.

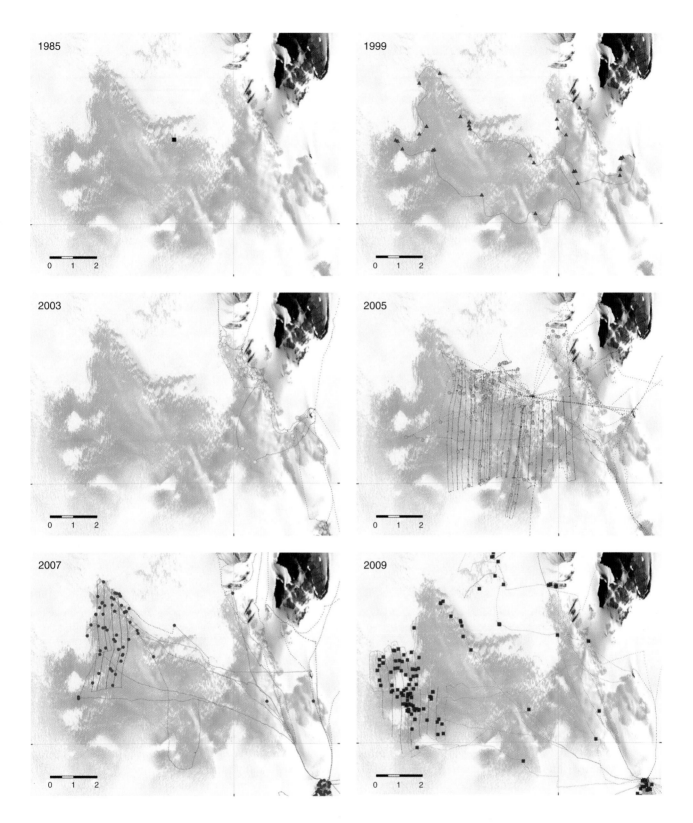

Figure 2.2a. Six seasons of ANSMET activities at a single icefield demonstrating reconnaissance and systematic styles of searching. An Advanced Spaceborne Thermal Emission and Reflection Radiometer (ASTER) satellite image of the Miller Range northern icefield is shown with finds and snowmobile traces for each season as labeled. Boxes, dots, and so on show individual finds in various colors. In all seasons, the path of a single GPS-equipped snowmobile is shown to demonstrate search activity; one or more additional snowmobiles would have also been active. The first visit, in 1985 (upper left), was a single overflight by helicopter; one meteorite was recovered. A two-person reconnaissance visit in 1999 (upper right) led to the recovery of 30 specimens, with searching suspended after two days given clear signs of a major concentration. A four-person team conducted extensive reconnaissance throughout the Miller Range in 2003 (middle left), recovering meteorites and documenting the need for systematic meteorite recoveries. The weather-plagued first season of systematic recoveries in 2005 (middle right) concentrated primarily on the northern icefield, with most systematic searches on the eastern side and including a few reconnaissance trips to nearby ice patches. Systematic meteorite recoveries from the northern icefield continued during the 2007 field season (lower left) and were completed in 2009 (lower right) with the middle icefield (extreme lower right) as a main target of activity.

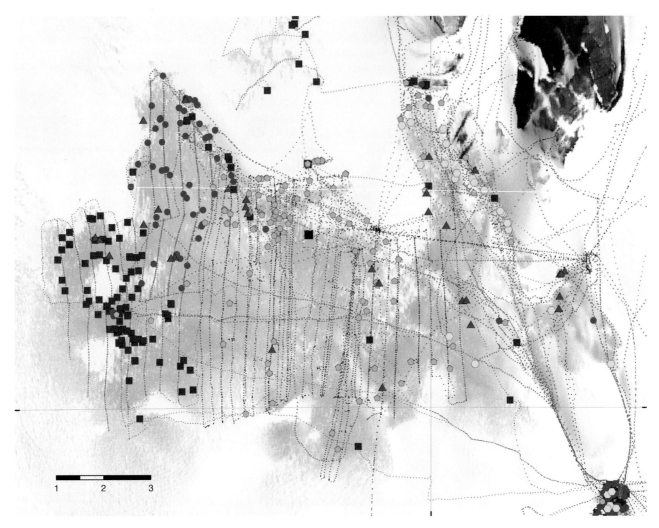

Figure 2.2b. (*Continued*) The 365 meteorite locations and snowmobile paths for all six seasons of ANSMET meteorite recoveries at the Miller Range northern icefield. Base map and symbols are as described in Figure 2.2a. Systematic recovery from the region continues to this day, primarily focused on other local icefields further to the south and smaller icefields in surrounding areas (not shown).

The payback comes in the form of future systematic meteorite recoveries that in many cases involve multiple seasons and thousands of new specimens.

2.3.1.3. Reconnaissance searching techniques. The imposed time limits and distinct goals of reconnaissance searching have led to search techniques that differ from those employed during more systematic meteorite recovery efforts. Reconnaissance searching is designed to cover large areas of blue ice quickly rather than to completely examine the surface in a systematic fashion. Reconnaissance searching typically takes the form of a loosely organized series of transects, with experienced personnel forming two semiparallel paths separated by several hundred meters and the other members of the team widely spaced between them. Overlap between searchers is neither encouraged nor controlled, and spacing between field party members may vary considerably as the team tries to accommodate local topography, hazards such as crevasses and moraines, and maintain continuous visual contact. The latter is important both for safety reasons (these are usually sites not previously visited) and to ensure that any meteorite discoveries are quickly noted. When a suspected meteorite is spotted, all members of the team converge on the find. Not only can the group as a whole establish whether or not it is a meteorite (and participate in subsequent collection activities), but the act of converging itself creates a locally dense search grid that often leads to further nearby discoveries.

Priorities can change dramatically during such a search. When a meteorite is found, the field team leaders may choose to redirect the transect in a new direction rather than continue on the previous bearings. If the find is in an area certain to be revisited and time is short, the specimen may be flagged and left for later recovery. When few or no meteorites are encountered, reconnaissance transects will explore as much of the icefield as possible, to eliminate (as far as possible) the possibility that a concentration was missed. When more meteorites are encountered, spacing within the transects may be narrowed to better define the scope of the concentration; and when abundant meteorites are encountered, the reconnaissance team may choose to conduct fully systematic recoveries using overlapping, highly controlled transects (as described in the next section). Typically, this happens when the scale of the concentration appears too small to warrant a future visit by a larger team, and enough time remains in the season to complete the work. Alternatively, the reconnaissance team may choose to move on to a new site as soon as possible and leave the rest of the recoveries to a larger, better-equipped team. Reconnaissance at any scale is considered complete when the value of possible future visits to the site by ANSMET is known.

2.3.1.4. Systematic searching procedures. Systematic searching is among the most basic of ANSMET field activities. It involves the methodical recovery of meteorites from a stranding surface where a meteorite concentration is known to exist and the potential for large numbers of recoveries is high. Typically, systematic searching field teams consist of eight individuals, but there can be more or fewer depending on factors such as logistical availability and the area of ice to be searched. Systematic search teams are normally only sent to sites that have been explored in some detail by prior reconnaissance teams, allowing priorities for a given season to be set in advance and logistical demands to be well constrained.

ANSMET search strategies typically follow the transect sampling model in use by natural scientists for hundreds of years [e.g., *Anderson et al.*, 2002; *Barabesi et al.*, 2002; *Chen et al.*, 2002; *Hammond et al.*, 2002]. During these transects the field team forms a line, each member a few tens of meters to several tens of meters apart. The team then proceeds to cross the meteorite stranding surface in a direction perpendicular to this line. After each pass is completed, the team changes direction and a new transect is started, covering new ground and exploring new areas of exposed ice (Figure 2.2). The orientation and pattern of the traverses are adapted to local geographical features, hazards, and weather conditions (wind, Sun angle, and snow cover) to maximize the coverage and efficiency of the search. The spacing, amount of overlap, and method of travel (foot or snowmobile) may also vary depending on frequency of meteorite encounter and density of terrestrial rock.

During early recovery efforts, single transects may be used as sampling tools to prioritize among several search areas. Because each field party member is capable of independent mobility (everyone has their own snowmobile), it is not unusual for an ANSMET party to split into temporary subparties to cover immediate needs (such as distinct GPS surveying and sample recovery groups when a large number of specimens has been found in a confined area). The location and path of at least one team member's snowmobile is continuously recorded to establish the geographical location of the transects and provide a record of the field team's progress (Figure 2.2). This record of traverses, available with the advent of high-resolution satellite imagery and GPS, has dramatically improved ANSMET's ability to track our own progress both during and between field seasons.

2.4. METEORITE RECOVERY TECHNIQUES

2.4.1. Minimizing Biases

The recovery of scientific samples always involves sampling biases related to the techniques used to acquire the samples and the choices made by the scientists during sampling. Throughout its history, ANSMET has chosen to use a simple and inexpensive but very effective meteorite detection system: the human vision system. For areas where the background of terrestrial rock is very low or absent, the innate human ability to rapidly differentiate a scene into key elements and recognize those that are unique or out of place allows field party members to scan enormous areas of blue ice quickly and immediately notice any rocks upon its surface. This ability is limited only by the seeing conditions and the resolution of the human eye, which typically allows a dark, centimeter-sized meteorite to be resolved at distances of up to 100 meters on the light-colored ice [*Harvey*, 2003]. Given that ANSMET searches typically involve much shorter distances, we routinely recover meteorite specimens much smaller than this; catalogs of Antarctic specimens contain many rare types recovered in the subcentimeter size range.

Meteorite recovery tasks become more difficult and the risk of biases rises when terrestrial rocks become abundant, such as on icefields adjacent to nunataks and moraines, or in the moraines themselves. In the earliest years of ANSMET fieldwork, moraine searches were avoided because many regions free of terrestrial rock were available for searching and the difficulty of distinguishing terrestrial rocks from meteorite could be easily avoided. ANSMET has tried several different meteorite detection strategies and techniques in such environments, and we have found none more effective than simply trusting the human eye-brain combo to identify the rocks that "don't belong" after a period of familiarization with local lithologies. Recovering all the rocks from such areas has been suggested and even tested, but as the number density of terrestrial rock increases, the scale of such an effort becomes impractical, even absurd. For example, in 1997 and 1998 ANSMET marked off a 100 × 100 m region of the informally named Mare Meteoriticus icefield in the Foggy Bottom region of the Walcott Névé (the major source of QUE specimens), an area subjectively considered representative of the average numerical surface density of rocks. One hundred twenty-five rocks were recovered during this exercise, but no meteorites. This same exercise, if scaled up to the entire Mare Meteoriticus icefield, would require the collection of more than 500 million rocks in the <4-g range alone, of which roughly one in 250,000 would probably be a meteorite. Sorting meteorites from terrestrial rocks in some fashion must inevitably be considered more effective.

A number of technologically sophisticated sorting tools have been suggested and tested by ANSMET, including everything from simple metal detectors to a meteorite-hunting robot (NOMAD) equipped with multiple sensors and intelligent processing algorithms [e.g., *Apostolopoulos et al.*, 2001]. In our experience, such technological sensors have inevitably proven both slow and prone to unintentional sorting. For example, while well-calibrated metal detectors can efficiently sort iron, stony iron, and ordinary chondrite meteorites from terrestrial rock due to the presence of metal in the former, many of the most scientifically valuable Antarctic meteorites contain little or no metal and are effectively indistinguishable from common Antarctic igneous rocks. Equally important is that operation of such detectors divides the operator's attention between their eyes and the signals from the detection device; all too often, the latter takes precedence because it seems less subjective and involves conscious recognition of a signal. In fact, it *is* simpler, but primarily because it is a less data-rich detection technique, focused on the ferromagnetic properties of a rock and ignoring other key variables such as size, shape, texture, patina, and color. Second, while the speed of modern computer processors and robotic systems is growing exponentially, it has not yet come close to the human mind's ability to integrate a scene and pick out key elements. Our experiments with NOMAD suggest that a trained individual with innate positioning, path-choosing, and visual synthesis skills may be several hundred times more efficient than a robot (at least from that era) [*Harvey*, 2003]. Finally, there is ample indication that the human visual system is effective even in confusing environments. Of the 5,900 specimens recovered from meteorite stranding surfaces in the Walcott Névé region (LEW, QUE, and MAC specimens), all but a few hundred were recovered from regions rich in terrestrial rock. These specimens include many notable samples easily confused with terrestrial

rocks, including two martian specimens, five lunar specimens, and several rare igneous specimens such as angrites and brachinites. Certainly some proportion of meteorites were not recovered, particularly those lacking diagnostic fusion crust; but the overall success of meteorite recovery in such confusing environments suggests losses are not high enough to warrant dramatic changes to our current operational procedures.

Another proposed sorting strategy is "high-grading": purposefully targeting recoveries on achondrites or large specimens that are of the most interest to science and ignoring more mundane discoveries such as small ordinary chondrites [*Harvey*, 2003]. Some amount of this does in fact take place during reconnaissance searches, when unique specimens are encountered by sheer chance, time is limited, and any recoveries that do take place must be prioritized, given the risk that a site might not be revisited. Unfortunately, the potential loss of interesting specimens during high-grading is very high. As noted earlier, rare specimens are not always easily recognized from among other meteorites; the differences in their lithologies may be subtle at the hand-specimen level of examination, and fusion crust typically hides their interior. Many unique specimens in the existing Antarctic collections were not recognized as such while in the field (Figure 2.3). It is also not clear that searching specifically for rare specimens would significantly reduce the amount of time it takes to find them, given that the geographical distribution of meteorites on each icefield shows no distinction among meteorite types. Getting to the meteorite concentration site for even the most cursory examination is the major logistical cost faced by ANSMET, and with actual collection times that are short, the value of high-grading decreases. ANSMET field searches take the opposite approach, choosing to recover everything that is clearly a meteorite or has the potential to be a meteorite. By doing so, we accept some level of false positives but increase the likelihood that unusual specimens will not be overlooked.

2.4.2. Recognizing Meteorites

Many of the meteorite stranding surfaces explored by ANSMET are far enough inland of the Transantarctic Mountains that they are devoid of terrestrial rocks; any rock found at such sites almost certainly fell from the sky, essentially making recognition of them as meteorites a trivial pursuit. At the remaining sites, however, meteorites are often mixed with terrestrial rocks, either blown out onto the ice by the katabatic winds or carried in by glacial movement to form moraines. Recognizing meteorites in such settings is thus a crucial task for ANSMET field parties. The capability of the human visual system as an innate and not-entirely-conscious tool for meteorite

Figure 2.3. Field portraits of meteorites illustrating some diagnostic characteristics. The counter shows the field number used to identify each specimen while in the field (and is not the formal sample number later assigned by the Antarctic meteorite curator at NASA's Johnson Space Center). (a) LAR 06266, A typical (albeit large) find in a moraine showing the distinctive fusion crust and rust staining associated with an H5 ordinary chondrite. (b) A large rounded CV3 carbonaceous chondrite (LAR 12002) showing prominent chondrules and evaporite growth on its downwind and sunnier northern side. (c) GRO 06059, an achondrite displaying the glossy fusion crust commonly associated with feldspar-rich eucrites. (d) LAR 12320, a diogenite with multicolored fusion crust ranging from black to yellow-green. (e) Reasons not to high-grade during searches, example one: this mundane-looking specimen is MIL 11207, an amphibole-bearing R6 chondrite. (f) Reasons not to high-grade during searches, example two: MIL 07259, an acapulcoite / lodranite of nondescript appearance.

detection can be improved through training designed to let the conscious mind play a supervisory role. Over the years we've made efforts to deconstruct the meteorite recognition process, and we now recognize two "trainable" factors: the visual clues provided by the meteorite itself, and development of an internal catalog of local terrestrial lithologies.

Improving the latter for ANSMET field party members is fairly simple and follows the old maxim "The best geologist is the one who has seen the most rocks" (attributed originally to H.H. Read; see *Young* [2003]). The first few days of ANSMET fieldwork are routinely dedicated to looking at lots of rocks during searches at sites rich in local lithologies. Typically, the search site will be a moraine where previous work has suggested not only a thorough representation of local lithologies but also the likely presence of a few "example" meteorites (Figure 2.3). During such searches field party members are strongly encouraged to consciously examine every rock that catches their eye and bring any rock they are curious about to the attention of the team as a whole and the veterans in particular for identification. False positives are par for the course early on and accepted as a crucial part of the training. Anecdotal evidence suggests that this early exposure to a very complex lithological environment quickly trains the brain; it is not unusual for an individual's meteorite finds to increase at a nearly exponential rate during this training period. It sometimes leads to a phenomenon we affectionately call a feeding frenzy, where the team's rapidly increasing power to recognize meteorites overwhelms leadership's attempts at managing systematic progress during the search. There are worse problems to have given our goals.

When meteorites are encountered during these early searches, focus shifts to the other trainable factor (recognition of the features of Antarctic meteorite finds). Most field party members have some prior experience with meteorites in hand sample. During training in McMurdo, they are asked to familiarize themselves with hundreds of images of previous finds. Meteorites in the wild can look very different than those images, due to lighting and background conditions, and even experienced veterans benefit from a refresher course on the features that distinguish Antarctic meteorites.

The most distinctive feature of meteorites and the one that most often distinguishes them from terrestrial rocks is fusion crust. On their way to the ground, meteorites develop a thin shell of melt as 10–20 km/s of velocity is converted into thermal energy within the Earth's atmosphere. The resulting layer of melt, once chilled to a glass, is called fusion crust. With notable exceptions, fusion crust is distinct from a meteorite's interior and much darker than the weathering rind common on native Antarctic rocks. It often shows flow lines and fluid features characteristic of a semi-liquid state and is rarely more than a few mm in thickness. Fusion crusts can range from a matte black, polygonally fractured surface reminiscent of a charcoal briquette to a smooth glassy black resembling furnace slag. Fusion crust is almost always black but can vary in color depending on the minerals being melted; gray, green, and even yellowish fusion crust has been noted on some unusual specimens (Figure 2.3). Only a very small percentage of Antarctic meteorites show no fusion crust whatsoever, usually due to physical weathering.

In the absence of visible fusion crust, other clues can help one recognize meteorites. Meteorites are often well rounded and equant in comparison to their terrestrial neighbors; their fiery plunge through the atmosphere tends to take off any sharp corners, and structural controls on their shape (such as bedding, jointing, etc.) are virtually absent in meteorites and common in terrestrial rock. Meteorites often are different in size than the local rocks, particularly in settings where aeolian sorting has occurred; they can be either larger than the wind-sorted rocks around them simply because they were delivered there by different means, or smaller because their higher density and rounded shape sorts them differently. The density of meteorites, and their ability to absorb solar energy when fusion crusted, can also lead to them sitting differently at the ice surface (often slightly sunken in). Because most meteorites contain native metal that oxidizes very easily, they can show significant spots of rust when weathered; this highly localized distribution of rust is quite distinct from the broader coloring associated with terrestrial oxidation of FeO in oxides and silicates (Figure 2.3). The presence of native metal is also readily detected by examination with a hand magnet, a test used by ANSMET field party members when other clues suggesting a meteoritic origin are not convincing enough. Chondrules can also be very diagnostic when exposed. Finally, most meteorite lithologies are distinct from most terrestrial lithologies, so any rock that just "looks different" has potential, whether or not you're a trained geologist. During ANSMET fieldwork we strive to recover any rock suspected of having fusion crust or that just seems exceptionally out of place, accepting some level of false positives and trusting the curatorial process that follows to weed these out.

In summary, ANSMET meteorite searches are an economic compromise. Maximizing recoveries for any given season means balancing currently available logistical access to a site with our understanding of local meteorite density, a site's propensity for foul weather, recent snow cover, the density of local terrestrial rock coverage, and even the expertise of a given year's field team. Our visual searches are prey to all the failings of the flesh, as well as the quirks of wind, snow cover, terrestrial rock camouflaging, and so on. For now, with no shortage of places where meteorite recovery can be effective, human visual searches continue to be extremely successful and economical. As technological advances occur, we will explore the ways these might improve the efficiency of our searches, but we will do so in ways that do not interfere with our current procedures for fast and efficient meteorite recovery.

2.4.3. Maximizing Scientific Return during Recovery

From the earliest days of fieldwork, Antarctic meteorite specimens were recognized as much more pristine chemically than most finds from the civilized continents, and the U.S. and Japanese programs worked quickly to establish collection and curation protocols. These protocols, while originally not strict or enforced in any legislative sense, have been recognized to be of immense value and have become the *de facto* standard for meteorite recovery efforts in many locales. A unique feature of these protocols is that they do more than preserve specimen integrity; they can also ensure early, unbiased access to the samples by members of the planetary materials research community. The U.S. governmental agency responsible for activity in the Antarctic has produced enforceable regulations regarding meteorite protection that guide current Antarctic meteorite recovery by U.S. citizens [*Federal Register*, 2004].

The ANSMET program files a meteorite sample recovery plan for each field season that describes how we intend to meet or exceed NSF Regulation 45 CFR Part 674 concerning the collection and curation of Antarctic meteorites, and fieldwork does not proceed without prior USAP approval of that sample plan. As described in our sample plan, the typical procedure for recovery of a find is as follows. Upon discovering a meteorite, the finder signals the remainder of the team, who mark their current positions within the transect and converge on the find site. A single GPS-equipped snowmobile is brought near the find, down- or side-wind to minimize contamination, while all other vehicles remain several meters away (Figure 2.4). While the location of the find is being accurately determined, several field party members begin the collection procedure. The first step is the assignment of a field number that is used as a unique identifier for the sample throughout all subsequent recovery procedures. The specimen is photographed using a digital camera while detailed notes are entered into the field notebook to record the measured size of the find (along three ordinal directions), the percentage of the specimen covered by fusion crust, the presumed type (chondrite, iron, etc.), and any distinguishing characteristics of the find, such as fractures, nearby fragments, contact with snow or terrestrial rocks, or accidental human contact. During this

Figure 2.4. A typical ANSMET collection scene. Two field party members (C. Corrigan and J. Pierce) assist each other in placing a meteorite in its protective bag while J. Schutt (above right) notes distinguishing characteristics of the find such as fusion crust coverage, size, and presumed type. The GPS-equipped snowmobile shown approached from downwind and carefully parked to the right (side-wind) of the specimen, placing the GPS antenna closest to the find and the exhaust on the other side of the vehicle.

process the field number is continuously cross-checked among all records.

The sample itself is placed into a clean Teflon or nylon bag, and all contact with skin, clothing, or "dirty" implements is avoided (Figure 2.4). A clean aluminum tag punched or anodized with the field number is then inserted into a fold of this bag, arranged to prevent contact with the meteorite. The bag is then securely sealed with Teflon freezer tape. The samples are then collectively put into a larger bag or dedicated sample container where they remain during the workday. Upon return to camp, the samples are sorted by sample number into further labeled bags to aid daily and weekly inventories and put into a dedicated storage and shipping container, which is left outdoors to keep the meteorites frozen. These containers are locked before shipping from the field, and (still frozen) accompany the field team back to McMurdo at the end of the field season. While in McMurdo the storage containers are kept closed whenever possible, and stored in a clean -20° C freezer. In late February the specimens are transported by ship (still frozen) to Port Hueneme, California, and upon arrival are forwarded by freezer truck to the Johnson Space Center in Houston, Texas, where they can be thawed under controlled dry conditions to minimize interaction with liquid water. Curatorial and characterization activities associated with ANSMET-collected meteorites are described in Chapter 3 (Righter et al., this volume).

Note that our sample protocols are designed to fully document possible anthropogenic contamination of samples rather than totally eliminate such contact. Prior studies conducted during ANSMET fieldwork have shown that imposing dramatic "cleanroom-style" constraints on meteorite recovery does little to reduce such contamination [*Fries et al.*, 2012]. In fact, the specimens have typically been immersed in the Antarctic environment for thousands of years; terrestrial contamination is completely unavoidable, and some part of that baseline is already anthropogenic.

2.5. CONCLUSIONS: THE FUTURE OF ANSMET METEORITE RECOVERIES

To date, ANSMET has conducted meteorite searches on nearly 200 icefields at 75 different sites in the Transantarctic Mountains and nearby regions (Figure 2.1). These icefields have ranged in size from parts of a square kilometer to several hundred square kilometers, and the number of meteorites recovered from these icefields has ranged from zero (in many cases) to several thousand (in just a few). In spite of ANSMET's long history, many icefields remain targets for both reconnaissance and systematic searching, both within and outside the Transantarctics. Given the large numbers of scientific mysteries that remain in planetary materials research, many of which can only be solved by new specimens, this is a good thing for both science and the future of the U.S. Antarctic meteorite program. ANSMET's field methods will continue to evolve with technological and logistical advances and as our understanding of meteorite concentrations improve; but we expect the baseline procedures described here, having served us so well for decades, will remain fundamental to those future operations.

Acknowledgements. The authors would also like to thank the 170+ volunteers who have participated in ANSMET as field party members. Bill Cassidy, the founder of ANSMET, called the program the poor person's space probe, making you the world's most underpaid astronauts. We also want to express our sincere thanks to the many thousands of folks of the McMurdo Station community who have supported and sacrificed for science in general and for ANSMET specifically.

REFERENCES

Anderson, J. B., A. L. Lowe, A. C. Mix, A. B. Mosola, S. S. Shipp, and J. S. Wellner (2002), The Antarctic ice sheet during the last glacial maximum and its subsequent retreat history: A review, *Quaternary Sci Rev*, *21*, 49–70.

Apostolopoulos, D., B. Shamah, M. Wagner, K. Shillcutt, and W. L. Whittaker (2001), Robotic search for Antarctic meteorites: Outcomes, *Proceedings of the 2001 IEEE International Conference on Robotics and Automation* (ICRA01), Seoul, Korea. 35–42.

Barabesi, L., L. Greco, and S. Naddeo (2002), Density estimation in line transect sampling with grouped data by local least squares, *Environmetrics*, *13*, 167–176.

Bayly, P. G. W., and F. L. Stillwell (1923), The Adelie Land meteorite. Australasian Antarctic Expedition, 1911-14, *Sci Rpts, Ser A*, *4*, 1–13.

Chen, S. X., P. S. F. Yip, and Y. Zhou (2002), Sequential estimation in line transect surveys, *Biometrics*, *58*, 263–269.

Duke, M. B. (1965), Discovery of Neptune Mountains iron meteorite, Antarctica, *Meteoritical Bull*, *34*, 2–3.

Federal Register (2004), Federal Regulations regarding Antarctic Meteorite Recovery (NSF regulation 45 CFR Part 674, RIN 3145-AA40), *Federal Register*, *68*(61), 15378.

Ferrigno, J. G., J. L. Mullins, J. A. Stapleton, P. S. Chavez Jr., M. G. Velasco, R. S. Williams, G. F. Delinski, and D. Lear (1996), *Satellite Image Map of Antarctica, US Geological Survey Miscellaneous Investigation Series*, map *I-2560*.

Ford, A. B., and R. W. Tabor (1971), The Thiel Mountains pallasite of Antarctica, *US Geol Survey Prof Paper*, *750-D*, 56–60.

Fries, M., R. Harvey, A. J. T. Jull, and N. Wainwright, ANSMET 07-08 Team (2012), The microbial contamination state of as-found Antarctic meteorites, in *Conference on life detection in Extraterrestrial Samples* Abstract 6036, Lunar and Planetary Institute, Houston.

Haack, H., J. Schutt, A. Meibom, and R. Harvey (2008), Results from the Greenland Search for Meteorites Expedition, *Meteoritics and Planetary Science*, *42*, 345–366.

Hammond, P. S, P. Berggren, H. Benke, D. L. Borchers, A. Collet, M. P. Heide-Jorgensen, S. Heimlich, A. R. Hiby, M. F. Leopold, and N. Oien (2002), Abundance of harbour porpoise and other cetaceans in the North Sea and adjacent waters. *J Appl Ecol*, *39*, 361–376.

Harvey, R. P. (2003). The origin and significance of Antarctic Meteorites. *Chemie der Erde*, *63*, 93–147.

Justice, C. O., E. Vermote, J. R. G. Townshend, R. Defries, D. P. Roy, D. K. Hall, V. V. Salomonson, J. L. Privette, G. Riggs, A. Strahler, W. Lucht, R. B. Myneni, Y. Knyazikhin, S. W. Running, R. R. Nemani, Z. Wan, A. R. Huete, W. van Veeuwen, R. E. Wolfe, L. Giglio, J.-P. Muller, P. Lewis, and M. J. Barnsley (1998), The Moderate Resolution Imaging Spectroradiometer (MODIS): Land remote sensing for global change research. *IEEE TRans Geosci Rem Sens*, *36*, 1228–1249.

Lucchitta, B. K., J. Bowell, K. Edwards, E. Eliason, and H. M. Ferguson (1987), Multispectral Landsat images of Antarctica, *US Geol Survey Bull, 8755–531X; B*, 1696.

Marvin, U. B. (2014), The origin and early history of the U.S. Antarctic Search for Meteorites program (ANSMET), in *35 Seasons of U.S. Antarctic Meteorites (1976-2011): A Pictorial Guide to the Collection, Special Publication 68*, edited by K. Righter, C. M. Corrigan, R. P. Harvey, and T. J. McCoy, American Geophysical Union/John Wiley & Sons, Washington, D. C.

Mawson, D. (1915), *The Home of the Blizzard 2*, Heinemann, London.

Ravich, M. G., and B. I. Revnov (1963), Lazarev iron meteorite, *Meteoritika*, *23*, 30–35 (In Russian), English translation in *Meteoritica* (1965), *23*, 38–43.

Righter, K., C. E. Satterwhite, K. M. McBride, and C. M. Corrigan (2014), Curation and allocation of samples in the U.S. Antarctic meteorite collection, in *35 seasons of U.S. Antarctic Meteorites (1976–2011): A Pictorial Guide to the Collection, Special Publication 68*, edited by K. Righter, C. M. Corrigan, R. P. Harvey, and T. J. McCoy, American Geophysical Union/John Wiley & Sons, Washington, D. C.

Schutt, J. (1989), The expedition to the Thiel Mountains and Pecora Escarpment, 1982–1983, *Smithsonian Contr Earth Sci*, *28*, 9–16.

Tolstikov, E. (1961), Discovery of Lazarev iron meteorite, Antarctica. *Meteoritical Bull*, *20*, 1.

Turner, M. D. (1962), Discovery of Horlick Mountains stony-iron meteorite, Antarctica. *Meteoritical Bull*, *24*, 1.

Yoshida, M. (2010), Discovery of the Yamato meteorites in 1969. *Polar Sci*, *3*, 272–284.

Yoshida, M., H. Ando, K. Omoto, R. Naruse, and Y. Ageta (1971), Discovery of meteorites near Yamato Mountains, East Antarctica. *Japanese Antarctic Record*, *39*, 62–65.

Young, D. A. (2003), *Mind Over Magma: The Story of Igneous Petrology*. Princeton University Press, Princeton, NJ.

3

Curation and Allocation of Samples in the U.S. Antarctic Meteorite Collection

Kevin Righter[1], Cecilia E. Satterwhite[1], Kathleen M. McBride[1], Catherine M. Corrigan[2], and Linda C. Welzenbach[2]

3.1. INTRODUCTION

This chapter provides an overview of the entire curation and allocation process for the U.S. program. Parts of this have been covered in publications that date back to the late 1980s and early 1990s [e.g., *Bogard and Annexstad*, 1980; *Graham and Annexstad*, 1989], but an in-depth and historical review has not been published. This chapter will give an overview of the early part of the program, describe the curation facilities, provide a summary of samples allocated from the collection, and give case studies of five important meteorite samples: three small and rare samples that have required special preservation efforts, and two larger samples that have provided amply to the scientific community.

3.2. CURATION FACILITIES AND APPROACHES

The long and difficult process of obtaining funding to collect meteorites in Antarctica was successful due to Bill Cassidy's persistence and tenacity. The successes of the field efforts then raised a new issue: How would the samples be stored and curated? Although Cassidy initially had a plan to establish a curation and processing facility at the University of Pittsburgh, his interactions with the community upon returning samples from the field drew much attention from scientific groups who were interested in being involved in this next phase. In this same time frame, sample processing facilities were available (*Apollo* 11 cabinets) [*Annexstad*, 2001] in Building 31 (B31) at NASA Johnson Space Center (JSC) after construction of

the new lunar sample handling facility in B31N. In fact, JSC Planetary Materials Branch Chief Larry Haskin had asked one of his research scientists, Don Bogard, to write a report on the possible use of the available lunar sample handling facilities for meteorites. Knowing of this report and of the meteorite recoveries of Cassidy, John Annexstad approached Bogard and they realized that the U.S. Antarctic meteorites could be curated at JSC [*Sears*, 2012; *Annexstad*, 2001]. Bogard then approached Cassidy and the arrangements began to be discussed. At the same time, Smithsonian meteoriticists were naturally interested in the Antarctic meteorites because they had been collected by the U.S. government. Representatives from these two agencies as well as NSF and other members of the university community met in 1977 and 1978 to make a plan for the short- and long-term storage, processing, and handling of the samples. In 1980, these meetings ultimately led to the establishment of a formal agreement between NSF, NASA, and the Smithsonian (often referred to as "the three-agency agreement").

Some aspects of the early years deserve explanation because there were some differences in handling approaches and protocols as experience with these new samples was gained. The 1976–1977 season recovered a total of nine samples [*Olsen et al.*, 1978] that were held under a special agreement between the United States and Japan [*Marvin*, 2014 (this volume)]. Samples were split at McMurdo Station in Antarctica using a radial rock saw. In fact, images of the 1976 season's samples show the radial pattern of the saw cuts; samples cut in subsequent years at JSC were done using a band saw. The samples from this first season are stored and available at the National Institute of Polar Research (NIPR; Tokyo, Japan), the Field Museum of Natural History (Chicago, IL, USA),

[1] *NASA Johnson Space Center*
[2] *Smithsonian Institution*

35 Seasons of U.S. Antarctic Meteorites (1976–2010): A Pictorial Guide to the Collection, Special Publication 68,
First Edition. Edited by Kevin Righter, Catherine M. Corrigan, Timothy J. McCoy and Ralph P. Harvey.
© 2015 American Geophysical Union. Published 2015 by John Wiley & Sons, Inc.

JSC (Houston, TX, USA), and the Smithsonian Institution (SI; Washington, DC, USA).

Samples from the subsequent two years (1977–1978 and 1978–1979 seasons) were collected jointly by the United States and Japan, with the arrangement for the division of samples set in place to cut samples >300 g in half (cutting to take place at JSC) and then dividing samples <300 g on a one-by-one basis. Thin sections were being made by the Smithsonian Institution of all of the large samples, and JSC, SI, and NIPR received one section from each meteorite. A subset of the smaller samples (referred to then as "pebbles") were classified by principal investigators (PIs) as determined or recommended by the Meteorite Working Group (MWG). For example, K. Keil at the University of New Mexico proposed to help classify 145 of the small equilibrated ordinary chondrites [*McGinley and Keil*, 1984]. Similarly, C. B. Moore (ASU), M. Rhodes (UMass), J. Fitzgerald (Univ. Adelaide), and S. J. B. Reed (Cambridge, UK) all received pebble-sized samples [*Fulton and Rhodes*, 1984]. This approach was not continued past this season, as it was not the most efficient way to classify large numbers of samples.

A small number of samples from the 1977–1978 field season were kept frozen, primarily for thermoluminescence studies. However, interest waned in keeping samples frozen after this initial subset of samples; nonetheless, the samples are stored frozen and remain available for study.

In the first few years of Antarctic meteorite handling at JSC, the staff experimented with the idea of storing and handling samples frozen, using a cold processing plate in a cabinet, and using a cold storage room. Although some hardware was assembled to do this, it became clear after detailed tests that this was not an effective way to handle samples due to the difficulty of keeping samples cold while still allowing dexterity of the sample processor, length of time required to process individual samples, and overall expense. The cold processing approach was abandoned in 1979, after review and discussion by the MWG. At the same time, JSC expanded their own thin-section-making laboratory to include meteorites (they were already making *Apollo* lunar sample thin sections). Also, realizing the importance of minimizing contamination of organic compounds contained in carbonaceous chondrites, processing protocols were established for subdividing and handling carbonaceous chondrites in a dedicated GN2 cabinet. The GN2 used for all Antarctic meteorite handling and storage at JSC is of high purity with levels not exceeding 20 ppm argon, 10 ppm oxygen, hydrogen, carbon dioxide, carbon monoxide, and moisture. Sample handling and processing techniques in the meteorite lab at JSC were firmly established by 1980–1981.

The growth of the collection over the ensuing decades posed new challenges and problems for the curation staff, and the roles of staff at JSC and SI (Table 3.1) will be highlighted where possible and relevant.

Table 3.1. Curators associated with U.S. Antarctic meteorite collection.

NASA Johnson Space Center Curators	Smithsonian Institution Curators
Donald Bogard, 1978–1984	Brian Mason, 1978–1987
James Gooding, 1984–1986	Glenn Macpherson, 1987–1996
Marilyn Lindstrom, 1987–2000	Tim McCoy, 1997–2007
David Mittlefehldt, 2000–2001	Cari Corrigan, 2008–present
Carl Allen, acting curator 2001–2002	
Kevin Righter, 2002–present	

3.2.1. Transport from McMurdo to Houston

Samples collected by ANSMET (U.S. Antarctic Search for Meteorites program) teams are kept at ambient conditions in isopods in the field and during transport back to McMurdo, and then stored in freezers at McMurdo Station (Crary Science Lab). The samples are shipped to Port Hueneme, California, and then transported to Houston, remaining frozen all along the journey. The samples were flown to Houston until approximately 1991, when they switched to truck transport via frozen cargo trucks. The samples are kept frozen in order to minimize interaction with water from melted ice/snow within bags, and also to prevent rusting that might be accelerated at room temperatures.

3.2.2. Staging

Once samples arrive in Houston they are immediately transferred into one of two freezers in the Antarctic Meteorite Processing Lab (MPL) at JSC, a class 10000 cleanroom with HEPA filtered air handler system that effectively makes the cleanroom operational at Class 1000 or better. The newly arrived samples are temporarily stored in the freezers which are set at −10° to −4° F (-23 to -20 °C). Storage under frozen conditions allows them to remain in their collected state until the processing personnel are ready to study them under the controlled conditions of the processing lab. Samples are assigned a random five-digit number in the field and are renumbered at JSC and given their official name, a three-letter prefix abbreviating the geographical area in which they were found, followed by two digits for the collection year (being the year the team began the field season), followed by a three- or four-digit number uniquely identifying the sample from that season (i.e., "ALH 84001" for Allan Hills 84001 or "MIL 090001" for Miller Range 090001). Four digits must be used for collection years in which greater than 999 samples have been recovered from one area. These are dense collections areas; therefore, all new sample prefixes and numbering schemes must be approved by the Nomenclature Committee of the Meteoritical Society. All names used by the ANSMET program and the U.S. Antarctic meteorite program have been formally approved by this process.

3.2.3. Initial Processing and Classification

When samples are ready for initial characterization, they are thawed in a GN2 cabinet for at least 24 hours. This step minimizes the possible oxidation or alteration of the meteorites from moisture (snow or ice) that can be trapped in the bags. Thawing in the nitrogen glove box allows the moisture to be released without reacting with the samples. Once thawed, the samples are then weighed, measured, photographed (larger samples in six orthogonal directions), and macroscopic features described, such as weathering, fracturing, notable inclusions, percentage fusion crust, and metal content. This process is carried out on a laminar flow workbench that has a stainless steel surface (Figure 3.1, e). Then samples are broken to obtain a small chip for classification. Depending on the specific sample geometry, size, or hardness, a stainless steel rock splitter (Figure 3.1, c) or a steel chisel and chipping bowl (Figure 3.1, f) is used to obtain a chip.

Classifications of all meteorites (except irons and pallasites—see below) are carried out in two ways: (a) oil immersion approaches combined with visual observation of fresh surface to classify equilibrated ordinary chondrites (EOCs) by H, L, or LL and petrologic type [e.g., *Lunning et al.*, 2012], and (b) electron microprobe analyses of characteristic mineral phases in all other chondrites and achondrites, allowing a detailed classification to be made. In the early years of the program, all samples were thin sectioned and classified using microprobe data and detailed petrography. In the mid 1980s, due to large numbers of samples being collected each season, Smithsonian meteorite curator Brian Mason initiated the classification of many of the EOCs by using a combination of oil immersion to estimate the olivine composition (and thus H, L, or LL) and optical microscopy to estimate petrologic grade (4, 5, or 6). This approach (recently reviewed by *Lunning et al.* [2012]), led to the classification of larger numbers of EOC samples and allowed the curators to classify every sample collected within two years of its being returned from Antarctica.

Iron meteorites are classified using standard metallographic techniques as originally laid out by Roy Clarke. In fact, Roy Clarke's classification work on some of the early irons from the U.S. collection includes those from the Allan Hills 1976, 1977, and 1978 seasons; Purgatory Peak 1977 and Reckling Peak 1979 and 1980 samples, and the huge Derrick Peak paired sample group [*Clarke*, 1982]. The basic approach utilized for classification of these early irons has been used to the present with 111 iron meteorites and 26 pallasites recovered so far.

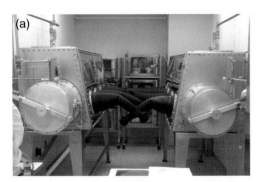

Figure 3.1. Facilities at JSC in the Class 10,000 (ISO 7) cleanroom Antarctic Meteorite Processing Laboratory, including (a) special and large sample stage cabinets, (b) intermediate-size sample storage cabinets, (c) rock splitter, (d) martian meteorite processing cabinet, (e) laminar flow bench, and (f) chisel and chipping bowl.

The macroscopic descriptions made at JSC and the microscopic descriptions made at SI are combined and announced in the *Antarctic Meteorite Newsletter*, which is released twice a year. These announcements are also coordinated with the Meteoritical Society's Nomenclature committee, which approves newly proposed meteorites.

The classification and announcement process has always provided the minimum amount of information to formulate a definitive classification, allowing the scientific community to make informed requests for meteorite samples. This also ensures that the scientific investigations are left to the science community rather than undertaken (either in whole or in part) by the curators overseeing the collection. For example, oxygen isotopic measurements, although frequently used for classification of meteorite samples, are not usually necessary because a variety of other mineral and textural information can be used to make an accurate classification. Therefore, oxygen isotopic data are only obtained for samples for which textural and mineral data do not allow a definitive classification to be made. For many years Robert N. Clayton and Tosh Mayeda [*Clayton and Mayeda*, 1996] provided oxygen isotopic analyses to support the collection in this way, and in more recent years some analyses have been contributed by James Farquhar (Univ. Maryland), Doug Rumble (Carnegie-DTM), and Zachary Sharp (UNM). When new data are published making it clear that a sample has been misclassified, reclassifications are announced in the *Antarctic Meteorite Newsletter* and also updated with the Meteoritical Society meteorite database.

3.2.4. Storage

Once thawed during initial processing (see above), the samples at JSC are bagged in nylon or Teflon bags and are stored in several different types of cabinets. Achondrites, carbonaceous chondrites, and special ordinary chondrites are stored in GN2 cabinets (Figure 3.1). The remaining samples (mostly EOCs) are stored in three different settings: large (>500 g) samples are stored in large-volume GN2 cabinets (Figure 3.1, a), medium-sized (200–500 g) samples are stored in two large GN2 cabinets with multiple trays (Figure 3.1, b), and small samples (<200 g) are stored in stainless steel cabinets with no nitrogen flow. Some exceptionally large samples are triple bagged and stored on open shelves in the lab. One exceptional sample is LEW 85320 (Plate 2), collected in the Lewis Cliffs Ice Tongue in 1985; because of its large size (110 kg), a customized GN2 plexiglass storage box was constructed and it resides in the meteorite lab at JSC. In the event of a hurricane threat, all GN2 cabinets that store the non-EOC meteorite samples are moved to a watertight vault in the lunar sample facility (B31N).

Iron meteorites and pallasites are stored at the Smithsonian Institution in the National Museum of Natural History in Washington, D.C. They are stored in dedicated auto-desiccators (Figure 3.2) near facilities uniquely equipped to cut and prepare irons and predominantly-iron-bearing samples.

3.2.5. Requests, MWG, MSG, and Allocations

Sample requests and allocations were designed to be less stringent than those developed for *Apollo* samples to increase accessibility to the collection for the scientific community. JSC curator Don Bogard and his staff strove to make the Antarctic meteorite allocations process more streamlined and efficient compared to the *Apollo* sample allocations. The JSC curator can approve sample requests unless they fall into one of the following categories: a new PI, a newly announced sample, a piece of a small (<20 g) meteorite, a larger piece (> 5 g) of a single meteorite, or a meteorite on the special list (see below). Sample requests from these five categories must be reviewed by the Meteorite Working Group (MWG), a panel of meteorite experts providing advice to the curators, and the Meteorite Steering Group (MSG), a body composed of member one from each of the three agencies involved (SI, NASA, and NSF) that approve the recommendations of MWG members. Early MWG meetings were long and focused on infrastructure needs, planning, funding, and rules, while later meetings have focused more on brevity and resolving decisions on the many requests received annually.

A great challenge facing curators Glenn MacPherson and Marilyn Lindstrom in the 1980s and early 1990s was to oversee a collection that was quickly growing in size, using an unchanged amount of personnel support and resources. During this time frame, the allocation guidelines changed as the collection matured in size and diversity (lunar and martian meteorites, many new chondrite groups, and some rare and unusual achondrites). They managed to strike a balance between the popularity and great demand for the samples being characterized and made available, while classifying ever-increasing numbers of samples from the new expeditions. This unwieldy task was done well under their direction and with the support of NASA headquarters.

3.2.5.1. Special samples. The special list was created to protect very unusual, scientifically unique and important samples. Requests for these samples must be evaluated by MWG and approved by MSG. This list is reassessed at every MWG meeting and has changed over the years. Some samples added to the list many years ago may have been removed as the collection has been augmented with similar samples. Some samples have moved onto the list as their remaining mass has become much smaller than

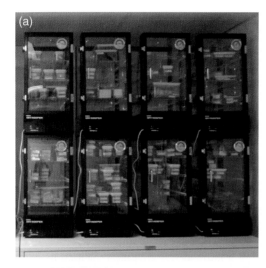

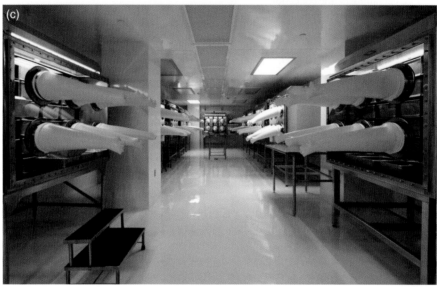

Figure 3.2. (a) Sanplatec auto-desiccators are used to house Antarctic iron meteorites, powdered stone types, and type sections of aubrites and enstatite chondrites. (b) Laboratory equipment and supplies are located within the clean room storage area. Sample allocations and study take place on a biological-grade laminar flow "clean" bench. (c) The Smithsonian Antarctic storage facility is a Class 10,000 clean room, which has a dedicated triple HEPA filtered air handling system. Fourteen stainless steel glove boxes are filled with 99.998% pure nitrogen gas supplied from a central LN tank. Paired glove boxes share an airlock with air shower. Each glove box has gas flow and relative humidity monitors.

the original mass yet scientific interest in the sample continues. A current listing of the special protected samples can be found on the JSC webpage: http://curator.jsc.nasa. gov/antmet/forms/.

Antarctic meteorite sample requests from individual scientists may involve one particular sample or dozens of samples, depending on the nature of the study being undertaken. The number of requests received by the program has steadily grown over the years, such that recent years have seen more than a hundred requests (and over a thousand individual meteorite samples) per year, whereas the program average is about 80 per year

(Figure 3.3). Interest in samples from the collection can be assessed by considering the history of requests for specific samples. In most cases, there is an initial high level of interest as a new sample is recovered and made available. This is followed by a long period of sustained interest as more is learned about specific samples and how they fit into the big picture of knowledge that might exist for a specific group or groups of samples. For example, the martian meteorite EET A79001 (Plate 70) was found in 1980, and even after several years of intense study it remains of great interest and has received more than 300 requests over the years of

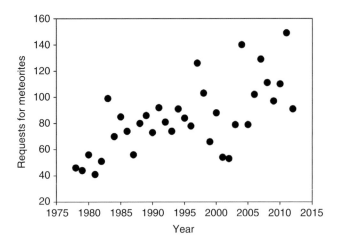

Figure 3.3. Requests for meteorites from the U.S. Antarctic meteorite program since the beginning of the collection in 1978. The peaks in 1983, 1997, and 2004 correspond to the announcement of the first lunar meteorite (ALH A81005), the first meeting after the announcement of possible fossil life in ALH 84001, and the announcement of the first U.S. Antarctic nakhlite, MIL 03346, respectively.

the program (Figure 3.4). The Miller Range nakhlite (Plate 74), in comparison, was found in 2004, with pairs discovered in 2009, and it has already received more than 100 requests (Figure 3.4). Interest in lunar meteorites has also been sustained over decades with the MAC 88105 regolith breccia (Plate 68) receiving a steady request rate over 25 years. The LaPaz Icefield (LAP) 02205 basaltic lunar meteorite (and its pairs) has received the most requests of any lunar meteorite to date (Plate 67). Small and rare lunar meteorites such as QUE 93069 and EET 87521 (Plate 65) continue to receive requests from the community; because they are small samples and more precious, their allocation is monitored closely by the MWG and MSG to preserve material for future generations.

Several chondrites have received large numbers of requests over the years and continue to provide important materials for those studying primitive and early solar system materials. For example, the CR2 chondrite GRA 95229 has been requested at a greater rate than others in the past decade due to the diverse materials (presolar grains, organics, chondrules, metals grains) found within the CR2s [see *Weisberg and Righter*, 2014 (this volume); *Righter*, 2013] (Figure 3.4). Enstatite and other carbonaceous chondrites (CH, CO3, and CM1/2) are also heavily requested as scientists examine the origin of organics, zoned metal grains, and aqueous alteration on carbonaceous parent bodies (asteroids).

Finally, various achondrites have been heavily requested over the years of the program, with some examples illustrated in Figure 3.4. The diogenites and ureilites (ALH

A77256 and 77257, respectively) continue to be of great interest scientifically. The brachinite (ALH 84025; Plate 48) has been steadily requested since 1984 but also has received new interest with the discovery of the related ungrouped achondrites GRA 06128 and 06129 (Figure 3.4; Plate 49). The two small angrites (LEW 86010 and 87051; Plates 53 and 54) have been requested at steady levels because they are two members of the small group of angrite meteorites and thus remain essential to understanding the origin of angrites. The lodranites (MAC 88177 and GRA 95209; Plate 41) were highly requested when first found because there were so few lodranites and acapulcoites in the 1990s. They have remained of great interest because new samples are continuously being found and compared to these important samples that originally helped to establish the groups.

3.2.6. Sample Handling, Preparation, and Documentation

At JSC, the meteorite samples are prepared several different ways and can be requested accordingly. Most samples are prepared as chips, obtained either by use of stainless steel chisels in a chipping bowl or using rock splitters. Tools, sample containers, and bags are made of stainless steel, aluminum, Teflon, or nylon. Polyvials (polystyrene vials with polyethylene lids) are also used as sample containers. In cases where a meteorite will be subdivided into many pieces for scientific studies, the samples are cut dry in a GN2 cabinet with a band saw, producing a 1- to 2-cm-thick slab. A listing of meteorites that have been cut or slabbed at JSC is available at http://curator.jsc.nasa.gov/antmet/PDFFiles/1aa-JSC-bandsawList.pdf. More details about sawing and possible effects on samples are at http://curator.jsc.nasa.gov/antmet/bandsaws.cfm.

3.2.6.1. Cabinet processing: carbonaceous chondrites and martian meteorites. Since 1978, carbonaceous chondrites have had a dedicated cabinet for processing due to the possibility that these samples contain unique organic compounds. Since 1997, martian meteorites have been processed in a dedicated cabinet (Figure 3.1, d), again due to the possibility that there are organic or other compounds that are specific to these samples, and to minimize the effects of handling under less clean conditions. On occasion, an additional cabinet will be cleaned for targeted processing of specific meteorites, such as lunar meteorites.

3.2.6.2. Flow bench processing. Three clean-room laminar flow benches are used at JSC to process all other samples (Figure 3.1, e), and also for initial processing of new meteorites.

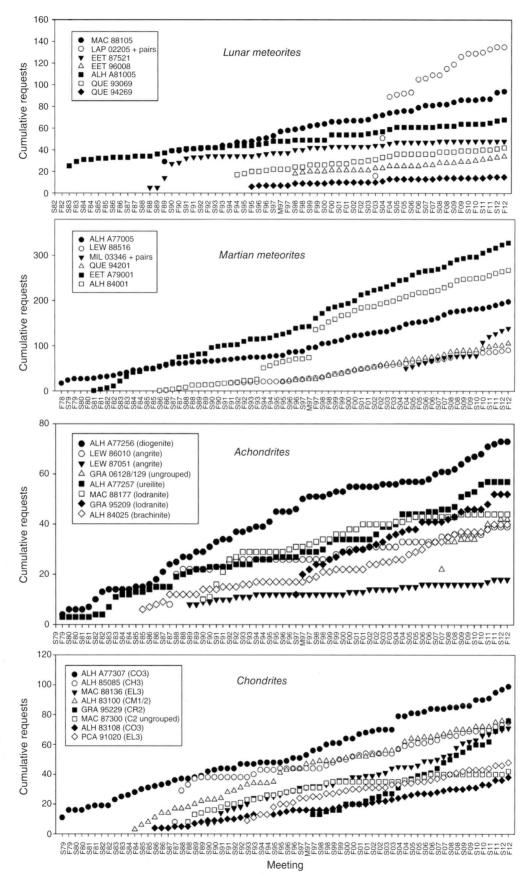

Figure 3.4. Cumulative requests versus Meteorite Working Group meeting year (S = spring and F = fall) for specific meteorites since the beginning of the U.S. Antarctic meteorite program. See text for detailed explanations and discussion.

3.2.6.3. Iron meteorite and pallasite preparation at Smithsonian. Iron meteorites and pallasites are stored and processed at the Smithsonian. These meteorites, unlike other meteorite types, are stored at the National Museum of Natural History. The majority of these are stored in auto-desiccators to reduce oxidation (Figure 3.2). Sample preparation facilities are specially equipped for cutting, polishing, and sectioning irons and predominantly iron-bearing meteorites. Iron meteorites and pallasites are cut using specialized composite (a hardened rubberized compound impregnated with abrasives) blades. Blade composition and sources can be provided upon request. Polished sections are made with standard epoxy and polished with diamond paste and aluminum oxide.

3.2.6.4. Thin and thick section preparation. The meteorite thin section labs at JSC and the Smithsonian can prepare standard 30-micrometer thin sections, thick sections of variable thickness (100 to 150 micrometers are common for LA-ICP-MS or microdrilling), or demountable sections using superglue for use in TEM studies. At JSC, water is not used in any part of the thin-sectioning process; instead, light mineral oil and alcohol are used for the cutting, grinding, and polishing stages; thin sectioning at SI utilizes water unless otherwise specified. Diamond grinding wheels are used on some samples, as well as a SiC slurry for lapping. Polishing compounds are typically diamond paste or diamond fluids, but alumina can be used as a final polish as well. Pure silica slides are available as a substitute for standard petrographic glass slides, if necessary.

3.2.6.5. JSC and SI "library" materials (thin sections and powders). A library of thin sections of all non-EOC (and some EOC) specimens is maintained at SI and JSC (those at JSC for meteorites with original mass >10 g). All enstatite meteorite (aubrite, EH, and EL chondrite) thin sections are stored in electronic desiccators to avoid breakdown of water-soluble minerals such as oldhamite. MWG does not in general advocate the loan of SI library sections, in order to maintain one relatively complete library. In special cases, however, the SI library section may be loaned for a brief period (up to 2 weeks) by the SI curator with the consent of one other member of MWG. The JSC library sections are not generally loaned out to PIs for research purposes except in rare cases and for no more than 3 months. PIs are free to visit both facilities to examine thin sections at any time after their announcement in the Antarctic Meteorite Newsletter.

For many large meteorites collected during the early years of the ANSMET program, homogenized powders were prepared by Gene Jarosewich and archived at the Smithsonian. A full listing is reported in *Jarosewich*

[1990], with an update published in 2006. This small formal contribution to the program is only one example of Gene Jarosewich's important involvement with the Antarctic meteorite collection [e.g., *Clarke et al.*, 2006].

3.2.7. Cleaning and Contamination

Collection and curation of Antarctic meteorites for 35 years have resulted in identification of potential contaminants, both organic and inorganic, to meteorite samples. Organic matter (windblown microorganisms or organic compounds) may be present in the ice [*Stroeven et al.*, 1998; *Burckle and Delaney*, 1999; *Botta et al.*, 2006] in which meteorites sit for ~100 Ka, and even though storage in nitrogen minimizes oxidation and reactivity of samples, the environment is not totally free of organisms, as anaerobic bacteria can survive in some samples [*Fries et al.*, 2005]. Nonetheless, curation facilities can be designed to guard against additional contamination. There are various ways in which the environment is kept as clean as possible, including special cleaning procedures, air filtering, and monitoring for potential contamination. Some of these practices changed substantially following the 1996 announcement of evidence for possible fossil life and biochemicals in ALH 84001 [*McKay et al.*, 1996] (see also Plate 69). The increased awareness of organic contamination of the Antarctic meteorites prompted several changes to the Antarctic meteorite lab at JSC: establishment of a dedicated cabinet for processing martian meteorites, separate storage of martian meteorites and most carbonaceous (once they are classified), and installation of a new HEPA filtered air system (see below). In addition, organics testing took place in clean rooms, including air samples, exposed witness plates in cabinets, and the ultrapure water (UPW) used to clean tools and bags. The cleaning and monitoring approaches are discussed below.

Cleaning of cabinets, heat sealers, balances, bags, tools, containers, and equipment used in the curatorial labs currently follows specific guidelines (*Johnson Space Center*, 2006; *Calaway et al.*, 2013a,b). For many years, cleaning of curatorial items was done using a Freon rinse (until the mid-1990s). In the 1970s and 1980s, the cleaning was done with Freon, and the cleanliness of the final flush was checked by particle counts, nonvolatile residue (NVR), and total hydrocarbon (THC) tests. Freon cleaning was done using three venues but all the same procedure: a Freon still in the high bay of Building 31, White Sands Test Facility, and the JSC Building 9 facility. Curatorial staff used the in-house Freon still whenever possible, but sometimes demand became too high for the size of the still, and at these times, White Sands or Building 9 would provide supplemental support. When Freon cleaning was discontinued, it was replaced with a

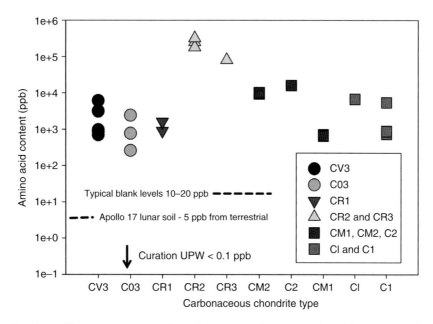

Figure 3.5. Summary of amino acid measurements made on Antarctic carbonaceous chondrites, compared to levels of amino acids measured in Apollo samples [*Brinton and Bada*, 1996], in UPW [*Bada et al.*, 1998], and blank levels measured during amino acid measurements [*Bada et al.*, 1998]. All carbonaceous chondrite data are from *Glavin et al.* [2006, 2011, 2012], *Glavin and Dworkin* [2009], *Burton et al.* [2012], and *Martins et al.* [2007].

hot UPW rinse for cabinets, tools, and other items. The total organic content of the UPW system is checked regularly and is typically near 10 ppb.

The Antarctic meteorite lab was designed as a Class 10,000 lab (ISO 7), at which it operated through early 2001, until HEPA filtered air and a new air handler unit was installed. The lab also has an air shower in its entrance, which takes air from the main room, passes it through a HEPA filter, and then recycles it within the shower. The purpose of the recycling is to remove any loose dirt from the gowning process in the change room, which potentially stirs up particles. This design has undoubtedly led to a cleaner environment. Particles that are >0.5 microns are kept to low levels, and this effectively cuts out mold, fungi, pollen, human hair, algae, and dusts. Since the installment of the HEPA filtered air system, particle counts in the various processing areas of the MPL have been consistently between 100 and 1000 counts.

The curation practices have led to a clean environment, and it is periodically monitored with testing of air, water, and glove boxes. For example, studies of the organic compounds in the meteorite lab main room and the martian meteorite cabinet revealed the presence of low levels of siloxanes, which can be from sealants, or lubricants in motors, or elastomers in gaskets. Small amounts of fluorocarbons detected in the meteorite room air may be derived from refrigerants. Isopropanol, butanol, propoxypropanol, 2-butoxyethanol, butoxypropanol,

dipropylene glycol, found in the room sample may be from solvents. Finally, aromatic compounds such as toluene, ethylbenzene, xylenes, alkylbenzenes, and low to medium boiling hydrocarbons are also used in solvents. None of these species or the levels at which they are present have been of concern to investigators who study meteorite samples, but knowledge of their presence is very important.

There are also concerns about some materials used during curation, and some examples are given here. For example, in some cases water can pyrolize nylon and produce an amino acid contaminant [*Shimoyama et al.*, 1985]; use of Teflon bags can eliminate this problem [*Glavin et al.*, 2006]. Heat sealing of bags can produce caprolactum, a monomer used in the production of Nylon 6; most cabinet processing uses heat sealing to seal storage bags, so it is important to know about the presence of this compound. Cabinet gloves can offgas various species such as CO (butyl rubber), isopropyl alcohol and C10-C14 hydrocarbons (viton), COS (neoprene), and SO_2 (hypalon) [*Righter et al.*, 2008], so gloves can be selected accordingly; the JSC Antarctic meteorite lab has used neoprene gloves. Fabrication and construction of processing cabinets (or other equipment) can utilize lubricants and other chemicals; an example is xylan, a complex amide, used in JSC processing cabinets in the 1980s; discovery of this material in cabinets led to elimination of the use of xylan in early 1990s [*Wright et al.*, 1991, 1992].

Band sawing frequently leaves dark streaks on the cut face of a meteorite, and the identity of this dark material was unknown for many years. To learn more about the effects of band sawing and to identify the components of the dark streaks, a band-sawed pristine *Apollo* sample known to have low levels of organics [*Allton*, 1999; *Clemett et al.*, 2005] was studied using a variety of microscopy techniques. Little organic contamination was found, but significant Ni and C contamination from the saw blade was found on the cut surface (C was also found by *Wright et al.* [1993]). These are some known examples of contaminants introduced during the handling of meteorites. There may be others depending upon materials and procedures used by various groups. The curators of the collection have always been open to modifying procedures and approaches when new information comes to light about potential contaminants.

Focusing on a specific organic compound (amino acids) illustrates that the clean room environment for Antarctic meteorites has proven acceptable for detailed studies of carbonaceous chondrites. Because Antarctic meteorites are recovered from blue ice in Antarctica and have a finite residence time (up to 1,200,000 years) on Earth before recovery, they may obtain organic contamination from their terrestrial environment. Therefore, it might be prudent to compare the total organic carbon (TOC) contents of *Apollo* lunar sample cabinets to those in the Antarctic meteorite lab. In 2000 the TOC content of an *Apollo* 16 pristine sample cabinet in B31N (PSL-38) was 6.28 ng/cm^2, and in 2007, TOC of the returned sample processing cabinet in B31N was measured at 5.3 ng/ cm^2 (*J. Allton*, pers. communication). In both of these measurements, a silicon wafer was exposed in the cabinet for 48 hours. These measurements are in line with the low levels of organic contamination estimated from the early *Apollo* days (1-10 ng/cm^2) as well as the very low levels of organics detected in *Apollo* samples especially after *Apollo* 12 [*Allton*, 1999]. Finally, *Brinton and Bada* [1996] measured 15 ppb of amino acids in an *Apollo* 17 sample, which was stored for ~23 years at JSC before analysis. Some of the measured amino acids were clearly indigenous to the sample, which means the amount due to terrestrial contamination was much lower, perhaps 5 ppb. In comparison to the lunar cabinets, the carbonaceous chondrite cabinet TOC was measured at 0.4 ng/cm^2, even smaller than that measured in *Apollo* lunar sample processing cabinets. It was from this environment in the carbonaceous chondrite processing cabinet that samples of the CM2 chondrite LEW 90500 and ungrouped carbonaceous chondrite LON 94102 were processed and sent out to various investigators, including *Glavin and Dworkin* [2009], who measured from 2 to 65 ppm total amino acids in these and many other carbonaceous chondrite samples and types, the majority of which were extraterrestrial in origin (Figure 3.5). In addition, amino acids in lunar meteorite MAC 88105 (Plate 68), processed on an open

flow bench in the Antarctic meteorite lab, were below blank levels, <10–20 ppb [*Brinton and Bada*, 1996].

Thus, it is clear that the environments maintained in JSC GN2 cabinets have acceptably low levels of TOC such that contamination of amino acids does not interfere with measurements of the low concentrations of martian meteorites, or the much higher concentrations in carbonaceous chondrites.

3.2.8. Long-Term Storage

By the mid-1980s, the collection had grown to >5,000 samples and JSC had reached capacity in terms of storage space. The MWG initial vision provided guidelines for managing the long-term storage of meteorites at the Smithsonian. Scientific interest in EOCs is lower than other groups, and because there are so many EOCs (~90%) in the U.S. Antarctic meteorite collection, procedures were defined for formal transfer of such samples on a regular basis to the Smithsonian to make more space at JSC for the active research collection. Currently, if an EOC specimen at JSC has not been requested or allocated for ~4 to 5 years, it is a candidate for transfer. A representative sampling of meteorites from each icefield, however, is kept at JSC for long-term storage, so that there is a set of samples available for study that have been stored in the same environment over a long period of time. Long-term storage of all meteorites at the Smithsonian (except irons and pallasites) is located at the Museum Support Center in Suitland, MD, USA (established in 1983). Nearly 20,000 meteorites are now stored in a Class 10,000 clean room (completed in 2011) with 14 stainless steel GN2 glove boxes, outfitted with hypalon gloves (Figure 3.2). Meteorites are segregated into cabinets and pans by class. Samples, which are only removed for allocations, are enclosed in the same style and materials as those used at JSC. Sampling takes place on a clean room laminar flow bench (Figure 3.2). Tools, sample containers, and bags are made of stainless steel, aluminum, Teflon, nylon, or polyethylene zip lock bags. Polyvials are also used as sample containers (polystyrene vials with polyethylene lids).

3.3. SAMPLES: CASE STUDIES

The value of detailed curation of samples in a controlled minimal contamination environment becomes clear when considering several case histories. The program had just been established when the collection received one of its most famous samples: the first recognized lunar meteorite ALH A81005 (Plate 64). The challenges this sample posed to the MWG have been best described elsewhere [*Cassidy*, 2003], but the detailed subdivision of this lunar meteorite is described later in this chapter. Although collected before ALH A81005, EET A79001 (Plate 70) was not proposed to be a piece of

Figure 3.6. Stage I processing of ALH A81005 which generated splits 1, 2, and 5. Top photo is NASA S82-35865, and bottom is S82-35867.

Mars until the early 1980s. This sample was originally 7.9 kg, contains multiple lithologies, and has been available to allocate to the scientific community with many kilograms of material still remaining. The challenges of truly small and rare samples are perhaps best illustrated by the case of the small shergottite QUE 94201 (Plate 71). This sample was originally 12 g and of a rare and interesting lithology, and thus it has not been possible to allocate many subsamples to the community. It has been protected by a regular and strict review process whose aim is to preserve this sample for posterity as best as possible. The first U.S. Antarctic nakhlite, MIL 03346 (Plate 74), was announced in 2004 and immediately received an enormous amount of interest. This large sample was subdivided for allocation to 50 individual studies and continues to provide to a large Mars science community much as EET A79001 has, except that the interest has been in a much more compressed period of time. Finally, the CR2 chondrites have a growing interest because they harbor a diverse collection of components informing a broad range of early solar system research. The CR2 chondrites have become more and more popular with time, even since their earliest appearance in the collection in the 1980s, and now are perhaps the most popular sample type in the collection, with requests sur-

passing those for lunar and martian meteorites. These five case studies illustrate the benefits of curating a diverse collection with the approaches developed and refined since 1980.

3.3.1. ALH A81005: First Lunar Meteorite (Plate 64)

On January 17, 1982, an interesting 31.39-g achondrite was found in the Allan Hills icefield (Figure 3.6). It had a 50% thin, tan-green fusion crust, and the exposed interior exhibited numerous white to grey breccia fragments. The ANSMET team had found what later became Allan Hills (ALH) A81005, the first recognized meteorite from the Moon (Figure 3.6). This sample was of historic significance not only because it was the first lunar meteorite, but it became a great piece of evidence in favor of dynamic arguments that fragments of the Moon and Mars could be delivered to the Earth after being ejected from their parent bodies during an impact event [e.g., *Marvin*, 1983; *Marvin*, 2014 (this volume); *Korotev and Zeigler*, 2014 (this volume)]. The possibility that this meteorite represented material not sampled by the *Luna* or *Apollo* missions led many scientists to request pieces for detailed study. The meteorite was fairly small, so the MWG delayed their recommendation to ensure that distribution

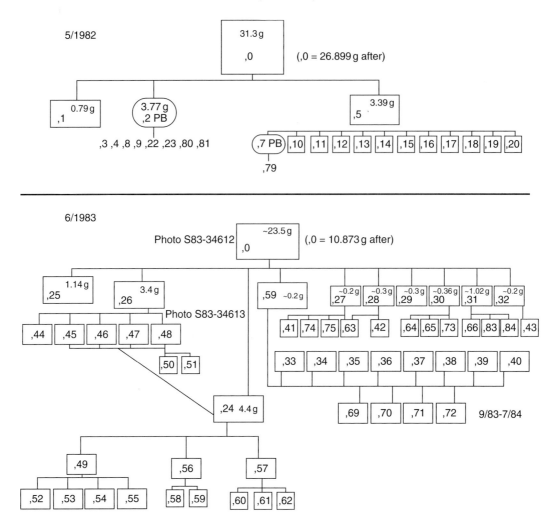

Figure 3.7. Genealogy of ALH A81005 showing processing in two main stages (I and II), as well as later (post-1984) processing. Ovals are potted butts used to make thin sections; square is a chip.

would be fair yet expedient. This process is described in some detail by *Cassidy* [2003] and set a precedent for how precious meteorite samples would be allocated. A summary of how the sample was divided (in two main stages) is presented in Figures 3.6, 3.7, and 3.8.

3.3.2. Allocation History of EET A79001 Lithology B: An Update and Synthesis (Plate 70)

Martian meteorite Elephant Moraine (EET) A79001, weighing 7.9 kg, was discovered in the Elephant Moraine region of the Transantarctic Mountains during the 1979–1980 field season. Having three distinct lithologies, olivine megacryst-bearing main lithology (A), a basaltic textured minor lithology (B), and many small glassy inclusions (C), it was recognized to have some affinity with Shergotty, which is texturally similar to Lithology B [*Score et al.*, 1982], and it became one of the important links between Mars and martian meteorites [*Bogard and Johnson*, 1983; *Bogard et al.*, 1984]. The three distinct lithologies con-

tinue to provide important information about the geologic history of Mars, and EET A79001 is the most highly requested sample in the U.S. Antarctic meteorite collection, having received close to 300 requests since 1980.

Now we will focus on the more rare Lithology B, estimated to have originally been ~350 g of the total mass of the meteorite (<5% of total mass of the meteorite). It is of interest to those studying the magmatic history of Mars and the connection between the various basaltic rocks represented in the world's martian meteorite collections. Lithology B illustrates the challenges of curating a more rare lithology in the face of a multitude of requests for scientific studies. Using this meteorite's data packs, photographic information, and published studies, an assessment will be made of how much Lithology B material has been utilized to enable new discoveries, and how much remains for future use.

Lithology B is best recognized with respect to the main lithology of EET A79001: Lithology A. Lithology A consists of a fine-grained matrix of plagioclase (maskelynite)

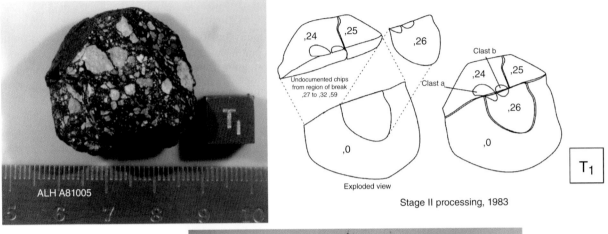

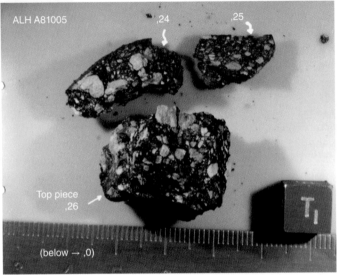

Figure 3.8. Stage II processing which generated splits 24, 25, 26, 59 and several smaller pieces for allocations (NASA photo S83-34612 and S83-34613).

and pyroxene, with about 10% of very large crystals of olivine. In hand sample, the light green to yellow to white olivine crystals stand out clearly. In comparison, lithology B is coarser grained, contains no megacrysts, and has elongate plagioclase (maskelynite) crystals as well as pigeonite. The differences are illustrated even more so when looking at the contact between the two in either hand sample or thin section (Figure 3.9; Plate 70). Confusing Lithology A for Lithology B is possible if one is looking at only one lithology without a reference piece of both lithologies. Also, if olivine is not present on any given piece of Lithology A, it is possible to confuse these in hand sample, given the similar color and overall texture of the matrix.

Lithology B is only present at one end of the meteorite, and more specifically when looking at the oriented photos from the meteorite laboratory, it is on the end designated "East" [see *McSween et al.*, 2014 (this volume)]. An estimate may be made of the mass of Lithology B by assuming an

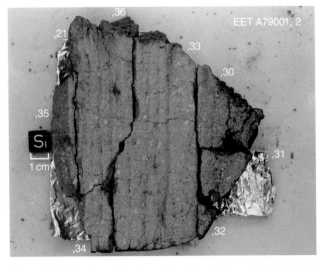

Figure 3.9. Photograph of a slab ,2 cut from EET A79001, exposing the contact between Lithology A and B (NASA photo S81-25271).

average thickness of 1 cm, overall height of 9 cm, and width of 15 cm. Assuming a density of 2.64 g/cm³ (calculated for this composition using *Ghiorso and Sack*, 1995) results in a mass of ~360 g. This estimate does not take into account the irregularities in the surface, such as the two visible dimples in many images (see Meyer, Mars Meteorite Compendium at http://curator.jsc.nasa.gov/antmet/mmc/index.cfm), and so may have some uncertainty (+/−10 g).

Two issues contribute to a major difference between the mass that is present, and the mass that is usable. First, the fusion crust is present over much of the exterior of Lithology B. Most researchers do not want to study pieces with fusion crust attached, so any piece of B that has fusion crust adhered is generally not usable (though still kept in the collection). Second, the contact between A and B is problematic. Not only is the contact gradational, it also has a large area, roughly 9 cm × 15 cm. There are many pieces of EET A79001 that were generated during subdivision that are mixtures of both lithologies. Separating A and B on small pieces is difficult if not impossible. To complicate the situation further, there was originally a natural fracture plane present at an oblique angle to the contact. During band sawing, this fracture created many small pieces that are a mixture of both Lithology A and B, such that the actual usable mass of Lithology B is reduced considerably from the estimated total mass. We can get an idea of what percentage of material is usable (70%–75%; see below) by looking at allocation data from a few individual pieces.

The first round of major allocations of Lithology B occurred in 1980, when the sample was band sawed into what became splits ,1 and ,2 and a 1-cm slab (,22). Band sawing in a GN2 cabinet was done frequently, and the cutting through Lithology B can be seen in Figure 3.10, with the final slab and ,2. The second major subdivision of Lithology B came with the band sawing of ,1 to

produce ,216 and ,1. Many allocations of Lithology B were subsequently made from ,216 (NASA photo S86-26478), and a substantial portion of Lithology B remains on ,1 (NASA photo S86-26477) and well as on ,2. A summary of the subdivision of EET A79001 was prepared by C. Meyer for the Mars Meteorite Compendium (http://curator.jsc.nasa.gov/antmet/mmc/index.cfm).

Based on the photographic information and data packs, we can summarize how much material from Lithology B has been used to estimate how much of the available mass is actually usable mass. For slab ,1 the original mass was ~24 g, and allocated mass is 18.1 g (13.6 g chips + 4.5 g potted butt). The remaining material from ,1 is either fine mixture of A and B, has fusion crust attached, or is attrition, which yields a usable mass of ~75%. Similarly, for split ,216, the original mass was ~46 g, and allocated mass is 32.1 g (all chips). The remaining material from ,216 is either fine mixture of A and B, has fusion crust attached, or is attrition, which yields a usable mass of ~70%. This number (70%–75%) can be used as a guide for future estimates of usable remaining mass by the curation group. For example, if 75% of the original mass of Lithology B is usable then that leaves 270 g. So far, 0.7 g + 32.1 g + 18.1 g have been allocated. Therefore, it seems that approximately 20% of the usable mass of Lithology B has been allocated to scientists for study, leaving about 80% or just over 200 g. Although 200 g is a reasonable amount of material, it is recommended that allocations of this sample, due to its scarceness, be limited to studies that cannot be completed using other available basaltic shergottites or martian meteorites.

3.3.3. QUE 94201: Small Rare Martian Meteorite (Plate 71)

In 1994, a 12.02-g meteorite, now called QUE 94201, was found in the Queen Alexandra Range of the Transantarctic Mountains. This sample is a basaltic rock (shergottite) that represents a melt of reduced and depleted martian mantle [see *McSween et al.*, 2014 (this volume)], contains hydrous phosphates, and has been an important sample to the Mars geoscience community. As such, it has been in high demand, and the Meteorite Working Group has been an important body in preserving this sample for the most effective research use. Since 1994, this sample has been subdivided into 63 splits, including 27 bulk samples (4.416 g) for destructive analysis, and 13 thin sections (using 2.2 g). To date, 23 principal investigators have studied the first set of splits (subsamples), and 29 principal investigators have examined splits that were created subsequently. In addition, 5.16 g of material is still available for study using new techniques or by a new generation of scientists. The manner in which QUE 94201 was subdivided and the number of investigators involved provides a case study for what we might expect if small samples like this are collected during a Mars Sample

Figure 3.10. Band sawing of main mass (slab = ,22) of EET A79001 (NASA photo S81-25260).

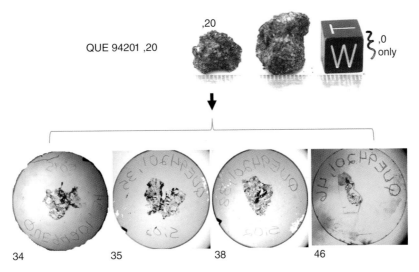

Figure 3.11. Images of QUE 94201 chip ,20 which was used to make four thin sections, one of which was studied by multiple investigators over a 15-year period.

Return mission: a rock sample could be divided into sub-portions that are subsequently divided for various analyses [e.g., *Beaty et al.*, 2008]. As an example of how material has been used sparingly and efficiently by the curators, a 2-g chip was used to make four thin sections, one of which (,46) has been allocated six times, allowing measurements to be made of such diverse properties as oxidation state, mineralogy, volatile content, rare earth element geochemistry, and clinopyroxene mineral zoning (Figure 3.11).

3.3.4. MIL 03346: A Large Nakhlite (Plate 74)

The 2003–2004 ANSMET team recovered a 715-g nakhlite from the Miller Range (MIL) region of the Transantarctic Mountains (MIL 03346). This was the first nakhlite found by the U.S. Antarctic meteorite program, and after the announcement in 2004 [*Satterwhite and Righter*, 2004], JSC received more than 50 requests for this sample for the Fall 2004 Meteorite Working Group meeting. Since then, it has been subdivided into >200 splits and distributed to ~70 scientists around the world for study. The 2009–2010 ANSMET team recovered three additional, paired masses of this nakhlite: MIL 090030 (452.6 g), MIL 090032 (532.2 g), and MIL 090136 (171.0 g) [*Satterwhite and Righter*, 2010], making the total amount of mass 1.871 kg. Given that the original find (MIL 03346) has been heavily studied and these new masses are available, we will present a comprehensive overview of the subdivision of the original mass as well as the scientific findings to date.

Nakhlites are a group of eight coarse-grained clinopy-roxene-rich rocks that contain minor amounts of olivine and mesostasis. They are thought to represent a series of cumulate igneous rocks from a shallow magma chamber on Mars [*Treiman*, 2005]. Studies of these rocks have thus led

to a better understanding of a wide variety of processes on Mars such as magma genesis and subsequent surficial processes (weathering and alteration) at the martian surface [see *McSween et al.*, 2014 (this volume)].

MIL 03346 was subdivided in three main stages: initial processing (13.15 g including 5 thin sections), slab allocations (46.46 g including 49 chips and 7 thin sections), and NE butt end allocations (32.05 g including 8 chips and 1 thin section butt). Small pieces were derived for the initial characterization of the sample, including a 2.1-g chip that was potted to make thin sections (Figure 3.12). Due to the large number of individual samples requested, band saw slabbing was considered the best way to preserve as much of the original mass as possible for future study and also to document the individual meteorite chips allocated (Figure 3.13). After the slab was totally subdivided and allocated (Figure 3.14), the NE butt end was subdivided (Figure 3.15). The total and allocated mass of the main mass, the slab, and the NE butt end are summarized in Table 3.2.

After the allocation of 57 thin/thick sections and 63 chips to individual researchers, much has been learned about MIL 03346. The scientific findings from the MIL nakhlites were summarized by *Righter and McBride* [2011]. A 447-g piece of the main mass remains for future studies, as well as a total of ~600 g of material that has not been allocated. Clearly, many additional studies are possible with this much mass left.

3.3.5. CR2 Chondrites: High Demand for Diverse Materials of Broad Interest

CR2 chondrites have been steadily recovered from Antarctica and have enabled exceptional discoveries on a wide variety of subfields including chondrule formation,

Figure 3.12. MIL 03346 main mass (top) and subsequent initial processing (bottom) which ultimately led to five potted butts for thin and thick sections. Red dots indicate locations or samples which were used to make potted butts for thin sections.

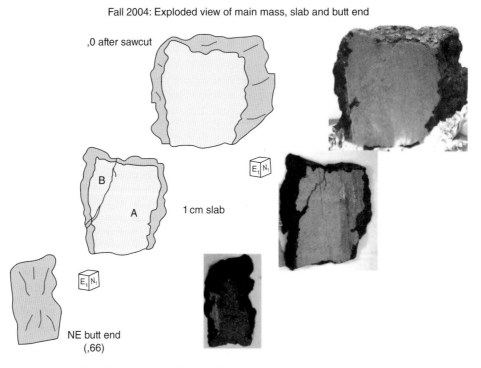

Figure 3.13. Expanded view of MIL 03346 in sketch form (left) and photo form (right), showing the derivation of three main splits after band sawing.

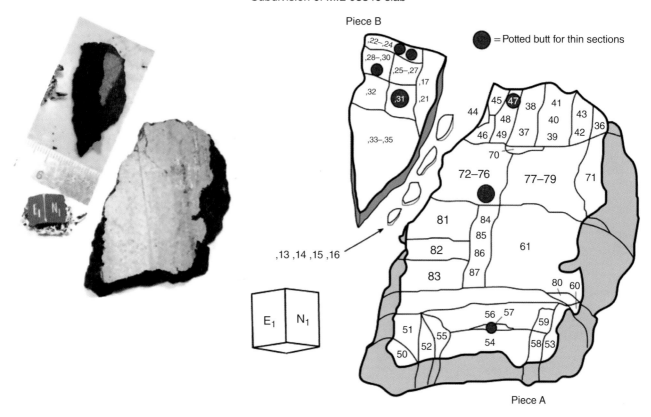

Subdivision of MIL 03346 slab

Figure 3.14. Detailed subdivision of the two pieces of MIL 03346 generated during the slabbing (A and B, which broke along a major fracture).

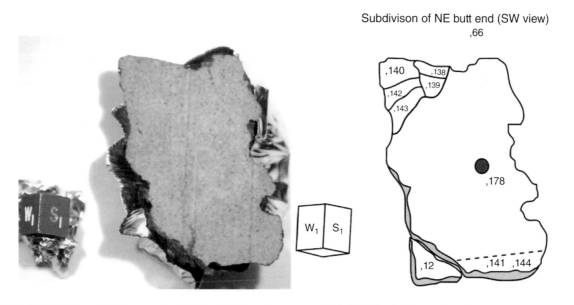

Subdivison of NE butt end (SW view)

Figure 3.15. Subdivision of NE butt end of MIL 03346 as of summer 2008. Red dot indicates location of sample used to make potted butt for thin sections.

Table 3.2. Summary of major subdivisions of MIL 03346.

Portion	Allocated mass (g)	PI splits	TS butts
Initial processing	13.15	0	5
Slab	46.46	49	7
NE butt end	32.05	8	1

PI = principal investigator; TS = thin section; NE = northeast.

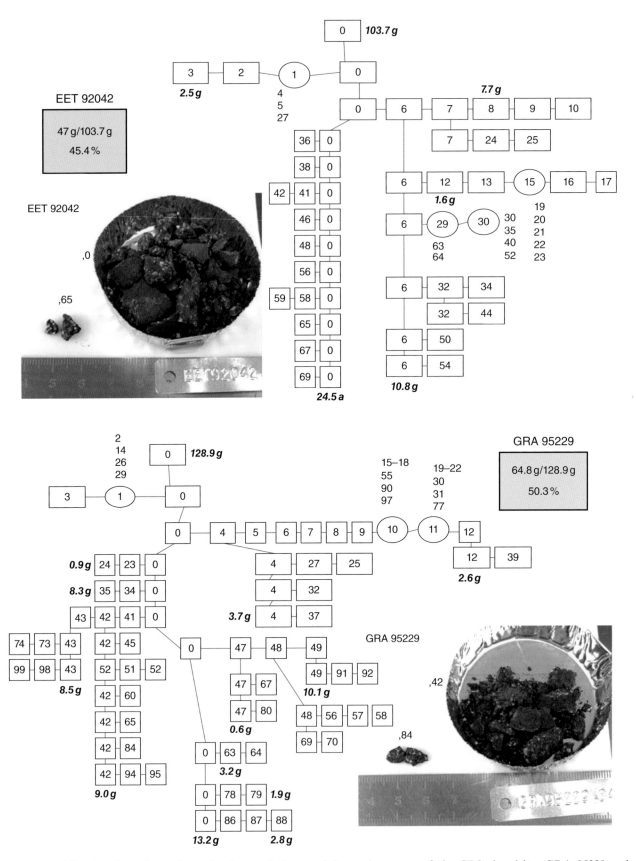

Figure 3.16. Allocation histories and sample photos of the remaining main masses of the CR2 chondrites GRA 95229 and EET 92042. Grams listed next to a sample indicate available mass; ovals are potted butts used to make thin sections; square is a chip. Shaded boxes indicate remaining mas available compared to the original mass..

matrix CAI/AOA inclusions, metal, presolar grains, isotopic studies, and organics [see *Weisberg and Righter*, 2014 (this volume)]. Many Antarctic CR2s have provided materials for a variety of studies; however, they have been heavily requested and are of limited size, and thus availability of material has become steadily smaller for some samples. Two examples are given: 128.9-g GRA 95229 and pairing group EET 92042 (Figure 3.16; Plate 25).

GRA 95229 has been an important sample for organic geochemists, with a diversity of compounds being discovered and analyzed including amino acids, hydroxyl acids, and ammonia [*Pizzarello et al.*, 2008, 2010, 2011]. Although it was originally >100 g, its popularity has resulted in ~50% of the original mass remaining for study (Figure 3.16).

Studies of EET 92042 (Plate 25), which is a highly primitive sample, have produced interesting results on organics, presolar grains, chondrules, and inclusions [*Busemann et al.*, 2006; *Glavin et al.*, 2010; *Schrader et al.*, 2011; *Makide et al.*, 2009]. Even though it is 45 g with only 45% left from the original mass, it is paired with nearly 50 other samples as part of a large pairing group spanning several collection seasons (1987, 1990, and 1992), with a combined mass of 1342.6 g. Collectively, there are 304.2 g of material remaining at JSC (from 22 specimens or 22.7% of available mass) and 941.3 g of material at SI (from 46 specimens or 70.1% of available mass), with 97.1 g allocated to scientists for various studies (7.2% of available mass).

3.4. SUMMARY

The U.S. Antarctic meteorite collection has grown more than anyone could have predicted in 1980. The curation of the samples has benefitted from the combined resources and genial cooperation of the three managing agencies. Rapid characterization of the hundreds of meteorites collected over more than 35 seasons has allowed community access to samples of high scientific interest. Detailed documentation of their handling and processing, as well as the safe and secure storage, has enabled researchers to access samples from Earth's Moon, Mars, and asteroids, to know their detailed history and to have material available for future studies using approaches and techniques that may not even have been developed yet.

REFERENCES

Allton, J. H. (1999), A brief history of organic contamination monitoring of lunar sample handling, *Lunar Planet. Sci.*, *XXIX*(1857).

Annexstad, J. O. (2001), *Oral History Transcript, JSC Oral History Project*, 15 March 2001, 93 pp.

Bada, J. L., D. P. Glavin, G. D. McDonald, and L. Becker (1998), A search for endogenous amino acids in martian meteorite ALH84001, *Science*, *279*, 362–365.

Beaty, D. W., L. E. Borg, D. J. DesMarais, O. Aharonson, S. A. Benner, D. D. Bogard, J. C. Bridges, C. J. Rudney, W. M. Calvin, B. C. Clark, J. L. Eigenbrode, M. M. Grady, J. W. Head, S. R. Hemming, N. W. Hinners, V. Hipkin, G. J. MacPherson, L. Marinangeli, S. M. McLennan H. Y., McSween, J. E. Moersch, K. H. Nealson, L. M. Pratt, K. Righter, S. W. Ruff, C. K. Shearer, A. Steele, D. Y. Sumner, S. J. Symes, J. L. Vago, and F. Westall (2008), Science priorities for Mars sample return, *Astrobiology*, *8*, 489–536.

Bogard, D. D., and J. O. Annexstad (1980), Curation and allocation procedures, *Smithsonian Contributions to the Earth Sciences*, *23*, 8–11.

Bogard, D. D., and P. Johnson (1983), Martian gases in an Antarctic meteorite? *Science*, *221*, 651–654.

Bogard, D. D., L. E. Nyquist, and P. Johnson (1984), Noble gas contents of shergottites and implications for the martian origin of SNC meteorites, *Geochim. Cosmochim. Acta*, *48*, 1723–1739.

Botta, O., Z. Martins, C. Emmennegger, J. P. Dworkin, D. P. Glavin, R. P. Harvey, R. Zenobi, J. L. Bada, and P. Ehrenfreund (2006), Reassessing the organic content of Antarctic ice and meteorites, *Lunar Planet. Sci.*, *XXXVII*(1464).

Brinton, K. L. F., and J. L. Bada (1996), A reexamination of amino acids in lunar soils: Implications for the survival of exogenous organic material during impact delivery, *Geochim. Cosmochim. Acta*, *60*, 349–354.

Burckle, L. H., and J. S. Delaney (1999), Terrestrial microfossils in Antarctic ordinary chondrites, *Meteoritics & Planetary Science*, *34*, 475–478.

Burton, A. S., J. E. Elsila, M. P. Callahan, M. G. Martin, D. P. Glavin, N. M. Johnson, and J. P. Dworkin (2012), A propensity for n-ω-amino acids in thermally altered Antarctic meteorites, *Meteoritics & Planetary Science*, *47*, 374–386.

Busemann, H., A. F. Young, C. O'D. Alexander, P. Hoppe, S. Mukhopadhyay, and L. R. Nittler (2006), Interstellar chemistry recorded in organic matter from primitive meteorites, *Science*, *312*, 727–730.

Calaway, M. J., J. H. Allton, C. C. Allen, and P. J. Burkett (2013a), *Organic Contamination Baseline Study on NASA JSC Astromaterial Curation Gloveboxes*. 44th Lunar and Planetary Science Conference, held March 18–22, 2013, in The Woodlands, Texas. LPI Contribution No. 1719, p. 1242.

Calaway, M. J., P. J. Burkett, J. H. Allton, and C. C. Allen (2013b), *Ultra Pure Water Cleaning Baseline Study on NASA JSC Astromaterial Curation Gloveboxes*. 44th Lunar and Planetary Science Conference, held March 18–22, 2013 in The Woodlands, Texas. LPI Contribution No. 1719, p. 1241.

Cassidy, W. A. (2003), *Meteorites, Ice, and Antarctica: A Personal Account*, Cambridge Univ. Press, Cambridge, UK, 349 pp.

Clarke, R. S., Jr. (1982), Overview of Antarctic irons, in *Catalog of Meteorites from Victoria Land, Antarctica, 1978–1980*, edited by U. B. Marvin and B. Mason, *Smithsonian Contributions to the Earth Sciences 24*, 49–57.

Clarke, R. S. Jr, H. Plotkin, and T. J. McCoy (2006), Meteorites and the Smithsonian Institution, in *The History of Meteorites and Key Meteorite Collections: Fireballs, Falls and Finds*, edited by G. J. H. McCall, A. J. Bowden, and R. J. Howarth, The Geological Society, London, UK, pp. 237–266.

Clayton, R. N., and T. K. Mayeda (1996), Oxygen isotopic studies of achondrites, *Geochim. Cosmochim. Acta, 60*, 1999–2017.

Clemett, S. J., L. P. Keller, and D. S. McKay (2005), Lunar organic compounds: Search and characterization, *Meteoritics & Planetary Science, 40*(5300).

Fries, M. D., G. Cody, M. Fogel, R. P. Harvey, T. Jull, L. Nittler, D. Rost, A. Steele, J. Toporski, E. Vicenzi, and N. Wainwright (2005), Contamination in meteorites stored since 1977: Preliminary results of Antarctic meteorite contamination study (ACMS), *Meteoritics & Planetary Science, 40*(5201).

Fulton, C. R., and J. M. Rhodes (1984), The chemistry and origin of the ordinary chondrites: implications from refractory-lithophile and siderophile elements, *J. Geophys. Res., 89*(Suppl), B543–B558.

Ghiorso, M. S., and R. O. Sack (1995), Chemical mass transfer in magmatic processes: IV. A revised and internally consistent thermodynamic model for the interpolation and extrapolation of liquid-solid equilibria in magmatic systems at elevated temperatures and pressures, *Contributions to Mineralogy and Petrology, 119*, 197–212.

Glavin, D. P., and J. P. Dworkin (2009), Enrichment of the amino acid L-isovaline by aqueous alteration on CI and CM meteorite parent bodies, *Proc. Nat. Acad. Sci., 106*, 5487–5492.

Glavin, D. P., J. P. Dworkin, A. Aubrey, O. Botta, J. H. Doty III, Z. Martins, and J. L. Bada (2006), Amino acid analyses of Antarctic CM2 meteorites using liquid chromatography—time of flight—mass spectrometry, *MAPS, 41*, 889–902.

Glavin, D. P., M. P. Callahan, J. P. Dworkin, and J. E. Elsila (2010), The effects of parent body processes on amino acids in carbonaceous chondrites, *Meteoritics & Planetary Science, 45*, 1948–1972.

Glavin, D. P., J. E. Elsila, A. S. Burton, M. P. Callahan, D. P. Dworkin, R. W. Hilts, and C. D. K. Herd (2012), Unusual nonterrestrial L-proteinogenic amino acid excesses in the Tagish Lake meteorite, *Meteoritics & Planetary Science, 47*, 1347–1364.

Graham, A. L., and J. O. Annexstad (1989), Antarctic meteorites. *Antarctic Science, 1*, 3–14.

Jarosewich, E. (1990), Chemical analyses of meteorites: A compilation of stony and iron meteorite analyses, *Meteoritics, 25*, 323–337.

Jarosewich, E. (2006), Chemical analyses of meteorites at the Smithsonian Institution: An update, *Meteoritics & Planetary Science, 41*, 1381–1382. doi: 10.1111/j.1945-5100.2006.tb00528.x.

Johnson Space Center (2006), *Astromaterials Curation Facility Cleaning Procedures for Contamination Control, Revision D* (JSC Document 03243), 34 pp.

Korotev, R. L., and R. A. Zeigler (2014), ANSMET meteorites from the Moon, in *35 Seasons of U.S. Antarctic Meteorites (1976–2011): A Pictorial Guide to the Collection, Special Publication 68*, edited by K. Righter, C. M. Corrigan, R. P. Harvey, and T. J. McCoy, American Geophysical Union/John Wiley & Sons, Washington, D. C.

Lunning, N. G., C. M. Corrigan, L. C. Welzenbach, and T. J. McCoy (2012), *Using Immersion Oils to Classify Equilibrated Ordinary Chondrites from Antarctica.* 43rd Lunar and Planetary Science Conference, held March 19–23, 2012, at The Woodlands, Texas. LPI Contribution No. 1659, id. 1566.

Makide, K., K. Nagashima, A. N. Krot, G. R. Huss, I. D. Hutcheon, A. Bischoff, (2009), Oxygen- and magnesium-isotope compositions of calcium-aluminum-rich inclusions from CR2 carbonaceous chondrites, *Geochim. Cosmochim. Acta, 73*, 5018–5050.

Martins, Z., C. M. O.'D. Alexander, G. E. Orzechowska, M. L. Fogel, and P. Ehrenfreund (2007), Indigenous amino acids in primitive CR meteorites, *Meteoritics & Planetary Science, 42*, 2125–2136.

Marvin, U. B. (1983), The discovery and initial characterization of Allan Hills 81005: The first lunar meteorite, *Geophysical Research Letters, 10*(9), 775–778.

Marvin, U. B. (2014), The origin and early history of the U.S. Search for Antarctic Meteorites (ANSMET), in *35 seasons of U.S. Antarctic Meteorites (1976–2011): A Pictorial Guide to the Collection, Special Publication 68*, edited by K. Righter, C. M. Corrigan, R. P. Harvey, and T. J. McCoy, American Geophysical Union/John Wiley & Sons, Washington, D. C.

McGinley, S. G., and K. Keil (1984), Petrology and classification of 145 small meteorites from the 1977 Allan Hills collection, *Smithsonian Contributions to the Earth Sciences, 26*, 55–71.

McKay, D. S., E. K. Gibson Jr., K. L. Thomas-Keprta, H. Vali, C. S. Romanek, S. J. Clemett, X. D. F. Chillier, C. R. Maechling, and R. N. Zare (1996), Search for past life on Mars: Possible relic biogenic activity in martian meteorite ALH84001, *Science, 273*, 924–930.

McSween, H. Y., R. P. Harvey, and C. M. Corrigan (2014), Meteorites from Mars, via Antarctica, in *35 seasons of U.S. Antarctic Meteorites (1976–2011): A Pictorial Guide to the Collection, Special Publication 68*, edited by K. Righter, C. M. Corrigan, R. P. Harvey, and T. J. McCoy, American Geophysical Union/John Wiley & Sons, Washington, D. C.

Olsen, E. J., A. Noonan, K. Fredriksson, E. Jarosewich, and G. Moreland (1978), Eleven new meteorites from Antarctica, 1976–1977, *Meteoritics, 13*, 209–225.

Pizzarello, S., Y. Huang, and M. R. Alexandre (2008), Molecular asymmetry in extraterrestrial chemistry: Insights from a pristine meteorite, *Proc. Nat. Acad. Sci., 105*, 3700–3704.

Pizzarello, S., Y. Wang, and G. M. Chaban (2010), A comparative study of the hydroxy acids from the Murchison, GRA 95229 and LAP 02342 meteorites, *Geochim. Cosmochim. Acta, 74*, 6206–6217.

Pizzarello, S., L. B. Williams, J. Lehman, G. P. Holland, and J. L. Yarger (2011), Abundant ammonia in primitive asteroids and the case for a possible exobiology, *Proc. Nat. Acad. Sci., 108*, 4303–4306.

Righter, K. (2013), New constraints on the size of chondrite parent bodies, *American Mineralogist, 98*, 1379–1380.

Righter, K., and K. M. McBride (2011), *The Miller Range Nakhlites: A Summary of the Curatorial Subdivision of the Main Mass in Light of Newly Found Paired Masses*. 42nd Lunar and Planetary Science Conference, held March 7–11, 2011, at The Woodlands, Texas. LPI Contribution No. 1608, p. 2161.

Righter, K., C.Meyer, J.Allton, J.Warren, C.Schwarz, S.Clemett, S.Wentworth, and K. McNamara (2008), Contamination sources associated with collection and curation of Antarctic meteorites: JSC lessons learned since 1976, *Meteoritics & Planetary Science*, 43, A189.

Satterwhite, C. E., and K. Righter (2004), *Antarct. Met. Newsletter*, 27(2), JSC Curator's Office, Houston.

Satterwhite, C. E., and K. Righter (2010), *Antarct. Met. Newsletter*, 33(2), JSC Curator's Office, Houston.

Schrader, D. L., I. A. Franchi, H. C. Connolly, R. C. Greenwood, D. S. Lauretta, and J. M. Gibson (2011), The formation and alteration of the Renazzo-like carbonaceous chondrites I: Implications of bulk-oxygen isotopic composition, *Geochim. Cosmochim. Acta*, 75, 308–325.

Score, R., T. V. V. King, C. M. Schwarz, A. M. Reid, and B. Mason (1982), Descriptions of stony meteorites, *Smithson. Contrib. Earth Sci.*, 24, 44–60.

Sears, D. W. (2012), Oral histories in meteoritics and planetary science: XVI. Donald D. Bogard, *Meteoritics & Planetary Science*, 47(3), 416–433.

Shimoyama, A., K. Harada, and K. Yanai (1985), Amino acids from the Yamato-791198 carbonaceous chondrite from Antarctica, *Chemistry Letters*, 1183–1186.

Stroeven, A. P., L. H. Burckle, J. Kleman, and M. L. Prentice (1998), Atmospheric transport of diatoms in the Antarctic Sirius Group: Pliocene deep freeze, *GSA Today*, 8, 1–8.

Treiman, A. H. (2005), The nakhlite meteorites: Augite-rich igneous rocks from Mars, *Chemie der Erde [Geochemistry]*, 65, 203–270.

Weisberg, M. K., and K. Righter (2014), Primitive asteroids: Expanding the range of known primitive materials, in *35 seasons of U.S. Antarctic Meteorites (1976–2011): A Pictorial Guide to the Collection, Special Publication 68*, edited by K. Righter, C. M. Corrigan, R. P. Harvey, and T. J. McCoy, American Geophysical Union/John Wiley & Sons, Washington, D. C.

Wright, I. P., C. P. Hartmetz, S. S. Russell, S. R. Boyd, and C. T. Pillinger (1991), On the properties of Xylan, a lubricant paint used in the dry-nitrogen sample handling cabinets at NASA-JSC, *Lunar Planet. Sci.*, XXII, 1523–1524.

Wright, I. P., S. S. Russell, S. R. Boyd, C. Meyer, and C. T. Pillinger (1992), Xylan: a potential contaminant for lunar samples and Antarctic meteorites, *Proc. Lunar Planet. Sci. Conf.*, 22, 449–458.

Wright, I. P., C. P. Hartmetz, and C. T. Pillinger (1993), An assessment of the nature and origins of the carbon-bearing components in fines collected during the sawing of EET A79001, *J. Geophys. Res.*, 98, 3477–3482.

PLATE PREFACE

The following plates highlight 80 samples from the collection that have influenced either the general field of meteoritics or the definition and understanding of specific meteorite groups. They are listed starting with chondrites (ordinary, carbonaceous, enstatite, Rumuruti, and K chondrites along with associated ungrouped chondrites), followed by achondrites including major achondrite groups, and lunar and martian meteorites.

For each sample the following information is presented:

Sample names are abbreviated as explained in chapter 2 of this book, and sample localities are presented in map Figures 2.1a to 2.1e from that chapter.

Classifications, masses (in grams), dimensions, find dates (by ANSMET teams), and weathering grades are taken from the NASA-JSC classification database (www.curator.nasa.gov) and associated field and curation data.

For each sample we present a combination of field photos, lab images, and thin section images. For some samples field photos are not available, and lab and/or thin section photos may not illustrate features clearly. Therefore, all three are not always presented for each sample. Cubes present in meteorite images are 1 cm on each side.

For each sample, we present a general overview of each meteorite type, a summary of the mineralogy, and the significance of that sample within the field of meteoritics or within that particular meteorite group.

Additional images for the plates were chosen to highlight some significance aspect of the sample, and are taken from peer reviewed literature published for the samples. References for each sample are presented at the end of the plate section. These references are meant to reflect a broad range of research carried out on the Antarctic meteorite collection but are not necessarily comprehensive. Specific references are called out by number in the mineralogy and significance sections.

35 Seasons of U.S. Antarctic Meteorites (1976–2010): A Pictorial Guide to the Collection, Special Publication 68,
First Edition. Edited by Kevin Righter, Catherine M. Corrigan, Timothy J. McCoy and Ralph P. Harvey.
© 2015 American Geophysical Union. Published 2015 by John Wiley & Sons, Inc.

Pictorial Guide to Selected Meteorites

Wisconsin Glacier (WSG) 95300
H3.3 chondrite
2733 g
Found December 24, 1995
14.0 × 10.0 × 9.0 cm
Weathering = A/B

Low-petrologic-grade ordinary chondrites are rare but contain many interesting features due to their unequilibrated nature. Because the level of postnebular thermal heating is low, many minerals and chemical features are preserved in these samples, making them a treasure trove of astromaterials. Presolar grains, pristine chondrules, isotopic heterogeneities, and organic compounds are among the rare components preserved in low-grade chondrites.

Plate 1

MINERALOGY

WSG 95300 contains numerous chondrules (up to 1.8 mm across), chondrule fragments, and mineral grains in a dark matrix containing a moderate amount of nickel-iron and troilite. The meteorite is essentially unweathered. Olivine and pyroxene is of variable composition: olivine, Fa_{1-21}; pyroxene, Fs_{2-17}.

SIGNIFICANCE

WSG 95300 is a low-petrologic-grade H chondrite (3.3), which are rare and contain a variety of very important components (e.g., organics (left, [3]), presolar grains) that may not survive metamorphism to higher temperatures that are so common among ordinary chondrites. Pristine chondrule textures led to using WSG 95300 as a basis for experimental studies aimed at duplicating chondrule textures. These studies propose that Type 1B chondrule textures (right, top [1]) can be explained by 20%–30% partial melting of chondritic materials (right bottom, [1]), with main variables being peak temperature, time of heating, and cooling rate.

References [1–3]

Lewis Cliffs (LEW) 85320
H5 chondrite
110,224 g
Found January 21, 1986
61.0 × 48.0 × 27.0 cm
Weathering = Be

H5 chondrites are perhaps the most common meteorite type in the U.S. Antarctic meteorite collection and most likely come from a parent body that was layered, producing unmetamorphosed shallow layers and metamorphosed deeper layers. There is, however, ongoing debate on whether H chondrites formed in a layered body or had a more complex history in which material was accreted, cooled, fragmented, and reaccreted.

Plate 2

MINERALOGY

Chondrules (~21%) are evident (with fuzzy borders) up to 500 microns in size, while the matrix (67%) consists of recrystallized grains 50 to 100 microns in size. There is a clear lineation texture, with metal (~10%) and sulfide (~2%) grains weakly aligned.

SIGNIFICANCE

LEW 85320 (affectionately known as "Big LEW") is one of the largest specimens recovered by the ANSMET program (110 kg), and it has a terrestrial exposure age of <0.10 Ma (right, [8]). It is a striking sample with beautiful regmaglypts, and efforts have been made to preserve the external features. Most attention has focused on the terrestrial weathering products on Big LEW, which consist of nesquehonite, magnesium sulfates, and other salts (left, [7]).

References [4–9]

LaPaz Icefield (LAP) 02240
H chondrite impact melt
28.163 g
Found December 21, 2002
3.9 × 2.5 × 2.5 cm
Weathering = C

A small subset of ordinary chondrites exhibit melted portions that have likely been formed during impact. These rare samples offer a glimpse at the pressure and temperature conditions and temporal relations among the most common chondritic materials known. Collectively, Ar-Ar ages of H, L, and LL chondrites suggest the occurrence of multiple large impact events on the ordinary chondrite parent bodies.

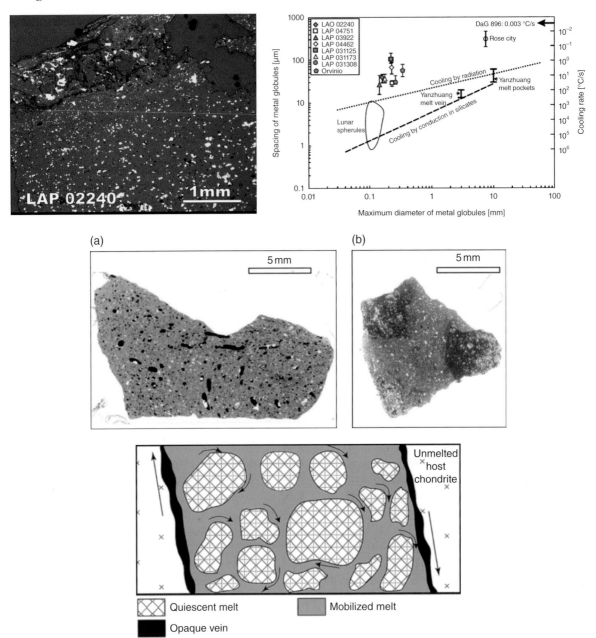

Plate 3

MINERALOGY

LAP 02240 consists of ~10% coarse-grained relict phases and ~90% fine-grained olivine and pyroxene (1–10 microns). The mineral compositions are homogenous; olivine is Fa_{18} and orthopyroxene is Fs_{16}.

SIGNIFICANCE

LAP 02240 impact melts were rapidly cooled as indicated by the homogeneous compositions of glass and opaque veins (upper right, [14]). Impact melt from LAP 02240 has yielded an Ar-Ar age of 3.9 Ga. It is one of perhaps a dozen Antarctic H chondrite impact melts that have an irregular distribution of ages <1.3 Ga and 3.4 to 4.5 Ga. These ages help to unravel the impact and geologic history (bottom, [15]) of ordinary chondrites, including some very recent impacts that are <500 Ma.

References [10–15]

Queen Alexandra Range (QUE) 97008
L3.0 chondrite
452.6 g
Found December 27, 1997
9.0 × 5.0 × 5.5 cm
Weathering = A

Low-petrologic-grade ordinary chondrites are rare but contain many interesting features due to their unequilibrated nature. Because the level of postnebular thermal heating is low, many minerals and chemical features are preserved in these samples, making them a treasure trove of astromaterials. Presolar grains, pristine chondrules, isotopic heterogeneities, and organic compounds are among the rare components preserved in low-grade chondrites.

Plate 4

MINERALOGY

QUE 97008 exhibits numerous large, well-defined chondrules (up to 2 mm) in a black matrix of fine-grained silicates, metal and troilite. Weak shock effects are present. Polysynthetically twinned pyroxene is extremely abundant. Silicates are unequilibrated with olivines ranging from Fa_{3-33} and pyroxenes from Fs_{3-21}.

SIGNIFICANCE

QUE 97008 preserves presolar grains of all four types: groups 1 and 3 form in red giant stars, and groups 2 and 4 form in low-mass asymptotic giant branch (AGB) stars (left, [21]). Its pristine chondrues have been used to define textbook examples of porphyritic and granular chondrules (right, [19]). The preservation of organic matter, as well as high δD ratios has been highlighted in survey studies of chondrite types.

References [16–23]

MacAlpine Hills (MAC) 87302
L4 chondrite breccia
1094.57 g
Found January 9, 1988
10.0 × 8.0 × 8.0 cm
Weathering = A/B

Studies of ordinary chondrites from large meteorite collections reveal that nearly 20% of all ordinary chondrites are breccias, and brecciation is often revealed only during study of the largest masses of some meteorites. Most asteroids have regolith at their surface and regolith compacts to breccia, so brecciated meteorites can lead to a better understanding of processes affecting near-surface samples from asteroids.

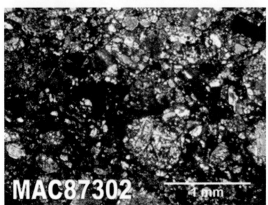

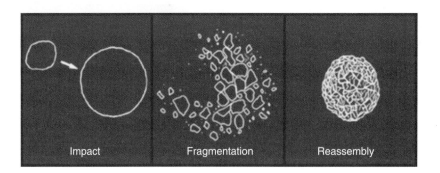

Plate 5

MINERALOGY

MAC 87302 consists of olivine with a range of compositions (Fa_{5-24}, with a distinct peak in compositions at Fa_{23-24}) and low-Ca pyroxene (Fs_{1-21}, with a peak at Fs_{20-21}) with minor plagioclase, metal, and troilite. Most clasts are similar in composition, although a few exhibit compositions more typical of H chondrites, and some olivine grains within the melt clasts are reversely zoned.

SIGNIFICANCE

MAC 87302 was the first sample recognized from a pairing group of L chondrite regolith breccias recovered during the 1987–1988 and 2002–2003 field seasons. Bearing solar wind–implanted gases, these meteorites contain a range of cognate xenoliths that include differing metamorphic types as well as impact melts [26]. The meteorite may also sample H chondritic material, and thus may represent a multilithologic asteroid body (rubble pile) formed from fragmented and reassembled material (bottom, [25]). Terrestrial age dating suggests that these meteorites fell within the last 50 kyr.

References [24–26]

Allan Hills (ALH) 85017
L6 chondrite
2361.43 g
Found December 13, 1985
15.0 × 10.0 × 11.5 cm
Weathering = A

One of the most common meteorite types in world collections are L6 chondrites, thought to be from an S-type asteroid parent body that had undergone metamorphism to near 900 °C. Many L chondrites show evidence for shock events and shock heating, so a more in-depth understanding of shock effects would be relevant to most equilibrated ordinary chondrites.

Plate 6

MINERALOGY

ALH 85017 is a polymict breccia containing lithic and monomineralic clasts, set in a fine-grained matrix of ~15% of the rock. The lithic clasts are commonly polymict, clastic breccias of varying grain sizes and modes. There are also irregular fragments of metal and sulfide, and a few relict chondrules. Olivine is Fa_{24}, orthopyroxene is Fs_{20}, clinopyroxene is $En_{47}Fs_7Wo_{46}$, feldspar is An_{14}, and there is minor Fe-Ni metal (4%), troilite (3%), chromite, and apatite.

SIGNIFICANCE

A detailed study of the fragmentation dynamics of ALH 85017 was carried out in the high-velocity impact facility at JSC. A 500-g sample was fragmented by multiple impacts down to a grain size of 125–250 μm fines (right, [27]), which were subsequently subjected to reverberation shock stresses of 14.5–67 GPa to understand the melting behavior of porous, unconsolidated, chondritic asteroid surfaces during meteorite impact (left, [28]). The results have implications for both shock melting and the production of nano-phase iron during space weathering.

References [27, 28]

Allan Hills (ALH) A78003
L6 chondrite
124.8 g
Found November 28, 1978
3.0 × 2.5 × 2.0 cm
Weathering = C

Many ordinary chondrites exhibit melt veins that have formed by shock processes. The veins can be composed mainly of silicates like the host, or of sulfide and metal veins. Such samples offer information about the impact processes and conditions experienced on small asteroid-sized bodies, which in turn offers information about the dynamics of cratering.

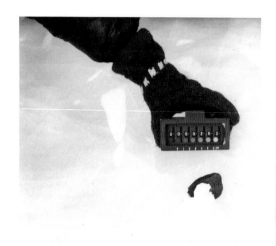

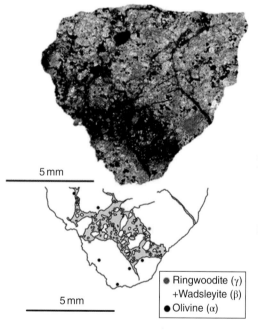

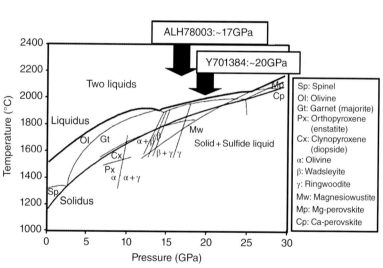

Plate 7

MINERALOGY

ALH A78003 contains olivine (Fa$_{24}$), low-Ca pyroxene (Fs$_{20}$), plagioclase, troilite, and iron nickel metal.

SIGNIFICANCE

Like many L6 chondrites, ALH A78003 is cross-cut by a series of shock melt veins formed during impact on the parent asteroid. These shock veins consist of micro-melted silicates and, in this case, much of the vein is filled with troilite. Veins in ALH A78003 contain high-pressure polymorphs of olivine and pyroxene, such as ringwoodite and wadsleyite (left, [29]), of the type found in the mantle of the Earth. These high pressure phases help constrain the shock pressures experienced by the sample on its parent body (right, [29]).

References [29, 30]

Patuxent Range (PAT) 91501
L impact melt
8550.621 g
Found December 20, 1991
19.1 × 14.3 × 14.5 cm
Weathering = B

A small subset of ordinary chondrites exhibit melted portions that have likely been formed during impact. These rare samples offer a glimpse at the pressure and temperature conditions and temporal relations among the most common chondritic materials known. Collectively, Ar–Ar ages of H, L, and LL chondrites suggest the occurrence of multiple large impact events on the ordinary chondrite parent bodies.

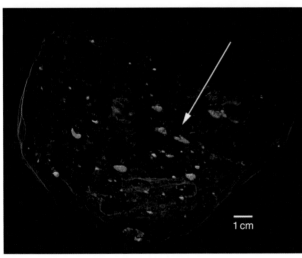

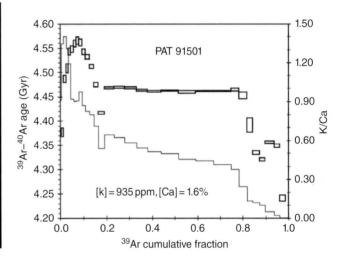

Plate 8

MINERALOGY

PAT 91501 shows an equigranular (~0.4 mm grain size) aggregate of anhedral to subhedral olivine (Fa_{24}) and low-Ca pyroxene (Fs_{20}), with minor plagioclase (An_{12}) and accessory nickel-iron and troilite. Plagioclase laths are larger than olivine and pyroxene (up to 3 mm long) and poikilitically enclose these minerals.

SIGNIFICANCE

PAT 91501 is unique among impact-melted ordinary chondrites for the distinct vesicle-metal-sulfide nodules it contains. While the silicate melt texture suggests incomplete melting, the cm-sized vesicles are obvious on the exterior of this large meteorite and common throughout the mass. High-resolution X-ray computed tomography reveals that the vesicles are often associated with metal-sulfide nodules that have a distinctive alignment of the dense metal and lighter sulfide (left, [31]). This orientation can be used to infer the gravitational field on the asteroid at the time of impact melting, which may have occurred at 4.47 Ga (right, [31]). The associated vesicles likely formed by volatilization of sulfur, and the compound vesicle-metal-sulfide particles may have been neutrally buoyant within the silicate melt.

References [31–33]

Queen Alexandra Range (QUE) 90201 and pairs
LL5 chondrite shower
1282.5 g
Found January 5, 1991
12 × 9 × 6 cm
Weathering = A/B

Meteorite showers have been documented in a number of cases and have allowed estimates to be made of impactor size, age, and dynamics. Antarctic meteorite showers are not common but have been documented on a small scale. Any new examples can provide important information about pairing, statistical evaluation of dense collection areas

Plate 9

MINERALOGY

Members of this shower have properties typical of L (e.g., pyroxene composition) and LL chondrites (e.g., metal abundance and composition), as well as properties intermediate between the L and LL groups (e.g., olivine composition), and are thus best described as L/LL5 chondrites.

SIGNIFICANCE

Samples span six field seasons: 1990, 1994, 1995, 1997, 1999, and 2002. The measured radionuclide concentrations of ^{10}Be, ^{36}Cl, and ^{41}Ca (bottom, [34]) in the metal and stone fractions of QUE 90201 indicate an age of 100–150 Ka and irradiation in an object with a preatmospheric radius of approximately 150 cm, representing one of the largest chondrites known so far. The QUE 90201 shower includes up to 2000 fragments with a total recovered mass of 60–70 kg, <1% of the preatmospheric mass of approximately 50,000 kg.

References [34]

La Paz Icefield (LAP) 04757
ungrouped chondrite
12.8 g
Found December 23, 2004
2.5 × 2.0 × 1.0 cm
Weathering = C

Some ordinary chondrites (OC) possess lower than typical olivine Fa content than has been established for the H chondrites (<17 mol%). These samples, as well as others from the higher FeO end of ordinary chondrites, suggest that ordinary chondrite-like material might be present in the asteroid belt and be part of a wider compositional group that is poorly represented in our collections.

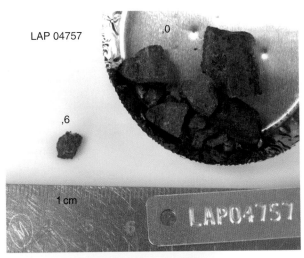

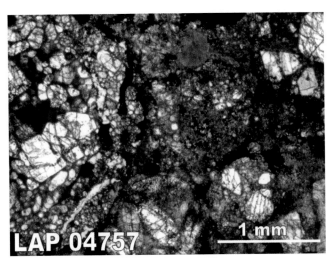

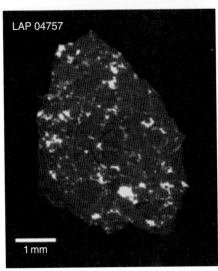

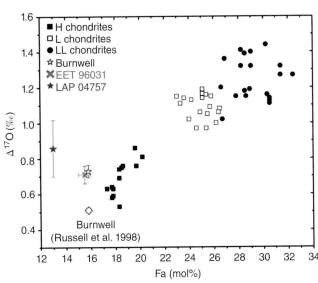

Plate 10

MINERALOGY

The meteorite consists of abundant chondrules set in a metal- and sulfide-bearing matrix. Olivine is homogeneous at Fa_{13} and pyroxene exhibits a small range at $Fs_{13–16}$. Shock effects are present, including metal-sulfide melting at grain boundaries.

SIGNIFICANCE

Overall, bulk trace element abundances and bulk oxygen isotopic analysis suggest a common origin for the low-FeO chondrites Burnwell, EET 96010, and LAP 04757 and the H chondrites (right, [35]). These three samples could represent extreme members of a redox process where a limiting nebular oxidizing agent, probably ice, reacted with material containing slightly higher amounts of metal than typically seen in the H chondrites (left, [35]).
References [35, 36]

Queen Alexandra Range (QUE) 97990
CM2 chondrite
67.297 g
Found December 15, 1997
5.5 × 3.5 × 3.0 cm
Weathering = Be

The CM2 chondrites are the most common carbonaceous group and have been at the center of many studies focused on aqueous alteration on their parent bodies, and on the abundance and species of organic matter. They are distinguished by their small (~300 micron) chondrules, accretionary rims around chondrules, common calcium-aluminum-rich inclusions (CAI), and amoeboid olivine aggregates (AOA), with little to no FeNi metal and a high matrix (~70%) that contains phyllosilicates, carbonate, sulfides, and magnetite. Alteration postdates primary features such as chondrules and CAIs, producing some impressive textures in CM2 chondrites.

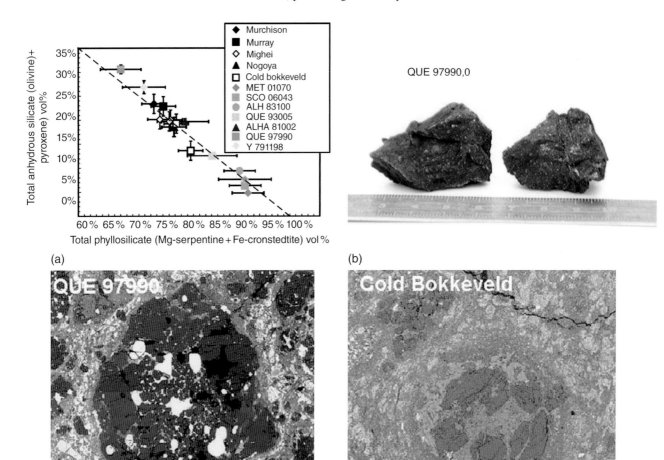

(a)

(b)

Plate 11

MINERALOGY

This meteorite consists of a few small chondrules (up to 1 mm), mineral grains and CAIs set in a black matrix. The matrix contains coarse serpentine-tochilinite intergrowths, 1–2 vol% of 10–30 mm-size carbonate grains, 1.4 vol% pyrrhotite and pentlandite, and ~1 vol% metallic Fe-Ni within porphyritic chondrules and as relatively coarse grains in the matrix. Olivine compositions are Fa_{0-61}, with many Fa_{0-2}; orthopyroxene is Fs_{0-1}.

SIGNIFICANCE

Among the CM chondrites, QUE 97990 exhibits evidence for the least amount of aqueous alteration (top left, [38]), has a suggested petrologic type of 2.6 [39], and is thus an end member for understanding the full range of textures and mineralogies that are representative of aqueous alteration on asteroid parent bodies. The degree of alteration is minor compared to the rather altered Cold Bokkeveld (bottom, [40]).

References [37–41]

Allan Hills (ALH) A81002
CM2 chondrite
14.01 g
Found January 6, 1982
2.5 × 2.5 × 2.5 cm
Weathering = Ae

The CM2 chondrites are the most common carbonaceous group and have been at the center of many studies focused on aqueous alteration on their parent bodies, and on the abundance and species of organic matter. They are distinguished by their small (~300 micron) chondrules, accretionary rims around chondrules, common calcium-aluminum-rich inclusions (CAI), and amoeboid olivine aggregates (AOA), with little to no FeNi metal and a high matrix (~70%) that contains phyllosilicates, carbonate, sulfides, and magnetite. Alteration postdates primary features such as chondrules and CAIs, producing some impressive textures in CM2 chondrites.

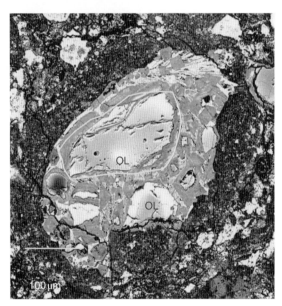

Plate 12

MINERALOGY

Chondrules and CAIs (40%), fine-grained matrix (35%), tochilinite/serpentine-rich (25%).

SIGNIFICANCE

ALH A81002 shows abundant evidence for aqueous alteration in CM2 chondrites, such as aqueously altered porphyritic and radial chondrules (left, right; [45]). It has been a critical sample in defining mineralogic and textural changes that occur during the alteration.
References [42–50]

Allan Hills (ALH) 83100
CM1/2 chondrite
3019 g
Found January 1, 1984
8.0 × 7.0 × 6.0 cm
Weathering = Be

The CM2 chondrites are the most common carbonaceous group and have been at the center of many studies focused on aqueous alteration on their parent bodies, and on the abundance and species of organic matter. They are distinguished by their small (~300 micron) chondrules, accretionary rims around chondrules, common calcium-aluminum-rich inclusions (CAI), and amoeboid olivine aggregates (AOA), with little to no FeNi metal and a high matrix (~70%) that contains phyllosilicates, carbonate, sulfides, and magnetite. Alteration postdates primary features such as chondrules and CAIs, producing some impressive textures in CM2 chondrites.

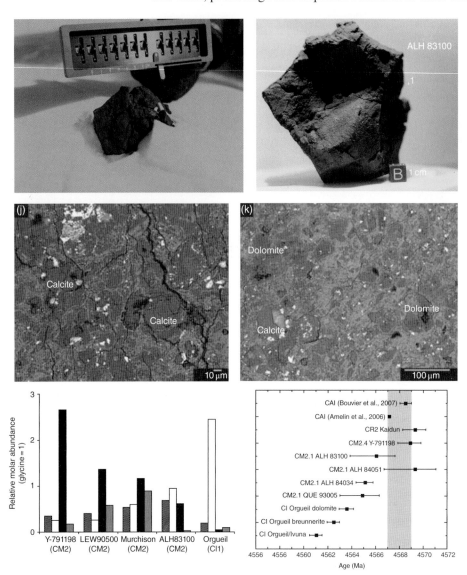

Plate 13

MINERALOGY

Mg-rich serpentine (62%), Fe-cronstedite (24%), olivine (Fa$_{0-10}$; 9%), magnetite (2%), sulfide (1%), calcite (1%).

SIGNIFICANCE

ALH 83100 is a large sample and has been classified as transitional between CM1 and CM2. Therefore, it has received much attention and been the subject of many studies. In particular, calcite grains in the matrix (middle, [53]) have been dated using Mn-Cr chronology showing the calcite formed early, but just after chondrules and CAIs (right, [53]). Amino acid studies show that ALH 83100 contains a similar load of amino acids to the CM fall, Murchison (left, where alanine = stripes, β-alanine = white, α-aminoisobutyric acid = black, and isovaline = gray; [56]).

References [51–63]

Meteorite Hills (MET) 01070
CM1 chondrite
40.585 g
Found December 17, 2000
4.5 × 2.5 × 3.5 cm
Weathering = Be

CM carbonaceous chondrites are perhaps the most abundant kind of CCs and have been at the center of many studies focused on aqueous alteration on their parent bodies, and on the abundance and species of organic matter. Alteration postdates primary features such as chondrules and CAIs, producing some impressive textures in CM2 chondrites. CM1 chondrites are rarer and provide evidence for the most extreme aqueous alteration known on asteroids.

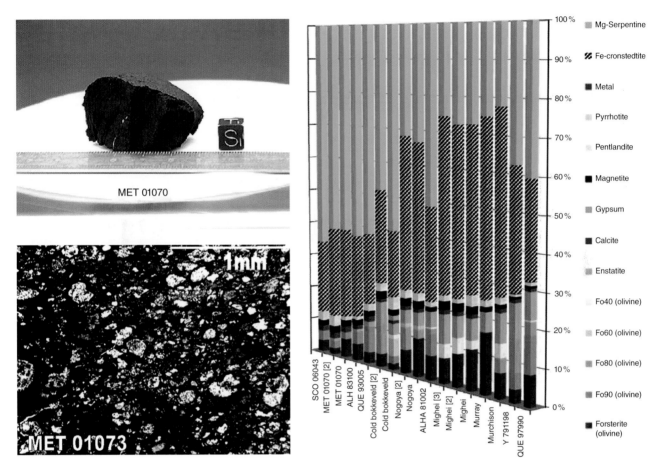

Plate 14

MINERALOGY

Mg-rich serpentine (61%), Fe-rich cronstedtite (26%), sulfide (2%), magnetite (2%), calcite (2%), olivine (7%; Fa_{0-10}).

SIGNIFICANCE

CM1 chondrites are rare and almost exclusively defined by Antarctic samples. Even so, there are many very small Antarctic samples (<5 g), so larger samples like MET 01070 can be studied by more scientists using more techniques than most of the samples in the collection. The fine-grained nature of CM1s has led to the development and application of micro x-ray diffraction analysis to identify the major mineralogy of samples (right, [66]). Discovery that MET 01070 represents a most aqueously altered member of the CM groups, as well as its unique amino acid content, makes it precious scientific material.
Reference [64–71]

Allan Hills (ALH) A77307
CO3 chondrite
181.3 g
Found January 7, 1978
4.5 × 4.0 × 5.0 cm
Weathering = Ae

The CO chondrites contain abundant (35%–45%) but small (150 micron) chondrules and mostly Type II metal-rich chondrules. CAIs and AOAs are present but not abundant (10%). The volume of matrix material is low, with secondary alteration minerals including nepheline, sodalite, fayalite, hedenbergite, and andradite, but no phyllosilicates. There is a metamorphic sequence present in CO chondrites, from unequilibrated 3.0 to partially equilibrate 3.7.

Plate 15

MINERALOGY

ALH A77307 shows a closely-packed aggregate of mineral grains (up to 0.2 mm), mineral aggregates (up to 0.8 mm), and rather sparse small (0.1–0.5 mm) chondrules, set in a dark brown to black opaque matrix. The matrix makes up 40%–50% of the section, and the mineral grains, aggregates, and chondrules consist of olivine and polysynthetically twinned clinopyroxene in approximately equal amounts. Most of the olivine is $Fa_{~1}$, with a few grains ranging up to Fa_{30}. Scattered grains of (1%–2%) iron-nickel and sulfide are present in the matrix.

SIGNIFICANCE

ALH A77307 is one of the least metamorphosed of the CO chondrites (type 3.05; [79]), and has been an integral sample in understanding the variation of oxygen isotopic values in chondritic components (chondrules, matrix, inclusions; left, [86]), as well as for identifying presolar grains. The latter include diamond, SiC, silicates, and oxides and are believed to come from red giant branch (RGB) stars, asymptotic giant branch (AGB) stars, stars with higher-than solar metallicity, or a few even of supernova origin (right, [92]).

References [72–94]

Lewis Cliffs (LEW) 85332
Ungrouped chondrite
113.692 g
Found January 15, 1996
5.5 × 4.0 × 3.5 cm
Weathering = B/C

Ungrouped chondrites are rare but offer an opportunity to document diversity of materials outside the established groups. Chondrites that have been ungrouped in the past have eventually led to the establishment of robust groups (like Renazzo [CR], Karoonda [CK], and Rumuruti [R] chondrites). Ungrouped chondrites constantly push the boundaries of classification and keep our classification schemes in check.

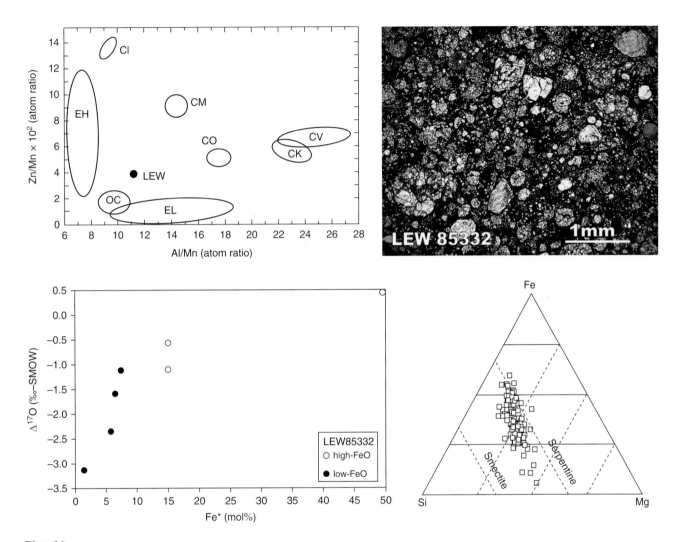

Plate 16

MINERALOGY

LEW 85332 shows an aggregate of small granular or porphyritic olivine chondrules (most <0.5 mm, but can be up to 1.2 mm across), chondrule fragments, and irregular granular masses set in a translucent yellow-brown matrix. Minor amounts of nickel-iron and sulfide are present as small grains in the matrix and on chondrule rims. Olivine shows a wide composition range, Fa_{1-20}, mean Fa_9; and less abundant pyroxene has composition range Fs_{1-16}.

SIGNIFICANCE

The bulk composition is clearly outside of established chondrite groups (top, [96]), and additional studies revealed some of the first evidence for a link between oxygen isotopic variation and FeO content in silicates (left, [98]). In addition, it contains a hydrated matrix that illustrates its presence in carbonaceous chondrite groups other than CR or CM (right, [97]).

References [95–98]

MacAlpine Hills (MAC) 87300,301
Ungrouped chondrite
167.535, 110.899 g
Found January 9, 1987
6.0 × 5.0 × 3.5; 5.5 × 5.0 × 3.5 cm
Weathering = B

Ungrouped chondrites are rare but offer an opportunity to document diversity of materials outside the established groups. Chondrites that have been ungrouped in the past have eventually led to the establishment of robust groups (like Renazzo [CR], Karoonda [CK], and Rumuruti [R] chondrites). Ungrouped chondrites constantly push the boundaries of classification and keep our classification schemes in check.

Plate 17

MINERALOGY

Chondrules are abundant but small (~0.6 mm across) and, together with chondrule fragments and mineral grains, are set in a dark brown to black matrix. Chondrules consist of granular or porphyritic olivine (almost pure Mg_2SiO_4, but mean composition is Fa_7) and olivine-pyroxene (Fs_{1-9}); mineral grains are mainly olivine.

SIGNIFICANCE

MAC 87300 (and 87301) represents carbonaceous chondrite material that does not fall into an established carbonaceous chondrite group but has Fe-rich matrix and evidence for hydration. It is an ungrouped C2 chondrite that provides evidence for aqueous alteration (right, [102]) on a parent body that differed from the more common (in our collections) CM2 parent body.

References [99–102]

MacAlpine Hills (MAC) 88107
Ungrouped chondrite
192.77 g
Found January 7, 1989
8.5 × 4.5 × 3.0 cm
Weathering = Be

Ungrouped chondrites are rare but offer an opportunity to document diversity of materials outside the established groups. Chondrites that have been ungrouped in the past have eventually led to the establishment of robust groups (like Renazzo [CR], Karoonda [CK], and Rumuruti [R] chondrites). Ungrouped chondrites constantly push the boundaries of classification and keep our classification schemes in check.

Plate 18

MINERALOGY

MAC 88107 has abundant small chondrules (~0.3 mm in diameter), numerous mineral aggregates, and mineral grains set in a black matrix. Most of the chondrules consist of granular olivine ($Fa_{0.5-39}$, with a mean of Fa_8) and olivine-pyroxene ($Fs_{0.8-9}$). Accessory amounts of finely dispersed nickel-iron and sulfide are present.

SIGNIFICANCE

MAC 88107 represents material that is intermediate in bulk composition between CM and CO chondrites (right, [107]) and contains oxygen isotopic variation that also spans the gaps between CO and CM chondrites (left, [103]). The infrared spectrum of this sample has also been compared to spectra of asteroid 1 Ceres, suggesting that it may represent material like that found on our largest asteroid.

References [103–108]

Allan Hills (ALH) 84028
CV3 chondrite
735.9 g
Found December 14, 1984
9.0 × 8.0 × 6.5 cm
Weathering = Ae

The CV chondrites contain 1-mm-sized chondrules that are mainly porphyritic. They have a high matrix to chondrule ratio, as well as abundant CAI and AOA contents. There are also abundant salite-hedenbergite-andradite nodules present. The CV chondrites have been subdivided into reduced and oxidized groups. The oxidized subgroup contains two kinds of secondary alteration: magnetite, fayalite, nepheline, or sodalite (A) or phyllosilicates, magnetite, fayalite, sulfide or hedenbergite (B). The reduced subgroup is unaltered.

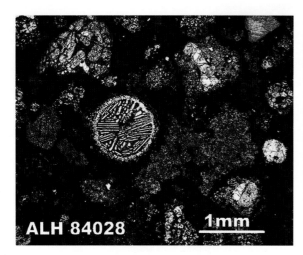
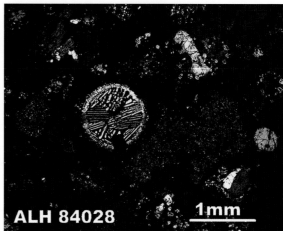

Plate 19

MINERALOGY

Chondrules up to ~2 mm in diameter, clasts, and inclusions up to ~4 mm in maximum dimension are distributed in a pristine matrix consisting of (at least) abundant minute olivine plates of Fa_{45-50} composition, troilite, and awaruite. The olivine in chondrules, and larger matrix grains, has a wide range of composition, from Fa_{0-30}, but most are Fa_{0-10}. Pyroxene grains having a composition close to Wo_1 En_{97} Fs_2 were found. Fine- and coarse-grained refractory inclusions are present, including large Type A with gehlenitic melilite, spinel, and Ti-rich fassaitic pyroxene.

SIGNIFICANCE

ALH 84028 contains fayalite-rich regions that replace sulfide and oxide nodules in the matrix and portions of chondrules. A satisfactory explanation for these Fa-rich regions has been elusive but essentially opens the age-old questions of whether they formed in a nebular process or later in parent body processes. This sample will continue to help address this question.

References [109–116]

Elephant Moraine (EET) 92002
CK5 chondrite
1040.95 g
Found December 19, 1992
13.6 × 11.2 × 4.7 cm
Weathering = A/Be

CKs are the only carbonaceous chondrite group to exhibit evidence for widespread thermal metamorphism. CK chondrites are perhaps the most oxidized meteorite group among the chondrites exhibiting values of fO_2 higher than FMQ. Their lack of metal and higher forsterite-content olivines suggest that they have been oxidized beyond any other group, and their oxide and sulfide mineralogy suggests that they equilibrated at temperatures close to 600 °C. Shock features in the olivine include small inclusions thought to cause darkening; this may aid in identifying the parent asteroid(s) for these chondrites.

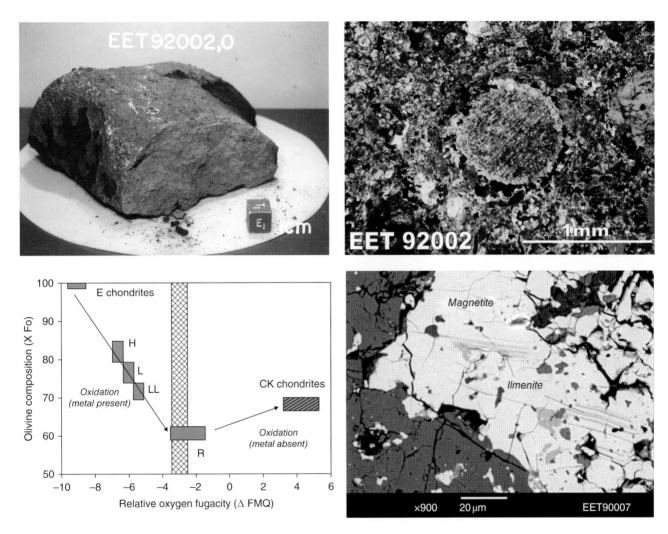

Plate 20

MINERALOGY

EET 90007 shows an aggregate of small (average about 0.02 mm) olivine grains and minor opaque material which is mainly magnetite with a little troilite and pentlandite. Occasional chondrules, up to 1.2 mm across, are present. Olivine composition is Fa_{29}; minor plagioclase of variable composition, An_{33-45}, minor diopside, $Wo_{40}Fs_{11}$.

SIGNIFICANCE

As with other groups, CK chondrites were a small group of samples in 1994 but have rapidly expanded into >100 samples with some diversity of mineralogy, with Antarctic members of the group being influential in defining the group. Detailed studies of shock features, bulk composition, and opaque minerals (sulfide, oxides; right, [122]) have allowed quantification of degree of oxidation (left, [122]) and metamorphism compared to other groups such as H, L, LL, and R chondrites.

References [117–125]

Queen Alexandra Range (QUE) 94411
CB$_a$ chondrite
39.6 g
Found December 14, 1994
3.2 × 2.1 × 1.4 cm
Weathering = B

Bencubbinites (CB chondrites) are metal rich and related to CH and CR chondrites. They contain high modal contents of metal, as well as CAIs and chondrules. Between the coarser-grained components is a thin layer of intimately intermixed glass and metallic quench liquids that formed upon shock of the meteorite. The hydrated matrix that is found in most other carbonaceous chondrites is largely missing; the only remnant is interstitial glass and very rare hydrated clasts. CB$_a$ chondrites are generally fine grained in texture with ~100 μm chondrules and metal grains.

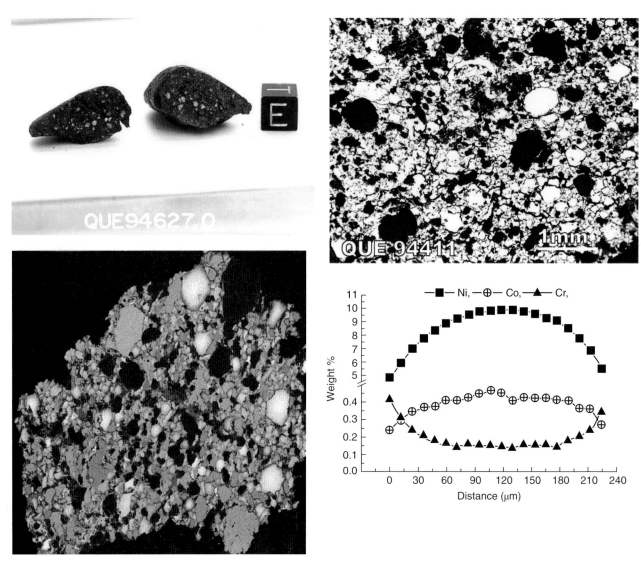

Plate 21

MINERALOGY

Zoned FeNi grains, CAIs, small cryptocrystalline chondrules, and fine-grained interstitial shock melt.

SIGNIFICANCE

QUE 94411 was the first bencubbinite recovered from Antarctica and has been the subject of intense study, especially on the origin of the zoned metal grains (zoned w.r.t. all siderophile elements, including PGEs) (right, [129]), and a weak foliation (left, [129]) attributed to shock.

References [126–130]

Miller Range (MIL) 05082
CB$_b$ chondrite
11.98 g
Found December 6, 2005
2.5 × 1.8 × 2.3 cm
Weathering = B

Bencubbinites (CB chondrites) are metal rich and related to CH and CR chondrites. They contain high modal contents of metal, as well as CAIs and chondrules. Between the coarser-grained components is a thin layer of intimately intermixed glass and metallic quench liquids that formed upon shock of the meteorite. The hydrated matrix that is found in most other carbonaceous chondrites is largely missing; the only remnant is interstitial glass and very rare hydrated clasts. CB$_b$ chondrites are generally coarse grained in texture with up to ~5 mm chondrules and metal grains.

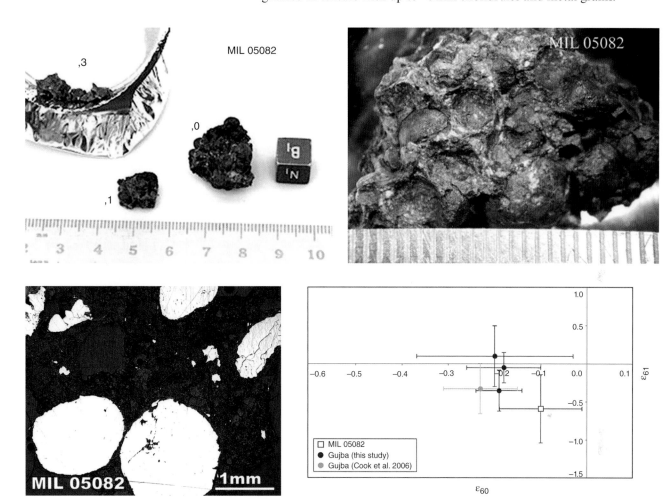

Plate 22

MINERALOGY

Subequal amounts of metal and sulfide, with rounded metal particles up to 4 mm in diameter. Chondrules and chondrule fragments up to 0.5 mm are dominated by radiating pyroxene textures. Silicates are magnesian (Fa$_{1-3}$, Fs$_{1-3}$).

SIGNIFICANCE

The bulk composition of MIL 05082 indicates that there was little or no metal-silicate fractionation between the major components of the meteorite (nor was there modification from original nebular origin; left). Amino acids in CB chondrites are abundant and structurally distinct from other carbonaceous chondrite groups. MIL 05082 comprises primitive chondritic materials, little modified since nebular times, and isotopic studies reveal that ^{60}Fe was not distributed homogeneously throughout the early solar system (right, [133]).

References [131–133]

Grosvenor Mountains (GRO) 95551
Ungrouped chondrite
213.389 g
Found January 15, 1996
6.5 × 4.0 × 4.0 cm
Weathering = C

Metal-rich chondrites have mainly been associated with CR carbonaceous chondrites, such as CH and CB chondrites, but there are a few meteorites that are metal rich yet have affiliation with ordinary chondrites. The processes that led to the formation of metal-rich chondrites were therefore not confined to just carbonaceous materials but instead must be more widespread. Therefore, understanding these chondrites can lead to a more general understanding of nebular processes.

Plate 23

MINERALOGY

The meteorite is a breccia with chondritic, carbonaceous, and enstatite-rich clasts. The chondritic clasts (up to 15+ mm) consist of chondrules and chondrule fragments (up to 1.8 mm across; mainly Fa_{1-2}; pyroxene, Fs_1) in a matrix of nickel-iron metal with minor troilite. The enstatite-rich clasts are up to 11 mm across and consist of highly shocked enstatite (or clinoenstatite), and the fine-grained carbonaceous clasts are as much as 3.6 mm across.

SIGNIFICANCE

The meteorite is anomalous and resembles bencubbinites. NWA 5492 and GRO 95551 likely represent a new type of chondrite with affinities to both E and H chondrites. Oxygen isotope compositions that plot near the TF line (right, [136]), combined with lack of any evidence of hydrous alteration, suggest their formation in the inner solar system, making them possible analogues of the materials that accreted to form the inner planets.

References [134–137]

Allan Hills (ALH) 85085
CH chondrite
11.922 g
Found December 19, 1985
1.5 × 2.0 × 0.8 cm
Weathering = A/B

CH chondrites contain very small (~20 micron) chondrules (radial and cryptocrystalline), a significant amount of metal, some small CAIs, and little to no matrix. The zoned metal grains in CH chondrites are thought to have formed from condensation from a gas of solar nebular composition. In addition, some CH chondrites contain nitrogen isotopic anomalies that are associated with insoluble organic matter.

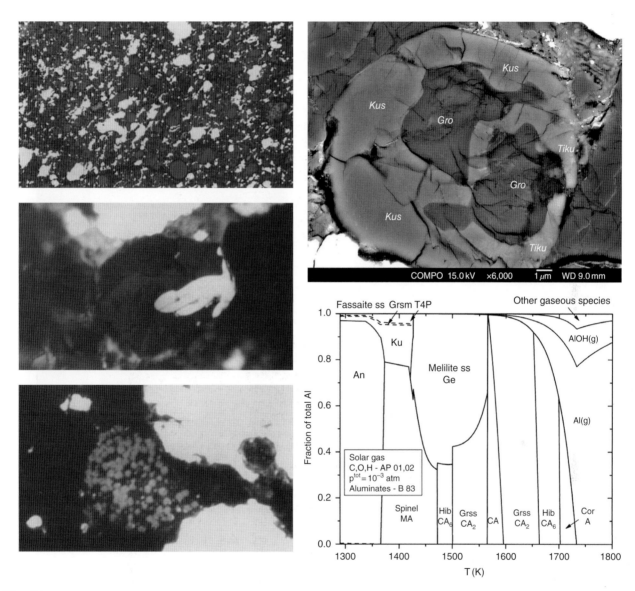

Plate 24

MINERALOGY

FeNi metal (36%), lithic and silicate mineral fragments (56%), chondrules (4%), and matrix lumps (4%). Olivine fayalite component is mainly 1% to 6% but can be as high as 30%.

SIGNIFICANCE

ALH 85085 was the first of this kind of carbonaceous chondrite to be recognized, but by 2012 there were close to 23 CH chondrites in world collections. Osbornite was documented here (left, center panel, [151]) and also in Stardust sample collection, hence spurring links between meteorites and cometary particles. Also, the new mineral kushiroite was discovered in ALH 85085 (upper right, [147]). It is a Ca,Al-silicate ($CaAl_2SiO_6$) that is part of condensation sequence (see "Ku") in nebular condensation (lower right, [145]).

References [138–153]

Elephant Moraine (EET) 92042
CR2 chondrite
103.67 g
Found December 22, 1992
5 × 4 × 2.5 cm
Weathering = B

CR chondrites are one of the most primitive carbonaceous chondrite groups, but they also record moderate secondary aqueous alteration. Unusual characteristics are their high (50%–60%) content of relatively large (0.7 mm) chondrules, high metal contents (7%–8%), and hydrated (phyllosilicate-rich) fine-grained matrix. Most chondrules are porphyritic types, and many chondrules have metal and sulfide rims. The matrix contains a wide range of minerals such as carbonates, sulfides, and magnetite, as well as serpentine and saponite.

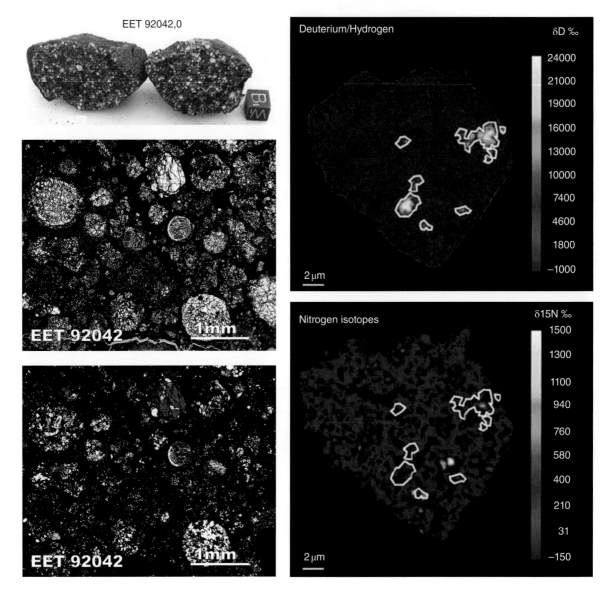

EET 92042,0

Deuterium/Hydrogen — δD ‰

24000
21000
19000
16000
13000
10000
7400
4600
1800
−1000

2 µm

EET 92042 — 1 mm

EET 92042 — 1 mm

Nitrogen isotopes — δ15N ‰

1500
1300
1100
940
760
580
400
210
31
−150

2 µm

Plate 25

MINERALOGY

Olivine and orthopyroxene are Mg rich (Fa, Fs <4%), solar FeNi metal, and metal rich (10%). The matrix contains hydrous silicates, carbonates, and magnetite. Plagioclase-bearing chondrules are also reported (An_{91-96}).

SIGNIFICANCE

EET 92042 is part of a large pairing group of CR2s found in the Elephant Moraine region of Antarctica. High resolution isotopic measurements of H and N in the matrix of this meteorite have revealed large isotopic anomalies that are associated with insoluble organic matter (right, [156]). Other studies have revealed diverse and interesting features such as amino acids, presolar grains, and metal and sulfide rims on chondrules.

References [154–174]

Queen Alexandra Range (QUE) 99177
CR2 chondrite
43.555 g
Found December 8, 1999
4.0 × 3.0 × 2.0 cm
Weathering = Be

CR chondrites are one of the most primitive carbonaceous chondrite groups, but they also record moderate secondary aqueous alteration. Unusual characteristics are their high (50%–60%) content of relatively large (0.7 mm) chondrules, high metal contents (7%–8%), and hydrated (phyllosilicate-rich) fine-grained matrix. Most chondrules are porphyritic types, and many chondrules have metal and sulfide rims. The matrix contains a wide range of minerals such as carbonates, sulfides, and magnetite, as well as serpentine and saponite.

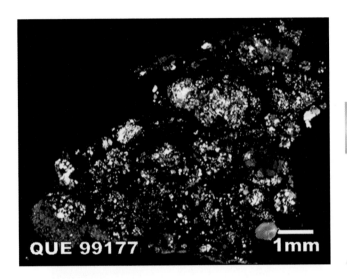

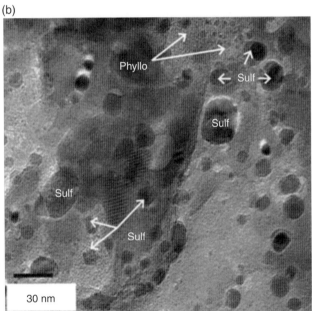

(a)

(b)

Plate 26

MINERALOGY

QUE 99177 exhibits well-defined, metal-rich chondrules up to 2 mm in diameter in a dark matrix of FeO-rich phyllosilicate and metal. Silicates are unequilibrated; olivines range from Fa_{1-31}, with most Fa_{0-2}, and pyroxenes from $Fs_{1-7}Wo_{1-5}$. The meteorite is probably a CR2 chondrite.

SIGNIFICANCE

The matrix of QUE 99177 is unlike many other CR2 chondrites in that it contains FeO-rich amorphus silicates and only rare phyllosilicates and sulfides that are usually common in matrices of CR2 (above images, [176]). In addition, QUE 99177 contains an unusual graphite-amphibole clast that records conditions of the primitive solar nebula and also contains a variety of presolar grains as in ALH 77307 and DOM 08006.

References [175–179]

Grosvenor Mountains (GRO) 95577
CR1 chondrite
106.2 g
Found January 10, 1996
6.4 × 3.6 × 3.3 cm
Weathering = B

CR chondrites are one of the most primitive carbonaceous chondrite groups, but they also record moderate secondary aqueous alteration. Unusual characteristics are their high (50%–60%) content of relatively large (0.7 mm) chondrules, high metal contents (7%–8%), and hydrated (phyllosilicate-rich) fine-grained matrix. Most chondrules are porphyritic types, and many chondrules have metal and sulfide rims. The matrix contains a wide range of minerals such as carbonates, sulfides, and magnetite, as well as serpentine and saponite. CR1 chondrites have undergone extensive aqueous alteration and contain no anhydrous silicates. There are not many CR1 chondrites known, so the properties of this group are based on very few samples.

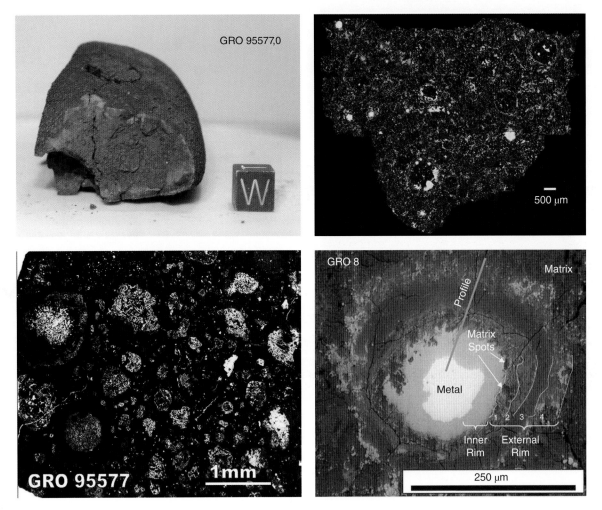

Plate 27

MINERALOGY

Remarkably, all of the chondrules in GRO 95577 are completely hydrated, consisting almost entirely of phyllosilicates, magnetite, and sulfides. Despite the mineralogic conversion, the initial chondrule textures and chondrule-matrix boundaries are well preserved (above right, [187]).

SIGNIFICANCE

GRO 95577 was the first classified as a CR1, and remains one of the most highly studied examples. It records effects of alteration on many of the unique features of CR2 such as organic compounds, chondrules, CAIs, and nebular metal (lower right, [186]).

References [180–188]

Grosvenor Mountains (GRO) 95517
EH3 enstatite chondrite
574 g
Found December 19, 1995
7.5 × 7.5 × 6.0 cm
Weathering = C

Enstatite chondrites are the most reduced chondritic materials we have in our meteorite collections and are composed of nearly pure Mg end-member silicate minerals such as enstatite and metallic FeNi metal that contains Si. EH chondrites include petrologic types 3 through 6, with unequilibrated type 3 chondrites providing insight into nebular processes in a more reduced portion of the solar nebula. Enstatite chondrites also share the same oxygen isotopic signature as the Earth, the Moon, and aubrites.

GRO 95517,0

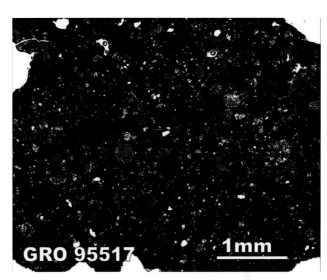

GRO 95517   1mm

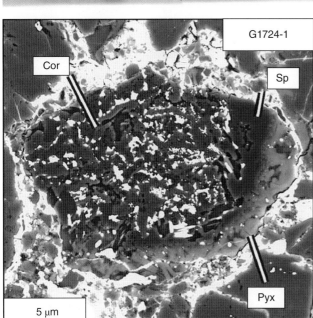

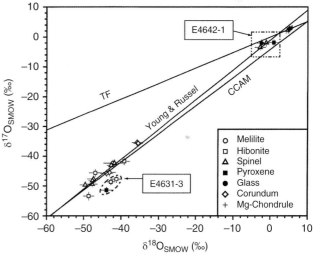

Plate 28

MINERALOGY

GRO 95517 contains numerous small chondrules (up to 0.6 mm diameter), CAIs, some irregular aggregates, and many small silicate grains in a black matrix. Trace amounts of nickel-iron metal and troilite are present as minute grains. The silicate grains are almost entirely olivine near Mg_2SiO_4 in composition, with a few more iron-rich grains. A little pyroxene near $MgSiO_3$ in composition is present. The matrix appears to consist largely of iron-rich serpentine.

SIGNIFICANCE

Studies of the CAIs from GRO 95517 showed that they are similar in mineralogy, composition, and age to CAIs from other more oxidized meteorites like carbonaceous and ordinary chondrites (left and right, [190]). Therefore, the EH CAIs either formed early in an oxidized environment and were later reduced, or they were formed in a different (oxidized) part of the solar nebula and then later transported to a more reduced region where enstatite chondrites were formed. This sample has thus provided fundamental constraints on the timing and sources of materials making up primitive meteorites.

References [189–192]

Allan Hills (ALH) A81189
EH3 enstatite chondrite
2.63 g
Found January 20, 1982
2.0 × 1.1 × 0.5 cm
Weathering = C

Enstatite chondrites are the most reduced chondritic materials we have in our meteorite collections, and are composed of nearly pure Mg end-member silicate minerals such as enstatite and metallic FeNi metal that contains Si. EH chondrites include petrologic types 3 through 6, with unequilibrated type 3 chondrites providing insight into nebular processes in a more reduced portion of the solar nebula. Enstatite chondrites also share the same oxygen isotopic signature as the Earth, the Moon, and aubrites.

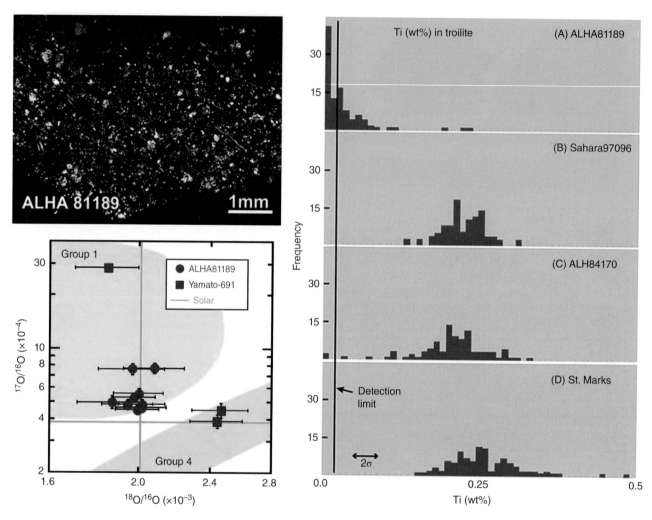

Plate 29

MINERALOGY

ALH A81189 contains an aggregate of chondrules, chondrule fragments, and mineral grains set in an opaque matrix. The chondrules range up to 0.9 mm in diameter; most of them consist of granular pyroxene (sometimes with a little olivine), but a few are made up of nickel-iron and troilite. The matrix consists largely of nickel-iron and troilite, with a considerable amount of limonite formed by weathering of the metal. Microprobe analyses show that the pyroxene is close to $MgSiO_3$ in composition. Most of the olivine grains are close to Mg_2SiO_4 in composition.

SIGNIFICANCE

ALH A81189 may be the most primitive EH3 chondrite. However, variations in chondrule types suggest that it is not the metamorphic protolith of other EH3 chondrites. ALH A81189 formed at slightly higher oxygen fugacities than other EH3s, based on the lower Ti content of the troilite (right, [196]). It also contains presolar grains that record stellar processes just like a few other meteorites in our collection (QUE 99177, QUE 97008; left, [193]).

References [193–196]

Pecora Escarpment (PCA) 91020
EL3 enstatite chondrite
1748.6 g
Found January 19, 1992
14.4 × 12.0 × 4.8 cm
Weathering = Ce

Enstatite chondrites are the most reduced chondritic materials we have in our meteorite collections and are composed of nearly pure Mg end-member silicate minerals such as enstatite and metallic FeNi metal that contains Si. EL chondrites also include petrologic types 3 through 6, with unequilibrated type 3 chondrites providing insight into nebular processes in a more reduced portion of the solar nebula. Enstatite chondrites also share the same oxygen isotopic signature as the Earth, the Moon, and aubrites.

PCA91020,0
1 cm

PCA91020,24
1 cm

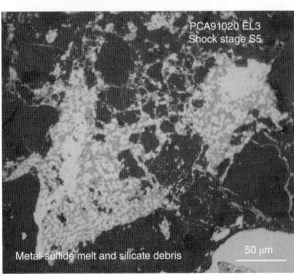
PCA91020 EL3
Shock stage S5
Metal-sulfide melt and silicate debris
50 µm

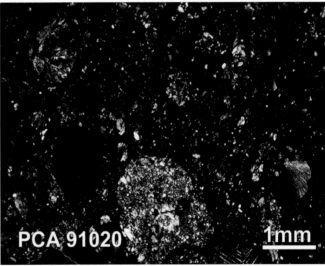
PCA 91020
1mm

Plate 30

MINERALOGY

PCA 91020 contains abundant chondrules, ranging up to 1.8 mm in diameter; they consist of granular or radiating pyroxene, sometimes with minor olivine. The matrix consists of pyroxene grains, with 25%–30% nickel-iron metal and a lesser amount of sulfides. Most of the pyroxene is close to $MgSiO_3$ in composition: mean Fs_1, range $Fs_{0.2-3}$, with CaO 0.1%–0.8%; olivine is forsterite (FeO 0.1%, CaO 1.5%). The nickel-iron metal contains 0.3%–0.7% Si.

SIGNIFICANCE

PCA 91020 is one of the most highly shocked enstatite chondrites. It contains maskelynite, metal-troilite textures (left, [201]) typical of rapid cooling from super-liquidus temperatures, and also elliptical and elongated chondrules and inclusions that are part of a foliation present in the meteorite. Given all of these unusual features, PCA 91020 records important information about shock processes that operated on the enstatite chondrite parent bodies.

References [197–204]

MacAlpine Hills (MAC) 88136
EL3 enstatite chondrite
74.4 g
Found January 14, 1989
4.0 × 4.0 × 2.5 cm
Weathering = A

Enstatite chondrites are the most reduced chondritic materials we have in our meteorite collections and are composed of nearly pure Mg end-member silicate minerals such as enstatite and metallic FeNi metal that contains Si. EL chondrites also include petrologic types 3 through 6, with unequilibrated type 3 chondrites providing insight into nebular processes in a more reduced portion of the solar nebula. Enstatite chondrites also share the same oxygen isotopic signature as the Earth, Moon, and aubrites.

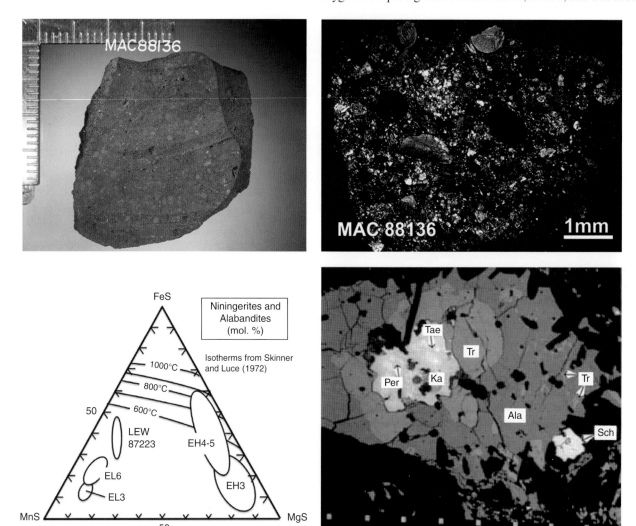

Plate 31

MINERALOGY

MAC 88136 contains abundant chondrules and chondrule fragments, ranging up to 2.4 mm diameter; they consist of granular or cryptocrystalline pyroxene. The matrix consists largely of small pyroxene grains, with some nickel-iron metal and minor sulfides. One ellipsoid metal grain, 1.1 × 1.5 mm, is present in the section. Microprobe analyses show that the pyroxene is clinoenstatite, with variable but low iron content (FeO 0.1%–2.1%); the metal grains show a low and variable silicon content, ranging up to 0.5%.

SIGNIFICANCE

MAC 88136 was the first known primitive (EL3) chondrite; prior to its discovery, only highly metamorphosed examples of EL chondrites were known. The MAC 88136 enstatite chondrite contains accessory olivine particularly rich in chromium. During metamorphism, chromium diffuses out of olivine. With Cr_2O_3 concentrations up to 0.6 wt.%, MAC 88136 contains the most chromium-rich olivines found in enstatite chondrites, perhaps indicating that it is one of the most primitive members of the enstatite chondrite group. Perryite, $(Ni,Fe)_5(Si,P)_2$, and alabandite, $(Mn,Fe,Mg)S$, are among the unusual opaque minerals found in MAC 88136 (left [220] and right [213]).

References [205–220]

Queen Alexandra Range (QUE) 94368
EL4 enstatite chondrite
1.2 g
Found January 24, 1995
1.5 × 1.0 × 0.5 cm
Weathering = C

Enstatite chondrites are the most reduced chondritic materials we have in our meteorite collections and are composed of nearly pure Mg end-member silicate minerals such as enstatite and metallic FeNi metal that contains Si. EL chondrites include petrologic types 3 through 6, with equilibrated type 6 chondrites providing constraints on the metamorphic history of the EL chondrite. Enstatite chondrites also share the same oxygen isotopic signature as the Earth, the Moon, and aubrites.

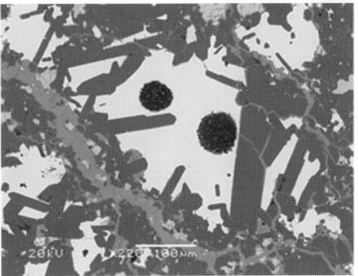

Plate 32

MINERALOGY

QUE 94368 shows numerous chondrules and irregular aggregates, up to 0.6 mm across, and mineral grains in a moderate amount of dark matrix. The matrix contains a minor amount of nickel-iron and sulfides. The chondrules, aggregates, and mineral grains appear to consist entirely of pyroxene. The pyroxene is close to $MgSiO_3$ in composition, but a few more iron-rich grains were analyzed. The nickel-iron contains 0.5%–0.7% Si and sometimes has graphite associated or included within (left, [222]).

SIGNIFICANCE

QUE 94368 was the first recognized EL4 chondrite, thus helping to define the metamorphic history of the EL chondrite parent body. Subsequently, sinoite (right, [221]) was discovered in QUE 94368, indicative of highly reduced conditions, by the equilibrium $SiO_2 + 3Si + 2N_2(g) = 2Si_2N_2O$.

References [221, 222]

LaPaz Icefield (LAP) 02225
EH chondrite impact melt
313.50 g
Found December 14, 2002
7.0 × 7.0 × 3.5 cm
Weathering = B

A very small subset of enstatite chondrites exhibit melted portions that were likely formed during impact. These rare samples offer a glimpse of the pressure and temperature conditions existing during impact of the EH parent body, as well as an opportunity to study mineral transformations for these unusual reduced meteorite groups.

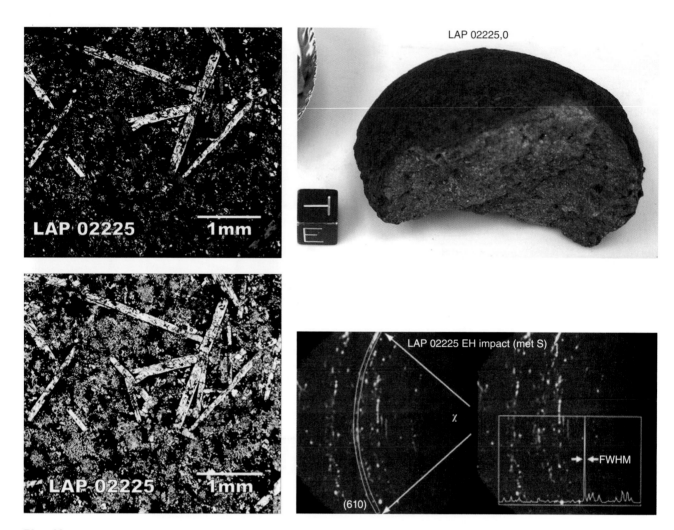

Plate 33

MINERALOGY

LAP 02225 consists of a matrix of small (~0.2 mm long) pyroxene laths with interstitial plagioclase, nickel-iron metal, troilite, daubreelite, Mg,Mn,Fe-sulfides, and perryite. The latter is often exsolved on the {111} axes of the metal. Also present are ~10 vol.% large enstatite laths that can exceed 3 mm in long dimension. Enstatite is $_{Fs0-1}$; plagioclase is An_0Or_{2-3}; and metal contains 3 wt.% Si.

SIGNIFICANCE

LAP 02225 is one of only a few EH impact melts, recording the conditions and effects of severe impacts on the EH parent body. Its fine-grained matrix has been characterized using micro x-ray diffraction (lower right, [223]), which has revealed plagioclase and enstatite together with FeNi metal, troilite, daubreelite, keilite, and perryite.

References [223]

Queen Alexandra Range (QUE) 94204
Ungrouped enstatite chondrite
2427.9 g
Found December 19, 1994
16.5 × 10.0 × 9.5 cm
Weathering = C

Ungrouped chondrites are not common, and they can sometimes form the foundation for a new group of chondrites, once more are discovered. However, there are several that have remained one-of-a-kind samples even after close to 25 years in the collection. Among the enstatite meteorites there are a few unusual ungrouped samples such as Itqiy and QUE 94204 and its pairs. These unusual samples offer insight into the diversity of materials in the solar system.

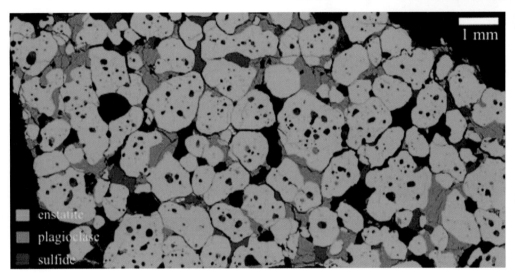

Plate 34

MINERALOGY

QUE 94204 is composed of a granular aggregate of about 75% pyroxene (average grain size 1.5 mm), 20% nickel-iron metal and troilite, and 5% intergranular plagioclase. Some of the metal grains contain numerous inclusions of graphite. Microprobe analyses show that the pyroxene is close to $MgSiO_3$ in composition, with a few more iron-rich grains. Plagioclase composition is An_{19-26}. The nickel-iron metal contains 2.4%–2.9% Si.

SIGNIFICANCE

The distinctive texture and mineralogy of QUE 94204 suggest both formation by impact melting and a poorly understood mechanism of crystallization. With its millimeter-sized polysynthetically twinned enstatite crystals and abundant interstitial metal, sulfide, and feldspar (above, [232]), it is compositionally similar to enstatite chondrites, yet its texture is distinctly igneous. Almost certainly formed as a result of impact melting, the large, rounded enstatites may suggest crystallization from a superheated melt nearly devoid of relict silicate nuclei.

References [224–232]

Pecora Escarpment (PCA) 91002
R3.8-6 chondrite
210.16 g
Found December 23, 1991
7.8 × 4.9 × 4.7 cm
Weathering = A/B

R chondrites (named for the fall Rumuruti) share some properties with ordinary chondrites, such as chondrule types, rare refractory inclusions, and similar chemical compositions. There are, however, significant differences between R chondrites and ordinary chondrites, such as a much higher proportion of matrix, more oxidized (high Ni-olivine and commonly absence of metallic FeNi), and enrichments in ^{17}O over that of ordinary chondrites. Many R chondrites are brecciated and may originate in an asteroid regolith.

Plate 35

MINERALOGY

PCA 91002 contains numerous polycrystalline silicate clasts (up to 1.2 mm in diameter), few small chondrules (up to 0.5 mm), and mineral grains in a finely granular matrix with minute sulfides (pentlandite with minor troilite) disseminated throughout. No nickel-iron metal was seen. Olivine compositions show a prominent peak at Fa_{39}. The pyroxene is almost entirely low-Ca: $Wo_{0.3-5}$, Fs_{1-28}, and maskelynite is present in a few clasts.

SIGNIFICANCE

PCA 91002 was one of the first R chondrites recognized, and it helped to define the distinguishing properties of the group. The oxygen isotopic values of the R chondrites fall well above the terrestrial fractionation and ordinary chondrite line (right, [234]), suggesting a larger role for water in their source materials. R chondrites also exhibit a diverse mineralogy, including metal, graphite, and magnetite (not always together) recording a wide range of oxygen fugacities within the group.

References [233–239]

LaPaz Icefield (LAP) 04840
R6 chondrite
50.405 g
Found January 20, 2005
5.0 × 3.0 × 1.75 cm
Weathering = A/B

R chondrites (named for the fall Rumuruti) share some properties with ordinary chondrites, such as chondrule types, rare refractory inclusions, and similar chemical compositions. There are, however, significant differences between R chondrites and ordinary chondrites, such as a much higher proportion of matrix, more oxidized (high Ni-olivine and low metallic FeNi contents), and enrichments in ^{17}O over that of ordinary chondrites. Many R chondrites are brecciated and may originate in an asteroid regolith.

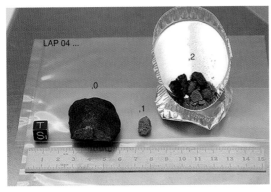

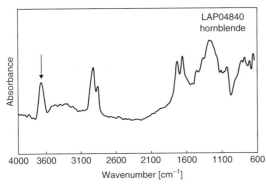

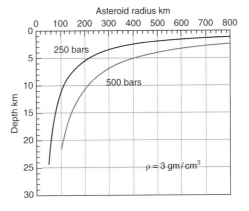

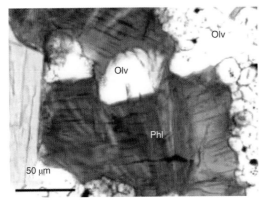

Plate 36

MINERALOGY

LAP 04840 is texturally heterogeneous, containing relict chondrules up to 1 mm, isolated mineral grains 100–200 microns, and microcrystalline areas reaching 1 mm with mafic silicate grain sizes of 5–10 microns. Shock effects are pervasive, particularly in plagioclase. The rock consists of olivine (Fa_{38}), orthopyroxene ($Fs_{30}Wo_1$), plagioclase (An_7Or_3), calcic hornblende and minor biotite (bottom left; [241]). Opaque minerals include troilite, pentlandite, and chromite.

SIGNIFICANCE

LAP 04840 is a one-of-a-kind meteorite: at the time it was classified it was the only such hornblende- and biotite-bearing chondrite known. Its OH-bearing minerals (middle left, arrow is OH peak; [243]), together with silicate and oxide mineralogy, define pressures of 250 to 500 bars (could be as deep as 25 km, right; [241]) and high oxygen fugacities, above the FMQ oxygen buffer.

References [240–244]

Lewis Cliffs (LEW) 87232
K chondrite
23.13 g
Found December 9, 1987
3.0 × 2.5 × 2.0 cm
Weathering = B

K chondrites are a small group of samples that share some common characteristics, such as reduced silicate mineralogy, and bulk compositions that fall outside of established chondrite groups. For this reason they have been grouped together with the prospect of establishing a new group of meteorite with some common features.

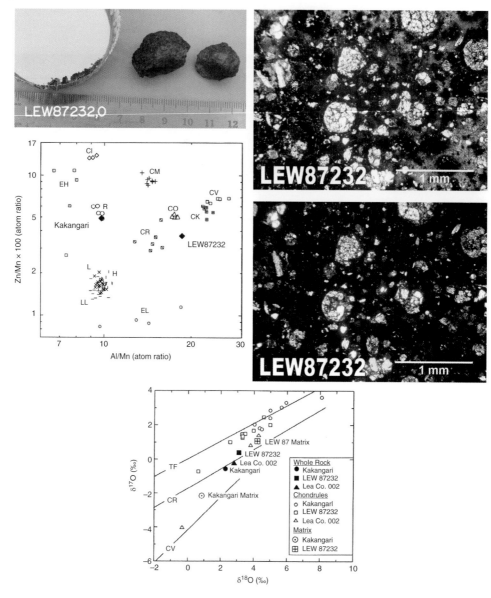

Plate 37

MINERALOGY

LEW 87232 shows numerous chondrules (up to 1.8 mm across), chondrule fragments, and mineral grains in a translucent brown matrix. Chondrules are mainly granular or porphyritic olivine (($Fa_{0.6-2}$, mean $Fa_{1.0}$) and olivine-pyroxene ($Fs_{0.5-9}$). Minor amounts of nickel-iron and a little sulfide are present as small grains in matrix or on chondrule rims. The matrix appears to consist largely of phyllosilicates.

SIGNIFICANCE

More detailed studies of LEW 87232 have identified some subtle differences from the other members of the trio (left and right, [249]). Whether LEW 87232 and the others will remain grouped after more extensive studies is not clear. However, the unique features of these samples nonetheless illustrate the diversity of materials in the meteorite collections, diversity that may have gone unrecognized without the collection of large numbers of samples from Antarctica.

References [245–249]

Queen Alexandra Range (QUE) 94535
Winonaite
11.3 g
Found December 16,1994
2.4 × 2.3 × 0.8 cm
Weathering = C

Winonaites are a small group of primitive achondrites with roughly chondritic mineralogy and major element chemistry. Some winonaites contain relict chondrules, and the group is related to silicate inclusions in IAB iron meteorites as suggested by their oxygen isotope compositions.

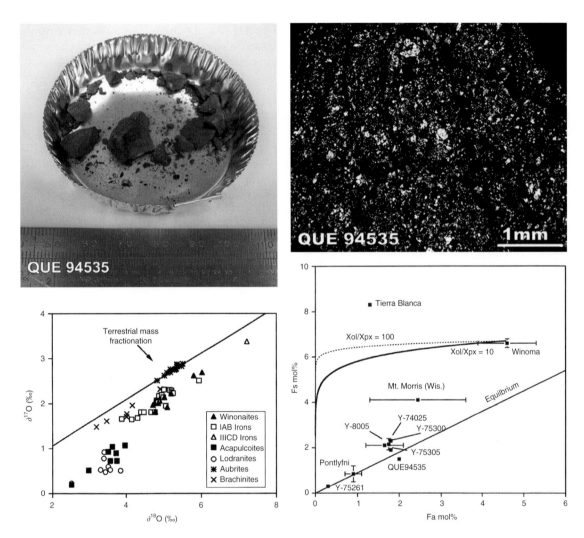

Plate 38

MINERALOGY

QUE 94535 is composed of a fine granular aggregate (0.02 mm grain size) of olivine (Fa_{1-3}) and pyroxene (Fs_{1-2}) with a few coarser areas (possibly vestigial chondrules); about 20% nickel-iron metal and troilite are present, with much of the metal as thin veinlets throughout the section. The nickel-iron metal contains 0.1%–0.2% Si.

SIGNIFICANCE

QUE 94535 is the only winonaite in the U.S. collection, and because the number of meteorites defining this group is small, it has been an important sample. Although it is quite different in texture and mineral composition than Winona, it shares some common features with the Yamato Antarctic winonaites from the 1974, 1975, and 1980 seasons (right, [250]), and together with other winonaites provided a link to the IAB and IIICD irons (left, [250]). Like lodranites and acapulcoites, the winonaites represent partial melting products of a chondrite precursor that is unknown among our collections. Therefore, this small group lies at the center of our understanding of melting processes in the early solar system.

References [250, 251]

Lewis Cliffs (LEW) 86220
Acapulcoite
25.04 g
Found December 18, 1986
4.0 × 1.0 × 1.5 cm
Weathering = ?

Acapulcoites have mafic silicate compositions intermediate between E and H chondrites, roughly chondritic mineralogies, achondritic, equigranular textures, micrometer- to centimeter-sized veins of Fe,Ni-FeS that cross-cut silicate phases, rapid metallographic cooling rates at ~600–400°C (10^3–10^5°C/Myr) and trapped noble gas abundances comparable to type 3–4 ordinary chondrites. They likely formed from a precursor chondrite that differs from known chondrites in mineral and oxygen isotopic compositions. Heating to ~950–1000°C resulted in melting at the Fe,Ni-FeS cotectic, but silicates did not melt. Silicate textures resulted from extensive solid-state recrystallization. Acapulcoites share a common mineralogy, mineral composition, and oxygen isotopic composition with lodranites and are believed to have originated on a common parent body.

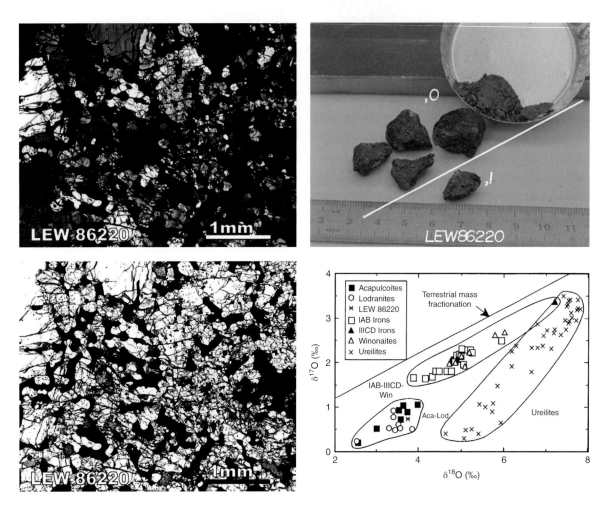

Plate 39

MINERALOGY

LEW 86220 is dominated by nickel-iron metal with a 0.3–1.2 mm diameter granular aggregate of olivine (Fa_7), pyroxene (Wo_2Fs_9), plagioclase (An_{15}), and minor diopside ($Wo_{43}Fs_4$).

SIGNIFICANCE

LEW 86220 is unique among the acapulcoite-lodranite clan (left, [254]) in sampling a fine-grained, granoblastic matrix cross-cut by a coarse-grained vein of pyroxene, augite, metal, and sulfide. This remarkable juxtaposition of lithologies likely formed when partial melts extracted from the lodranite source region intruded into a slightly cooler acapulcoite host and solidified. These remain our only significant samples of the melts complementary to the residual acapulcoites and lodranites.

References [252–259]

Allan Hills (ALH) A77081
Acapulcoite
8.59 g
Found December 27, 1977
2.0 × 2.0 × 2.5 cm
Weathering = B

Acapulcoites have mafic silicate compositions intermediate between E and H chondrites, roughly chondritic mineralogies, achondritic, equigranular textures, micrometer- to centimeter- sized veins of Fe,Ni-FeS that cross-cut silicate phases, rapid metallographic cooling rates at ~600–400°C (10^3–10^5°C/Myr), and trapped noble gas abundances comparable to type 3-4 ordinary chondrites. They likely formed from a precursor chondrite that differs from known chondrites in mineral and oxygen isotopic compositions. Heating to ~950–1000°C resulted in melting at the Fe,Ni-FeS cotectic, but silicates did not melt. Silicate textures resulted from extensive solid-state recrystallization. Acapulcoites share a common mineralogy, mineral composition, and oxygen isotopic composition with lodranites and are believed to have originated on a common parent body.

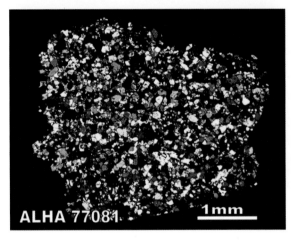

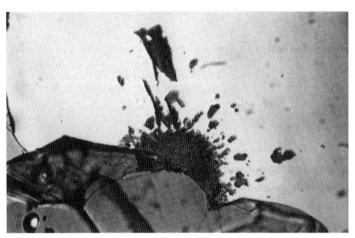

Plate 40

MINERALOGY

ALH A77081 is a moderately coarse equigranular (grains 0.1–0.3 mm diameter) aggregate of approximately equal amounts of olivine (Fa_{11}) and orthopyroxene ($Wo_{1.7}Fs_{11}En_{87}$), with minor amounts of diopside ($Wo_{45}Fs_5En_{50}$) and plagioclase (An_{15}) with 0.8% K_2O, nickel-iron metal, troilite, graphite (above, growing into kamacite, [267]), and accessory chromite.

SIGNIFICANCE

ALH A77081 was the first U.S. Antarctic meteorite grouped with the acapulcoites and lodranites. Originally described as H7 chondrites, careful petrographic examination revealed micron-sized veins of metal and sulfide indicative of the earliest partial melting of an asteroid. These veins, formed when low-degree partial melts produced excess pressures and minute fractures, record the onset of transformation of chondritic bodies to the differentiated worlds that dominate the inner solar system today.

References [260–269]

Graves Nunatak (GRA) 95209
Lodranite
948.8 g
Found January 31, 1996
10.5 × 8.0 × 5.0 cm
Weathering = B

Lodranites share a common mineralogy, mineral composition, and oxygen isotopic composition with acapulcoites and are believed to have originated on a common parent body. Lodranites experienced melting and loss of the silicate component, with peak temperatures of ~1250°C. They provide our best glimpse of the early melting history of asteroids.

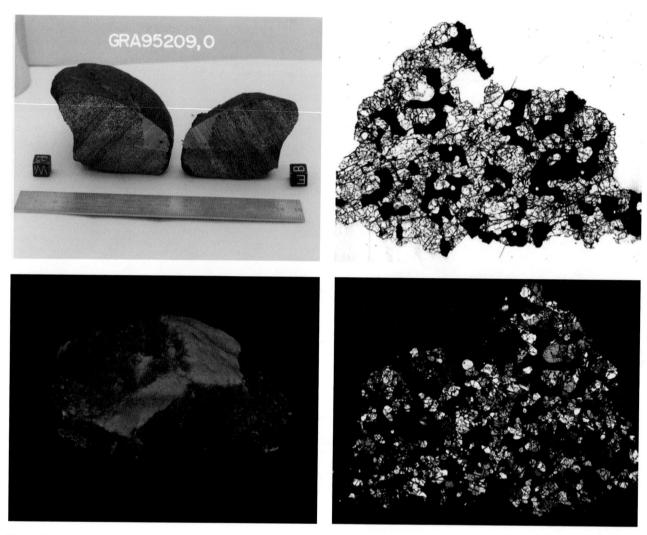

Plate 41

MINERALOGY

GRA 95209 is an equigranular aggregate (grains averaging 0.3 mm) of subequal amounts of nickel-iron, olivine (Fa$_7$), and pyroxene (Wo$_3$Fs$_7$), with minor plagioclase (An$_{19}$) and clinopyroxene, (Wo$_{41}$Fs$_5$). The mixed metal-silicate-sulfide matrix is cut by metal-rich, graphite-bearing veins exceeding 1 cm in width and grades into a volumetrically minor metal-poor region. Silicate compositions and modal abundances are typical for lodranites, while the mineralogy of the metal-sulfide component is complex and differs among the three lithologies.

SIGNIFICANCE

GRA 95209 is the largest and, by far, exhibits the greatest lithologic diversity among the acapulcoites and lodranites. At almost 1 kg in mass, it exhibits metal-sulfide-phosphate veins that formed during the earliest partial melting of an asteroid (left, [275]). The pairings of phosphates and phosphides, Ni-rich and Ni-poor metal, and graphite suggest that oxidation-reduction reactions accompanied partial melting, testifying to the complex chemical and physical processes occurring during the earliest stages of differentiation of an asteroid.

References [270–277]

Lewis Cliffs (LEW) 88280
Lodranite
5.97 g
Found December 30, 1988
1.5 × 1.5 × 1.1 cm
Weathering = B

Lodranites share a common mineralogy, mineral composition, and oxygen isotopic composition with acapulcoites and are believed to have originated on a common parent body. Lodranites experienced melting and loss of the silicate component, with peak temperatures of ~1250°C. They provide our best glimpse of the early melting history of asteroids.

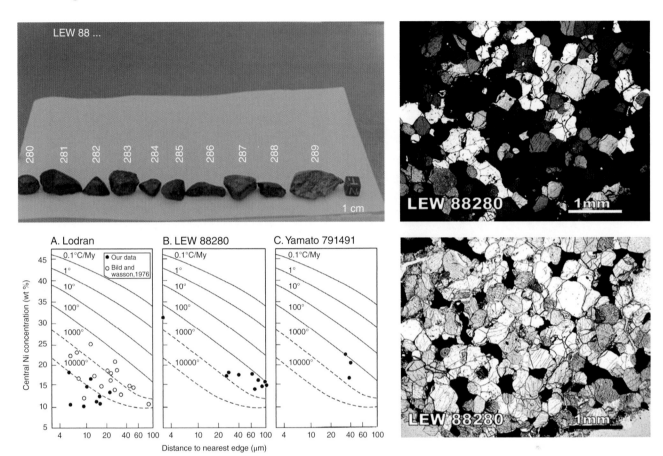

Plate 42

MINERALOGY

LEW 88280 contains a coarse equigranular aggregate of anhedral to subhedral olivine (Fa_{13}) and pyroxene (Wo_3Fs_{12}) grains, 0.6–1.2 mm across, with about 20% of intergranular nickel-iron and a little troilite.

SIGNIFICANCE

LEW 88280 has metal lamellae and compositions indicating a very rapid cooling rate, as do several other lodranites and acapulcoites (left, [283]). It has high FeO olivine, and abundant FeS, and represents one of the highest $\delta^{18}O$ lodranites known. All of these qualities make it an important sample representing one end of the spectrum of lodranite characteristics. However, it also has a distinctly older cosmic ray exposure age, so it is possibly from a different parent body than many other lodranites.

References [278–285]

Lewis Cliffs (LEW) 88763
Ungrouped achondrite
4.12 g
Found December 17, 1988
2.0 × 1.6 × 0.6 cm
Weathering = B

Ungrouped achondrites are not common, and they can sometimes form the foundation for a new group of achondrites, once more are discovered. However, there are several that have remained one-of-a-kind samples even after close to 25 years in the collection. These unusual samples offer insight into the diversity of materials in the solar system.

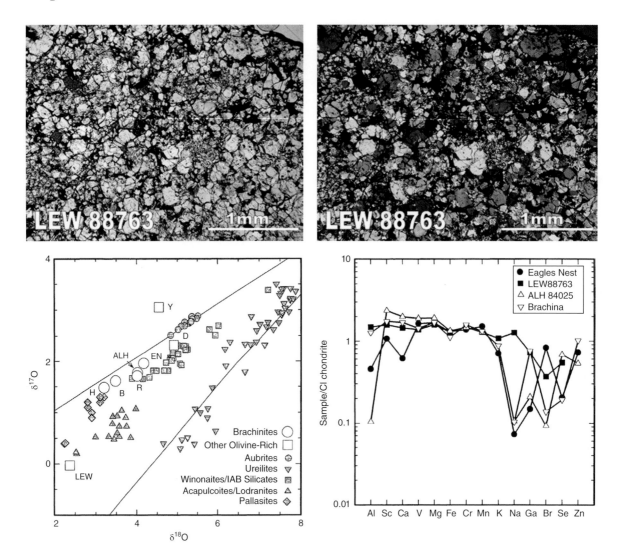

Plate 43

MINERALOGY

LEW 88763 is a fine-grained (<1 mm crystals), granular textured, olivine-rich (70% Fo_{63}) rock with minor plagioclase ($Ab_{55-76}An_{44-19}Or_{2-7}$), augite, pigeonite, chromite, whitlockite, ilmenite, trolite, and taenite.

SIGNFICANCE

Although initially thought to be a brachinite, an achondrite group that formed by igneous processes on a differentiated body, it has distinct oxygen isotope and bulk compositional characteristics (left and right, [290]). Instead, its similarity in bulk composition to chondritic meteorites suggests that it may represent material that has been heated and recrystallized, perhaps even lost a very small portion of sulfide to a melting event. This sample lies at the boundary between chondritic and achondritic meteorites and may provide important distinctions between these groups as more such samples are found.

References [286–290]

Allan Hills (ALH) A78019
Ureilite
30.27 g
Found November 23, 1978
3.0 × 2.5 × 3.0 cm
Weathering = B/C

Ureilites are coarse-grained ultramafic rocks containing olivine and pigeonite. Graphite is also present in many ureilites, and the Fe-Mg-O relation in this group of meteorites indicates equilibration at high temperatures in the presence of C at some pressure (50 bar) within an asteroid-sized body. Because ureilites also define an unusual oxygen isotopic composition that has a mass-independent slope of 1:1 on a three-isotope diagram for oxygen, clues to their origin lie in an explanation that includes silicates, graphite, and oxygen.

Plate 44

MINERALOGY

ALH A78019 contains Fo_{77} olivine and En_{72} pigeonite and minor amounts of metal, sulfide, and graphite. Olivine and pigeonite form triple-junction boundaries. Graphite occurs as laths interstitial to the silicates (left, [291]). Metal-sulfide-graphite veins occur in fractures and cleavage planes.

SIGNIFICANCE

ALH A78019 represents a very low-shock-state ureilite, as graphite is well preserved. Many ureilites contain diamond, which is thought to have formed by shock processes. The carbon-bearing phases and Fe-Ni-Si metallic alloys found in ALH A78019 allow a detailed understanding of the redox history of a portion of the ureilite parent body (right, [295]).

References [291–311]

Pecora Escarpment (PCA) 82506
Ureilite
5316 g
Found December 28, 1982
22.0 × 16.0 × 9.0 cm
Weathering = A/Be

Ureilites are coarse-grained ultramafic rocks containing olivine and pigeonite. Graphite is also present in many ureilites, and the Fe-Mg-O relation in this group of meteorites indicates equilibration at high temperatures in the presence of C at some pressure (50 bar) within an asteroid-sized body. Because ureilites also define an unusual oxygen isotopic composition that has a mass-independent slope of 1:1 on a three-isotope diagram for oxygen, clues to their origin lie in an explanation that includes silicates, graphite, and oxygen.

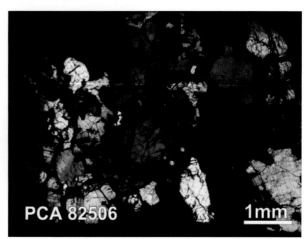

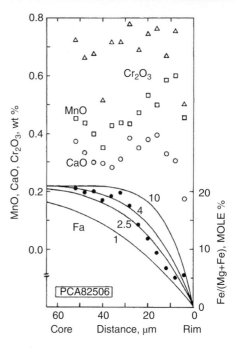

Plate 45

MINERALOGY

PCA 82506 is composed of an aggregate of anhedral to subhedral grains (0.6–3 mm in diameter) of olivine (Fa$_{21}$), high CaO content (0.3%), and pyroxene (Wo$_6$Fs$_{18}$En$_{76}$). Individual grains are rimmed by carbonaceous material, which includes thin stringers of troilite. Trace amounts of nickel-iron metal originally present have been largely weathered to limonite. Some grains show undulose extinction, but apart from that the meteorite appears to be relatively unshocked compared to most ureilites.

SIGNIFICANCE

Along with ALH A77257, PCA 82506 is a large ureilite collected in the early years of the program. As such, it has influenced many systematic studies of ureilites, and its large size has allowed wide distribution within the science community. Cooling rate studies of PCA 82506 olivines showed that ureilites likely cooled rapidly (right, [320]), consistent with the idea that ureilites formed during a disruptive impact of a large parent body.

References [312–329]

Elephant Moraine (EET) 83309
Polymict ureilite
60.76 g
Found December 27, 1983
4.0 × 4.0 × 2.5 cm
Weathering = C

A rarer kind of ureilite is a polymict ureilite, which, like howardites or polymict eucrites, contain lithic clasts of at least two kinds of meteorites. A stunning case of a polymict ureilite is EET 83309, in which are found clasts of R chondrites, identified as such based on oxygen isotopic measurements and mineral compositions of silicates. These observations show that an asteroid's regolith can be a mixture of quite diverse materials.

EET83309,0

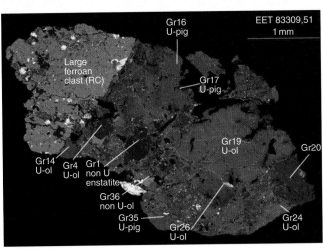

Plate 46

MINERALOGY

Most polymict ureilites contain olivine grains that span the range of olivine seen in monomict ureilites, but the range in EET 83309 is wider, from Fo_{62} to Fo_{98}. EET 83309 also contains plagioclase feldspar, and clasts composed of pyroxene, feldspar, and ulvöspinel. There are also commonly clasts of R chondrite material in EET 83309.

SIGNIFICANCE

Careful petrographic and geochemical studies of EET 83309 have revealed that it contains multiple lithologies in addition to the purely ureilitic clasts. The other "exotic" clasts indicate that the ureilite parent body had a heterogeneous regolith containing pieces of R chondrites (left, [331]), possibly enstatite chondrites (or achondrites), and perhaps even some material from unknown meteorite groups. This rarer kind of ureilite can provide information about impact gardening and regolith development on asteroids.

References [330–342]

Lewis Cliffs (LEW) 88774
Anomalous ureilite
3.068 g
Found December 11, 1988
1.2 × 1.3 × 0.8 cm
Weathering = B/C

Among large meteorite groups there are unusual or anomalous members such as in the pallasites or mesosiderites. Among the ureilites there is a pigeonite ureilite (MET 01085) that has been studied by a few, and an even more unusual anomalous ureilite LEW 88774, which contains chromite and other Cr-silicates, as well as pyroxene with exsolution lamellae, indicating a slower cooling history than most ureilites.

Plate 47

MINERALOGY

LEW 88774 contains exsolved pyroxene (78%), olivine (12%), chromite (6%), and minor brezinaite (Cr_3S_4), eskolaite (Cr_2O_3), C polymorphs, carbides, and Si-Al-rich glass. Most ureilites do not contain chromite, and most have unexsolved pyroxene.

SIGNIFICANCE

LEW 88774 has an unusual Cr-rich bulk composition that has led to the stabilization of a large number of Cr-bearing minerals (carbides, sulfide, oxides, pyroxenes; right [345]). It has also undergone cooling from 1280 to 1160°C, followed by a breakup of the parent body similar to what most other ureilites experienced. The unusual bulk composition can be extended to Ca and Al and gives LEW 88774 some characteristics that are chondritic or primitive, making it unique among ureilites and sharing similarities with NWA 766.

References [343–353]

Allan Hills (ALH) 84025
Brachinite
4.58 g
Found December 26, 1984
2.0 × 1.5 × 0.8 cm
Weathering = A/Be

Brachinites are olivine-rich achondrites with a range of grain sizes, and having minor amounts of metal, sulfide, chromite, clinopyroxene, and plagioclase. They are thought be from the deeper part of an asteroid-sized body, perhaps representing a residue from partial melting or an olivine-rich cumulate from early differentiation. These are rare meteorites with some affinity to the HED meteorites (similar oxygen isotopes) but not much coherency in the membership.

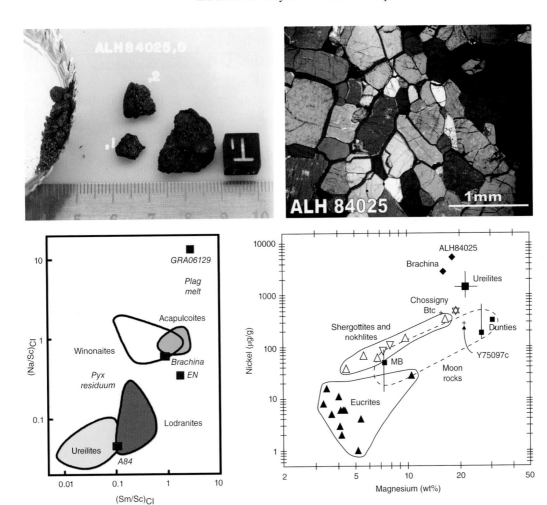

Plate 48

MINERALOGY

Allan Hills 84025 has a coarse-grained (0.5 to 1.5 mm olivines), equigranular recrystallized texture with 80%–90% Fo$_{68}$ olivine, 4%–15% clinopyroxene, chromite, sulfide, and metal. The olivine contains no melt inclusions as do many other olivine-bearing achondrites.

SIGNIFICANCE

As the second member of the small brachinite group of achondrites, ALH 84025 has helped define the characteristics of this group. Brachinites are distinct from other achondrite groups such as acapulcoite/lodranites, eucrites, and shergottites, and they have distinct chemical features such as high Ni/Mg (right, [357]). ALH 84025 may be linked to the ungrouped achondrite GRA 06128 by melting relations (left, [364]), where ALH 84025 represents a residue from melting. The parent asteroid is a differentiated body and not a primitive achondrite that might still contain some relict chondritic material or residues. The differentiation event was very ancient, just a few Ma after the start of the solar system.

References [354–357]

Graves Nunataks (GRA) 06128, 06129
Ungrouped achondrite
447.6, 196.45 g
Found January 12 and 8, 2006
8.5 × 4.0 × 7.5; 8.0 × 5.0 × 2.5 cm
Weathering = Ce

Antarctica has been a steady source of new meteorites that do not initially fall into a standard classification but rather define new groups. One such sample is the ungrouped achondrite pair Graves Nunataks 06128 and 06129. These samples have been studied intensely by many scientists and their grouping is still unknown, but they are apparently related to another small group of achondrites, the brachinites, and represent differentiated material.

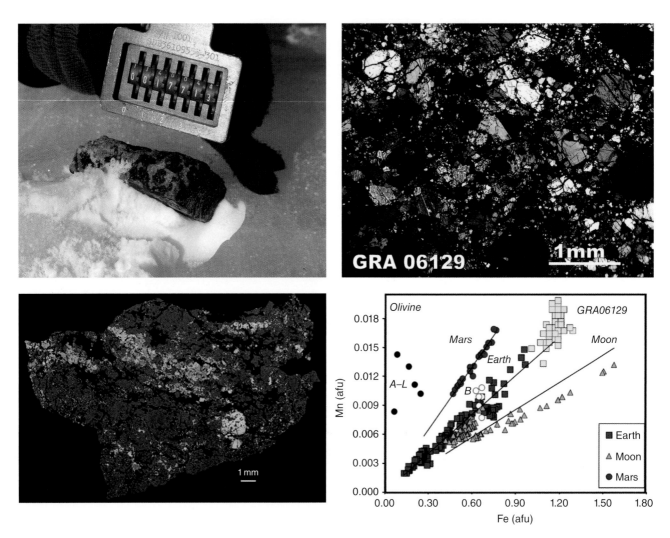

Plate 49

MINERALOGY

The unusal mineralogy is typified by 60%–70% sodic feldspar (Ab, An, Or), with smaller amount of olivine, pyroxenes, phosphates, sulfide, and chromite. The mafic phases are found associated together, and overall there is a lineation texture to the meteorites (above left with plagioclase = purple, pyroxene = green, olivine = orange-yellow, phosphates = blue-green, and iron oxides, sulfides, and metals = red; [364]), along with the platy fracture seen in both stones.

SIGNIFICANCE

The relation to the brachinites has been made by O isotopes and mineralogy (above right, [365]), and suggests that the brachinites may come from a melted chondritic parent body. The differentiation age is old and may provide additional constraints on the thermal conditions in small parent bodies in the early solar system. Connections to other ungrouped achondrites that are related to brachinites remain an unclear but exciting possibility that could lead to a larger group of samples representing this parent body.

References [358–367].

Allan Hills (ALH) A78113
Aubrite
298.6 g
Found December 22, 1978
11.0 × 10.0 × 8.0 cm
Weathering = A/Be

Aubrites are enstatite pyroxene–bearing achondrites that share a common oxygen isotopic value with Earth, Moon, and enstatite chondrites, and are the most reduced basalts known. They contain unusual sulfides, low FeO silicates, and sodic plagioclase feldspar. Some aubrites contain a basaltic vitrophyre lithology that may represent a basaltic component from the aubrite parent body. Their low FeO contents and small amount of metal would give these samples a high albedo for a parent asteroid.

Plate 50

MINERALOGY

ALH A78113 consists almost entirely of orthopyroxene ($Wo_{0.6} En_{99}$) clasts up to 2 mm in a groundmass of comminuted pyroxene. The pyroxene is an iron-free enstatite (FeO < 0.1%) with minor and variable amounts of CaO (0.2%–0.6%, average 0.5%). Accessory amounts of olivine (Fo_{99}), sulfides (troilite, and trace amounts of oldhamite, alabandite, and daubreelite), and nickel-iron are present as small grains in the groundmass.

SIGNIFICANCE

ALH A78113 is the first Antarctic meteorite to be identified as an aubrite. It contains chondritic clasts and also FeO-rich pyroxene, indicating that perhaps material from another more oxidized parent body has become incorporated into aubrites. This meteorite has also been used to compare to infrared spectra of E-type asteroids (left, [378]) to determine possible parent bodies of enstatite chondrites.

References [368–382]

LaPaz Icefield (LAP) 03719
Aubrite
62.02 g
Found January 1, 2004
5.5 × 4.0 × 2.5 cm
Weathering = B

Aubrites are enstatite pyroxene–bearing achondrites that share a common oxygen isotopic value with Earth, Moon, and enstatite chondrites, and are the most reduced basalts known. They contain unusual sulfides, low FeO silicates, and sodic plagioclase feldspar. Some aubrites contain a basaltic vitrophyre lithology that may represent a basaltic component from the aubrite parent body. Their low FeO contents and small amount of metal would give these samples a high albedo for a parent asteroid.

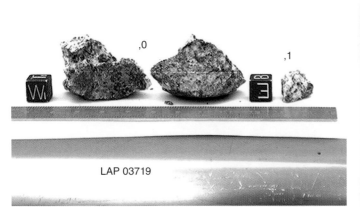

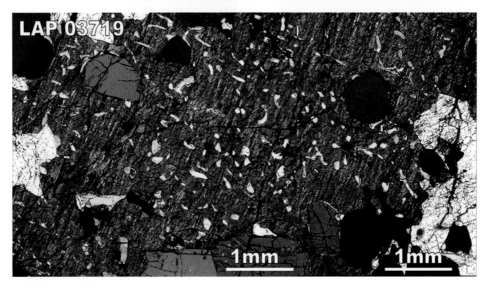

Plate 51

MINERALOGY

LAP 03719 is an unbrecciated aubrite consisting of essentially FeO-free enstatite, low CaO diopside, and olivine with minor nickel-iron metal, troilite, daubreelite, and alabandite. Enstatites reach 1 cm diameter, and contacts between enstatite grains are often interfingering. Diopside exsolution blebs occur within enstatite and at enstatite-enstatite grain boundaries. Olivine ($Fo_{99.7}$), which occupies ~20% of the section, reaches 3 mm. Although the pyroxene textures are reminiscent of some aubrite clasts, the unbrecciated nature, texture, and olivine abundance in this aubrite are unique.

SIGNIFICANCE

LAP 03719 is unique among the aubrites in being both unbrecciated and olivine-rich. This meteorite allows us to test the proposed link between the enstatite chondrites and the aubrites. Unlike enstatite chondrites, aubrites tend to contain significant proportions of olivine. Models for olivine formation during melting of the aubrite parent body range from reduction of silica to metallic silicon to co-crystallization with rare alkaline phases like roedderite. The absence of these proposed phases in LAP 03719 may suggest that aubrites could not have formed from known enstatite chondrites.

References [383–384]

Larkman Nunatak (LAR) 04316
Aubrite
1163 g
Found January 6, 2005
12.0 × 7.0 × 7.0 cm
Weathering = A

Aubrites are enstatite pyroxene–bearing achondrites that share a common oxygen isotopic value with Earth, Moon, and enstatite chondrites, and are the most reduced basalts known. They contain unusual sulfides, low FeO silicates, and sodic plagioclase feldspar. Some aubrites contain a basaltic vitrophyre lithology that may represent a basaltic component from the aubrite parent body. Their low FeO contents and small amount of metal would give these samples a high albedo for a parent asteroid.

Plate 52

MINERALOGY

LAR 04316 is composed of comminuted matrix of essentially FeO-free enstatite (Fs_{0-1}) and diopside (Fs_1Wo_{45}) with grain sizes reaching 3 mm in diameter with rare nickel-iron metal, schreibersite, troilite, daubreelite, and alabandite. Two clasts described in the macroscopic description, and apparently in contact in the piece, are an aubrite basalt vitrophyre composed of enstatite (Fs_2) and forsterite (Fa_2) in a matrix of feldspathic glass and a metal-sulfide quench-textured clast (above right) with a single 0.5 mm alabandite grain.

SIGNIFICANCE

LAR 04316 is remarkable in containing a pyroxene-plagioclase-metal-sulfide clast approximately 2 cm in diameter that may record the earliest melting of an asteroidal body. Volcanoes on asteroids should have produced sprays of droplets (pyroclasts) whose eruption velocity exceeded the escape velocity of the parent asteroid. Only the largest such droplets, those in excess of 1 cm in diameter, might have been preserved. The mineralogy and texture of the clast in LAP 04316 suggests it might be a rare example of these large pyroclasts. Modeling the physics of eruption places upper and lower bounds on the size of the parent asteroid (above left, [386]).

References [385–387]

Lewis Cliffs Icefield (LEW) 86010
Angrite
6.91 g
Found January 12, 1987
1.5 × 1.5 × 1.5 cm
Weathering = A/B

Angrites are a rare group of oxidized basaltic meteorites with oxygen isotopic values similar to, but distinct from, the HED group of meteorites. Their FeO-rich and volatile depleted characteristics are reflected in the FeO-rich clinopyroxene, fayalitic olivine, and calcic plagioclase. The low concentrations of core-loving elements such as Ni, Co, and W suggest they are from a small body that underwent core formation. Some have proposed a link between angrites and the IVB irons, originating from a relatively oxidized asteroidal parent body.

Plate 53

MINERALOGY

LEW 86010 consists of plagioclase (An_{100}), clinopyroxene, and olivine, with a trace of opaques, in a granular texture of 0.5 to 2.5 mm grain size. Reddish purple clinopyroxene ($Wo_{56}Fs_{19}$) is fassaitic with 10.4 wt.% Al_2O_3 and 2.4 wt.% TiO_2. Olivine (Fa_{63}) has 1.5 to 2.6 wt.% CaO and contains exsolution lamellae of kirschsteinite ($Ca[Mg,Fe]SiO_4$).

SIGNIFICANCE

LEW 86010 records an age of approximately 4558 Ma (left, [390]), which helps define the age limits of this group of meteorites that formed by ancient differentiation in a planetesimal. The kirschsteinite exsolution lamellae record cooling rates of ~40–50°C/year, implying depth of formation of 75 m (right, [392]).

References [388–393]

Lewis Cliffs Icefield (LEW) 87051
Angrite
0.61 g
Found December 13, 1987
1.0 × 0.7 × 0.5 cm
Weathering = A

Angrites are a rare group of oxidized basaltic meteorites with oxygen isotopic values similar to, but distinct from, the HED group of meteorites. Their FeO-rich and volatile depleted characteristics are reflected in the FeO-rich clinopyroxene, fayalitic olivine, and calcic plagioclase. The low concentrations of core-loving elements such as Ni, Co, and W suggest they are from a small body that underwent core formation. Some have proposed a link between angrites and the IVB irons, originating from a relatively oxidized asteroidal parent body.

Plate 54

MINERALOGY

The texture of LEW 87051 is dominated by a subparallel arrangement of plagioclase laths, 0.02 mm wide and up to 0.3 mm long. The pyroxene is weakly pleochroic with a purplish tint. Major phases are plagioclase (An_{100}), olivine (Fa_{19}), a titanian fassaite pyroxene (with up to 8.5% Al_2O_3 and 4.6% TiO_2), and kirschsteinite.

SIGNIFICANCE

The fine-grained texture of LEW 87051 indicates it cooled rapidly, whereas at the same time it contains zoned olivine crystals, the cores of which are not in equilibrium with the surrounding fine-grained matrix (left and right, [395]). This sample and its unique texture among angrites likely formed by shallow melting on the parent body, perhaps by impact processes.

References [394–399]

Elephant Moraine (EET) 90020
Eucrite
154.02 g
Found January 4, 1991
5.3 × 4.5 × 4.8 cm
Weathering = A

A small number of eucrites experienced metamorphism deeper in the crust of their parent body (likely 4 Vesta) and thus have a recrystallized texture that formed during this more intense heating. In addition, the crust experienced intense heating from impact events onto the surface. Distinguishing between these two different kinds of heating events can lead to a detailed understanding of the thermal evolution of differentiated asteroid crusts.

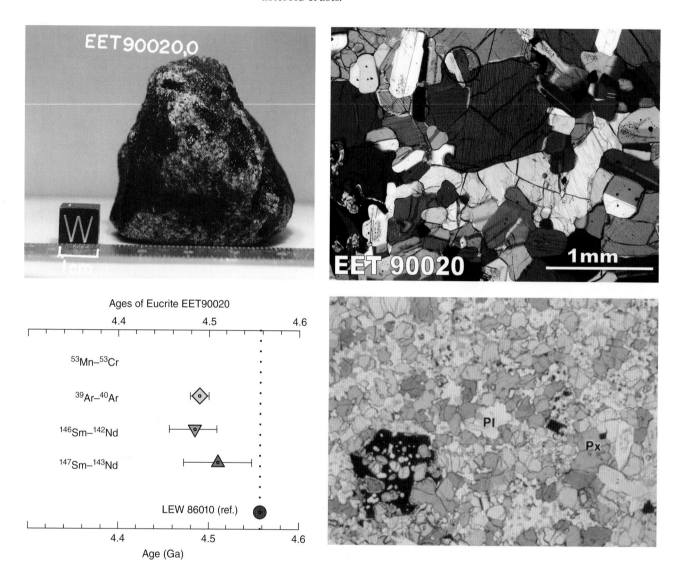

Plate 55

MINERALOGY

EET 90020 shows an equigranular aggregate (0.6 mm avg. grain size) of plagioclase laths and anhedral to subhedral pyroxene grains, with minor tridymite, Ti-chromite, ilmenite, and metal. There are coarser grained and finer grained portions of the sample, with the coarser grained regions containing igneous sub-ophitic texture, and the finer grained region exhibiting recrystallized granulitic texture. The pyroxene is pale brown and weakly pleochroic, and ranges from Wo_5Fs_{55} to $Wo_{23}Fs_{40}$, with relatively uniform En content. Plagioclase composition is An_{86-92}.

SIGNIFICANCE

The undisturbed Ar-Ar age date for EET 90020 indicates that it did not experience intense younger reheating that many eucrites did between 3.1 and 4.4 Ga (left, [407]). The concurrence of other dating systems (Rb-Sr, Sm-Nd, and Mn-Cr, Hf-W), as well as mineralogic evidence from the chromite-ilmenite equilibria (right, [407]), all indicate this sample experienced an impact event at 4.5 Ga.

Reference [400–407]

Allan Hills (ALH) A81001
Unbrecciated eucrite
52.93 g
Found January 7, 1982
4.5 × 4.0 × 4.5 cm
Weathering = Ae

Eucrites are basaltic igneous rocks that comprise a range of textural types from ophitic to subophitic to variolitic, and contain plagioclase feldspar and pyroxene. The range of textures indicates formation conditions from shallow intrusive to extrusive. There are even a few examples of vesicular basalt. Two compositional types may define an equilibrium melting trend and a fractional crystallization trend. Eucrites are thought to be from the asteroid 4 Vesta, which has a similar surficial composition and mineralogy.

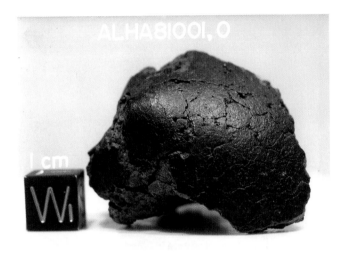

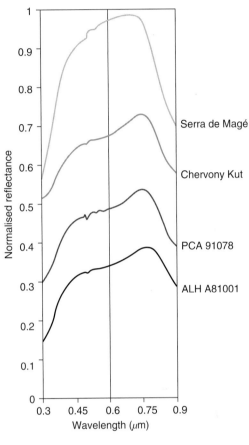

Plate 56

MINERALOGY

Allan Hills A81001 is an unbrecciated eucrite with very fine grained, variolitic-spherulitic texture, and also contains rare elongate microphenocrysts of pyroxene. Plagioclase (An_{96}) is approximately 61%, and pyroxene is 39%; orthopyroxene has rather uniform composition, averaging $Wo_{1.6}En_{40}Fs_{59}$, and minor clinopyroxene has $Wo_{43.7}En_{31.4}Fs_{24.8}$.

SIGNIFICANCE

ALH A81001 has the appearance of a quench-textured melt, and thus its origin has been proposed to be that of an impact melt or from the quenched portion of an early magma ocean. The latter hypothesis is based on the high Cr content of the pyroxenes, which may indicate a more primitive composition, coupled with the evidence for a 0.6 micron absorption feature of IR spectra associated with the high Cr (right, [410]). Studies of 4 Vesta by the *Dawn* spacecraft may help evaluate this idea.

References [408–412]

Allan Hills (ALH) 76005 and pairs
Polymict eucrite
1425 g
Found January 20, 1977
6.6 × 7.8 × 4.3 cm
Weathering = A

Many eucrites contain two or more lithologies and have been brecciated from being near the surface of the asteroid. Clasts can represent a variety of unrelated basaltic rocks of a broad range of cooling histories. The matrix contains fragments of all lithologies and can contain glassy regions and dark fine-grained lithologies. These eucrites are called polymict due to the presence of multiple igneous lithologies.

Plate 57

MINERALOGY

The ALH polymict eucrites contain clasts of plagioclase-pigeonite basalt (variety of textures), dark fine-grained clasts, and mineral fragments of plagioclase and pyroxene. Clast plagioclase ranges from An_{83} to An_{92}, and the pyroxenes are orthopyroxene, pigeonite, and augite with Mg# between 0.3 and 0.7. Mineral fragments in the matrix include plagioclase, pyroxene, tridymite, chromite, ilmenite, and FeNi metal.

SIGNIFICANCE

The ALH pairing group comprises 14 different samples with a combined mass of 4.2 kg. It represents a large well-characterized mass of material that is representative of the brecciated basaltic crust (left, [427]) of the eucrites' parent body (asteroid 4 Vesta). The material from this group helped to shape our understanding of the differentiation of 4 Vesta. It has been studied by dozens of scientists using a broad range of analytical techniques and approaches.

References [413–427]

LaPaz Icefield (LAP) 91900
Diogenite
786.87 g
Found January 14, 1992
13.0 × 6.9 × 6.0 cm
Weathering = A/B

Diogenites are orthopyroxenites that also contain chromite, metal, and rarely olivine. Most common are brecciated diogenites. The diogenites are thought to represent either pyroxenite cumulates from an early magmatic system on an asteroid, or partial melt residue. In either case, there has been subsequent thermal metamorphism producing somewhat uniform Mg# among the diogenite group.

Plate 58

MINERALOGY

LAP 91900 and its pairs contain >99% orthopyroxene, with minor to trace amounts of clinopyroxene, chromite, troilite, FeNi metal, and silica.

SIGNIFICANCE

LAP 91900 has nine paired masses and represents close to 3 kg of diogenitic material. As such, it has contributed to evaluating many aspects of diogenite formation such as magmatic trends of major and trace elements, cooling rates, and isotopic heterogeneity in the inner solar system. Early studies utilized LAP 91900 to strengthen the infrared spectral link to asteroid 4 Vesta (left, [428]). Orthopyroxene crystallographic studies indicate that many diogenites equilibrated at low temperatures between 300 and 400°C (right, [431]).

References [428–431]

Grosvenor Mountains (GRO) 95555
Unbrecciated diogenite
250.6 g
Found January 13, 1996
6.0 × 6.0 × 5.0 cm
Weathering = A/B

Most diogenites are brecciated and have undergone some modification due to impact-related surficial processes in the asteroid regolith. Rarely there are unbrecciated diogenites that preserve the prebrecciation texture. GRO 95555 is such a sample and has been the focus of several detailed studies highlighting its unique character.

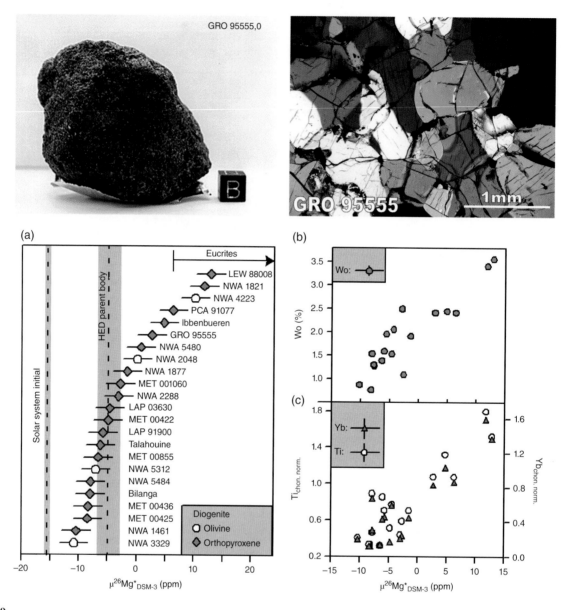

Plate 59

MINERALOGY

GRO 95555 contains 96.7% orthopyroxene, 1.5% spinel, 1.4 % silica, and trace amounts of metal and troilite. It has no olivine or plagioclase like some other diogenites.

SIGNIFICANCE

Despite its unbrecciated texture, its mineral compositions have been homogenized by metamorphic re-equilibration. Nonetheless, this sample has provided important constraints on the history of the diogenites, such as evidence for excess ^{26}Mg (left, [436]) (from decay of ^{26}Al) and trace element characteristics that allow evaluation of magma ocean (right, [436]), plutonic, and crustal assimilation models for formation.

References [432–438]

Graves Nunataks (GRA) 98108
Olivine diogenite
12.68 g
Found January 11, 1999
2.0 × 2.0 × 2.0 cm
Weathering = B

Many diogenites contain only orthopyroxene, but a growing number have been recognized that contain significant (>10%) olivine in addition to the orthopyroxene as a major silicate phase. GRA 98108 is such a sample and is coarse grained with as much as 20% olivine.

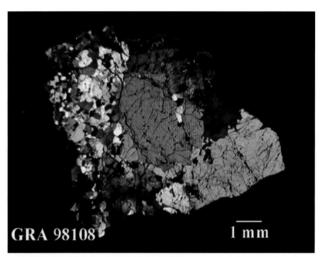

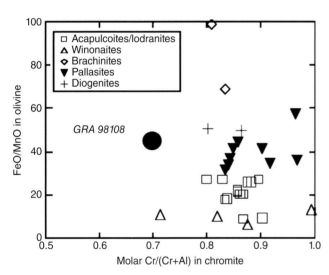

Plate 60

MINERALOGY

GRA 98108 contains 80% orthopyroxene, nearly 20% olivine, and minor amounts of chromite, plagioclase, clinopyroxene, FeNi metal, troilite, and silica. The orthopyroxene has Mg# of 0.77, olivine of Mg# = 0.73, and plagioclase is An_{91}.

SIGNIFICANCE

GRA 98108 has one of the highest MgO contents of diogenites and may represent the deeper or more primitive part of a differentiated asteroid where olivine may have been part of the fractionating assemblage. The FeO/MnO ratio of GRA 98108 is like other diogenites but distinct from other olivine-rich achondrites (above, [441]). Because diogenites typically have a small compositional range, and we do not have samples of the mantle of the diogenite parent body, these olivine-rich diogenites may provide compositional leverage that can help to better constrain models for early differentiation.

References [439–443]

**Elephant Moraine (EET) 87503
and pairing group
Howardite
1734.5 g
Found December 21, 1987
16.5 cm × 10 cm × 8.5 cm,
Weathering = A**

Eucrites and diogenites are known to come from the same parent body because fragments of both are preserved in the brecciated achondritic meteorites known as howardites. Howardites contain a minimum of >10% of either eucritic or diogenitic material in order to be classified as such. They are also polymict breccias, and sometimes contain glassy spherules that presumably formed in the regolith process. Some howardites contain clasts of carbonaceous chondrite material, indicating a mixture of lithologies in some asteroid regolith.

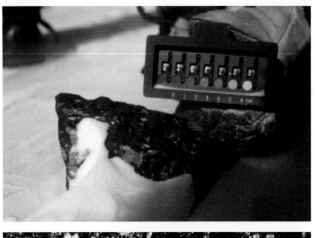

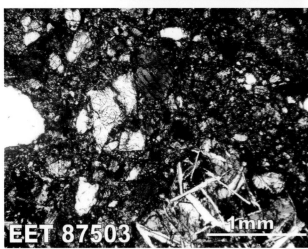

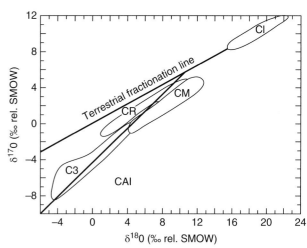

Plate 61

MINERALOGY

EET 87503 displays a variety of fine- to coarse-grained diogenite, eucrite (multiple varieties), impact melt/breccia, glassy, and even carbonaceous chondrite clasts in a light gray matrix.

SIGNIFICANCE

EET 87503 contains a great diversity of materials that are likely from the regolith of asteroid 4 Vesta. Among the materials found are clasts of carbonaceous chondrites (right, [449]). In some other howardites the portion of carbonaceous chondrite material can be very high, as much as 50% of some sections. Careful studies of the clasts in howardites can lead to a detailed understanding of the processes acting on the surface (or regolith) of an asteroid, and also offer insight into magmatic history.

References [444–454]

Miller Range (MIL) 03443
HED dunite
46.25 g
Found December 14, 2003
4.0 × 3.0 × 2.0 cm
Weathering = B

If HED meteorites represent crustal rocks from a differentiated asteroid parent body, there must be some olivine-rich (dunite) material preserved in the parent body as well. Therefore, many meteoriticists have looked carefully for dunitic material among achondrites that may come from the HED parent body. A handful of dunitic HEDs have been recognized, one of which is Miller Range 03443.

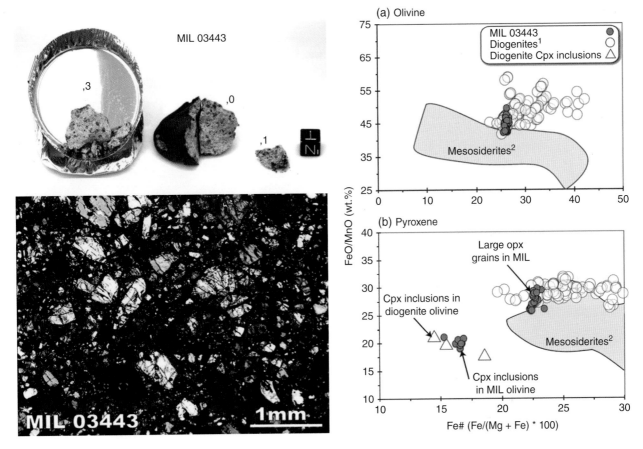

Plate 62

MINERALOGY

MIL 03443 is a brecciated dunite composed of mainly 200–400 mm angular olivines, but some as large as 2.5 mm are present. Olivine comprises 91% of the sample (Fa_{26}), whereas minor amounts of orthopyroxene, troilite, and chromite are also present.

SIGNIFICANCE

Although originally classified as an olivine-rich clast from a mesosiderite, this sample clearly is part of the HED meteorite group and distinct from mesosiderites (right, [455]). Thus, it offers a rare glimpse into igneous processes on asteroid 4 Vesta that may form dunites (in the crust or mantle), and it helps to refine models for the origin of the eucrites and diogenites.

References [455–457]

Queen Alexandra Range (QUE) 93148
Ungrouped achondrite
1.09 g
Found December 22, 1993
1.0 × 0.8 × 0.5 cm
Weathering = B

The HED clan (howardite, diogenites, and eucrites) comprise more than 750 meteorites that have been linked to asteroid 4 Vesta. These samples represent crustal and shallow mantle lithologies, whereas deeper mantle pieces (dunite and harzburgite) have been lacking in our collections. There are only a few anomalous ultramafic meteorites that have been linked to 4 Vesta and the HED, and each newly found piece offers more insights into the differentiation of this large asteroid.

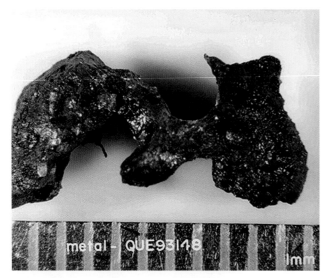

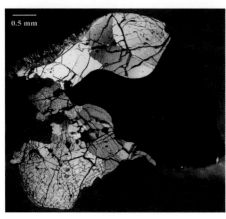

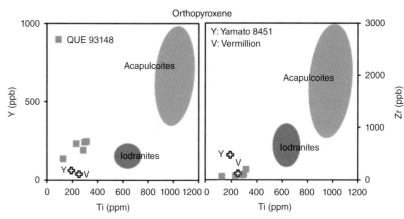

Plate 63

MINERALOGY

FeNi metal, coarse olivine (Fa$_{14}$) and orthopyroxene (Fs$_{13}$), as well as minor augite, and symplectities containing augite, orthopyroxene, chromite, troilite, and phosphate.

SIGNIFICANCE

QUE 93148 is a very small meteorite, but it has offered much information nonetheless due to the variety of techniques that can be applied to small samples. Mineralogic studies using SEM, TEM, and EBSD have provided important constraints on its formation history (left, [459]), as have trace element measurements using SIMS and the electron microprobe (right, [458]). QUE may have a connection to pyroxene pallasites.

References [458–462].

Allan Hills (ALH) A81005
Lunar anorthositic
regolith breccia
31.39 g
Found January 18, 1982
3.0 × 2.5 × 3.0 cm
Weathering = A/B

Anorthositic portions of the Moon are found in the lunar highlands, or the rough mountainous regions between the smoother mare. These are some of the oldest portions of the Moon but were only sampled extensively by a few of the *Apollo* missions. Additional samples of the highlands can provide better and more comprehensive understanding of the geologic history of this fundamental province on the Moon.

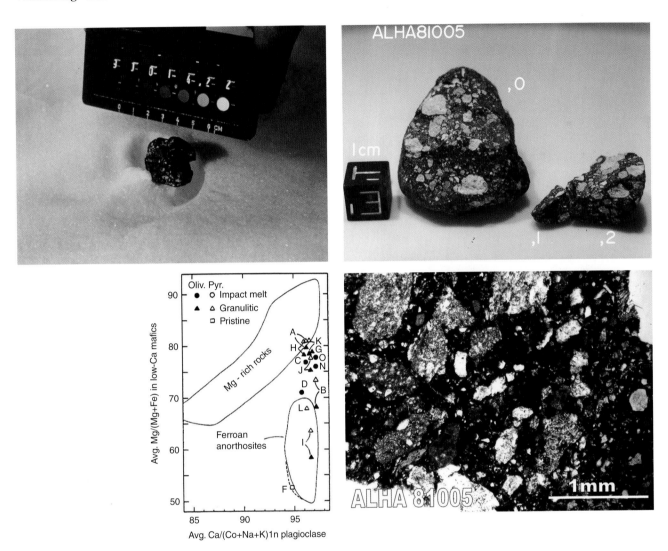

Plate 64

MINERALOGY

ALH A81005 is a polymict regolith breccia that contains clasts of low Ti mare basalt, high Ti mare basalt, granulitic breccia, cumulate breccia, impact melt, anorthosite, norite, and troctolite. It also contains many soil components (regolith breccia and agglutinate), and mineral and glass fragments.

SIGNIFICANCE

ALH A81005 was the first recognized lunar meteorite. It contains clasts that bridge the gap between the magnesium suite samples (MGS) and ferroan anorthosites (FAN), illustrating that lunar meteorites provide information that complements the *Apollo* and *Luna* collections (left, [487]).
References [436–489]

Elephant Moraine (EET) 87521, 96008
Lunar basaltic regolith breccia
30.7, 53.0 g
Found December 20, 1987 and
December 14, 1996
3.7 × 2.5 × 2.0; 4.5 × 3.5 × 1.5 cm
Weathering = A

Although lunar geology is dominated by two end members (mare basalt and anorthositic crust), there are polymict breccias that represent more complex rock units and provide information about lunar geologic processes that cannot be discovered in *Apollo/Luna* collections. Polymict lunar breccias are more rare and provide information about diversity of materials in Moon rocks.

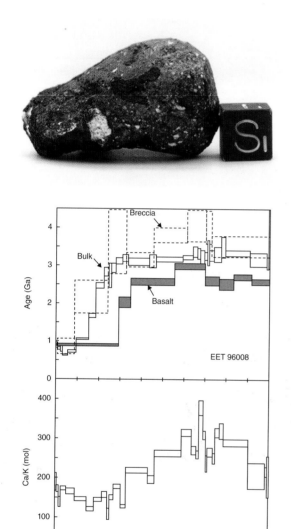

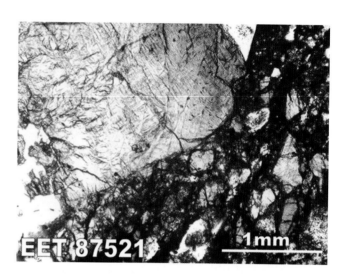

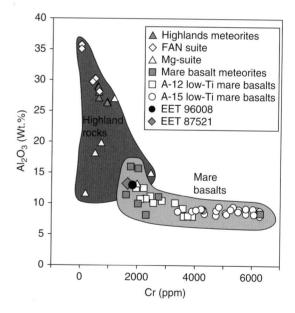

Plate 65

MINERALOGY

EET 87521 is a polymict regolith breccia that contains clasts of low Ti mare basalt, high Ti mare basalt, granulitic breccia, cumulate breccia, impact melt, anorthosite (An_{90-92}), norite, and troctolite. It also contains many soil components (regolith breccia and agglutinate) and mineral and glass fragments.

SIGNIFICANCE

The basaltic clasts found in EET 87521 and 96008 exhibit young ages that overlap with those of *Apollo* basalts (left, [494]). However, the overall composition of the breccia falls intermediate between the basalt and feldspathic (anorthositic) end members of the lunar geology (right, [490]), illustrating the diversity of materials and thus the rich geologic information present in mingled lunar breccias.

References [490–499]

Miller Range (MIL) 05035
Unbrecciated lunar gabbro
142.2 g
Found December 11, 2005
4.5 × 4.0 × 3.5 cm
Weathering = A/B

Lunar basaltic meteorites are samples of mare regions on the Moon that may not have been sampled by *Apollo* or *Luna* missions and therefore can provide additional constraints on basaltic magmatism and volcanism on the Moon in general.

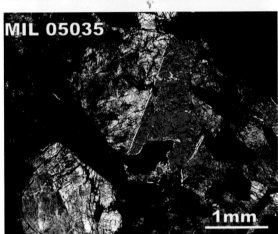

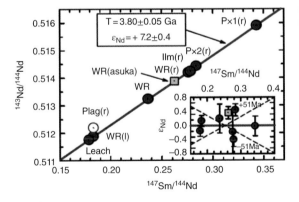

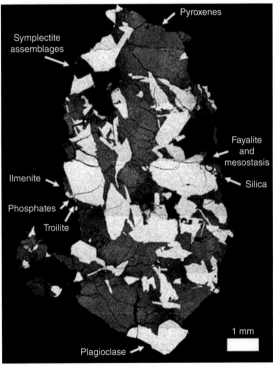

Plate 66

MINERALOGY

This coarse-grained gabbroic textured rock contains ~55%–60% calcic to subcalcic pyroxene, maskelynitized plagioclase (~30%), and minor amounts of silica, troilite, phosphates, ilmenite, and symplectites (right, [501]).

SIGNIFICANCE

MIL 05035 contains maskelynitized plagioclase that is very clear and glassy looking, giving the rock a shiny or glossy appearance. This, combined with the older age of 3.8 Ga (left, [503]), give this sample unique characteristics that complement the *Apollo/Luna* collections.

References [500–504]

LaPaz Icefield (LAP) 02205
Unbrecciated lunar basalt
1226.3 g
Found December 18, 2002
10.0 × 8.5 × 5.5 cm
Weathering = B

Lunar basaltic meteorites are samples of mare regions on the Moon that may not have been sampled by *Apollo* or *Luna* missions and therefore can provide additional constraints on basaltic magmatism and volcanism on the Moon in general.

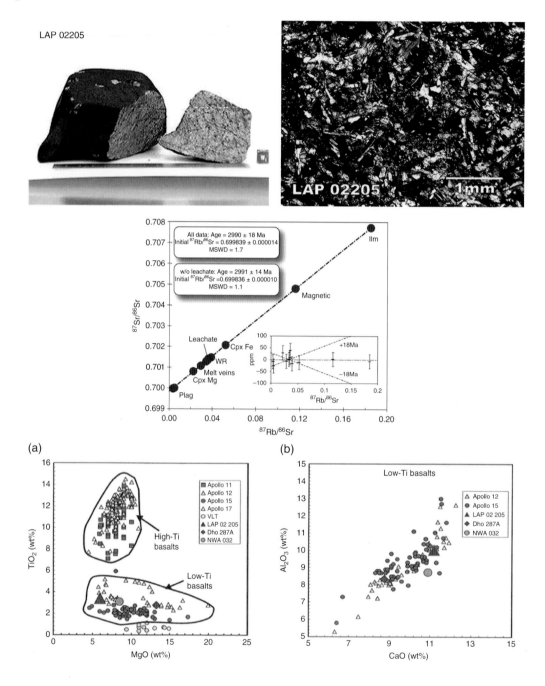

Plate 67

MINERALOGY

Olivine phenocrysts (~2%), pyroxene (~50%), plagioclase (~35%), ilmenite (~4%), spinel, phosphate, troilite, and fayalitic symplectite.

SIGNIFICANCE

LAP 02205 has a distinctively young age of 3.0 Ga (upper right, [512]), and is similar to the basaltic series defined by the Apollo 12 and 15 low Ti basalts (lower figure, [505]). It is paired with five other pieces for a combined mass close to 2 kg.

References [505–516]

MacAlpine Hills (MAC) 88104, 88105
Lunar anorthositic regolith breccia
61.2, 662.5 g
Found January 13, 1989
4.0 × 4.5 × 2.5; 11.0 × 7.0 × 6.5 cm
Weathering = A/Be

Anorthositic portions of the Moon are found in the lunar highlands, or the rough mountainous regions between the smoother mare. These are some of the oldest portions of the Moon but were sampled extensively by only a few of the *Apollo* missions. Additional samples of the highlands can provide better and more comprehensive understanding of the geologic history of this fundamental provenance on the Moon.

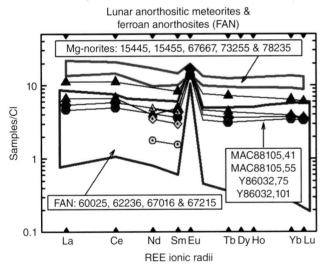

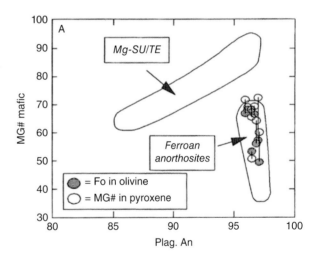

Plate 68

MINERALOGY

The modal mineralogy of MAC 88105 is dominated by lithic clasts of mafic and feldspathic melt breccias, mafic and feldspathic rocks, and granulites (75%–80%), mineral fragments (15%–20%), and minor glass veins and clasts.

SIGNIFICANCE

MAC 88105 contains ancient ferroan anorthosite (FAN) clasts (left [529] and right [527]) and is a mature regolith breccia. It was one of the first meteorites of this kind found and documented in detail to allow comparison to *Apollo* samples.

References [517–536]

Allan Hills (ALH) 84001
Martian orthopyroxenite
1930.9 g
Found December 27, 1994
17 × 9.5 × 6.5 cm
Weathering = A/B

Linked to shergottites, nakhlites, and chassignites by oxygen and noble gas isotopes, the orthopyroxenite is thought to represent the ancient crust of Mars. The age of the orthopyroxenite is ~4.1 Ga, and overall much older than any of the other martian meteorites. These meteorites provide the only samples of Mars available for study in our laboratories, and ALH 84001 is one of the oldest known pieces of Mars.

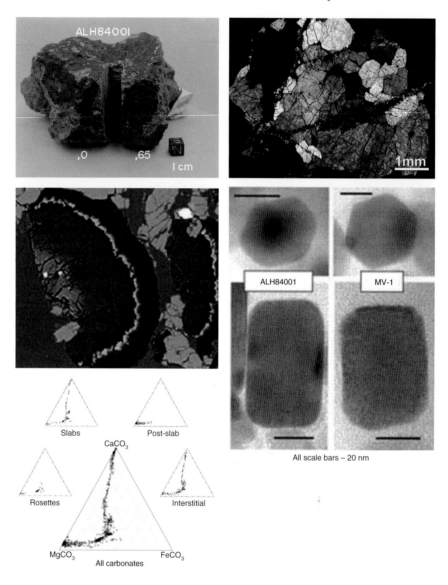

Plate 69

MINERALOGY

ALH 84001 consists mainly of coarse-grained orthopyroxenite (97% orthopyroxene $En_{70}Wo_3$), with 2% chromite inclusions within the orthopyroxene, and ~1% maskelynite ($An_{35}Or_3$) with 0.15% apatite between grain boundaries. Minor augite ($En_{45}Wo_{43}$), olivine (Fo_{65}), magnetite (right, [574]), and pyrite are found as clusters around globules of zoned Fe-Mg-Mn-Ca carbonates (top, [559]).

SIGNIFICANCE

Originally classified as a diogenite, ALH 84001 is the first and only orthopyroxenite among the martian meteorites. It contains veins of feldspathic glass melted by shock and carbonate globules (some are in rosette or slab form, some are less defined). The occurrence of these carbonate minerals was one of the first solid, mineralogical clues that liquid water occurred on Mars at some point in its history (left, [559]). The carbonates were also at the center of the paper by *McKay et al.* (1996, [599]) that asserted that they had found ancient life in the meteorite. In addition to being the only orthopyroxenite among the martian meteorites, ALH 84001 is also one of the oldest martian meteorites, dating at ~4.1 billion years old [591].

References [537–638]

Elephant Moraine (EET) A79001
Shergottite
7942 g
Found January 13, 1980
22 × 17 × 14 cm
Weathering = Ae

Shergottites are FeO-rich basalts that were identified as pieces of Mars by the composition of trapped gases that match those measured by the *Viking* spacecraft at the surface of Mars. Their young ages (~180 Ma) indicate they also came from a planet with young volcanism on its surface, like Mars. In addition, plagioclase feldspar has been entirely converted to the shock glass, maskelynite, demonstrating the high shock pressures required to eject them from the surface of Mars. Studies of shergottites have also revealed information about the interior (mantle and crust) of Mars, thus helping constrain its conditions of formation.

Plate 70

MINERALOGY

Lithology A is composed of a basaltic host (pyroxene, maskelynite, high-Ti chromite, merrillite, minor Cl apatite, ilmenite, pyrrhotite and mesostasis) containing xenocrysts of olivine, Cr spinel, and low-Ca pyroxene (left). Lithology B is homogeneous basalt containing augite laths in a matrix of pigeonite-augite, maskelynite, ulvöspinel-ilmenite intergrowth, whitlockite, Cl apatite, and mesostasis (right). Lithology C is an assemblage of glass "pods" and thin, interconnecting, glass veins.

SIGNIFICANCE

EET A79001 is one of the largest martian meteorites and contains three different lithologies, all of which have made important contributions to our understanding of martian geology. The black glassy pods (Lithology C) are the source of trapped gas that provided a match and link to Mars. The basaltic lithology (B), which makes up about 10% of the meteorite's volume, is similar to other basaltic shergottites and exhibits a direct contact with the main lithology of the meteorite (A). Lithology A is olivine-phyric textured but also contains many individuals and clusters of xenocrysts of olivine (distinct from the phenocrysts) and orthopyroxene. The latter provide some additional information about the source magmatic history of the parent melt.

References [639–700]

Queen Alexandra Range (QUE) 94201
Basaltic shergottite
12.02 g
Found December 16, 1994
2.3 × 2.0 × 1.5 cm
Weathering Be

Basaltic shergottites are more evolved basalts such as Shergotty and Zagami that are linked to Mars by their age, O isotopes, and noble gas content. Studies of basaltic shergottites have revealed information about the production of magmas on Mars, and efforts have been made to understand the links between olivine phyric shergottites, basaltic shergottites, and lherzolitic shergottites to better understand the magmatic history of Mars.

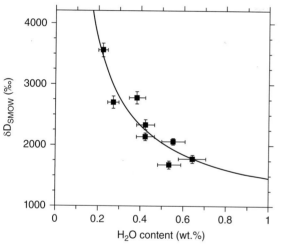

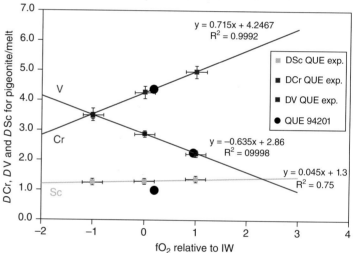

Plate 71

MINERALOGY

QUE 94201 exhibits a medium-grained (1–5 mm) texture dominated by plagioclase and pyroxene. Plagioclase is normally zoned (An_{68} to An_{57}) and pyroxene is adjacent to and sometimes enclosed by plagioclase. Pyroxene consists of pigeonite and augite; pigeonite exhibits normal Mg-Fe-Ca igneous zoning (from $Wo_9En_{60}Fs_{31}$ to $Wo_{16}En_1Fs_{83}$), whereas the augite compositions are bimodal $Wo_{30}En_{45}Fs_{25}$ and $Wo_{40}En_2Fs_{58}$.

SIGNIFICANCE

QUE 94201 is one of the most reduced basaltic rocks among the martian meteorites, near the IW buffer (right, [710]), and represents an end member of the known oxidation states within the martian interior. QUE 94201 is also a member of the isotopically depleted shergottites, which define an important interior reservoir of the mantle and crust. In addition, this sample contains hydrous minerals (phosphate) and evaporites. These minerals provide information about martian volatiles (H_2O, CO_2, and S_2) (left, [714]) and igneous processes in martian magmas, as well as alteration processes at the martian surface.

References [701–722]

Allan Hills (ALH) A77005
Lherzolitic shergottite
482.5 g
Found December 29, 1977
9.5 × 7.5 × 5.25 cm
Weathering = A

Lherzolitic shergottites are coarse-grained ultramafic rocks consisting of olivine, orthopyroxene, chromite, and plagioclase (maskelynite). A distinctive characteristic of lherzolitic shergottites is the abundant evidence for shock, including maskelynite, mosaicised olivine, and pockets of melted rock in the interior. Lherzolitic shergottites may represent the residue from melting of the martian mantle during production of basaltic rocks. They also are among the most heavily shocked of all the martian meteorites. Thus, they hold information about the interior and differentiation history of Mars as well as constraints on its impact history.

Plate 72

MINERALOGY

ALH A77005 is a cumulate achondrite comprising olivine (~Fo_{74}), low- and high-Ca pyroxene, maskelynite (An_{54}–An_{45}), and minor amounts of chromite, ilmenite, troilite, and whitlockite.

SIGNIFICANCE

ALH A77005 was the first recognized lherzolitic shergottite and provides unique information about melting and the martian mantle over and above what we can learn from nakhlites, basaltic shergottites, and chassignites. The trapped melt inclusions (right, [731]) provide information about possible parental liquids. Shock minerals and chemical characteristics indicate that this meteorite (and LEW 88516, another lherzolitic shergottite from Antarctica) have been severely shocked. The two distinct lithologies, poikilitic and nonpoikilitic, offer important constraints on the detailed magmatic history of the lherzolitic shergottites (left, [770]). Combined with RBT 04261/262, these samples may represent the bottommost and topmost, respectively, units of a martian magmatic complex.

References [723–763]

Roberts Massif (RBT) 04261, 262
Lherzolitic shergottite
78.8, 204.6 g
Found December 25 and 28, 2004
4.0 × 3.5 × 2.5; 6.5 × 5.5 × 3.5 cm
Weathering = B

Lherzolitic shergottites are coarse-grained ultramafic rocks consisting of olivine, orthopyroxene, chromite, and plagioclase (maskelynite). A distinctive characteristic of lherzolitic shergottites is the abundant evidence for shock, including maskelynite, mosaicised olivine, and pockets of melted rock in the interior. Lherzolitic shergottites may represent the residue from melting of the martian mantle during production of basaltic rocks. They also are among the most heavily shocked of all the martian meteorites. Thus, they hold information about the interior and differentiation history of Mars as well as constraints on its impact history.

Plate 73

MINERALOGY

RBT 04262 is composed of poikilitic and nonpoikilitic areas. The poikilitic areas (up to 5 mm in size) are comprised of pyroxene oikocrysts enclosing small olivine grains (~0.5 mm). Pyroxene oikocrysts in poikilitic areas are low-Ca pyroxene ($En_{73}Wo_5$ to $En_{60}Wo_{10}$) in the cores and augite in the rims ($En_{52}Wo_{30}$–$En_{45}Wo_{40}$). Kaersutite-bearing magmatic inclusions are found within pyroxene oikocrysts. The nonpoikilitic areas comprise olivine, maskelynite, and pyroxene, the latter mostly pigeonite with small amounts of augite and slightly higher in Fe contents than those in poikilitic areas ($En_{62}Wo_7$–$En_{58}Wo_{12}$ and $En_{50}Wo_{30}$–$En_{44}Wo_{39}$). The olivine grains in poikilitic areas are more magnesian and show wider compositional variation (Fa_{28-39}) than those in nonpoikilitic areas (Fa_{32-41}). Maskelynite is $An_{58}Or_3$ to $An_{30}Or_7$. K-rich feldspar is $An_{10}Or_{25}$ to An_2Or_{70}. Minor amounts of Ca phosphates, chromite, ilmenite, and FeNi sulfide are also present.

SIGNIFICANCE

The presence of large pyroxene oikocrysts in RBT 04261/262 and its overall mineralogy show that it is a member of the lherzolitic shergottite group (left, [770]). However, the relatively high abundance of plagioclase (10%–15%) indicates that it has affinity to the basaltic shergottites as well. The nonpoikilitic areas have the most ferroan olivine composition, highest modal abundance among lherzolitic shergottites, and more albitic maskelynite than the others. Thus, RBT 04262 may be located near the upper region of the igneous complex in which the lherzolitic shergottites formed (right, [769]), where cumulus phases were less abundant.

References [764–781]

Miller Range (MIL) 03346
Nakhlite
715.2 g
Found December 15, 2003
10.0 × 6.0 × 5.5 cm
Weathering = B

Nakhlites are clinopyroxene-rich cumulate rocks thought to be from Mars, based on the composition of trapped noble gases, their young ages (1.3 Ga), and their O isotopic composition. The eight different nakhlites recognized in our collections are apparently from one magmatic body that shows textural variation from deeper cumulate-rich to shallower mesostasis-rich lithologies. Studies of nakhlites have provided constraints on the differentiation history of Mars, the high oxidation state of the mantle, surficial volatiles such as S, C, and Cl, and organic geochemistry.

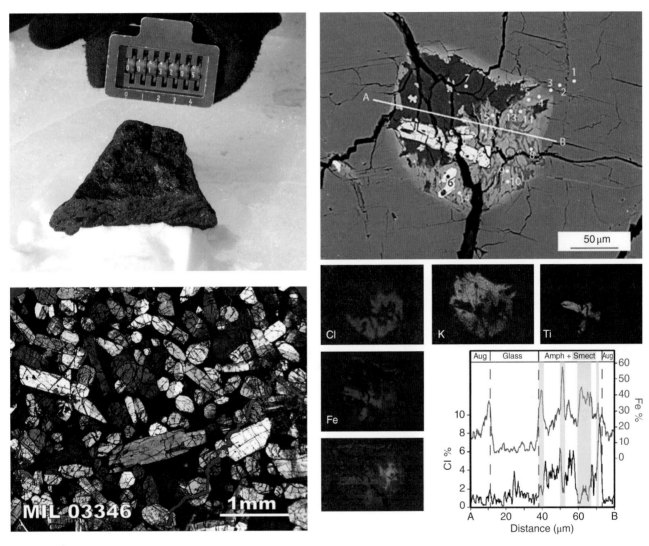

Plate 74

MINERALOGY

MIL03346 is composed of ~80% clinopyroxene, ~1% olivine, and ~20% vitrophyric intercumulus material (skeletal Fe-Ti oxide, fayalite, and sulfide), which suggests rapid cooling in a shallow intrusion or a lava flow. Olivines contain trapped melt inclusions (top right, [801]).

SIGNIFICANCE

MIL 03346 crystallized 1.3 Ga ago, from a shallow oxidized magma that may have had a deeper origin. The textures and chemical compositions suggest it cooled quickly in a thin flow or dike. After emplacement, it was shocked only mildly based on its density and magnetism. Weathering at the surface of Mars produced Cl-amphibole (bottom right, [801]), jarosite, and other oxidized minerals. The pyroxenes have been used to better understand remote sensing at the surface of Mars.

References [782–803]

Elephant Moraine (EET) 87500
Mesosiderite
8132 g
Found December 29, 1987
25.0 × 16.0 × 12.5 cm
Weathering = B

Mesosiderites are brecciated mixtures of metal and silicate and represent a diverse group of meteorites. The silicate portions of mesosiderites are very similar to eucritic and diogenitic materials, with gabbroic and basaltic clasts as well as large orthopyroxenes. In addition, there are many additional mineral fragments and lithologies. The mesosiderite metal is similar in composition to that in the pallasite and IIIAB iron meteorites. However, a link between mesosiderites and these other groups has not been definitively made.

EET 87500

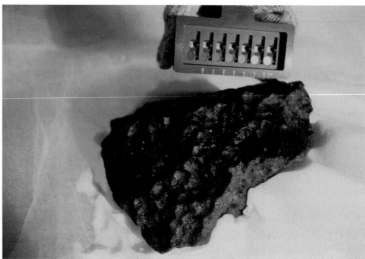

EET 87500   1 mm

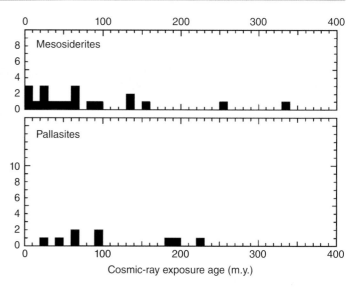

Cosmic-ray exposure age (m.y.)

Plate 75

MINERALOGY

EET 87500 is composed of a granular aggregate of approximately 50% pyroxene, 30% nickel-iron, and 20% plagioclase, with accessory merrillite and an SiO_2 polymorph, probably tridymite. The grain size is relatively coarse, with individual pyroxene (mean of Wo_6Fs_{30}) and plagioclase (An_{91-96}) grains up to 2 mm across. Many pyroxenes are partly or completely converted into a mosaic of small granules.

SIGNIFICANCE

The mesosiderites exhibit a wide range of textures and mineral compositions, and EET 87500 has contributed to documenting the range of these characteristics within the group. It also has the oldest cosmic ray exposure age (340 Ma) of the mesosiderites (right, [807]), making it an important sample in establishing the origin and parent body or bodies of the mesosiderites.

References [804–807]

Reckling Peaks (RKP) A79015
Mesosiderite
10,022 g
Found January 11, 1980
26.0 × 18.5 × 13.0 cm
Weathering = A/B

Some mesosiderites have anomalous or unusual features and thus deviate from the more common type of mesosiderite. This may be due to diverse processes on a single parent asteroid, or there may be multiple parent bodies for the mesosiderites, thus giving rise to more compositional and textural variation.

Plate 76

MINERALOGY

RKP A79015 is an iron with a slightly weathered exterior in which are set silicate fragments about 5 mm in average size. The most common fragments of this type are greenish single crystals showing cleavage that are probably orthopyroxenes, which appear to be present throughout the meteorite. The major portion of and exposed interior surface shows kamacite with lesser taenite. Approximately 10% of the meteorite consists of ellipsoidal to subangular inclusions which are dominantly troilite, silicate, or troilite plus silicate. The silicate has the composition of a magnesian orthopyroxene (estimated at $Wo_2En_{73}Fs_{25}$) in the largest inclusion; a second phase may be intergrown with the pyroxene. Other phases present include chromite and schreibersite.

SIGNIFICANCE

RKP A79015 exhibits metal composition and high modal metal content that is distinct from most mesosiderites, thus making it a more anomalous member of the group (right, [810]). However, it does exhibit a very slow cooling rate based on the exsolution textures in the metal (left, [814]).

References [808–814]

Cumulus Ridge (CMS) 04071
Main group pallasite
2110.1 g
Found December 20, 2004
13.0 × 9.0 × 8.0 cm
Weathering = B/C

Pallasites are perhaps the most striking of all meteorites. Composed of approximately half metal and half large (up to 3 cm) olivine crystals, pallasites are generally thought to be from near the core-mantle interface in a small asteroid. Other phases such as pyroxene, phosphate, chromite, phosphide, and sulfide, as well as textural varieties with angular and rounded olivine, give the pallasites some diversity as a group. It is thought they may represent as many as 10 different parent bodies, and are not associated with another major meteorite group, except for possibly the IIIAB irons with which they share a similar bulk metal composition. Some pallasites are thought to be formed by impact processes.

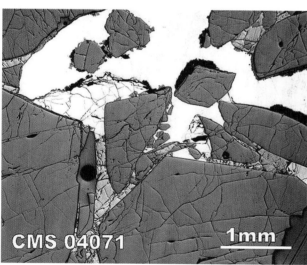

Plate 77

MINERALOGY

CMS 04071 exhibits fragmental, angular olivine (Fa_{12}) grains ranging in size from a hundred microns to 2 cm. Interstitial to these grains are euhedral and irregular chromites, as well as troilite and schreibersite. The texture, mineralogy, and even degree of weathering of CMS 04071 are reminiscent of Imilac.

SIGNIFICANCE

CMS 04071 is paired with CMS 04061–04079, one of 18 paired masses, comprising a total of 170.55 kg in recovered mass. Although there have been other pallasites discovered in Antarctica (Thiel Mountains, Queen Alexandra Range [QUE] 93544, Pecora Escarpment [PCA] 91004, Yamato 8451 and 74044), the CMS pairing group constitutes the largest pallasite recovered from Antarctica.

References [815]

Derrick Peak (DRP) 00200, 201
IIAB iron (coarse octahedrite)
10,000, 2689.4 g
Found December 19, 2000
1.0 × 0.8 × 0.5 cm
Weathering = B

The IIAB irons include hexahedrites and octahedrites, and are part of the third largest iron meteorite group. The IIAB iron meteorites are thought to have formed during the crystallization of an S- and P-bearing metallic core, at the center of a differentiated asteroid. These deep-formed asteroid samples thus offer a glimpse into differentiation processes in the early solar system.

DRP 00200,0

DRP 00201

5 cm

Plate 78

MINERALOGY

From a cut surface, DRP 00200 resembles typical members of the Derrick Peak iron shower (bottom, [817]). Only a thin layer of corrosion is found on the surface, and neither fusion crust nor heat-altered zone is found. Structurally, they are coarsest octahedrites with large areas of swathing kamacite enclosing elongate, skeletal schreibersite crystals and centimeter-sized round troilite inclusions.

SIGNIFICANCE

The Derrick Peak irons were found in 1978 and 2000 field seasons and comprise 27 members with a total mass of 408.4 kg. This pairing group has helped to define the sizable group of iron meteorites that constitutes the IIAB irons. The group has the largest known range in Ir concentrations (a factor of 4000) and slopes that are steeply negative on plots of Ir versus Au or As (or Ni). IIAB appears to have had the highest S contents of any magmatic group of iron meteorites, consistent with its high contents of other volatile siderophiles, particularly Ga and Ge.

References [816–817]

Meteorite Hills (MET) 00400 (+pairs)
IIIAB iron (medium octahedrite)
4583.8 g
Found December 20, 2000
17.5 × 11.0 × 7.0 cm
Weathering = B

The IIIAB irons include medium and coarse octahedrites and are part of the second largest iron meteorite group. The IIIAB iron meteorites are thought to have formed during the crystallization of an S- and P-bearing metallic core, at the center of a differentiated asteroid. The metal composition overlaps with that of the main group pallasite meteorites, suggesting that they may be from the same parent body. These deep-formed asteroid samples thus offer a glimpse into differentiation processes in the early solar system.

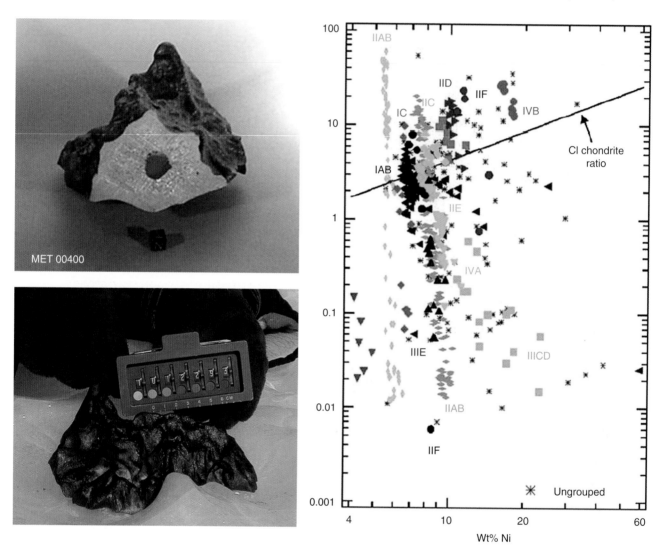

Plate 79

MINERALOGY

MET 00400 exhibits a medium octahedrite structure with original band widths of 1–1.5 mm and typically L/W of 15–20. Each mass exhibits a single orientation of the Widmanstätten pattern, suggesting the formation of each (and perhaps all) from a single austenite crystal. A large, polycrystalline troilite nodule (17 cm in diameter) is present. The interior structure is extensively heat-altered and dominated by recrystallized kamacite. Dimensions of recrystallized grains are typically 100–500 microns.

SIGNIFICANCE

MET 00400 is the largest mass of a 33-member pairing group, comprising a total of 5.3 kg. These meteorites range in mass from 4583.8 to 3.2 grams. They occurred within 200 m on either side of a line 5.8 km long connecting the largest and smallest mass. Although larger masses tend to lie at one end and smaller masses at the other, the distribution is imperfect and a mixed-mass clump is found in the center. The lineation of the distribution is not a result of either ice exposure or search strategy, as a much larger, rectangular area of blue ice was exposed and searched and meteorites were recovered throughout this larger area. The lineation may suggest a relatively recent fall.

References [818]

Mount Howe (HOW) 88403
Ungrouped iron
2480.7 g
Donated by science field team
11.5 × 7.0 × 7.0 cm
Weathering = ?

A large percentage of iron meteorites recovered from Antarctica are not part of the main magmatic groups of irons and are ungrouped. These samples offer a glimpse into other processes operating in the early solar system, from impact melting, condensation, or as-yet poorly defined magmatic groups. One such example is highlighted here, but there are many ungrouped irons that have been studied from the U.S. Antarctic meteorite collection.

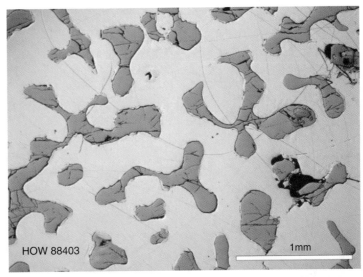

HOW 88403   1mm

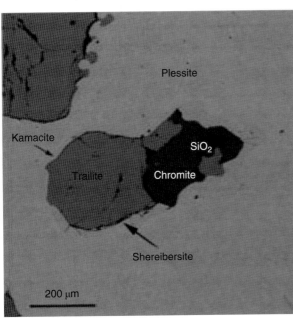

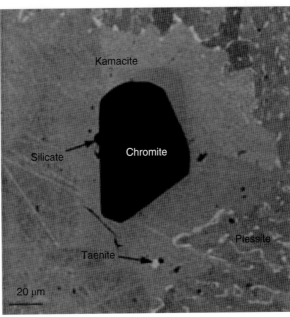

Plate 80

MINERALOGY

The appearance of HOW 88403 exterior surface suggests a metal-sulfide eutectic, with ~2/3 metal and ~1/3 sulfide. It is a fine-grained structure with individual sulfide cells averaging 0.3 mm in diameter. A thin fusion crust is present, underlain by a 2–3 mm heat-altered zone. The metal matrix is ataxitic, decomposing into a very fine-grained mixture of taenite and kamacite with abundant small schreibersites [819]. Troilite is surrounded by discontinuous rims of schreibersite, and an occasional chromite was observed in the sulfide [819].

SIGNIFICANCE

The meteorite was found by a research group under the direction of Dr. Charles Swithinbank on the blue icefield at Mt. Howe, Antarctica. HOW 88403 is more reduced than H chondrites, and likely formed by impact melting on a chondritic parent body. Bulk composition suggests that it was liquid above 1350°C, and estimated cooling rates suggest that HOW 88403 formed buried but near the surface of its parent body.

References [819–821]

PLATE REFERENCE LIST

1. WSG 95300—H3.3 CHONDRITE

[1] Lofgren, G. E., and Le, L. (2002), Experimental reproduction of Type IB chondrules, *Lunar Planet. Sci.*, *33*, 1746.

[2] Nittler, L. R., C. M. O'D. Alexander, R. Gallino, P. Hoppe, A. N. Nguyen, F. J. Stadermann, and E. K. Zinner (2008), Aluminum-, calcium- and titanium-rich oxide stardust in ordinary chondrite meteorites and erratum, *Astrophys. J.*, *682*, 1450–1478.

[3] Alexander, C. M. O'D., S. D. Newsome, M. L. Fogel, L. R. Nittler, H. Busemann, and G. D. Cody (2010), Deuterium enrichments in chondritic macromolecular material: Implications for the origin and evolution of organics, water and asteroids, *Geochim. Cosmochim. Acta*, *74*, 4417–4437.

2. LEW 85320—H5 CHONDRITE

[4] Grady, M. M., E. K. Gibson Jr., I. P. Wright, and C. T. Pillinger (1989), The formation of weathering products on the LEW 85320 ordinary chondrite: Evidence from carbon and oxygen stable isotope compositions and implications for carbonates in SNC meteorites, *Meteoritics*, *24*, 1–7.

[5] Jull, A. J. T., S. Cheng, J. L. Gooding, and M. A. Velbel (1988), Rapid growth of magnesium-carbonate weathering products in a stony meteorite from Antarctica, *Science*, *242*, 417–419.

[6] Velbel, M. A., and D. T. Long (1989), Meteoritic source of large-ion lithophile elements in terrestrial nesquehonite from Antarctic meteorite LEW 85320 (H5), *Meteoritics*, *24*, 334.

[7] Velbel, M. A., D. T. Long, and J. L. Gooding (1991), Terrestrial weathering of Antarctic stone meteorites: Formation of Mg-carbonates on ordinary chondrites, *Geochim. Cosmochim. Acta*, *55*, 67–76.

[8] Welten, K. C., L. Lindner, C. Alderliesten, and K. van der Borg (1999), Terrestrial ages of ordinary chondrites from the Lewis Cliff stranding area, East Antarctica, *Meteorit. Planet. Sci.*, *34*, 558–569.

[9] Velbel (2012), Terrestrial weathering of ordinary chondrites in nature and continuing during laboratory storage and processing: Review and implications for Hayabusa sample integrity, *Meteorit. Planet. Sci.*, Online Early.

3. LAP 02240—H CHONDRITE IMPACT MELT

[10] Cheek, L. C., and D. A. Kring (2008), Cooling rate determination for H chondrite impact melt breccia LAP 02240, *Lunar Planet. Sci. Conf.*, *39*(1391), 1169.

[11] Niihara, T., N. Imae, and H. Kojima (2008), Petrology and mineralogy of an impact melted H chondrite, LAP02240, *Lunar Planet. Sci. Conf.*, *39*(1391), 1856.

[12] Swindle, T. D., C. E. Isachsen, J. R. Weirich, and D. A. Kring (2009), ^{40}Ar-^{39}Ar ages of H-chondrite impact melt breccias, *Meteorit. Planet. Sci.*, *44*, 747–762.

[13] Swindle, T. D., D. A. Kring, and J. R. Weirich (2013), ^{40}Ar/^{39}Ar ages of impacts involving ordinary chondrite meteorites, in *Advances in ^{40}Ar/^{39}Ar dating from archaeology to planetary sciences*, edited by F. Jourdan, D. F. Mark, and C. Verati, Geological Society, London, Spec. Pub. 378, 10.1144/SP378.6.

[14] Wittmann, A., T. D. Swindle, L. C. Cheek, E. A. Frank, and D. A. Kring (2010), Impact cratering on the H chondrite parent asteroid, *J. Geophys. Res.*, *115*(E7), CiteID E07009.

[15] Niihara, T., N. Imae, K. Misawa, H. Kojima (2011), Petrology and mineralogy of the shock-melted H chondrites Yamato 791088 and LaPaz Ice Field 02240, *Polar Science*, *4*, 558–573.

4. QUE 97008—L3.05 CHONDRITE

[16] Alexander C. M.O'D., S. D. Newsome, M. L. Fogel, L. R. Nittler, H. Busemann, and G. D. Cody (2010), Deuterium enrichments in chondritic macromolecular material: Implications for the origin and evolution of organics, water and asteroids, *Geochim. Cosmochim. Acta*, *74*, 4417–4437.

[17] Busemann, H., C. M. O'D. Alexander, and L. R. Nittler (2007), Characterization of insoluble organic matter in primitive meteorites by microRaman spectroscopy, *Meteorit. Planet. Sci.*, *42*, 1387–1416.

[18] Grossman, J. N., and A. J. Brearley (2005), The onset of metamorphism in ordinary and carbonaceous chondrites, *Meteorit. Planet. Sci.*, *40*, 87–122.

[19] D. S. Lauretta, H. Nagahara, and C. M. O'D. Alexander (2006), Petrology and origin of ferromagnesian silicate chondrules, in *Meteorites and the Early Solar System II*, edited by D. S. Lauretta, and H. Y. McSween Jr., pp. 431–459, University of Arizona Press, Tucson.

[20] Nettles, J. W., G. E. Lofgren, W. D. Carlsom, and H. Y. McSween Jr. (2006), Extent of chondrule melting: Evaluation of experimental textures, nominal grain size, and convolution index, *Meteorit. Planet. Sci.*, *41*, 1059–1072.

[21] Nittler, L. R., C. M. O'D. Alexander, R. Gallino, P. Hoppe, A. N. Nguyen, F. J. Stadermann, and E. K. Zinne (2008), Aluminum-, calcium- and titanium-rich oxide

35 Seasons of U.S. Antarctic Meteorites (1976–2010): A Pictorial Guide to the Collection, Special Publication 68, First Edition. Edited by Kevin Righter, Catherine M. Corrigan, Timothy J. McCoy and Ralph P. Harvey.
© 2015 American Geophysical Union. Published 2015 by John Wiley & Sons, Inc.

stardustin ordinary chondrite meteorites, *Astrophys. J.*, *682*, 1450–1478.

[22] Qin, L., R. W. Carlson, and C. M. O'D. Alexander (2011), Correlated nucleosynthetic isotopic variability in Cr, Sr, Ba, Sm, Nd and Hf in Murchison and QUE 97008, *Geochim. Cosmochim. Acta 75*, 7806–7828.

[23] Rudraswami, N. G., and J. N. Goswami (2007), ^{26}Al in chondrules from unequilibrated L chondrites: Onset and duration of chondrule formation in the early solar system, *Earth Planet. Sci. Lett.*, *257*, 231–244.

5. MAC 87302—L4 CHONDRITE

[24] Bischoff, A., E. R. D. Scott, K. Metzler, and C. A. Goodrich (2006), Nature and origins of meteoritic breccias, in *Meteorites and the Early Solar System II*, edited by D. S. Lauretta and H. Y. McSween Jr., pp. 679–712, University of Arizona Press, Tucson.

[25] Rubin A. E., G. J. Taylor, E. R. D. Scott, and K. Keil (1982), Petrologic insights into the fragmentation history of asteroids, *Workshop on Lunar Breccias and Soils and Their Meteoritic Analogs. LPI Technical Rpt. No. 82–20*, pp. 107–110.

[26] Welzenbach, L. C., T. J. McCoy, A. Grimberg, and R. Wieler (2005), Petrology and noble gases of the regolith breccia MAC 87302 and implications for the classification of Antarctic meteorites, *Lunar Planet. Sci. Conf. 36*(1425).

6. ALH 85017—L6 CHONDRITE

[27] Cintala, M. J., and F. Hörz (2008), Experimental impacts into chondritic targets: Part I. Disruption of an L6 chondrite by multiple impacts, *Meteorit. Planet. Sci.*, *43*, 771–803.

[28] Horz, F., Cintala, M. J., See, T. H., and Le, L. (2005), Shock melting of ordinarychondrite powders and implications for asteroidal regoliths, *Meteorit. Planet. Sci.*, *40*, 1329–1346.

7. ALH 78003—L6 CHONDRITE (WITH SHOCK MELT VEINS)

[29] Ohtani, E., Y. Kimura, M. Kimura, T. Kubo, and T. Takata (2006), High-pressure minerals in shocked L6-chondrites: Constraints on impact conditions, *Shock Waves*, *16*, 45–52.

[30] Miyahara, M., A. El Goresy, E. Ohtani, M. Kimura, S. Ozawa, T. Nagase, and M. Nishijima (2009), Fractional crystallization of olivine melt inclusion in shock-induced chondritic melt vein, *Physics of the Earth and Planetary Interiors*, *177*, 116–121.

8. PAT 91501—L CHONDRITE IMPACT MELT

[31] Benedix, G. K., R. A. Ketcham, L. Wilson, T. J. McCoy, D. D. Bogard, D. H. Garrison, G. F. Herzog, S. Xue, J. Klein, and R. Middleton (2008), The formation and chronology of the PAT 91501 impact-melt L chondrite with vesicle-metal-sulfide assemblages, *Geochim. Cosmochim. Acta*, *72*, 2417–2428.

[32] Mittlefehldt, D. W., and M. M. Lindstrom (2001), Petrology and geochemistry of Patuxent Range 91501, a clast-poor impact melt from the L-chondrite parent body and Lewis Cliff 88763, and L7 chondrite, *Meteorit. Planet. Sci.*, *36*, 439–457.

[33] Swindle, T. D., D. A. Kring, J. Bond, E. Olson, and C. Jones (2005), Petrological and Ar-Ar studies of shocked chondrites, *Meteorit. Planet. Sci.*, *40*, 5295.

9. QUE 90201—LL5 CHONDRITE STREWNFIELD

[34] Welten, K. C., M. W. Caffee, D. J. Hillegonds, T. J. McCoy, J. Masarik, K. Nishiizumi (2011), Cosmogenic radionuclides in L5 and LL5 chondrites from Queen Alexandra Range, Antarctica: Identification of a large L/LL5 chondrite shower with a preatmospheric mass of approximately 50,000 kg, *Meteorit. Planet. Sci.*, *46*, 177–196.

10. LAP 04757—UNGROUPED CHONDRITE

[35] Troiano, J., D. Rumble, M. L. Rivers, and J. M. Friedrich (2011), Compositions of three low-FeO ordinary chondrites: Indications of a common origin with the H chondrites, *Geochim. Cosmochim. Acta*, *75*, 6511–6519.

[36] Russell, S. S., T. J. McCoy, E. Jarosewich, and R. D. Ash (1990), The Burnwell, Kentucky, Low-FeO chondrite fall: Description, classification and origin, *Meteorit. Planet. Sci.*, *33*, 853–856.

11. QUE 97990—CM2 CHONDRITE

[37] Maeda, M., and K. Tomeoka (2008), Chondrules and rims in the least aqueously altered CM chondrite QUE 97990: Is this meteorite a primary accretionary rock? *Meteorit. Planet. Sci.*, *43*(Supp.), 5027.

[38] Howard, K. T., G. K. Benedix, P. A. Bland, and G. Cressey (2011), Modal mineralogy of CM chondrites by X-ray diffraction (PSD-XRD): Part 2. Degree, nature and settings of aqueous alteration, *Geochim. Cosmochim. Acta*, *75*, 2735–2751.

[39] Rubin, A. E. (2007), Petrography of refractory inclusions in CM2.6 QUE 97990 and the origin of melilite-free spinel inclusions in CM chondrites, *Meteorit. Planet. Sci.*, *42*, 1711–1726.

[40] Rubin, A. E., J. M. Trigo-Rodriguez, H. Huber, and J. T. Wasson (2007), Progressive aqueous alteration of CM carbonaceous chondrites, *Geochim. Cosmochim. Acta*, *71*, 2361–2382.

[41] Trigo-Rodriguez, J. M., A. E. Rubin, and J. T. Wasson (2006), Non-nebular origin of dark mantles around chondrules and inclusions in CM chondrites, *Geochim. Cosmochim. Acta*, *70*, 1271–1290.

12. ALH A81002—CM2 CHONDRITE

[42] Brearley, A. J. (2006), The action of water, in *Meteorites and the Early Solar System II*, edited by D. S. Lauretta and H. Y. McSween Jr., pp. 587–624, University of Arizona Press, Tucson.

[43] Chizmadia, L. J., and A. J. Brearley (2008), Mineralogy, aqueous alteration, and primitive textural characteristics of fine-grained rims in the Y-791198 CM2 carbonaceous chondrite: TEM observations and comparison to ALH A81002, *Geochim. Cosmochim. Acta, 72*, 602–625.

[44] Hanowski, N. P. and A. J. Brearley (2000), Iron-rich aureoles in the CM carbonaceous chondrites, Murray, Murchison and Allan Hills 81002: Evidence for in situ aqueous alteration, *Meteorit. Planet. Sci., 35*, 1291–1308.

[45] Hanowski, N. P., and A. J. Brearley (2001), Aqueous alteration of chondrules in the CM carbonaceous chondrite, Allan Hills 81002: Implications for parent body alteration, *Geochim. Cosmochim. Acta, 65*, 495–518.

[46] Hiroi, T., M. E. Zolensky, C. M. Pieters, and M. E. Lipschutz (1996), Thermal metamorphism of the C, G, B, and F asteroids seen from the 0.7 micron, 3 micron and UV absorption strengths in comparison with carbonaceous chondrites, *Meteorit. Planet. Sci., 31*, 321–327.

[47] Howard, K. T., G. K. Benedix, P. A. Bland, and G. Cressey (2009), Modal mineralogy of CM2 chondrites by X-ray diffraction (PSD-XRD): Part 1. Total phyllosilicate abundance and the degree of aqueous alteration, *Geochim. Cosmochim. Acta, 73*, 4576–4589.

[48] Hua, X., J. Wang, and P. R. Buseck (2002), Fine-grained rims in the ALH 81002 and LEW 90500 CM2 meteorites: Their origin and modification, *Meteorit. Planet. Sci., 37*, 229–244.

[49] Lauretta, D. S., X. Hua, and P. R. Buseck (2000), Mineralogy of fine-grained rims in the ALH A81002 CM chondrite, *Geochim. Cosmochim. Acta, 64*, 3263–3273.

[50] Zolensky, M. E., D. W. Mittlefehldt, M. E. Lipschutz, M. S. Wang, R. N. Clayton, T. K. Mayeda, M. M. Grady, C. Pillinger, and B. David (1997), CM chondrites exhibit the complete petrologic range from type 2 to 1, *Geochim. Cosmochim. Acta, 61*, 5099–5115.

13. ALH 83100—CM1/2 CHONDRITE

[51] Busemann H., C. M. O'D. Alexander, and L. R. Nittler (2007), Characterization of insoluble organic matter in meteorites by Raman spectroscopy, *Meteorit. Planet. Sci., 42*, 1387–1416.

[52] de Leuw, S., A. E. Rubin, and J. T. Wasson (2010), Carbonates in CM chondrites: Complex formational histories and comparison to carbonates in CI chondrites, *Meteorit. Planet. Sci., 45*(4), 513–530.

[53] de Leuw, S., A. E. Rubin, A. K. Schmitt, and J. T. Wasson (2009), Mn-53-Cr-53 systematics of carbonates in CM chondrites: Implications for the timing and duration of aqueous alteration, *Geochim. Cosmochim. Acta, 73*, 7433–7442.

[54] Fehr, M. A., M. Rehkämper, A. N. Halliday, U. Wiechert, B. Hattendorf, D. Günther, S. Ono, J. L. Eigenbrode, and D. Rumble III (2005), Tellurium isotopic composition of the early solar system: A search for effects resulting from stellar nucleosynthesis, ^{126}Sn decay, and mass-independent fractionation, *Geochim. Cosmochim. Acta, 69*, 5099–5112.

[55] Friedrich, J. M., M. S. Wang, and M. E. Lipschutz (2002), Comparison of the trace element composition of Tagish Lake with other primitive carbonaceous chondrites, *Meteorit. Planet. Sci., 37*, 677–686.

[56] Glavin, D. P., J. P. Dworkin, A. Aubrey, O. Botta, J. H. Doty, Z. Martins, J. L. Bada (2006), Amino acid analyses of Antarctic CM2 meteorites using liquid chromatography-time of flight-mass spectrometry, *Meteoritics, 41*, 889–902.

[57] Hiroi, T., M. E. Zolensky, C. M. Pieters, and M. E. Lipschutz (1996), Thermal metamorphism of the C, G, B, and F asteroids seen from the 0.7 micron, 3 micron and UV absorption strengths in comparison with carbonaceous chondrites, *Meteorit. Planet. Sci., 31*, 321–327.

[58] Howard, K. T., G. K. Benedix, P. A. Bland, G. Cressey (2011), Modal mineralogy of CM chondrites by X-ray diffraction (PSD-XRD): Part 2. Degree, nature and settings of aqueous alteration, *Geochim. Cosmochim. Acta, 75*, 2735–2751.

[59] Martins, Z., C. M. O'D. Alexander, G. E. Orzechowska, M. L. Fogel, and P. Ehrenfreund (2007), Indigenous amino acids in primitive CR meteorites, *Meteorit. Planet. Sci., 42*, 2125–2136.

[60] Nazarov, M. A., G. Kurat, F. Brandstaetter, T. Ntaflos, M. Chaussidon, and P. Hoppe (2009), Phosphorus-bearing sulfides and their associations in CM chondrites, *Petrology, 17*, 101–123.

[61] Russell, S. S., J. W. Arden, and C. T. Pillinger (1996), The carbon and nitrogen isotopic composition of chondritic diamond, *Meteorit. Planet. Sci., 31*, 343–355.

[62] Yokoyama, T., C. M. O'D. Alexander, and R. J. Walker (2011), Assessment of nebular versus parent body processes on presolar components present in chondrites: Evidence from osmium isotopes, *Earth Planet. Sci. Lett., 305*, 115–123.

[63] Zolensky, M. E., D. W. Mittlefehldt, M. E. Lipschutz, M. S. Wang, R. N. Clayton, T. K. Mayeda, M. M. Grady, C. Pillinger, and D. Barber (1997), CM chondrites exhibit the complete petrologic range from type 2 to 1, *Geochim. Cosmochim. Acta, 61*, 5099–5115.

14. MET 01070—CM1 CHONDRITE

[64] Botta, O., Z. Martins, and P. Ehrenfreund (2007), Amino acids in Antarctic CM1 meteorites and their relationship to other carbonaceous chondrites, *Meteorit. Planet. Sci. 42*, 81–92.

[65] Busemann H., C. M. O'D. Alexander, and L. R. Nittler (2007), Characterization of insoluble organic matter in meteorites by Raman spectroscopy, *Meteorit. Planet. Sci., 42*, 1387–1416.

[66] Howard, K. T., G. K. Benedix, P. A. Bland, and G. Cressey (2011), Modal mineralogy of CM chondrites by X-ray diffraction (PSD-XRD): Part 2. Degree, nature and settings of aqueous alteration, *Geochim. Cosmochim. Acta, 75,* 2735–2751.

[67] de Leuw, S., A. E. Rubin, A. K. Schmitt, and J. T. Wasson (2009), Mn-53-Cr-53 systematics of carbonates in CM chondrites: Implications for the timing and duration of aqueous alteration, *Geochim. Cosmochim. Acta, 73,* 7433–7442.

[68] Moriarty, G. M., D. Rumble, III, and J. M. Friedrich (2009), Compositions of four unusual CM or CM-related Antarctic chondrites, *Chemie der Erde, 69,* 161–168.

[69] Rubin, A. E., J. M. Trigo-Rodriguez, H. Huber, and J. T. Wasson (2007), Progressive aqueous alteration of CM carbonaceous chondrites, *Geochim. Cosmochim. Acta, 71,* 2361–2382.

[70] Trigo-Rodriguez, J. M., A. E. Rubin, and J. T. Wasson (2006), Non-nebular origin of dark mantles around chondrules and inclusions in CM chondrites, *Geochim. Cosmochim. Acta, 70,* 1271–1290.

[71] Yokoyama, T., C. M. O'D. Alexander, R. J. Walker (2011), Assessment of nebular versus parent body processes on presolar components present in chondrites: Evidence from osmium isotopes, *Earth Planet. Sci. Lett., 305,* 115–123.

15. ALH A77307—CO3 CHONDRITE

[72] Busemann, H., C. M. O'D. Alexander, and L. R. Nittler (2007), Characterization of insoluble organic matter in meteorites by Raman spectroscopy, *Meteorit. Planet. Sci., 42,* 1387–1416.

[73] Biswas, S., T. Walsh, H. Ngo, and M. Lipschutz (1981), Trace element contents of selected Antarctic meteorites: II. Comparison with non-Antarctic specimens, *Proc. 6th Symp. Ant. Met.* (T. Nagata, editor), 221–228.

[74] Bonal, L., M. Bourot-Denise, E. Quirico, G. Montagnac, and E. Lewin (2008), Organic matter and metamorphic history of CO chondrites, *Geochim. Cosmochim. Acta, 71,* 1605–1623.

[75] Brearley, A. J. (1993), Matrix and fine-grained rims in the unequilibrated CO3 chondrite, ALH A77307: Origins and evidence for diverse, primitive nebular dust components, *Geochim. Cosmochim. Acta, 57,* 1521–1550.

[76] Brearley, A. J., S. Bajt, and S. R. Sutton (1993), Distribution of moderately volatile trace elements in fine-grained chondrule rims in the unequilibrated CO3 chondrite, ALH A77307, *Geochim. Cosmochim. Acta, 59,* 4307–4316.

[77] Chizmadia, L. J., A. E. Rubin, and J. T. Wasson (2002), Mineralogy and petrology of amoeboid olivine inclusions in CO3 chondrites: Relationship to parent-body aqueous alteration, *Meteorit. Planet. Sci., 37,* 1781–1796.

[78] Gibson, E. K., and M. Andrawes (1980), The Antarctic environment and its effect upon the total carbon and sulfur abundances in recovered meteorites, *Lunar Planet. Sci. Conf., 11,* Proceedings 2, 1223–1234.

[79] Grossman, J. N., and A. J. Brearley (2005), The onset of metamorphism in ordinary and carbonaceous chondrites, *Meteorit. Planet. Sci., 40,* 87–122.

[80] Huss, G. R., and R. S. Lewis (1995), Presolar diamond, SiC, and graphite in primitive chondrites: abundances as a function of meteorite class and petrologic type, *Geochim. Cosmochim. Acta, 59,* 115–160.

[81] Huss, G. R., A. P. Meshik, J. S. Smith, and C. M. Hohenburg (2003), Presolar diamond, silicon carbide, and graphite in carbonaceous chondrites: Implications for thermal processing in the solar nebula, *Geochim. Cosmochim. Acta, 6,* 4823–4848.

[82] Huss, G. (2004), Implications of isotopic anomalies and presolar grains for the formation of the early solar system, *Antarct. Meteor. Res., 17,* 132–152.

[83] Ikeda, Y. (1984), Alteration of chondrules and matrices in the four Antarctic carbonaceous chondrites ALH-77307 (C3), Y-790123 (C2), Y-75293 (C2), and Y-74662 (C2), *Mem. Natl. Inst. Polar Res., 30,* 93–108.

[84] Jones, R. H. (1992), On the relationship between isolated and chondrule olivine grains in the carbonaceous chondrite ALH A77307, *Geochim. Cosmochim. Acta, 56,* 467–482.

[85] Jones, R. H. (1993), Effect of metamorphism on isolated olivine grains in CO3 chondrites, *Geochim. Cosmochim. Acta, 57,* 2853–2867.

[86] Jones, R. H., J. M. Saxton, I. C. Lyon, and G. Turner (2000), Oxygen isotopes in chondrule olivine and isolated olivine grains from the CO3 chondrite, Allan Hills A77307, *Meteorit. Planet. Sci., 35,* 849–857.

[87] Kallemeyn, G. W., and J. T. Wasson (1982), The compositional classification of chondrites: III. Ungrouped carbonaceous chondrites, *Geochim. Cosmochim. Acta, 46,* 2217–2228.

[88] Keck, B. D., and Sears, D. W. G. (1987), Chemical and physical studies of type 3 chondrites: VIII. Thermoluminescence and metamorphism in the CO chondrites, *Geochim. Cosmochim. Acta, 51,* 3013–3021.

[89] Moore, C., J. Cronin, S. Pizzarello, M.-S. Ma, and R. Schmitt (1981), New analyses of Antarctic carbonaceous chondrites, *Proc. 6th Symp. Ant. Met.* 29–32.

[90] Murae, T., A. Masuda, and T. Takahashi (1984), Pyrolytic studies of organic components in Antarctic carbonaceous chondrites Y-74662 and ALH-77307, *Mem. Natl. Inst. Polar Res., 35,* 250–259.

[91] Nagahara, H., and I. Kushiro (1982), Petrology of chondrules, inclusions and isolated olivine grains in ALH-77307 (CO3), chondrite, *Mem. Natl. Inst. Polar Res., 25,* 66–77.

[92] Nguyen, A. N., F. J. Stadermann, E. Zinner, R. M. Stroud, C. M. O'D. Alexander, and L. R. Nittler (2007), Characterization of presolar silicate and oxide grains in primitive carbonaceous chondrites, *Astrophys. J., 65,* 1223–1240.

[93] Rubin, A. E. (1989), Size-frequency distributions of chondrules in CO3 chondrites, *Meteoritics, 24,* 179–189.

[94] Sears, D. W. G., and M. Ross (1983), Classification of the Allan Hills A77307 meteorite, *Meteoritics, 18,* 1–7.

16. LEW 85332—UNGROUPED CHONDRITE

[95] Brearley, A. J. (1997), Phyllosilicates in the matrix of the unique carbonaceous chondrite Lewis Cliff 85332 and possible implications for the aqueous alteration of CI chondrites, *Meteorit. Planet. Sci.*, *32*, 377–388.

[96] Rubin, A. E., and G. W. Kallemeyn (1990), Lewis Cliff 85332: A unique carbonaceous chondrite, *Meteoritics*, *25*, 215–225.

[97] Tonui, E., M. E. Zolensky, and M. E. Lipschutz (2002), Petrography, mineralogy, and trace element chemistry of Yamato-86029, Yamato-793321, and Lewis Cliff 85332: Aqueous alteration and heating events, *Ant. Met. Res.*, *15*, 38–58.

[98] Wasson, J. T., G. W. Kallemeyn, and A. E. Rubin (2000), Chondrules in the LEW 85332 ungrouped carbonaceous chondrite: Fractionation processes in the solar nebula, *Geochim. Cosmochim. Acta*, *64*, 1279–1290.

17. MAC 87300,301—UNGROUPED CHONDRITE

[99] Russell, S. S., A. M. Davis, G. J. MacPherson, Y. Guan, and G. R. Huss (2000), Refractory inclusions from the ungrouped carbonaceous chondrites MAC 87300 and MAC 88107, *Meteorit. Planet. Sci.*, *35*, 1051–1066.

[100] Wang, M.-S. and M. E. Lipschutz (1998), Thermally metamorphosed carbonaceous chondrites from data for thermally mobile trace elements, *Meteorit. Planet. Sci.* *33*, 1297–1302.

[101] Zolensky, M. E. (1991), Mineralogy and matrix composition of "CR" chondrites Renazzo and EET 87770, and ungrouped chondrites Essebi and MAC 87300, *Meteoritics*, *26*, 414.

[102] Zolensky, M., R. Barrett, and L. Browning (1993), Mineralogy and composition of matrix and chondrule rims in carbonaceous chondrites, *Geochim. Cosmochim. Acta*, *57*, 3123–3148.

18. MAC 88107—UNGROUPED CHONDRITE

[103] Clayton, T. N., and T. K. Mayeda (1999), Oxygen isotope studies of carbonaceous chondrites, *Geochim. Cosmochim. Acta*, *63*, 2089–2104.

[104] Elsila, J. E., N. P. de Leon, P. R. Buseck, and R. N. Zare (2005), Alkylation of polycyclic aromatic hydrocarbons in carbonaceous chondrites, *Geochim. Cosmochim. Acta*, *69*, 1349–1357.

[105] Friedrich, J. M., M. S. Wang, and M. E Lipschutz (2002), Comparison of the trace element composition of Tagish Lake with other primitive carbonaceous chondrites, *Meteorit. Planet. Sci.*, *37*, 677–688.

[106] Krot, A. N., I. D. Hutcheon, A. J. Brearley, O. V. Pravdivtseva, M. I. Petaev, and C. M. Hohenberg (2006), Timescales and settings for alteration of chondritic meteorites, in *Meteorites and the Early Solar System II*, edited by D. S. Lauretta and H. Y. McSween Jr., pp. 525–553, University of Arizona Press, Tucson.

[107] Krot, A. N., A. J. Brearley, M. I. Petaev, G. W. Kallemeyn, D. W. G. Sears, P. H. Benoit, I. D. Hutcheon, M. E. Zolensky, and K. Keil (2000), Evidence for low-temperature growth of fayalite and hedenbergite in MacAlpine Hills 88107, an ungrouped carbonaceous chondrite related to the CM-CO clan, *Meteorit. Planet. Sci.*, *35*, 1365–1386.

[108] Russell, S. S., A. M. Davis, G. J. MacPherson, Y. Guan, and G. R. Huss (2000), Refractory inclusions from the ungrouped carbonaceous chondrites MAC 87300 and MAC 88107, *Meteorit. Planet. Sci.*, *35*, 1051–1066.

19. ALH 84028—CV3 CHONDRITE (OXIDIZED)

[109] Busemann, H., C. O'D. Alexander, and L. R. Nittler (2007), Characterization of insoluble organic matter in primitive meteorites by microRaman spectroscopy, *Meteorit. Planet. Sci.*, *42*, 1387–1416.

[110] Fehr, M. A., M. Rehkämper, A. N. Halliday, U. Wiechert, B. Hattendorf, D. Günther, S. Ono, J. L. Eigenbrode, and D. Rumble (2005), Tellurium isotopic composition of the early solar system: A search for effects resulting from stellar nucleosynthesis, ^{126}Sn decay, and mass-independent fractionation, *Geochim. Cosmochim. Acta*, *69*, 5099–5112.

[111] Guimon, R. K., S. J. K. Symes, D. W. G. Sears, and P. H. Benoit (1995), Chemical and physical studies of type 3 chondrites: XII. The metamorphic history of CV chondrites and their components, *Meteoritics*, *30*, 704–715.

[112] Krot, A. N., M. I. Petaev, E. R. D. Scott, B. G. Choi, M. E. Zolensky, and K. Keil (1998), Progressive alteration in CV3 chondrites: More evidence for asteroidal alteration, *Meteorit. Planet. Sci.*, *33*, 1065–1085.

[113] Krot, A. N., M. I. Petaev, M. E. Zolensky, K. Keil, E. R. D. Scott, and K. Nakamura (1998), Secondary Ca-Fe-rich minerals in the Bali-like and Allende-like oxidized CV3 chondrites and Allende dark inclusions, *Meteorit. Planet. Sci.*, *33*, 623–645.

[114] Krot, A. N., M. I. Petaev, and P. A. Bland (2004), Multiple formation mechanisms of ferrous olivine in CV carbonaceous chondrites during fluid-assisted metamorphism, *Antarctic Meteorite Research*, *17*, 153–175.

[115] Paque, J. M., G. E. Lofgren, and L. Le (2000), Crystallization of calcium-aluminum-rich inclusions: Experimental studies on the effects of repeated heating events, *Meteorit. Planet. Sci.*, *35*, 363–371.

[116] Weisberg, M. K., and M. Prinz (1998), Fayalitic olivine in CV3 chondrite matrix and dark inclusions: A nebular origin, *Meteorit. Planet. Sci.*, *33*, 1087–1099.

20. EET 90007—CK5 CHONDRITE

[117] Elsila, J. E., N. P. de Leon, P. R. Buseck, and R. N. Zare (2005), Alkylation of polycyclic aromatic hydrocarbons in carbonaceous chondrites, *Geochim. Cosmochim. Acta*, *69*, 1349–1357.

[118] Geiger, T., and A. Bischoff (1995), Formation of opaque minerals in CK chondrites, *Planet. Space Sci.*, *43*, 485–498.

[119] Hiroi, T., C. M. Pieters, M. E. Zolensky, and M. E. Lipschutz (1994), Possible thermal metamorphism on the C, G, B, and F asteroids detected from their reflectance spectra in comparison with carbonaceous chondrites, *Proceedings of the NIPR Symposium on Antarctic Meteorites*, 7, 230–243.

[120] Huber, H., A. E. Rubin, G. W. Kallemeyn, and J. T. Wasson (2006), Siderophile-element anomalies in CK carbonaceous chondrites: Implications for parent-body aqueous alteration and terrestrial weathering of sulfides, *Geochim. Cosmochim. Acta*, 70, 4019–4037.

[121] Oura, Y., C. Takahashi, and M. Ebihara (2004), Boron and chlorine abundances in Antarctic chondrites: A PGA study, *Antarctic Meteorite Res.*, 17, 172.

[122] Righter, K., and K. E. Neff (2007), Temperature and oxygen fugacity constraints on CK and R chondrites and implications for water and oxidation in the early solar system, *Polar Science*, 1, 25–44.

[123] Rubin, A. E. (1993), Magnetite-sulfide chondrules and nodules in CK carbonaceous chondrites: Implications for the timing of CK oxidation, *Meteoritics*, 28, 30–135.

[124] Sugiura, N., and S. Zashu (1995), Nitrogen isotopic composition of CK chondrites, *Meteoritics*, 30, 430–435.

[125] Wang, M.-S., and M. E. Lipschutz (1998), Thermally metamorphosed carbonaceous chondrites from data for thermally mobile trace elements, *Meteorit. Planet. Sci.*, 33, 1297–1302.

21. QUE 94411 AND 94627—CB$_a$ CHONDRITE

[126] Campbell, A. J., M. Humayun, A. Meibom, A. N. Krot, and K. Keil (2001), Origin of zoned metal grains in the QUE94411 chondrite, *Geochim. Cosmochim. Acta*, 65, 163–180.

[127] Greshake, A., A. N. Krot, A. Meibom, M. K. Weisberg, M. E. Zolensky, and K. Keil (2002), Heavily-hydrated lithic clasts in CH chondrites and the related, metal-rich chondrites Queen Alexandra Range 94411 and Hammadah al Hamra 237, *Meteorit. Planet. Sci.*, 37, 281–293.

[128] Krot, A. N., K. D. McKeegan, S. S. Russell, A. Meibom, M. K. Weisberg, J. Zipfel, T. V. Krot, T. J. Fagan, and K. Keil (2001), Refractory calcium-aluminum-rich inclusions and aluminum-diopside-rich chondrules in the metal-rich chondrites Hammadah al Hamra 237 and Queen Alexandra Range 94411, *Meteorit. Planet. Sci.*, 36, 1189–1216.

[129] Meibom, A., Righter, N. Chabot, G. Dehn, A. Antignano, T. J. McCoy, A. N. Krot, M. E. Zolensky, M. I. Petaev, and K. Keil (2005), Shock melts in QUE 94411, Hammadah al Hamra 237, and Bencubbin: Remains of the missing matrix, *Meteorit. Planet. Sci.*, 40, 1377.

[130] Petaev, M. I., A. Meibom, A. N. Krot, J. A. Wood, and K. Keil (2001), The condensation origin of zoned metal grains in Queen Alexandra Range 94411: Implications for the formation of the Bencubbin-like chondrites, *Meteorit. Planet. Sci.*, 36, 93–106.

22. MIL 05082—CB$_b$ CHONDRITE

[131] Burton, A. S., J. E. Elsila, J. E. Hein, D. P. Glavin, and J. P. Dworkin (2013), Extraterrestrial amino acids identified in metal-rich CH and CB carbonaceous chondrites from Antarctica, *Meteorit. Planet. Sci. 48*, 390–402.

[132] Lauretta, D. S., J. S. Goreva, D. H. Hill, and M. Killgore (2007), Bulk compositions of the CB chondrites Bencubbin, Fountain Hills, MAC 02675, and MIL 05082, *Lunar Planet. Sci. Conf.*, 38, LPI Contribution No. 1338, 2236.

[133] Quitté, G., A. Markowski, C. Latkoczy, A. Gabriel, and A. Pack (2010), Iron-60 heterogeneity and incomplete isotope mixing in the early solar system, *Astrophys. J.*, 720, 1215–1224.

23. GRO 95551—UNGROUPED CHONDRITE

[134] Campbell, A. J., and M. Humayun (2003), Formation of metal in Grosvenor Mountains 95551 and comparison to ordinary chondrites, *Geochim. Cosmochim. Acta*, 67, 2481–2495.

[135] Sugiura, N., S. Zashu, M. K. Weisberg, and M. Prinz (2000), A nitrogen isotope study of bencubbinites, *Meteorit. Planet. Sci.*, 35, 987–996.

[136] Weisberg, M. K., M. Prinz, R. N. Clayton, T. K. Mayeda, N. Sugiura, S. Zashu, and M. Ebihara (2001), A new metal-rich chondrite group, *Meteorit. Planet. Sci.*, 36, 401–418.

[137] Weisberg, M. K., T. E. Bunch, J. H. Wittke, D. Rumble, and D. S. Ebel (2012), Petrology and oxygen isotopes of NWA 5492, a new metal-rich chondrite, *Meteorit. Planet. Sci.*, 47, 363–373.

24. ALH 85085—CH CHONDRITE

[138] Bischoff, A., H. Palme, L. Schultz, D. Weber, H. W. Weber, and B. Spettel (1993), Acfer 182 and paired samples, an iron-rich carbonaceous chondrite: Similarities with ALH 85085 and relationship to CR chondrites, *Geochim. Cosmochim. Acta*, 57, 2631–2648.

[139] Campbell, A. J., and M. Humayun (2004), Formation of metal in the CH chondrites ALH 85085 and PCA 91467, *Geochim. Cosmochim. Acta*, 68, 3409–3422.

[140] Eugster, O., and S. Niedermann (1990), Solar noble gases in the unique chondritic breccia Allan Hills 85085, *Earth and Planet. Sci. Letters*, 101, 139–147.

[141] Goldstein, J. I., R. H. Jones, P. G. Kotula, and J. R. Michael (2007), Microstructure and thermal history of metal particles in CH chondrites, *Meteorit. Planet. Sci.*, 42, 913–933.

[142] Gosselin, D. C., and J. C. Laul (1990), Chemical characterization of a unique chondrite: Allan Hills 85085, *Meteoritics*, 25, 81–87.

[143] Grady, M. M., and C. T. Pillinger (1990), ALH 85085: Nitrogen isotope analysis of a highly unusual primitive chondrite, *Earth Planet. Sci. Letters*, 97, 29–40.

[144] Grossman, J. N., A. E. Rubin, and G. J. MacPherson (1988), ALH 85085: A unique volatile-poor carbonaceous chondrite with possible implications for nebular fractionation processes, *Earth Planet. Sci. Letters, 91*, 33–54.

[145] Grossman, L. (2010), Vapor-condensed phase processes in the early solar system, *Meteorit. Planet. Sci., 45*, 7–20.

[146] Kimura, M., A. El Goresy, H. Palme, and E. Zinner (1993), Ca-,Al-rich inclusions in the unique chondrite ALH 85085: Petrology, chemistry and isotopic compositions, *Geochim. Cosmochim. Acta, 57*, 2329–2359.

[147] Kimura, M., T. Mikouchi, A. Suzuki, M. Miyahara, E. Ohtani, and A. El Goresy (2009), Kushiroite, CaAlAlSiO$_6$: A new mineral of the pyroxene group from the ALH 85085 CH chondrite, and its genetic significance in refractory inclusions, *American Mineralogist, 94*, 1479–1482.

[148] Scott, E. R. D. (1988), A new kind of primitive chondrite, Allan Hills 85085, *Earth Planet. Sci. Letters, 91*, 1–18.

[149] Wasson, J. T., and G. W. Kallemeyn (1990), Allan Hills 85085: A subchondritic meteorite of mixed nebular and regolithic heritage, *Earth Planet. Sci. Letters, 101*, 148–161.

[150] Weber, D., E. K. Zinner, and A. Bischoff (1994), An ion microprobe study of an osbornite-bearing inclusion from ALH 85085, *Meteoritics, 29*, 547–548.

[151] Weisberg, M. K., M. Prinz, and C. E. Nehru (1988), Petrology of ALH 85085: A chondrite with unique characteristics, *Earth Planet. Sci. Letters, 91*, 19–32.

[152] Weisberg, M. K., C. E. Nehru, M. Prinz (1990), The Bencubbin chondrite breccia and its relationship to CR chondrites and the ALH 85085 chondrite, *Meteoritics, 25*, 269–279.

[153] Weisberg, M. K., M. Prinz, R. N. Clayton, T. K. Mayeda, M. M. Grady, and C. T. Pillinger (1995), The CR chondrite clan, Proc. NIPR Symp. *Antarctic Meteorites, 8*, 11–32.

25. EET 92042—CR2 CHONDRITE

[154] Abreu, N. M., and A. J. Brearley (2005), HRTEM and EFTEM Studies of phyllosilicate-organic matter associations in matrix and dark inclusions in the EET 92042 CR2 carbonaceous chondrite, *Lunar Planet. Sci. Conf., 36*, 1826.

[155] Busemann, H., C. M. O'D. Alexander, and L. R. Nittler (2007), Characterization of insoluble organic matter in meteorites by microRaman spectroscopy, *Meteorit. Planet. Sci., 42*, 1387–1416.

[156] Busemann, H., A. F. Young, C. M. O'D. Alexander, P. Hoppe, S. Mukhopadhyay, L. R. Nittler (2006), Interstellar chemistry recorded in organic matter from primitive meteorites, *Science, 312*, 727–730.

[157] Chaussidon, M., G. Libourel, and A. N. Krot (2008), Oxygen isotopic constraints on the origin of magnesian chondrules and on the gaseous reservoirs in the early solar system, *Geochim. Cosmochim. Acta, 72*, 1924–1938.

[158] Cody, G., and C. M. O'D. Alexander (2005), NMR studies of chemical structural variation of insoluble organic matter from different carbonaceous chondrite groups, *Geochim. Cosmochim. Acta, 69*, 1085–1097.

[159] Cody, G. D., C. M. O'D. Alexander, H. Yabuta, A. L. D. Kilcoyne, T. Araki, H. Ade, P. Dera, M. Fogel, B. Militzer, and B. O. Mysen (2008), Organic thermometry for chondritic parent bodies, *Earth Planet. Sci. Lett., 272*, 446–455.

[160] Connolly, H. C., G. R. Huss, and G. J. Wasserburg (2001), On the formation of Fe-Ni metal in Renazzo-like carbonaceous chondrites, *Geochim. Cosmochim. Acta, 65*, 4567–4588.

[161] Glavin, D. P., and J. P. Dworkin (2009), Enrichment of the amino acid L-isovaline by aqueous alteration on CI and CM meteorite parent bodies, *Proc. Nat. Acad. Sci., 106*, 5487–5492.

[162] Horan, M. F., R. J. Walker, J. W. Morgan, J. N. Grossman, and A. E. Rubin (2003), Highly siderophile elements in chondrites, *Chem. Geol., 196*, 27–42.

[163] Hutcheon, I. D., K. K. Marhas, A. N. Krot, J. N. Goswami, and R. H. Jones (2009), [26]Al in plagioclase-rich chondrules in carbonaceous chondrites: Evidence for an extended duration of chondrule formation, *Geochim. Cosmochim. Acta, 73*, 5080–5099.

[164] Kong, P. (1999), Distribution of siderophile elements in CR chondrites: Evidence for evaporation and recondensation during chondrule formation, *Geochim. Cosmochim. Acta, 63*, 2637–2652.

[165] Krot, A. N., C. M. Hohenberg, A. P. Meshik, O. V. Pravdivtseva, H. Hiyagon, M. I. Petaev, M. K. Weisberg, A. Meibom, and K. Keil (2002), Two-stage asteroidal alteration of the Allende dark inclusions, *Meteorit. Planet. Sci., 37*(Suppl.), A82.

[166] Krot, A. N., A. Meibom, M. K. Weisberg, and K. Keil (2002), The CR chondrite clan: Implications for early solar system processes, *Meteorit. Planet. Sci., 37*, 1451.

[167] Krot, A. N., and K. Keil (2002), Anorthite-rich chondrules in CR and CH carbonaceous chondrites: Genetic link between Ca, Al-rich inclusions and ferromagnesian chondrules, *Meteorit. Planet. Sci., 37*, 91–111.

[168] Krot, A. N., G. Libourel, and Chaussidon (2006), Oxygen isotope compositions of chondrules in CR chondrites, *Geochim. Cosmochim. Acta, 70*, 767–779.

[169] Makide, K., K. Nagashima, A. N. Krot, G. R. Huss, I. D. Hutcheon, and A. Bischoff (2009), Oxygen- and magnesium-isotope compositions of calcium-aluminum-rich inclusions from CR2 carbonaceous chondrites, *Geochim. Cosmochim. Acta, 73*, 5018–5050.

[170] Martins, Z., C. M. O'D. Alexander, G. E. Orzechowska, M. L. Fogel, and P. Ehrenfreund (2007), Indigenous amino acids in primitive CR meteorites, *Meteorit. Planet. Sci. 42*, 2125–2136.

[171] Oura, Y., C. Takahashi, and M. Ebihara (2004), Boron and chlorine abundances in Antarctic chondrites: A PGA study, *Antartic Meteorite Research, 17*, 172.

[172] Pearson, V. K., M. A. Sephton, I. A. Franchi, J. M. Gibson, and I. Gilmour (2006), Carbon and nitrogen in carbonaceous chondrites: Elemental abundances and stable isotopic compositions, *Meteorit. Planet. Sci. 41*, 1899–1918.

[173] Wang, Y., Y. Huang, C. M. O'D. Alexander, M. Fogel, and G. Cody (2005), Molecular and compound-specific hydrogen isotope analyses of insoluble organic matter from different carbonaceous chondrite groups, *Geochim. Cosmochim. Acta*, *69*, 3711–3721.

[174] Yokoyama, T., V. K. Rai, C. M. O'D. Alexander, R. S. Lewis, R. W. Carlson, S. B. Shirey, M. H. Thiemens, and R. J. Walker (2007), Osmium isotope evidence for uniform distribution of s- and r-process components in the early solar system, *Earth Planet. Sci. Lett.* *259*, 567–580.

26. QUE 99177—CR2 CHONDRITE

[175] Abreu, N. M. (2013), A unique omphacite, amphibole, and graphite-bearing clast in Queen Alexandra Range (QUE), 99177: A metamorphosed xenolith in a pristine CR3 chondrite, *Geochim. Cosmochim. Acta*, *105*, 56–72.

[176] Abreu, N. M., and A. J. Brearley (2010), Early solar system processes recorded in the matrices of two highly pristine CR3 carbonaceous chondrites, MET 00426 and QUE 99177, *Geochim. Cosmochim. Acta*, *74*, 1146–1171.

[177] Floss, C., and F. Stadermann (2009), Auger Nanoprobe analysis of presolar ferromagnesian silicate grains from primitive CR chondrites QUE 99177 and MET 00426, *Geochim. Cosmochim. Acta*, *73*, 2415–2440.

[178] Floss, C., and F. Stadermann (2009), High abundances of circumstellar and interstellar C-anomalous phases in the primitive CR3 chondrites QUE 99177 and MET 00426, *Astrophys. J.*, *697*, 1242–1255.

[179] Glavin, D. P., and J. P. Dworkin (2009), Enrichment of the amino acid L-isovaline by aqueous alteration on CI and CM meteorite parent bodies, *Proc. Nat. Acad. Sci.*, *106*, 5487–5492.

27. GRO 95577—CR1 CHONDRITE

[180] Alexander, M., H. Fogel, Yabuta, and G. D. Cody (2007), The origin and evolution of chondrites recorded in the elemental and isotopic compositions of their macromolecular organic matter, *Geochim. Cosmochim. Acta*, *71*, 4380–4403.

[181] Perronnet, M., and M. E. Zolensky (2006), Characterization and quantification of metallic and mineral phases in the highly hydrated Grosvenor Mountains 95577 CR1 chondrite, 37th Annual *Lunar Planet. Sci. Conf.*, abstract 2402.

[182] Busemann, H., C. M. O'D. Alexander, and L. R. Nittler (2007), Characterization of insoluble organic matter in meteorites by microRaman spectroscopy, *Meteorit. Planet. Sci.*, *42*, 1387–1416.

[183] Busemann, H., A. F. Young, C. M. O'D. Alexander, P. Hoppe, S. Mukhopadhyay, and L. R. Nittler (2006), Interstellar chemistry recorded in organic matter from primitive meteorites, *Science*, *312*, 727–730.

[184] Glavin, D. P., and J. P. Dworkin (2009), Enrichment of the amino acid L-isovaline by aqueous alteration on CI and CM meteorite parent bodies, *Proc. Nat. Acad. Sci.*, *106*, 5487–5492.

[185] Martins, Z., C. M. O'D. Alexander, G. E. Orzechowska, M. L. Fogel, and P. Ehrenfreund (2007), Indigenous amino acids in primitive CR meteorites, *Meteorit. Planet. Sci.*, *42*, 2125–2136.

[186] Morlok, A., and G. Libourel (2013), Aqueous alteration in CR chondrites: Meteorite parent body processes as analogue for long-term corrosion processes relevant for nuclear waste disposal, *Geochim. Cosmochim. Acta*, *103*, 76–103.

[187] Weisberg, M. K., and H. Huber (2007), The GRO 95577 CR1 chondrite and hydration of the CR parent body, *Meteorit. Planet. Sci.*, *42*, 1495–1503.

[188] Weisberg, M. K., and M. Prinz (2000), The Grosvenor Mountains 95577 CR1 chondrite and hydration of the CR chondrites, *Meteorit. Planet. Sci.*, *35*(Suppl.), A168.

28. GRO 95517—EH3 CHONDRITE

[189] Guan, Y., G. R. Huss, G. J. MacPherson, and G. J. Wasserburg (2000), Calcium-aluminum-rich inclusions from enstatite chondrites: Indigenous or foreign? *Science*, *289*, 1330–1333.

[190] Guan, Y., K. D. McKeegan, and G. J. MacPherson (2000), Oxygen isotopes in calcium-aluminum-rich inclusions from enstatite chondrites: New evidence for a single CAI source in the solar nebula, *Earth Planet. Sci. Lett.*, *181*, 271–277.

[191] Patzer, A., and L. Schultz (2001), Noble gases in enstatite chondrites I: Exposure ages, pairing, and weathering effects, *Meteorit. Planet. Sci.*, *36*, 947–961.

[192] Patzer, A., and L. Schultz (2002), Noble gases in enstatite chondrites II: The trapped component, *Meteorit. Planet. Sci.*, *37*, 601–612.

29. ALH 81189—EH3 CHONDRITE

[193] Ebata, S., K. Nagashima, S. Itoh, S. Kobayashi, N. Sakamoto, T. J. Fagan, and H. Yurimoto (2006), Presolar silicate grains in enstatite chondrites, *Lunar Planet. Sci.*, *37*, abstract 1619.

[194] Ebata, S., T. J. Fagan, and H. Yurimoto (2008), Identification of silicate and carbonaceous presolar grains by SIMS in the type-3 enstatite chondrite ALH A81189, *Applied Surface Science*, *255*, 1468–1471.

[195] Fagan, T. J., A. N. Krot, and K. Keil (2000), Calcium-aluminum-rich inclusions in enstatite chondrites, *Meteorit. Planet. Sci.*, *35*, 771–781.

[196] Fagan, T. J., S. Kataoka, A. Yoshida, and K. Norose (2010), Transition to low oxygen fugacities in the solar nebula recorded by EH3 chondrite ALH A81189, *Lunar Planet. Sci. Conf.*, *41*, LPI Contribution No. 1533, p. 1534.

30. PCA 91020—EL3 CHONDRITE

[197] Izawa, M. R. M., R. L. Flemming, N. R. Banerjee, and P. J. A. McCausland (2011), Micro-X-ray diffraction assessment of shock stage in enstatite chondrites, *Meteorit. Planet. Sci.*, *46*, 638–651.

[198] Kimura, M., M. K. Weisberg, Y. Lin, A. Suzuki, E. Ohtani, and R. Okazaki (2005), Thermal history of the enstatite chondrites from silica polymorphs, *Meteorit. Planet. Sci.*, *40*, 855.

[199] Kong, P., T. Mori, and M. Ebihara (1997), Compositional continuity of enstatite chondrites and implications for heterogeneous accretion of the enstatite chondrite parent body, *Geochim. Cosmochim. Acta*, *61*, 4895.

[200] Rubin, A. E., E. R. D. Scott, and K. Keil (1997), Shock metamorphism of enstatite chondrites, *Geochim. Cosmochim. Acta*, *61*, 847–858.

[201] Rubin, A. E., H. Huber, and J. T. Wasson (2009), Possible impact-induced refractory-lithophile fractionations in EL chondrites, *Geochim. Cosmochim. Acta*, *73*, 1523–1537.

[202] Schneider, D. M., P. H. Benoit, A. Kracher, and D. W. G. Sears (2003), Metal size distributions in EH and EL chondrites, *Geophys. Res. Lett.*, *30*, 1420.

[203] Schneider, D. M., S. J. K. Symes, P. H. Benoit, and D. W. G.Sears (2002), Properties of chondrules in EL3 chondrites, comparison with EH3 chondrites, and the implications for the formation of enstatite chondrites, *Meteorit. Planet. Sci.*, *37*, 1401–1416.

[204] Zhang, Y., P. H. Benoit, and D. W. G. Sears (1995), The classification and complex thermal history of the enstatite chondrites, *J. Geophys. Res.*, *100*, 9417–9438.

31. MAC 88136—EL3 CHONDRITE

[205] Bouvier, A., J. D. Vervoort, and P. J. Patchett (2008), The Lu Hf and Sm Nd isotopic composition of CHUR: Constraints from unequilibrated chondrites and implications for the bulk composition of terrestrial planets, *Earth Planet. Sci. Lett.*, *273*, 48–57.

[206] Crozaz, G., and L. L. Lundberg (1995), The origin of oldhamite in unequilibrated enstatite chondrites, *Geochim. Cosmochim. Acta*, *59*, 3817–3831.

[207] Fagan, T. J., A. N. Krot, and K. Keil (2000), Calcium-aluminum-rich inclusions in enstatite chondrites, *Meteorit. Planet. Sci.*, *35*, 771–781.

[208] Guan, Y., G. R. Huss, and L. A. Leshin (2007), 60Fe 60Ni and ^{53}Mn ^{53}Cr isotopic systems in sulfides from unequilibrated enstatite chondrites, *Geochim. Cosmochim. Acta*, *71*, 4082–4091.

[209] Hsu, W., and G. Crozaz (1998), Mineral chemistry and the origin of enstatite in unequilibrated enstatite chondrites, *Geochim. Cosmochim. Acta*, *62*, 1993–2004.

[210] Hsu, W. (1998), Geochemical and petrographic studies of oldhamite, diopside, and roedderite in enstatite meteorites, *Meteorit. Planet. Sci.*, *33*, 291–301.

[211] Lin, Y., and M. Kimura (1998), Petrographic and mineralogical study of new EH melt rocks and a new enstatite chondrite grouplet, *Meteorit. Planet. Sci.*, *33*, 501–511.

[212] Lin, Y. T., H. Nagel, L. L. Lundberg, and A. El Goresy (1991), MAC 88136: The first EL3 chondrite, *Lunar Planet. Sci. Conf.*, *22*, 811.

[213] Lin, Y., and A. El Goresy (2002), A comparative study of opaque phases in Qingzhen (EH3), and MacAlpine Hills 88136 (EL3): Representatives of EH and EL parent bodies, *Meteorit. Planet. Sci.*, *37*, 577–599.

[214] Lundberg, L. L., E. Zinner, and G. Crozaz (1994), Search for isotopic anomalies in oldhamite (CaS), from unequilibrated (E3), enstatite chondrites, *Meteoritics*, *29*, 384–393.

[215] Nakashima, D., T. Nakamura, and R. Okazaki (2006), Cosmic-ray exposure age and heliocentric distance of the parent bodies of enstatite chondrites ALH 85119 and MAC 88136, *Meteoritics*, *41*, 851–862.

[216] Patzer, A., and L. Schultz (2001), Noble gases in enstatite chondrites I: Exposure ages, pairing, and weathering effects, *Meteorit. Planet. Sci.*, *36*, 947–961.

[217] Patzer, A., and L. Schultz (2002), Noble gases in enstatite chondrites II: The trapped component, *Meteorit. Planet. Sci.*, *37*, 601–612.

[218] Quirico, E., M. Bourot-Denise, C. Robin, G. Montagnac, and P. Beck (2011), A reappraisal of the metamorphic history of EH3 and EL3 enstatite chondrites, *Geochim. Cosmochim. Acta*, *75*, 3088–3102.

[219] Wadhwa, M., E. K. Zinner, and G. Crozaz (1997), Mn-Cr systematics in sulfides of unequilibrated enstatite chondrites, *Meteoritics*, *32*, 281–292.

[220] Weisberg, M. K., and M. Kimura (2012), The unequilibrated enstatite chondrites, *Chemie der Erde*, *72*, 101–115.

32. QUE 94368—EL4 CHONDRITE

[221] Rubin (1997), Sinoite (Si$_2$N$_2$O): Crystallization from EL chondrite impact melts. *Amer. Mineral.*, *82*, 1001–1006.

[222] Rubin, A. E., E. R. D. Scott, and K. Keil (1997), Shock metamorphism of enstatite chondrites, *Geochim. Cosmochim. Acta*, *61*, 847–858.

33. LAP 02225—E CHONDRITE IMPACT MELT

[223] Izawa, M. R. M., R. L. Flemming, N. R. Banerjee, and S. Matveev (2011), QUE 94204: A primitive enstatite achondrite produced by the partial melting of an E chondrite-like protolith, *Meteorit. Planet. Sci. 46*, 1742–1753.

34. QUE 94204—UNGROUPED ENSTATITE METEORITE

[224] Izawa, M. R. M., R. L. Flemming, N. R. Banerjee, and S. Matveev (2011), QUE 94204: A primitive enstatite achondrite produced by the partial melting of an E chondrite-like protolith, *Meteorit. Planet. Sci. 46*, 1742–1753.

[225] Lin, Y., and M. Kimura (1998), Petrographic and mineralogical study of new EH melt rocks and a new enstatite chondrite grouplet, *Meteorit. Planet. Sci.*, *3*, 501–511.

[226] Newton, J., I. A. Franchi, and C. T. Pillinger (2000), The oxygen isotope record in enstatite meteorites, *Meteorit. Planet. Sci.*, *35*, 689–698.

[227] Patzer, A., and L. Schultz (2002), Noble gases in enstatite chondrites II: The trapped component, *Meteorit. Planet. Sci.*, *37*, 601–612.

[228] Patzer, A., and L. Schultz (2001), Noble gases in enstatite chondrites I: Exposure ages, pairing, and weathering effects, *Meteorit. Planet. Sci.*, 36, 947–961.

[229] Rubin, A. E., E. R. D. Scott, and K. Keil (1997), Shock metamorphism of enstatite chondrites, *Geochim. Cosmochim. Acta*, 61, 847–858.

[230] Rubin, A . E., and E. R. D. Scott (1997), Abee and related EH chondrite impact-melt breccias, Geochim. Cosmochim. Acta, 61, 425–435.

[231] Weisberg, M. K., M. Prinz, and C. E. Nehru (1997), QUE 94204: an EH-chondritic melt rock, *Lunar Planet. Sci. Conf.*, 28, 525.

[232] van Niekerk, D., and K. Keil (2012), Anomalous enstatite meteorites Queen Alexandra Range 94204 and pairs: The perplexing question of impact melts or partial melt residues, either way, unrelated to Yamato 793225. *Lunar Planet. Sci. Conf.*, 43, LPI Contribution No. 1659, 2644.

35. PCA 91002—R CHONDRITE

[233] Dixon, E. T., D. D. Bogard, and D. H. Garrison (2003), ^{39}Ar -^{40}Ar chronology of R chondrites, *Meteorit. Planet. Sci.*, 38, 341–355.

[234] Greenwood, J. P., A. E. Rubin, and J. T. Wasson (2000), Oxygen isotopes in R-chondrite magnetite and olivine: Links between R chondrites and ordinary chondrites, *Geochim. Cosmochim. Acta*, 64, 3897–3911.

[235] Kallemeyn, G. W., A. E. Rubin, and J. T. Wasson (1996), The compositional classification of chondrites: VII. The R chondrite group, *Geochim. Cosmochim. Acta*, 60, 2243–2256.

[236] Rubin, A. E., and G. W. Kallemeyn (1994), Pecora Escarpment 91002: A member of the new Rumuruti (R), chondrite group, *Meteoritics*, 29, 255–264.

[237] Rubin, A. E., and G. W. Kallemeyn (1993), Carlisle Lakes chondrites: Relationship to other chondrite groups, *Meteoritics*, 28, 424.

[238] Rubin, A. E., and H. Huber (2005), A weathering index for CK and R chondrites, *Meteorit. Planet. Sci.*, 40, 1123.

[239] Sugiura, N., and S. Zashu (1995), Isotopic composition of nitrogen in the PCA 91002 R group chondrite, *Proc. NIPR Symp.*, 8, 273–285.

36. LAP 04840—R CHONDRITE

[240] Dyar, M. D., M. C. McCanta, A. H. Treiman, E. C. Sklute, G. J. Marchand (2007), Mössbauer spectroscopy and oxygen fugacity of amphibole-bearing R-chondrite LAP 04840, *Lunar Planet. Sci. Conf.*, 38, LPI Contribution No. 1338, 2047.

[241] McCanta, M. C., A. H. Treiman, M. D. Dyar, C. M. O.'D. Alexander, D. Rumble III, and E. J. Essene (2008), The LaPaz Icefield 04840 meteorite: Mineralogy, metamorphism, and origin of an amphibole- and biotite-bearing R chondrite, *Geochim. Cosmochim. Acta*, 72, 5757–5780.

[242] Mikouchi, T., K. Ota, J. Makishima, A. Monkawa, K. Sugiyama (2007), Mineralogy and crystallography of

LAP04840: Implications for metamorphism at depth in the R chondrite parent body, *Lunar Planet. Sci. Conf.*, 38, LPI Contribution No. 1338, 1928.

[243] Ota, K., T. Mikouchi, and K. Sugiyama (2009), Crystallography of hornblende amphilbole in LAP 04840 R chondrite and implications for its metamorphic history, *J. Mineral. Petrol. Sci.*, 104, 215–225.

[244] Righter, K., and K. E. Neff (2007), Temperature and oxygen fugacity constraints on CK and R chondrites and implications for water and oxidation in the early solar system, *Nat. Inst. Polar Sci.*, 1, 25–44.

37. LEW 87232—K CHONDRITE

[245] Kallemeyn, G. W. (1994), Compositional study of Kakangari and LEW 87232, *Meteoritics*, 29, 479–480.

[246] Krot, A., H. Yurimoto, K. McKeegan, L. Leshin, M. Chaussidon, G. Libourel, M. Yoshitake, G. Huss, Y. Guan, and B. Zanda (2006), Oxygen isotopic compositions of chondrules: Implications for evolution of oxygen isotopic reservoirs in the inner solar nebula, *Chemie der Erde*, 66, 249–276.

[247] Nagashima, K., A. N. Krot, and G. R. Huss (2012), Oxygen-isotope compositions of chondrules and matrix grains in the LEW 87232, Kakangari-like chondrite. *Lunar Planet. Sci. Conf.*, 43, LPI Contribution No. 1659, 1768.

[248] Weisberg, M. K., M. Prinz, R. N. Clayton, T. K. Mayeda, Grady, M. M., and I. A. Franchi (1993), Petrology and stable isotopes of LEW 87232, a new Kakangari-type chondrite, *Meteoritics*, 28, 458.

[249] Weisberg, M. K., M. Prinz, R. N. Clayton, T. K. Mayeda, M. M. Grady, I. Franchi, C. T. Pillinger, and G. W. Kallemeyn (1996), The K (Kakangari), chondrite grouplet, *Geochim. Cosmochim. Acta*, 60, 4253–4263.

38. QUE 94535—WINONAITE

[250] Benedix, G. K., T. J. McCoy, K. Keil, D. D. Bogard, and D. H. Garrison (1998), A petrologic and isotopic study of winonaites: Evidence for early partial melting, brecciation, and metamorphism, *Geochim. Cosmochim. Acta*, 62, 2535–2553.

[251] Benedix, G. K., T. J. McCoy, K. Keil, and S. G. Love (2000), A petrologic study of the IAB iron meteorites: Constraints on the formation of the IAB-Winonaite parent body, *Meteorit. Planet. Sci.*, 35, 1127–1141.

39. LEW 86220—ACAPULCOITE

[252] Crozaz, G., C. Floss, and M. Wadhwa (2007), Chemical alteration and REE mobilization in meteorites from hot and cold deserts, *Geochim. Cosmochim. Acta*, 67, 4727–4741.

[253] Floss, C. (2002), Complexities on the acapulcoite-lodranite parent body: Evidence from trace element distributions in silicate minerals, *Meteorit. Planet. Sci.*, 35, 1073–1085.

[254] McCoy, T. J., K. Keil, D. W. Muenow, and L. Wilson (1997), Partial melting and melt migration in the acapulcoite-lodranite parent body, *Geochim. Cosmochim. Acta*, *61*, 639–650.

[255] McCoy, T. J., K. Keil, R. N. Clayton, T. K. Mayeda, D. D. Bogard, D. H. Garrison, G. R. Huss, I. D. Hutcheon, and R. Wieler (1996), A petrologic, chemical, and isotopic study of Monument Draw and comparison with other acapulcoites: Evidence for formation by incipient partial melting, *Geochim. Cosmochim. Acta*, *60*, 2681–2708.

[256] Mittlefehldt, D. W., M. M. Lindstrom, D. D. Bogard, D. H. Garrison, and S. W. Field (1996), Acapulco- and Lodran-like achondrites: Petrology, geochemistry, chronology, and origin, *Geochim. Cosmochim. Acta*, *60*, 867–882.

[257] Patzer, A., D. H. Hill, and W. V. Boynton (2004), Evolution and classification of acapulcoites and lodranites from a chemical point of view, *Meteorit. Planet. Sci.*, *39*, 61–85.

[258] Rai, V. K., T. L. Jackson, M. H. Thiemens (2005), Photochemical mass-independent sulfur isotopes in achondritic meteorites, *Science*, *309*, 1062–1065.

[259] Rubin, A. E. (2007), Petrogenesis of acapulcoites and lodranites: A shock-melting model, *Geochim. Cosmochim. Acta*, *71*, 2383–2401.

40. ALH A77081—ACAPULCOITE

[260] Biswas, S., T. Walsh, H. Ngo, and M. Lipschutz (1981), Trace element contents of selected Antarctic meteorites: II. Comparison with non-Antarctic specimens, *Proc. Sixth Symp. Antarc. Meteor.*, 221–228.

[261] Hiroi, T., J. F. Bell, H. Takeda, and C. M. Pieters (1993), Modeling of S-type asteroid spectra using primitive achondrites and iron meteorites, *Icarus*, *102*, 107–116.

[262] Kallemeyn, G. W., and J. T. Wasson (1985), The compositional classification of chondrites: IV. Ungrouped chondritic meteorites and clasts, *Geochim. Cosmochim. Acta*, *49*, 261–270.

[263] McCoy, T. J., K. Keil, R. N. Clayton, T. K. Mayeda, D. D. Bogard, D. H. Garrison, G. R. Huss, I. D. Hutcheon, and R. Wieler (1996), A petrologic, chemical, and isotopic study of Monument Draw and comparison with other acapulcoites: Evidence for formation by incipient partial melting, *Geochim. Cosmochim. Acta*, *60*, 2681–2708.

[264] McCoy, T. J., K. Keil, D. W. Muenow, and L. Wilson (1997), Partial melting and melt migration in the acapulcoite-lodranite parent body, *Geochim. Cosmochim. Acta*, *61*, 639–650.

[265] Mittlefehldt, D. W., M. M. Lindstrom, D. D. Bogard, D. H. Garrison, and S. W. Field (1996), Acapulco- and Lodran-like achondrites: Petrology, geochemistry, chronology, and origin, *Geochim. Cosmochim. Acta*, *60*, 867–882.

[266] Rubin, A. E. (2007), Petrogenesis of acapulcoites and lodranites: A shock-melting model, *Geochim. Cosmochim. Acta*, *71*, 2383–2401.

[267] Schultz, L., H. Palme, B. Spettel, H. W. Weber, H. Wanke, M. C. Michel-Levy, and J. C. Lorin (1982), Allan Hills 77081: An unusual stony meteorite, *Earth Planet. Sci. Lett.*, *61*, 23–31.

[268] Weigel, A., O. Eugster, C. Koeberl, R. Michel, U. Krahenbuhl, and S. Neumann (1999), Relationships among lodranites and acapulcoites: Noble gas isotopic abundances, chemical composition, cosmic-ray exposure ages, and solar cosmic ray effects, *Geochim. Cosmochim. Acta*, *63*, 175–192.

[269] Zipfel, J., H. Palme, A. K. Kennedy, and I. D. Hutcheon (1995), Chemical composition and origin of the Acapulco meteorite, *Geochim. Cosmochim. Acta*, *59*, 3607–3627.

41. GRA 95209—LODRANITE

[270] Busemann, H., and O. Eugster (2002), The trapped noble gas component in achondrites, *Meteorit. Planet. Sci.*, *37*, 1865–1891.

[271] Crowther, S. A., J. A. Whitby, A. Busfield, G. Holland, H. Busemann, and J. D. Gilmour (2009), Collisional modification of the acapulcoite/lodranite parent body revealed by the iodine-xenon system in lodranites, *Meteorit. Planet. Sci. 44*, 1151–1159.

[272] Floss, C. (2000), Complexities on the acapulcoite-lodranite parent body: Evidence from trace element distributions in silicate minerals, *Meteorit. Planet. Sci.*, *35*, 1073–1085.

[273] Grew, E. S., M. G. Yates, R. J. Beane, C. Floss, and C. Gerbi (2007), Chopinite-sarcopside solid solution, $[(Mg,Fe)_3](PO_4)_2$, in GRA 95209, a transitional acapulcoite: Implications for phosphate genesis in meteorites, *American Mineralogist, 95*, 260–272.

[274] Lee, D.-C. (2008), [182]Hf-[182]W chronometry and the early evolution history in the acapulcoite-lodranite parent body, *Meteorit. Planet. Sci.*, *43*, 675–684.

[275] McCoy, T. J., W. D. Carlson, L. R. Nittler, R. M. Stroud, D. D. Bogard, and D. H. Garrison (2006), Graves Nunataks 95209: A snapshot of metal segregation and core formation, *Geochim. Cosmochim. Acta*, *70*, 516–531.

[276] Patzer, A., D. H. Hill, and W. V. Boynton (2004), Evolution and classification of acapulcoites and lodranites from a chemical point of view, *Meteorit. Planet. Sci.*, *39*, 61–85.

[277] Terribilini, D., Eugster, O., Herzog, G. F., and Schnabel, C. (2000), Evidence for common break-up events of the acapulcoites/lodranites and chondrites, *Meteorit. Planet. Sci.*, *35*, 1043–1050.

42. LEW 88280—LODRANITE

[278] Busemann, H., and O. Eugster (2007), The trapped noble gas component in achondrites, *Meteorit. Planet. Sci.*, *37*, 1865–1891.

[279] Crowther, S. A., J. A. Whitby, A. Busfield, G. Holland, H. Busemann, and J. D. Gilmour (2009), Collisional modification of the acapulcoite/lodranite parent body revealed by the iodine-xenon system in lodranites, *Meteorit. Planet. Sci.*, *44*, 1151–1159.

[280] Crozaz, G., C. Floss, and M. Wadhwa (2007), Chemical alteration and REE mobilization in meteorites from hot and cold deserts, *Geochim. Cosmochim. Acta*, *67*, 4727–4741.

[281] Floss, C. (2000), Complexities on the acapulcoite-lodranite parent body: Evidence from trace element distributions in silicate minerals, *Meteorit. Planet. Sci.*, *35*, 1073–1085.

[282] Mittlefehldt, D. W., M. M. Lindstrom, D. D. Bogard, D. H. Garrison, and S. W. Field (1996), Acapulco- and Lodran-like achondrites: Petrology, geochemistry, chronology, and origin, *Geochim. Cosmochim. Acta*, *60*, 867–882.

[283] McCoy, T. J., K. Keil, R. N. Clayton, T. K. Mayeda, D. D. Bogard, D. H. Garrison, and R. Wieler (1997), A petrologic and isotopic study of lodranites: Evidence for early formation as partial melt residues from heterogeneous precursors, *Geochim. Cosmochim. Acta*, *61*, 623–637.

[284] Patzer, A., D. H. Hill, and W. V. Boynton (2004), Evolution and classification of acapulcoites and lodranites from a chemical point of view, *Meteorit. Planet. Sci.*, *39*, 61–85.

[285] Rubin, A. E. (2007), Petrogenesis of acapulcoites and lodranites: A shock-melting model, *Geochim. Cosmochim. Acta*, *71*, 2383–2401.

43. LEW 88763—UNGROUPED ACHONDRITE

[286] Clayton, R. N., and T. K. Mayeda (1996), Oxygen isotope studies of achondrites, *Geochim. Cosmochim. Acta*, *60*, 1999–2017.

[287] Gardner-Vandy, K. G., T. J. McCoy, and D. S. Lauretta (2009), Formation conditions of FeO-rich primitive achondrites, *Lunar Planet. Sci. Conf.*, *40*, 2520.

[288] Greenwood, R. C., I. A. Franchi, J. A. Gibson, and G. K. Benedix (2007), Oxygen isotope composition of the primitive achondrites, *Lunar and Planetary Science*, *28*, 2163.

[289] Mittlefehldt, D. W., D. D. Bogard, J. L. Berkley, and D. H. Garrison (2003), Brachinites: Igneous rocks from a differentiated asteroid, *Meteorit. Planet. Sci.*, *38*, 1601–1625.

[290] Swindle, T. D., D. A. Kring, M. K. Burkland, D. H. Hill, and W. V. Boynton (1998), Noble gases, bulk chemistry, and petrography of olivine-rich achondrites Eagles Nest and LEW 88763: Comparison to brachinites, *Meteorit. Planet. Sci.*, *33*, 31–48.

44. ALH 78019—UREILITE

[291] Berkley, J. L., and J. H. Jones (1982), Primary igneous carbon in ureilites: Petrological implications, *Lunar Planet. Sci. Conf.*, *13*, A353–A364.

[292] Clayton, R. N., and T. K. Mayeda (1996), Oxygen isotope studies of achondrites, *Geochim. Cosmochim. Acta*, *60*, 1999–2017.

[293] Farquhar, J., T. L. Jackson, and M. H. Thiemens (2000), A [33]S enrichment in ureilite meteorites: Evidence for a nebular sulfur component, *Geochim. Cosmochim. Acta*, *64*, 1819–1825.

[294] Gibson, E. K., Jr., and F. F. Andrawes (1980), The Antarctic environment and its effect upon the total carbon and sulfur abundances recovered in meteorites, *Lunar Planet. Sci. Conf.*, *11*(12), 1223–1234.

[295] Goodrich, C. A., E. R. D. Scott, and A. M. Fioretti (2004), Ureilitic breccias: Clues to the petrologic structure and impact disruption of the ureilite parent asteroid, *Chemie der Erde*, *64*, 283–327.

[296] Goodrich, C. A. (1992), Ureilites: A critical review, *Meteoritics*, *27*, 327–352.

[297] Miyamoto, M., H. Takeda, and H. Toyoda (1985), Cooling history of some Antarctic ureilites, *J. Geophys. Res.*, *90*, D116–D122.

[298] Nakamuta, Y., and Y. Aoki (2000), Mineralogical evidence for the origin of diamond in ureilites, *Meteorit. Planet. Sci.*, *35*, 487–494.

[299] Okazaki, R., T. Nakamura, N. Takaoka, and K. Nagao (2003), Noble gases in ureilites released by crushing, *Meteorit. Planet. Sci.*, *38*, 767–781.

[300] Rai, V. K., S. V. S. Murty, and U. Ott (2002), Nitrogen in diamond-free ureilite Allan Hills 78019: Clues to the origin of diamond in ureilites, *Meteorit. Planet. Sci.*, *37*, 1045–1055.

[301] Rai, V. K., S. V. S. Murty, and U. Ott (2003), Noble gases in ureilites: Next term cosmogenic, radiogenic, and trapped components, *Geochim. Cosmochim. Acta*, *67*, 4435–4456.

[302] Rankenburg, K., A. D. Brandon, and M. Humayun (2007), Osmium isotope systematics of ureilites, *Geochim. Cosmochim. Acta*, *71*, 2402–2413.

[303] Rubin, A. E. (2006), Shock, post-shock annealing, and post-annealing shock in ureilites, *Meteoritics*, *41*, 125–133.

[304] Spitz, A. H., and W. V. Boynton (1991), Trace element analysis of ureilites: New constraints on their petrogenesis, *Geochim. Cosmochim. Acta*, *55*, 3417–3430.

[305] Sinha, S. K., R. O. Sack, and M. E. Lipschutz (1997), Ureilite meteorites: Equilibration temperatures and smelting reactions, *Geochim. Cosmochim. Acta*, *61*, 4235–4242.

[306] Takeda, H., H. Mori, K. Yanai, and K. Shiraishi (1980), Mineralogical examination of the Allan Hills achondrites and their bearing on the parent bodies, *Mem. Natl. Inst. Polar Res.*, *17*(Spec. Issue), 119–144.

[307] Toyoda, H., N. Haga, O. Tachikawa, H. Takeda, and T. Ishii (1986), Thermal history of ureilite, Pecora Escarpment 82506 deduced from cation distribution and diffusion profile of minerals, *Mem. Natl. Inst. Polar Res.*, *41*(Spec. Issue) 206.

[308] Tribaudino, M., D. Pasqual, G. Molin, and L. Secco (2003), Microtextures and crystal chemistry in P21/c pigeonites, *Mineralogy and Petrology*, *77*(3–4), 161–176.

[309] Wacker, J. F. (1986), Noble gases in the diamond-free ureilite, ALHA78019: The roles of shock and nebular processes, *Geochim. Cosmochim. Acta*, *50*, 633–642.

[310] Wang, M.-S., and M. E. Lipschutz (1995), Volatile trace elements in Antarctic ureilites, *Meteoritics*, *30*, 319.

[311] Warren, P. H., F. Ulff-Møller, H. Huber, and G. W. Kallemeyn (2006), Siderophile geochemistry of ureilites: A record of early stages of planetesimal core formation, *Geochim. Cosmochim. Acta*, *70*, 2104–2126.

45. PCA 82502—UREILITE

[312] Berkley, J. L. (1986), Four Antarctic ureilites: Petrology and observations on ureilite petrogenisis, *Meteoritics*, *21*, 169–189.

[313] Burns, R. G., and S. L. Martinez (1991), Mossbauer spectra of olivine-rich achondrites: Evidence for preterrestrial redox reactions, *Proc. Lunar Planet. Sci.*, *21*, 331–340.

[314] Clayton, R. N., and T. K. Mayeda (1996), Oxygen isotope studies of achondrites, *Geochim. Cosmochim. Acta*, *60*, 1999–2017.

[315] Farquhar, J., T. L. Jackson, and M. H. Thiemens (2000), A ^{33}S enrichment in ureilite meteorites: Evidence for a nebular sulfur component, *Geochim. Cosmochim. Acta*, *64*, 1819–1825.

[316] Goodrich, C. A. (1999), Are ureilites residues from partial melting of chondritic material? The answer from MAGPOX = 20, *Meteorit. Planet. Sci.*, *34*, 109–119.

[317] Goodrich, C. A., P. J. Patchett, G. W. Lugmair, and M. J. Drake (1991), Sm-Nd and Rb-Sr isotopic systematics of ureilites, *Geochim. Cosmochim. Acta*, *55*, 829–848.

[318] Goodrich, C. A. (1992), Ureilites: A critical review, *Meteoritics*, *27*, 327–352.

[319] Leya, I., M. Schönbächler, U. Wiechert, U. Krähenbühl, and A. N. Halliday (2008), Titanium isotopes and the radial heterogeneity of the solar system, *Earth Planet. Sci. Lett.*, *266*, 233–244.

[320] Miyamoto, M., H. Takeda, and H. Toyoda (1985), Cooling history of some Antarctic ureilites, *J. Geophys. Res.*, *90*, D116–D122.

[321] Pasqual, D., G. Molin, and M. Tribaudino (2000), Single-crystal thermometric calibration of Fe-Mg order-disorder in pigeonites, *Am. Mineral.*, *85*, 953–962.

[322] Rankenburg, K., A. D. Brandon, and M. Humayun (2007), Osmium isotope systematics of ureilites, *Geochim. Cosmochim. Acta*, *71*, 2402–2413.

[323] Rubin, A. E. (2006), Shock, post-shock annealing, and post-annealing shock in ureilites, *Meteoritics*, *41*, 125–133.

[324] Sandford, S. A. (1993), The mid-infrared transmission spectra of Antarctic ureilites, *Meteoritics*, *28*, 579–585.

[325] Spitz, A. H., and W. V. Boynton (1991), Trace element analysis of ureilites: New constraints on their petrogenesis, *Geochim. Cosmochim. Acta*, *55*, 3417–3430.

[326] Toyoda, H., N. Haga, O. Tachikawa, H. Takeda, and T. Ishii (1986), Thermal history of ureilite, Pecora Escarpment 82506 deduced from cation distribution and diffusion profile of minerals, *Mem. Natl. Inst. Polar Res.*, *41*(Spec. Issue), 206.

[327] Tribaudino, M., D. Pasqual, G. Molin, and L. Secco (2003), Microtextures and crystal chemistry in P 2 1 / c pigeonites, *Mineralogy and Petrology*, *77*(3–4), 161–176.

[328] Wang, M.-S., and M. E. Lipschutz (1995), Volatile trace elements in Antarctic ureilites, *Meteoritics*, *30*, 319.

[329] Warren, P. H., F. Ulff-Møller, H. Huber, and G. W. Kallemeyn (2006), Siderophile geochemistry of ureilites: A record of early stages of planetesimal core formation, *Geochim. Cosmochim. Acta*, *70*, 2104–2126.

46. EET 83309—POLYMICT UREILITE

[330] Cohen, B. A., C. A. Goodrich, and K. Keil (2004), Feldspathic clast populations in polymict ureilites: Stalking the missing basalts from the ureilite parent body, *Geochim. Cosmochim. Acta*, *68*, 4249–4266.

[331] Downes, H., D. W. Mittlefehldt, N. T. Kita, and and J. W. Valley (2008), Evidence from polymict ureilite meteorites for a disrupted and re-accreted single ureilite parent asteroid gardened by several distinct impactors, *Geochim. Cosmochim. Acta*, *72*, 4825–4844.

[332] Farquhar, J., T. L. Jackson, and M. H. Thiemens (2000), A ^{33}S enrichment in ureilite meteorites: Evidence for a nebular sulfur component, *Geochim. Cosmochim. Acta*, *64*, 1819–1825.

[333] Goodrich, C. A., E. R. D. Scott, and A. M. Fioretti (2004), Ureilitic breccias: Clues to the petrologic structure and impact disruption of the ureilite parent asteroid, *Chemie der Erde*, *64*, 283–327.

[334] Grady, M. M., and C. T. Pillinger (1988), 15 N-enriched nitrogen in polymict ureilites and its bearing on their formation, *Nature*, *331*, 321–323.

[335] Guan, Y., and G. Crozaz (2000), LREE-enrichments in ureilites: A detailed ion microprobe study, *Meteorit. Planet. Sci.*, *35*, 131–144.

[336] Guan, Y., and G. Crozaz (2001), Microdistributions and petrogenetic implications of rare earth elements in polymict ureilites, *Meteorit. Planet. Sci.*, *36*, 1039–1056.

[337] Ikeda, Y., N. T. Kita, Y. Morishita, and M. K. Weisberg (2003), Primitive clasts in the Dar al Gani 319 polymict ureilite: Precursors of the ureilites, *Antarctic Meteorite Research*, *16*, 105–127.

[338] Kita, N. T., Y. Ikeda, S. Togashi, L. Yongzhong, Y. Morishita, and M. K. Weisberg (2004), Origin of ureilites inferred from a SIMS oxygen isotopic and trace element study of clasts in the Dar al Gani 319 polymict ureilite, *Geochim. Cosmochim. Acta*, *68*, 4213–4235.

[339] Rai, V. K., S. V. S. Murty, and U. Ott (2003), Noble gases in ureilites: Cosmogenic, radiogenic, and trapped components, *Geochim. Cosmochim. Acta*, *67*, 4435–4456.

[340] Rai, V. K., S. V. S. Murty, and U. Ott (2003), Nitrogen components in ureilites next term, *Geochim. Cosmochim. Acta*, *67*, 2213–2237.

[341] Warren, P. H., and G. W. Kallemeyn (1989), Geochemistry of polymict ureilite EET 83309, and a partially-disruptive impact model for ureilite origin, *Meteoritics*, *24*, 233–246.

[342] Wang, M.-S., and M. E. Lipschutz (1995), Volatile trace elements in Antarctic ureilites, *Meteoritics*, *30*, 319.

47. LEW 88774—UREILITE (ANOMALOUS)

[343] Chikami, J., T. Mikouchi, H. Takeda, and M. Miyamoto (1997), Mineralogy and cooling history of the Ca-Al-Cr enriched ureilite, LEW 88774, *Meteoritics, 32*, 343–348.

[344] Downes, H., D. W. Mittlefehldt, N. T. Kita, and Valley, J. W. (2008), Evidence from polymict ureilite meteorites for a disrupted and re-accreted single ureilite parent asteroid gardened by several distinct impactors, *Geochim. Cosmochim. Acta, 72*, 4825–4844.

[345] Goodrich, C. A., G. E. Harlow, and T. Mikouchi (2007), New investigations of "Knorringite-Uvarovite Garnet" and "Cr-Eskola Pyroxene" in ureilites LEW 88774 and NWA 766, *Lunar Planet. Sci. Conf., 38*, 1434.

[346] Guan, Y., and G. Crozaz (2000), LREE-enrichments in ureilites: A detailed ion microprobe study, *Meteorit. Planet. Sci., 35*, 131–144.

[347] Mikouchi, T., J. Chikami, H. Takeda, and M. Miyamoto (1995), Mineralogical study of LEW 88774, not so unusual ureilite, *Lunar Planet. Sci. Conf., 26*, 971.

[348] Prinz, M., M. K. Weisberg, and C. E. Nehru (1994), LEW 88774: A new type of Cr-rich ureilite, *Lunar Planet. Sci. Conf., 25*, 1107.

[349] Rankenburg, K., A. D. Brandon, and M. Humayun (2007), Osmium isotope systematics of ureilites, *Geochim. Cosmochim. Acta, 71*, 2402–2413.

[350] Warren, P. H., F. Ulff-Møller, H. Huber, and G. W. Kallemeyn (2006), Siderophile geochemistry of ureilites: A record of early stages of planetesimal core formation, *Geochim. Cosmochim. Acta, 70*, 2104–2126.

[351] Warren, P. H., H. Huber, and F. Ulff-Møller (2006), Alkali-feldspathic material entrained in Fe,S-rich veins in a monomict ureilite, *Meteorit. Planet. Sci., 41*, 835–849.

[352] Warren, P. H., and G. W. Kallemeyn (1994), Petrology of LEW 88774: An extremely chromium-rich ureilite, *Lunar Planet. Sci. Conf., 25*, 1465.

[353] Kallemeyn, G. W., and P. H. Warren (1994), Geochemistry of LEW 88774 and two other unusual ureilites, *Lunar Planet. Sci. Conf., 25*, 663.

48. ALH 84025—BRACHINITE

[354] Mittlefehldt, D. W., D. D. Bogard, J. L. Berkley, and D. H. Garrison (2003), Brachinites: Igneous rocks from a differentiated asteroid, *Meteorit. Planet. Sci., 38*, 1601–1625.

[355] Ott, U., H. P. Löhr, and F. Begemann (1987), Noble gases in ALH 84025: Like brachina, unlike chassigny, *Meteoritics 22*, 476.

[356] Swindle, T. D., D. A. Kring, M. K. Burkland, D. H. Hill, and W. V. Boynton (1998), Noble gases, bulk chemistry, and petrography of olivine-rich achondrites Eagles Nest and LEW 88763: Comparison to brachinites, *Meteorit. Planet. Sci., 33*, 31–48.

[357] Warren, P. H., and G. W. Kallemeyn (1989), Allan Hills 84025: The second brachinite, far more differentiated than brachina, and an ultramafic achondritic clast from L chondrite Yamato 75097, *Lunar Planet. Sci. Conf., 19*, 475–486.

49. GRA 06128, 06129—UNGROUPED ACHONDRITE

[358] Arai, T., T. Tomiyama, K. Saiki, H. Takeda (2008), Unique achondrites GRA 06128/06129: Andesitic partial melt from a volatile-rich parent body, *Lunar Planet. Sci. Conf., 39*, LPI Contribution No. 1391, 2465.

[359] Arculus, R., I. H. Campbell, S. M. McLennan, and S. R. Taylor (2009), Asteroids and andesites, *Nature, 459*, E1–1.

[360] Day, J. M. D., R. D. Ash, Y. Liu, J. J. Bellucci, D. Rumble III, W. F. McDonough, R. J. Walker, and L. A. Taylor (2009), Early formation of evolved asteroidal crust, *Nature, 457*, 179–182.

[361] Day, J. M. D., R. J. Walker, R. D. Ash, Y. Liu, D. Rumble III, A. J. Irving, C. A. Goodrich, K. Tait, W. F. McDonough, and L. A. Taylor (2012), Origin of felsic achondrites Graves Nunataks 06128 and 06129, and ultramafic brachinites and brachinite-like achondrites by partial melting of volatile-rich primitive parent bodies, *Geochim. Cosmochim. Acta, 81*, 94–128.

[362] Nyquist, L. E., C.-Y. Shih, and Y. D. Reese (2009), Early petrogenesis and late impact(?), metamorphism on the GRA 06128/9 parent body, *Lunar Planet. Sci. Conf., 40*, 1290.

[363] Shearer, C. K., P. V. Burger, J. J. Papike, Z. D. Sharp, and K. D. McKeegan (2013), Fluids on differentiated asteroids: Evidence from phosphates in differentiated meteorites GRA 06128 and GRA 06129, *Meteorit. Planet. Sci. 46*, 1345–1362.

[364] Shearer C. K., P. V. Burger, C. R. Neal, Sharp Z. D., L. E. Borg, L. Spivak-Birndorf, M. Wadhwa, J. J. Papike, J. M. Karner, A. M. Gaffney, J. Shafer, B. P. Weiss, J. Geissman, and V. A. Fernandes (2008), A unique glimpse into asteroidal melting processes in the early solar system from the Graves Nunatak 06128/06129 achondrites, *American Mineralogist, 93*, 1937–1940.

[365] Shearer, C. K., P. V. Burger, C. R. Neal, Z. Sharp, L. Spivack-Birndorf, L. E. Borg, V. A. Fernandes, J. J. Papike, J. M. Karner, M. Wadhwa, A. M. Gaffney, J. Shafer, J. Geissman, N. V. Atudorei, C. Herd, B. P. Weiss, P. L. King, S. A. Crowther, and J. A. Gilmour (2010), Non-basaltic asteroidal magmatism during the earliest stages of solar system evolution: A view from Antarctic achondrites Graves Nunatak 06128 and 06129, *Geochim. Cosmochim. Acta, 74*, 1172–1199.

[366] Usui, T., J. H. Jones, D. W. Mittlefehldt (2010), Low-degree partial melting experiments of CR and H chondrite compositions: Implications for asteroidal magmatism recorded in GRA 06128 and GRA 06129, *Lunar Planet. Sci. Conf., 41*, LPI Contribution No. 1533, 1186.

[367] Zeigler, R. A., B. L. Jolliff, R. L. Korotev, D. Rumble, P. K. Carpenter, A. Wang (2008), Petrology, geochemistry, and likely provenance of unique achondrite Graves Nunataks 06128, *Lunar Planet. Sci. Conf., 39,* LPI Contribution No. 1391, 2456.

50. ALH A78113—AUBRITE

[368] Casanova, I., K. Keil, and H. E. Newsom (1993), Composition of metal in aubrites: Constraints on core formation, *Geochim. Cosmochim. Acta*, *57*, 675–682.

[369] Floss, C., G. Crozaz, and M. M. Strait (1990), Rare earth elements and the petrogenesis of aubrites, *Geochim. Cosmochim. Acta*, *54*, 3553–3558.

[370] Fogel, R. A. (2005), Aubrite basalt vitrophyres: The missing basaltic component and high-sulfur silicate melts, *Geochim. Cosmochim. Acta 69*, 1633–1648.

[371] Gaffey, M. J., E. A. Cloutis, M. S. Kelley, and K. L. Reed (2002), Mineralogy of asteroids, in *Asteroids III*, edited by W. F. Bottke Jr., A. Cellino, P. Paolicchi, and R. P. Binzel (pp. 183–204), University of Arizona Press, Tucson.

[372] Keil, K. (1989), Enstatite meteorites and their parent bodies, *Meteoritics*, *24*, 195–208.

[373] Kimura, M., Y.-T. Lin, Y. Ikeda, A. El Goresy, K. Yanai, and H. Kojima (1992), Petrology and mineralogy of Antarctic aubrites, Y-793592 and ALH-78113, in comparison with non-Antarctic aubrites and E-chondrites, *17th Symposium on Antarctic Meteorites, NIPR*, 217–218.

[374] Kimura, M., Y. T. Lin, Y. Ikeda, A. El Goresy , K. Yanai, and H. Kojima (1993), Mineralogy of Antarctic aubrites, Yamato-793592 and Allan Hills-78113: Comparison with non-Antarctic aubrites and E-chondrites, *17th Symposium on Antarctic Meteorites, Proc. NIPR*, *6*, 186.

[375] Lipschutz, M. E., R. M. Verkouteren, D. W. G. Sears, F. A. Hasan, and M. Prinz (1988), Cumberland Falls chondritic inclusions: III. Consortium study of relationship to inclusions in Allan Hills 78113 aubrite, *Geochim. Cosmochim. Acta*, *52*, 1835–1848.

[376] Lorenzetti, S., O. Eugster, H. Busemann, K. Marti, T. H. Burbine, and T. McCoy (2003), History and origin of aubrites, *Geochim. Cosmochim. Acta*, *67*, 557–571.

[377] Miura, Y. N., H. Hidaka, K. Nishiizumi, and M. Kusakabe (2007), Noble gas and oxygen isotope studies of aubrites: A clue to origin and histories, *Geochim. Cosmochim. Acta*, *71*, 251–270.

[378] Nedelcu, D. A., M. Birlan, P. Vernazza, R. P. Binzel, M. Fulchignoni, M. A. Barucci (2007), E-type asteroid (2867), Steins: Flyby target for Rosetta, *Astronomy and Astrophysics*, *473*, L33–L36.

[379] Rosenshein, E. B., M. A. Ivanova, T. L. Dickinson, T. J. McCoy, D. S. Lauretta, Y. Guan, L. A. Leshin, and G. K. Benedix (2006), Oxide-bearing and FeO-rich clasts in aubrites, *Meteoritics*, *41*, 495–503.

[380] Watters, T. R., M. Prinz, E. R. Rambaldi, and J. T. Wasson (1980), ALH A78113, Mt. Egerton and the Aubrite Parent Body, *Meteoritics*, *15*, 386.

[381] Wheelock, M. M., K. Keil , C. Floss, G. J. Taylor, and G. Crozaz (1994), REE geochemistry of oldhamite-dominated clasts from the Norton County aubrite: Igneous origin of oldhamite, *Geochim. Cosmochim. Acta*, *58*, 449–458.

[382] Wilson, L., and K. Keil (1991), Consequences of explosive eruptions on small solar system bodies: The case of the missing basalts on the aubrite parent body. *Earth Planet. Sci. Lett.*, *104*, 505–512.

51. LAP 03719—UNBRECCIATED OLIVINE AUBRITE

[383] McCoy, T. J., A. Gale, and T. L. Dickinson (2005), The early crystallization history of the aubrite parent body, *Annual Meteoritical Society Meeting*, *68*, 5165.

[384] Rubin, A. E. (2009), Shock effects in EH6 chondrites and aubrites: Implications for collisional heating of asteroids, *Lunar Planet. Sci. Conf.*, *40*, 1353.

52. LAR 04316—AUBRITE (WITH BASALTIC VITROPHYRE CLAST)

[385] Fogel, R. A. (2005), Aubrite basalt vitrophyres: The missing basaltic component and high-sulfur silicate melts, *Geochim. Cosmochim. Acta*, *69*, 1633–1648.

[386] Keil, K., T. J. McCoy , L. Wilson, J.-A. Barrat, D. Rumble, M. M. M. Meier, R. Wieler, and G. R. Huss (2011), A composite Fe,Ni-FeS and enstatite-forsterite-diopside-glass vitrophyre clast in the Larkman Nunatak 04316 aubrite: Origin by pyroclastic volcanism, *Meteorit. Planet. Sci.*, *46*, 1719–1741.

[387] McCoy, T. J., and A. Gale (2007), Pyroclastic volcanism on the aubrite parent body: Evidence from an Fe,Ni-FeS clast in LAR 04316, *Meteorit. Planet. Sci.*, *41*, 5082.

53. LEW 86010—ANGRITE

[388] Amelin, Y. (2008), U-Pb ages of angrites, *Geochim. Cosmochim. Acta*, *72*, 221–232.

[389] Crozaz, G., and G. McKay (1990), Rare earth elements in Angra dos Reis and Lewis Cliff 86010, two meteorites with similar but distinct magma evolutions, *Earth Planet. Sci. Lett.*, *97*, 369–381.

[390] Kleine, T., U. Hans, A. J. Irving, and B. Bourdon (2012), Chronology of the angrite parent body and implications for core formation in protoplanets, *Geochim. Cosmochim. Acta*, *84*, 186–203.

[391] McKay, G., L. Le, J. Wagstaff, and G. Crozaz (1994), Experimental partitioning of rare earth elements and strontium: Constraints on petrogenesis and redox conditions during crystallization of Antarctic angrite Lewis Cliff 86010, *Geochim. Cosmochim. Acta*, *58*, 2911–2919.

[392] McKay, G. A., M. Miyamoto, T. Mikouchi, and T. Ogawa (1998), The cooling history of the LEW 86010 angrite as inferred from kirschsteinite lamellae in olivine, *Meteorit. Planet. Sci.*, *33*, 977–983.

[393] Mikouchi, T., H. Takeda, H. Mori, M. Miyamoto, and G. McKay (1993), Exsolved kirschsteinite in angrite LEW 86010 olivine, *Lunar Planet. Sci. Conf.*, *24*, 987–988.

54. LEW 87051—ANGRITE

[394] Floss, C., G. Crozaz, G. McKay, T. Mikouchi, and M. Killgore (2003), Petrogenesis of angrites, *Geochim. Cosmochim. Acta*, 67, 4775–4789.

[395] Mikouchi T., M. Miyamoto, and G. A. McKay (1996), Mineralogical study of angrite Asuka-881371: Its possible relation to angrite LEW 87051, *Proc. NIPR Symp. Antarct. Meteorites*, 9, 174–188.

[396] McKay G., G. Crozaz, T. Mikouchi, and M. Miyamoto (1995), Petrology of Antarctic angrites LEW 86010, LEW 87051 and Asuka 881371, *Antarct. Meteorites*, 20, 155–158.

[397] McKay G., G. Crozaz, T. Mikouchi, and M. Miyamoto (1995), Exotic olivine in Antarctic angrites LEW 87051 and Asuka 881371, *Meteoritics*, 30, 543–544.

[398] McKay G., G. Crozaz, J. Wagstaff, S.-R. Yang, and L. Lundberg (1990), A petrographic, electron microprobe and ion microprobe study of mini-angrite Lewis Cliff 87051, *Lunar Planet. Sci.*, 21, 771–772.

[399] Prinz M., M. K. Weisberg, and C. E. Nehru (1990), LEW 87051, a new angrite: Origin in a Ca-Al-enriched eucritic planetesimal, *Lunar Planet. Sci.*, 21, 979–980.

55. EET 90020– UNBRECCIATED EUCRITE

[400] Bogard, D. D., and D. H. Garrison (1997), ^{39}Ar-^{40}Ar ages of igneous non-brecciated eucrites, *Lunar Planet. Sci. Conf.*, 28, 127.

[401] Lee, D.-C. (2008), ^{182}Hf-^{182}W chronometry and the early evolution history in the acapulcoite-lodranite parent body, *Meteorit. Planet. Sci.*, 43, 675–684.

[402] Mittlefehldt, D. W., and M. M. Lindstrom (2003), Geochemistry of eucrites: genesis of basaltic eucrites, and Hf and Ta as petrogenetic indicators for altered Antarctic eucrites, *Geochim. Cosmochim. Acta*, 67, 1911–1939.

[403] Scott, E. R. D, R. C. Greenwood, I. A. Franchi, and I. S. Sander (2009), Oxygen isotopic constraints on the origin and parent bodies of eucrites, diogenites, and howardites, *Geochim. Cosmochim. Acta*, 73, 5835–5853.

[404] Srinivasan, G., M. J. Whitehouse, I. Weber, and A. Yamaguchi (2007), The crystallization age of eucrite zircon, *Science*, 317, 345–347.

[405] Stirling, C. H., A. N. Halliday, E. K. Potter, M. B. Andersen, and B. Zanda (2009), A low initial abundance of ^{247}Cm in the early solar system and implications for r-process nucleosynthesis, *Earth Planet. Sci. Lett.*, 251, 386–397.

[406] Warren, P. H., G. W. Kallemeyn, H. Huber, F. Ulff-Møller, and W. Choe (2009), Siderophile and other geochemical constraints on mixing relationships among HED-meteoritic breccias, *Geochim. Cosmochim. Acta*, 73, 5918–5943.

[407] Yamaguchi, A, G. J. Taylor, K. Keil, C. Floss, G. Crozaz, L. E. Nyquist, D. H. Garrison, Y. D., Reese, H. Wiesmann, C. Y. and Shih (2001), Post-crystallization reheating and partial melting of eucrite EET 90020 by impact into the hot crust of asteroid 4 Vesta ~4.50 Ga ago, *Geochim. Cosmochim Acta*, 65, 3577–3599.

56. ALH A81001—UNBRECCIATED EUCRITE

[408] Delaney, J. S., M. Prinz, C. E. Nehru, and C. P. Stokes (1984), Allan Hills A81001, cumulate eucrites and black clasts from polymict eucrites, *Lunar and Planetary Science Conf.*, 25, 212–213.

[409] Mayne, R. G., H. Y. McSween, T. J. McCoy, and A. Gale (2009), Petrology of the unbrecciated eucrites, *Geochim. Cosmochim. Acta*, 73, 794–819.

[410] Mayne, R. G., J. M. Sunshine, H. Y. McSween, T. J. McCoy , C. M. Corrigan, and A. Gale (2010), Petrologic insights from the spectra of the unbrecciated eucrites: Implications for Vesta and basaltic asteroids, *Meteorit. Planet. Sci.*, 45(7), 1074–1092.

[411] Ruzicka, A., G. A. Snyder, and L. A. Taylor (1997), Vesta as the howardite, eucrite and diogenite parents body: Implications for the size of a core and for large-scale differentiation, *Meteorit. Planet. Sci.*, 32, 825–840.

[412] Scott, E. R. D., R. C. Greenwood, I. A. Franchi, and I. S. Sanders (2009), Oxygen isotopic constraints on the origin and parent bodies of eucrites, diogenites, and howardites, *Geochim. Cosmochim. Acta*, 73, 5835–5853.

57. ALH A76005—POLYMICT EUCRITE

[413] Delaney, J. S., M. Prinz, and H. Takeda (1983), Modal comparison of Yamato and Allan Hills polymict eucrites, *Mem. Natl. Inst. Polar Res.*, 30(Spec. Issue), 206–223.

[414] Fuhrman, M., and J. J. Papike (1982), Howardites and polymict eucrites: Regolith samples from the eucrite parent body: Petrology of Bholgati, Bununu, Kapoeta, and ALH A76005, *12th Lunar Planet. Sci. Conf. Proc.* (pp. 1257–1279), New York and Oxford, Pergamon Press.

[415] Grossman, L., E. Olsen, A. M. Davis, T. Tanaka, and G. J. MacPherson (1981), The Antarctic achondrite ALH A76005: A polymict eucrite, *Geochim. Cosmochim. Acta*, 45, 1267–1279.

[416] Kunz, J., M. Trieloff, K. D. Bobe, K. Metzler, D. Stöffler, and E. K. Jessberger (1995), The collisional history of the HED parent body inferred from ^{40}Ar-^{39}Ar ages of eucrites, *Planetary and Space Science*, 43, 527–543.

[417] Labotka, T. C., and J. J. Papike (1980), Howardites: Samples of the regolith of the eucrite parent-body: Petrology of Frankfort, Pavlovka, Yurtuk, Malvern, and ALH A77302, *11th Lunar Planet. Sci. Conf. Proc.* (Vol. 2, pp. 1103–1130), New York, Pergamon Press.

[418] Metzler, K., K. D. Bobe, H. Palme, B. Spettel, and D. Stöffler (1995), Thermal and impact metamorphism on the HED parent asteroid, *Planetary and Space Science*, 43, 499–525.

[419] Mittlefehldt, D. W., and M. M. Lindstrom (2003), Geochemistry of eucrites: Genesis of basaltic eucrites, and Hf and Ta as petrogenetic indicators for altered Antarctic eucrites, *Geochim. Cosmochim. Acta*, 67, 1911–1934.

[420] Miyamoto, M., M. B. Duke, and D. S. McKay (1985), Chemical zoning and homogenization of Pasamonte-type pyroxene and their bearing on thermal metamorphism of a howardite parent body, *J. Geophys. Res.*, *90*(Suppl.), C629 – C635.

[421] Nakamura, N., and A. Masuda (1980), REE abundances in the whole rock and mineral separates of the Allan Hills-765 meteorite, *Mem. Natl. Inst. Polar Res.*, *17*(Spec. Issue), 159–167.

[422] Olsen, E. J., A. Noonan, K. Fredriksson, E. Jarosewich, and G. Moreland (1978), Eleven new meteorites from Antarctica, 1976–1977, *Meteoritics*, *13*, 209–225.

[423] Reid, A., and R. Score (1981), A preliminary report on the achondrite meteorites in the 1979 U.S. Antarctic meteorite collection, *Mem. Natl. Inst. Polar Res.*, *20*, 33–52.

[424] Smith, M. R., and R. A. Schmitt (1981), Preliminary chemical data for some Allan Hills polymict eucrites, *Lunar Planet. Sci.*, *12*, 1014–1016.

[425] Simon, S. B., and J. J. Papike (1983), Petrology of igneous lithic clasts from polymict eucrites ALH A76005 and ALH A77302, *Meteoritics*, *18*, 35–50.

[426] Treiman, A. H., and M. J. Drake (1985), Basaltic volcanism on the eucrite parent body: Petrology and chemistry of the polymict eucrite ALH A80102, *J. Geophys. Res.*, *90*(Suppl.), C619–C628.

[427] Warren, P. H., G. W. Kallemeyn, H. Huber, F. Ulff-Møller, and W. Choe (2009), Siderophile and other geochemical constraints on mixing relationships among HED-meteoritic breccias, *Geochim. Cosmochim. Acta*, *73*, 5918–5943.

58. LAP 91900—DIOGENITE

[428] Hiroi, T., C. M. Pieters, F. Vilas, S. Sasaki, Y. Hamabe, and E. Kurahashi (2001), The mystery of 506.5 nm feature of reflectance spectra of Vesta and Vestoids: Evidence for space weathering? *Earth Planets Space*, *53*, 1071–1075.

[429] Fowler, G. W., C. K. Shearer, J. J. Papike, and G. D. Layne (1995), Diogenites as asteroidal cumulates: Insights from orthopyroxene trace element chemistry, *Geochim. Cosmochim. Acta*, *59*, 3071–3084.

[430] Mittlefehldt, D. W. (1994), The genesis of diogenites and HED parent body petrogenesis, *Geochim. Cosmochim. Acta*, *58*, 1537.

[431] Zema, M., M. Chiara Domeneghetti, G. M. Molin, and V. Tazzoli (1997), Cooling rates of diogenites: A study of Fe^{2+}-Mg ordering in orthopyroxene by X-ray single-crystal diffraction, *Meteorit. Planet. Sci.*, *32*, 855–862.

59. GRO 95555—UNBRECCIATED DIOGENITE

[432] Barrat, J. A., A. Yamaguchi, R. C. Greenwood, M. Benoit, J. Cotten, M. Bohn, and I. A. Franchi (2009), Geochemistry of diogenites: Still more diversity in their parental melts, *Meteorit. Planet. Sci.*, *43*, 1759–1775.

[433] Hiroi, T., C. M. Pieters, and H. Takeda (1994), Grain size of the surface regolith of asteroid 4 Vesta estimated from its reflectance spectrum in comparison with HED meteorites, *Meteoritics*, *29*, 394–396.

[434] Hiroi, T., C. M. Pieters, F. Vilas, S. Sasaki, Y. Hamabe, and E. Kurahashi (2001), The mystery of 506.5 nm feature of reflectance spectra of Vesta and Vestoids: Evidence for space weathering? *Earth Planets Space*, *53*, 1071–1075.

[435] Papike, J. J., C. K. Shearer, M. N. Spilde, and J. M. Karner (2000), Metamorphic diogenite Grosvenor Mountains 95555: Mineral chemistry of orthopyroxene and spinel and comparisons to the diogenite suite, *Meteorit. Planet. Sci.*, *35*, 875–879.

[436] Schiller, M., J. Baker, J. Creech, C. Paton, M.-A. Millet, A. Irving, and M. Bizzarro (2011), Rapid timescales for magma ocean crystallization on the howardite-eucrite-diogenite parent body, *Astrophys. J. Lett.*, *740*, L22–L27.

[437] Scott, E. R. D., R. C. Greenwood, I. A. Franchi, and I. S. Sanders (2009), Oxygen isotopic constraints on the origin and parent bodies of eucrites, diogenites, and howardites, *Geochim. Cosmochim. Acta*, *73*, 5835–5853.

[438] Stimpfl, M., and J. Ganguly (2002), Thermal history of the unbrecciated diogenite GRO (Grosvenor Mountains), 95555: Constraints from inter- and intra-crystalline Fe-Mg exchange reactions, *Lunar Planet. Sci. Conf.*, *33*, No. 1966.

60. GRA 98108—OLIVINE DIOGENITE

[439] Beck, A. W., and H. Y. McSween Jr. (2010), Diogenites as polymict breccias composed of orthopyroxenite and harzburgite, *Meteorit. Planet. Sci.*, *45*, 850–872.

[440] Mittlefehldt, D. W. (2002), Geochemistry of new, unusual diogenites and constraints on diogenite genesis, *Meteorit. Planet. Sci.*, *37*, A100.

[441] Righter, K. (2001), Petrography, mineralogy and petrology of two new HED meteorites: Diogenite GRA98108 and howardite GRA98030, *Lunar Planet. Sci. Conf.*, *32*, No. 1765.

[442] Scott, E. R. D., R. C. Greenwood, I. A. Franchi, and I. S. Sanders (2009), Oxygen isotopic constraints on the origin and parent bodies of eucrites, diogenites, and howardites, *Geochim. Cosmochim. Acta*, *73*, 5835–5853.

[443] Shearer, C. K., P. Burger, and J. J. Papike (2010), Petrogenetic relationships between diogenites and olivine diogenites: Implications for magmatism on the HED parent body, *Geochim. Cosmochim. Acta*, *74*, 4865–4880.

61. EET 87503 AND PAIRS—HOWARDITE

[444] Mittlefehldt, D. W., and M. M. Lindstrom (2003), Geochemistry of eucrites: Genesis of basaltic eucrites, and Hf and Ta as petrogenetic indicators for altered antarctic eucrites, *Geochim. Cosmochim. Acta*, *67*, 1911–1934.

[445] Buchanan, P. C., and D. W. Mittlefehldt (2003), Lithic components in the paired howardites EET 87503 and EET 87513: Characterization of the regolith of 4 Vesta, *Antarc. Met. Res.*, *16*, 128–151.

[446] Buchanan, P. C., and A. M. Reid (1991), Eucrite and diogenite clasts in three Antarctic achondrites, Lunar Planet. *Sci. Conf.*, *22*, 149.

[447] Buchanan, P. C., and A. M. Reid (1992), Matrix pyroxenes in howardites and polymict eucrites, Lunar Planet. *Sci. Conf.*, *23*, 173.

[448] Buchanan, P. C., M. E. Zolensky, A. M. Reid, and R. Barrett (1993), The EET 87513 clast N: A CM2 fragment in an HED polymict breccias, *Lunar Planet. Sci. Conf.*, *24*(Part 1), 209–210.

[449] Buchanan, P. C., M. E. Zolensky, and A. M. Reid (1993), Carbonaceous chondrite clasts in the howardites Bholghati and EET 87513, *Meteoritics*, *28*, 659–669.

[450] Zolensky, M. E., M. K. Weisberg, P. C. Buchanan, and D. W. Mittlefehldt (1996), Mineralogy of carbonaceous chondrite clasts in HED achondrites and the Moon, *Meteorit. Planet. Sci.*, *31*, 518–537.

[451] Buchanan, P. C., D. J. Lindstrom, and Mittlefehldt, D. W (2000), Pairing among the EET 87503 group of howardites and polymict eucrites, *Workshop on Extraterrestrial Materials from Cold and Hot Deserts*, p. 21.

[452] Metzler, K., K. D. Bobe, H. Palme, B. Spettel, and D. Stöffler (1995), Thermal and impact metamorphism on the HED parent asteroid, *Planetary and Space Science*, *43*, 499–525.

[453] Mittlefehldt, D. W., and M. M. Lindstrom (1991), Generation of abnormal trace element abundances in Antarctic eucrites by weathering processes, *Geochim. Cosmochim. Acta*, *55*, 77–87.

[454] Zolensky, M. E., M. K. Weisberg, P. C. Buchanan, M. Prinz, A. Reid, and R. A. Barrett (1992), Mineralogy of dark clasts in CT chondrites, encrites and howardites, Lunar Planet. *Sci. Conf.*, *23*, 1587.

62. MIL 03443—HED DUNITE

[455] Beck, A. W., D. W. Mittlefehldt, H. Y. McSween Jr., D. Rumble III, C.-T. A. Lee, and R. J. Bodnar (2011), MIL 03443, a dunite from asteroid 4 Vesta: Evidence for its classification and cumulate origin, *Meteorit. Planet. Sci.*, *46*, 1133–1151.

[456] Krawczynski, M. J., L. T. Elkins-Tanton, and T. L. Grove (2008), Petrology of olivine-diogenite MIL 03443,9: Constraints on eucrite parent body bulk composition and magmatic processes, *Lunar Planet. Sci. Conf.*, *39*, LPI Contribution No. 1391, #1229.

[457] Mittlefehldt, D. W. (2008), Meteorite dunite breccia MIL 03443: A probable crustal cumulate closely related to diogenites from the HED parent asteroid, *Lunar Planet. Sci. Conf.*, *39*, LPI Contribution No. 1391, 1919.

63. QUE 93148—UNGROUPED ACHONDRITE

[458] Floss, C. (2002), Queen Alexandra Range 93148: A new type of pyroxene pallasite? *Meteorit. Planet. Sci.*, *37*, 1129–1139.

[459] Goodrich, C. A., and K. Righter (2000), Petrology of unique achondrite Queen Alexandra Range 93148: A piece of the pallasite (howardite-eucrite-diogenite?) parent body? *Meteorit. Planet. Sci.*, *35*, 521–535.

[460] Mikouchi, T., and K. Righter (2005), SEM-EBSD analysis on symplectic inclusion in the QUE 93148 olivine, *Meteorit. Planet. Sci.*, *40*, 5113.

[461] Righter M., T. J. Lapen, and K. Righter (2008), Relationships between HEDs, mesosiderites, ungrouped achondrites: Trace element analyses of mesosiderite RKPA 79015 and ungrouped achondrite QUE 93148, *Lunar Planet. Sci. Conf.*, *39*, LPI Contribution No. 1391, 2468.

[462] Shearer, C. K., J. J. Papike, and J. M. Karner (2001), Chemistry of olivine from planetary materials. Mn/Fe and trace element systematics in an unusual achondrite: QUE 93148, *Lunar Planet. Sci. Conf.*, *31*, 1634.

64. ALH A81005 –LUNAR ANORTHOSITIC BRECCIA

[463] Boynton, W. V., and D. H. Hill (1983), Composition of bulk fragments and a possible pristine clast from Allan Hills A81005, *Geophys. Res. Lett.*, *10*, 837–840.

[464] Delano, J. W. (1991), Geochemical comparison of impact glasses from lunar meteorites ALH A81005 and MAC 88105 and Apollo 16 regolith 64001, *Geochim. Cosmochim. Acta*, *55*, 3019–3029.

[465] Eugster, O., J. Geiss, U. Krähenbühl, and S. Niedermann (1986), Noble gas isotopic composition, cosmic ray exposure history, and terrestrial age of the meteorite Allan Hills A81005 from the Moon, Earth Planet. *Sci. Lett.*, *78*, 139–147.

[466] Goodrich, C. A., G. J. Taylor, K. Keil , W. V. Boynton, and D. H. Hill (1984), Petrology and chemistry of hyperferroan anorthosites and other clasts from lunar meteorite ALH A81005, *J. Geophys. Res.*, *89*(Suppl.), C87–C94.

[467] Gross J., and A. H. Treiman (2011), Unique spinel-rich lithology in lunar meteorite ALH A81005: Origin and possible connection to M3 observations of the farside highlands, *J. Geophys. Res.*, *116*, E10009, doi:10.1029/2011JE003858.

[468] Kallemeyn, G. W., and P. H. Warren (1983), Compositional implications regarding the lunar origin of the ALH A81005 meteorite. *Geophys. Res. Lett.*, *10*, 833–836.

[469] Korotev, R. L., M. M. Lindstrom, D. J. Lindstrom, and L. A. Haskin (1983), Antarctic meteorite ALH A81005: Not just another lunar anorthositic norite, *Geophys. Res. Lett.*, *10*, 829–832.

[470] Kurat, G., and F. Brandstätter (1983), Meteorite ALH A81005: Petrology of a new lunar highland sample, *Geophys. Res. Lett.*, *10*, 795–798.

[471] Laul, J. C., M. R. Smith, and R. A. Schmitt (1983), ALH A81005 meteorite: Chemical evidence for lunar highland origin, *Geophys. Res. Lett.*, *10*, 825–828.

[472] Marvin, U. B. (1983), The discovery and initial characterization of Allan Hills 81005: The first lunar meteorite, *Geophys. Res. Lett.*, *10*, 775–778.

[473] Mayeda, T. K., R. N. Clayton, and C. A. Molini-Velsko (1983), Oxygen and silicon isotopes in ALH A81005, *Geophys. Res. Lett.*, *10*, 799–800.

[474] Morris R. V. (1983), Ferromagnetic resonance and magnetic properties of ALH A81005, *Geophys. Res. Lett.*, *10*, 807–808.

[475] Nishiizumi, K., J. R. Arnold, J. Klein, D. Fink, R. Middleton, P. W. Kubik, P. Sharma, D. Elmore, and R. C. Reedy (1991), Exposure histories of lunar meteorites: ALH A81005, MAC 88104, MAC 88105, and Y791197, *Geochim. Cosmochim. Acta*, *55*, 3149–3155.

[476] Palme, H., B. Spettel, G. Weckwerth, and H. Wänke (1983), Antarctic meteorite ALH A81005, a piece from the ancient lunar crust, *Geophys. Res. Lett.*, *10*, 817–820.

[477] Pieters, C. M., B. R. Hawke, M. Gaffey, and L. A. McFadden (1983), Possible lunar source areas of meteorite ALH A81005: Geochemical remote sensing information. *Geophys. Res. Lett.*, *10*, 813–816.

[478] Robinson K. L., A. H. Treiman, and J. H. Joy (2012), Basaltic fragments in lunar feldspathic meteorites: Connecting sample analyses to orbital remote sensing, *Meteorit. Planet. Sci.*, *43*, 387–399.

[479] Ryder, G., and R. Ostertag (1983), ALH A81005: Moon, Mars, petrography, and Giordano Bruno, *Geophys. Res. Lett.*, *10*, 791–794.

[480] Simon, S. B., J. J. Papike, and C. K. Shearer (1983), Petrology of ALH A81005, the first lunar meteorite, *Geophys. Res. Lett.*, *10*, 787–790.

[481] Sutton, R. L., and G. Crozaz (1983), Thermoluminescence and nuclear particle tracks in ALH A81005: Evidence for a brief transit time, *Geophys. Res. Lett.*, *10*, 809–812.

[482] Sutton, S. R. (1986), Thermoluminesence of lunar meteorites Yamato-791197 and ALH A81005, *Mem. Natl. Inst. Polar Res.*, *41*(Spec. Iss.), 133–139.

[483] Treiman, A. H., and M. J. Drake (1983), Origin of lunar meteorite ALH A81005: Clues from the presence of terrae clasts and a very low-titanium mare basalt clast, *Geophys. Res. Lett.*, *10*, 783–786.

[484] Treiman A. H., A. K. Maloy, C. K. Shearer Jr., and J. Gross (2010), Magnesian anorthositic granulites in lunar meteorites Allan Hills A81005 and Dhofar 309: Geochemistry and global significance, *Meteorit. Planet. Sci.*, *45*, 163–180.

[485] Tuniz, C., D. K. Pal, R. K. Moniot, W. Savin, T. H. Kruse, G. F. Herzog, and J. C. Evans (1983), Recent cosmic ray exposure history of ALH A81005, *Geophys. Res. Lett.*, *10*, 804–806.

[486] Verkouteren, R. M., J. E. Dennison, and M. E. Lipschutz (1983), Siderophile, lithophile and mobile trace elements in the lunar meteorite Allan Hills A81005, *Geophys. Res. Lett.*, *10*, 821–824.

[487] Warren, P. H., G. J. Taylor, and K. Keil (1983), Regolith breccia Allan Hills A81005: Evidence of lunar origin and petrography of pristine and nonpristine clasts, *Geophys. Res. Lett.*, *10*, 779–782.

[488] Warren, P. H., and G. W. Kallemeyn (1986), Geochemistry of lunar meteorite Yamato-791197: Comparison with ALH A81005 and other lunar samples, *Mem. Natl. Inst. Polar Res.*, *41*(Spec. Iss.), 3–16.

[489] Warren, P. H., and G. W. Kallemeyn (1987), Geochemistry of lunar meteorite Yamato-82192: Comparison with Yamato-791197, ALH A81005, and other lunar samples, *Mem. Natl. Inst. Polar Res.*, *46*(Spec. Iss.), 3–20.

65. EET 87521/96008—LUNAR BASALTIC BRECCIA (POLYMICT)

[490] Anand, M., L. A. Taylor, C. R. Neal, G. A. Snyder, A. Patchen, Y. Sano, and K. Terada (2003), Petrogenesis of lunar meteorite EET 96008, *Geochim. Cosmochim. Acta*, *67*, 3499–3518.

[491] Delaney, J. S. (1989), Lunar basalt breccia identified among Antarctic meteorites, *Nature*, *342*, 889, 890.

[492] Eugster, O., Ch. Thalmann, A. Albrecht, G. F. Herzog, J. S. Delaney, J. Klein, and R. Middleton (1996), Exposure history of glass and breccia phases of lunar meteorite EET 87521, *Meteorit. Planet. Sci.*, *31*, 299–304.

[493] Eugster, O., E. Polnau, E. Salerno, and D. Terribilini (2000), Lunar surface exposure models for meteorites Elephant Moraine 96008 and Dar al Gani 262 from the Moon, *Meteorit. Planet. Sci.*, *35*, 1177–1181.

[494] Fernandes, V. A., R. Burgess, and A. Morris (2009), ^{40}Ar-^{39}Ar age determinations of lunar basalt meteorites Asuka 881757, Yamato 793169, Miller Range 05035, LaPaz Icefield 02205, Northwest Africa 479, and basaltic breccia Elephant Moraine 96008, *Meteorit. Planet. Sci.*, *44*, 805–821.

[495] Korotev R. L, R. A. Zeigler, B. L. Jolliff, A. J. Irving, and T. E. Bunch (2009), Compositional and lithological diversity among brecciated lunar meteorites of intermediate iron composition, *Meteorit. Planet. Sci.*, *44*, 1287–1322.

[496] Takeda, H., H. Mori, J. Saito, and M. Miyamoto (1992), Mineralogical studies of lunar mare meteorites EET 87521 and Y793274, *Proc. Lunar Planet. Sci.*, *22*, 275–301.

[497] Terada, K., T. Saiki, Y. Oka, Y. Hayasaka, and Y. Sano (2005), Ion microprobe U-Pb dating of phosphates in lunar basaltic breccia, Elephant Moraine 87521, *Geophys. Res. Lett.*, *32*, L20202.

[498] Vogt, S., G. F. Herzog, O. Eugster, T. Michel, S. Niedermann, U. Krahenbuhl, R. Middleton, B. Dezfouly-Arjomandy, D. Fink, and J. Klein (1993), Exposure history of the lunar meteorite, Elephant moraine 87521, *Geochim. Cosmochim. Acta*, *57*, 3793–3799.

[499] Warren, P. H., and G. W. Kallemeyn (1989), Elephant Moraine 87521: The first lunar meteorite composed of predominantly mare material, *Geochim. Cosmochim. Acta*, *53*, 3323–3300.

66. MIL 05035—UNBRECCIATED LUNAR GABBRO

[500] Arai, T., B. R. Hawke, T. A. Giguere, K. Misawa, M. Miyamoto, and H. Kojima (2010), Antarctic lunar meteorites Yamato-793169, Asuka-881757, MIL 05035, and MET 01210 (YAMM): Launch pairing and possible crypto-mare origin, *Geochim. Cosmochim. Acta, 74*, 2231–2248.

[501] Joy, K. H., I. A. Crawford, M. Anand, R. C. Greenwood, I. A. Franchi, and S. S. Russell (2008), The petrology and geochemistry of Miller Range 05035: A new lunar gabbroic meteorite, *Geochim. Cosmochim. Acta, 72*, 3822–3844.

[502] Liu, Y., C. Floss, J. M. D Day, E. Hill, and L. A. Taylor (2009), Petrogenesis of lunar mare basalt meteorite Miller Range 05035, *Meteorit. Planet. Sci., 44*, 261–284.

[503] Nyquist, L. E., C.-Y. Shih, and Y. D. Reese (2007), Sm-Nd and Rb-Sr ages for MIL 05035: Implications for surface and mantle sources, *Lunar Planet. Sci., 38*, abstract 1702.

[504] Zhang, A., W. Hsu, Q. Li, Y. Liu, Y. Jiang, and G. Tang (2010), SIMS Pb/Pb dating of Zr-rich minerals in lunar meteorites Miller Range 05035 and LaPaz Icefield 02224: Implications for the petrogenesis of mare basalt, *Science China Earth Sciences, 53*, 327–334. doi: 10.1007/s11430-010-0041-z.

67. LAP 02205—UNBRECCIATED LUNAR BASALT

[505] Anand, M., L. A. Taylor, C. Floss, C. R. Neal, K. Terada, and S. Tanikawa (2006), Petrology and geochemistry of LaPaz Icefield 02205: A new unique low-Ti mare-basalt meteorite, *Geochim. Cosmochim. Acta, 70*, 246–264.

[506] Day, J. M. D., and L. A. Taylor (2007), On the structure of mare basalt lava flows from textural analysis of the LaPaz Icefield and Northwest Africa 032 lunar meteorites, *Met. Planet. Sci., 42*, 3–18.

[507] Day, J. M. D., L. A. Taylor, C. Floss , A. D. Patchen, D. W. Schnare, and D. G. Pearson (2006), Comparative petrology, geochemistry, and petrogenesis of evolved, low-Ti lunar mare basalt meteorites from the LaPaz Icefield, Antarctica, *Geochim. Cosmochim. Acta, 70*, 1581–1600.

[508] Day, J. M. D., D. G. Pearson, and L. A. Taylor (2007), Highly siderophile element constraints on accretion and differentiation of the Earth-Moon system, *Science, 315*, 217–219.

[509] Fernandes, V. A., R. Burgess, and A. Morris (2009), [40]Ar-[39]Ar age determinations of lunar basalt meteorites Asuka 881757, Yamato 793169, Miller Range 05035, LaPaz Icefield 02205, Northwest Africa 479, and basaltic breccia Elephant Moraine 96008, *Meteorit. Planet. Sci., 44*, 805–821.

[510] Hill, E., L. A. Taylor, C. Floss, and Y. Liu (2009), Lunar meteorite LaPaz Icefield 04841: Petrology, texture, and impact-shock effects of a low-Ti mare basalt, *Meteorit. Planet. Sci., 44*, 87–94.

[511] Joy, K. H., I. A. Crawford, H. Downes, S. S. Russell, and A. T. Kearsley (2006), A petrological, mineralogical and chemical analysis of the lunar mare basalt meteorites LaPaz Icefield 02205, 02224 and 02226, *Meteorit. Planet. Sci., 41*, 1003–1025.

[512] Rankenburg, K., A. D. Brandon, and M. D. Norman (2007), A Rb-Sr and Sm-Nd isotope geochronology and trace element study of lunar meteorite LaPaz Icefield 02205, *Geochim. Cosmochim. Acta, 71*, 2120–2135.

[513] Righter, K., S. J. Collins, and A. D. Brandon (2005), Mineralogy and petrology of the LaPaz Icefield lunar mare basaltic meteorites, *Meteorit. Planet. Sci., 40*, 1703–1722.

[514] Spicuzza, M. J., J. M. D. Day, L. A. Taylor, and J. W. Valley (2007), Oxygen isotope constraints on the origin and differentiation of the Moon, *Earth and Planetary Science Letters, 253*, 254–265.

[515] Zeigler, R. A., R. L. Korotev , B. L. Jolliff, and L. A. Haskin (2005), Petrography and geochemistry of the LaPaz Icefield basaltic lunar meteorite and source crater pairing with Northwest Africa 032, *Meteorit. Planet. Sci. 40*, 1073–1101.

[516] Zhang, A., W. Hsu, Q. Li, Y. Liu, Y. Jiang, and G. Tang (2010), SIMS Pb/Pb dating of Zr-rich minerals in lunar meteorites Miller Range 05035 and LaPaz Icefield 02224: Implications for the petrogenesis of mare basalt, *Science China Earth Sciences, 53*, 327–334.

68. MAC 88105—LUNAR ANORTHOSITIC BRECCIA

[517] Cohen, B. A., T. D. Swindle, and D. A. Kring (2000), Support for the lunar cataclysm hypothesis from lunar meteorite impact melt ages. *Science, 290*, 1754–1756.

[518] Cohen, B. A., T. D. Swindle, and D. A. Kring (2005), Geochemistry and [40]Ar-[39]Ar geochronology of impact-melt clasts in feldspathic lunar meteorites: Implications for lunar bombardment history, *Meteorit. Planet. Sci., 40*, 755–777.

[519] Delano, J. W. (1991), Geochemical comparison of impact glasses from lunar meteorites ALH A81005 and MAC88105 and Apollo 16 regolith 64001, *Geochim. Cosmochim. Acta, 55*, 3019–3029.

[520] Eugster, O., M. Burger, U. Krähenbühl, Th. Michel, J. Beer, H. J. Hofmann, H. A. Synal, W. Woelfli, and R. C. Finkel (1991), History of the paired lunar meteorites MAC 88104 and MAC 88105 derived from noble gas isotopes, radionuclides, and some chemical abundances, *Geochim. Cosmochim. Acta, 55*, 3139–3148.

[521] Jolliff, B. L., R. L. Korotev, and L. A. Haskin (1991), A ferroan region of the lunar highlands as recorded in meteorites MAC 88104 and MAC 88105, *Geochim. Cosmochim. Acta, 55*, 3051–3071.

[522] Koeberl, C., G. Kurat, and F. Brandstätter (1991), MAC 88105—A regolith breccia from the lunar highlands: Mineralogical, petrological, and geochemical studies, *Geochim. Cosmochim. Acta, 55*, 3073-3087.

[523] Korotev, R. L., B. L. Jolliff, R. A. Zeigler, J. J. Gillis, and L. A. Haskin (2003), Feldspathic lunar meteorites and

their implications for compositional remote sensing of the lunar surface and the composition of the lunar crust, *Geochim. Cosmochim. Acta, 67,* 4895–4923.

[524] Lindstrom, M. M., C. Schwarz, R. Score, and B. Mason (1991), MacAlpine Hills 88104 and 88105 lunar highland meteorites: General description and consortium overview, *Geochim. Cosmochim. Acta, 55,* 2999–3007.

[525] Lindstrom, M. M., S. J. Wentworth, R. R. Martinez, D. W. Mittlefehldt, D. S. McKay, M.-S. Wang, and M. J. Lipschutz (1991), Geochemistry and petrography of the MacAlpine Hills lunar meteorites, *Geochim. Cosmochim. Acta, 55,* 3089–3103.

[526] Joy, K. H., I. A. Crawford, S. S. Russell, and A. T. Kearsley (2010), Lunar meteorite regolith breccias: An in situ study of impact melt composition using LA-ICP-MS with implications for the composition of the lunar crust, *Meteorit. Planet. Sci., 45,* 917–946.

[527] Neal, C. R., L. A. Taylor, Y. Liu, and R. A. Schmitt (1991), Paired lunar meteorites MAC 88104 and MAC 88105: A new "FAN" of lunar petrology, *Geochim. Cosmochim. Acta, 55,* 3037–3049.

[528] Nishiizumi, K., J. R. Arnold, J. Klein, D. Fink, R. Middleton, P. W. Kubik, P. Sharma, D. Elmore, and R. C. Reedy (1991), Exposure histories of lunar meteorites: ALH A81005, MAC 88104, MAC 88105, and Y791197, *Geochim. Cosmochim. Acta, 55,* 3149–3155.

[529] Nyquist, L. E., D. D. Bogard, C. Y. Shih, and H. Wiesmann (2002), Negative eNd in anorthositic clasts in Yamato 86032 and MAC 88105: Evidence for the LMO? *Lunar Planet. Sci., 33,* CD-ROM no. 1289, Lunar and Planetary Institute.

[530] Palme, H., B. Spettel, K. P. Jochum, G. Dreibus, H. Weber, G. Weckwerth, H. Wänke, A. Bischoff, and D. Stöffler (1991), Lunar highland meteorites and the composition of the lunar crust, *Geochim. Cosmochim. Acta, 55,* 3105–3122.

[531] Robinson, K. L., A. H. Treiman, and J. H. Joy (2012), Basaltic fragments in lunar feldspathic meteorites: Connecting sample analyses to orbital remote sensing, *Meteorit. Planet. Sci., 43,* 387–399.

[532] Sears, D. W. G., P. H. Benoit, H. Sears, J. D. Batchelor, and S. Symes (1991), The natural thermoluminescence of meteorites: III. Lunar and basaltic meteorites, *Geochim. Cosmochim. Acta, 55,* 3167–3180.

[533] Takeda, H., H. Mori, J. Saito, and M. Miyamoto (1991), Mineral-chemical comparisons of MAC 88105 with Yamato lunar meteorites, *Geochim. Cosmochim. Acta, 55,* 3009–3017.

[534] Taylor, G. J. (1991), Impact melts in the MAC 88105 lunar meteorite: Inferences for the lunar magma ocean hypothesis and the diversity of basaltic impact melts, *Geochim. Cosmochim. Acta, 55,* 3031–3036.

[535] Vogt, S., D. Fink, J. Klein, R. Middleton, B. Dockhorn, G. Korschinek, E. Nolte, and G. F. Herzog (1991), Exposure histories of the lunar meteorites: MAC88104, MAC88105, Y791197, and Y86032, *Geochim. Cosmochim. Acta, 55,* 3157–3165.

[536] Warren, P. H., and G. W. Kallemeyn (1991), The MacAlpine Hills lunar meteorite and implications of the lunar meteorites collectively for the composition and origin of the Moon, *Geochim. Cosmochim. Acta, 55,* 3123–3138.

69. ALH 84001—MARTIAN ORTHOPYROXENITE

[537] Antretter, M., M. Fuller, E. Scott, M. Jackson, B. Moskowitz, and P. Solheid (2003), Paleomagnetic record of martian meteorite ALH 84001, *J. Geophys. Res., 108,* 5049.

[538] Ash, R. D., S. F. Knott, and G. Turner (1996), A 4-Gyr shock age for a martian meteorite and implications for the cratering history of Mars, *Nature, 380,* 57–59.

[539] Bada, J. L., D. P. Glavin, G. D. McDonald, and L. Becker (1998), A search for endogenous amino acids in martian meteorite ALH 84001, *Science, 279,* 362–365.

[540] Barber, D. J., and E. R. D. Scott (2002), Origin of supposedly biogenic magnetite in the martian meteorite Allan Hills 84001, *Proc. Nat. Acad. Sci., 99,* 6556–6561.

[541] Barber, D. J., and E. R. D. Scott (2003), Transmission electron microscopy of minerals in the martian meteorite Allan Hills 84001, *Meteorit. Planet. Sci., 38,* 831–848.

[542] Barber, D. J., and E. R. D. Scott (2006), Shock and thermal history of martian meteorite Allan Hills 84001 from transmission electron microscopy, *Meteorit. Planet. Sci., 41,* 643–662.

[543] Barrat, J. A., and C. Bollinger (2010), Geochemistry of the martian meteorite ALH 84001, revisited, *Meteorit. Planet. Sci., 45,* 495–512.

[544] Beard, B. L., J. M. Ludois, T. J. Lapen, and C. M. Johnson (2013), Pre-4.0 billion year weathering on Mars constrained by Rb-Sr geochronology on meteorite ALH 84001, *Earth Planet. Sci. Lett., 361,* 173–182.

[545] Becker, L., D. P. Glavin, and J. L. Bada (1997), Polycyclic aromatic hydrocarbons (PAHs), in Antarctic martian meteorites, carbonaceous chondrites and polar ice, *Geochim. Cosmochim. Acta, 61,* 475–481.

[546] Becker, L., B. Popp, T. Rust, and J. L. Bada (1999), The origin of organic matter in the martian meteorite ALH 84001, Earth Planet. *Sci. Lett., 167,* 71–79.

[547] Bell, M. S. (2007), Experimental shock decomposition of siderite and the origin of magnetite in martian meteorite ALH 84001, *Meteorit. Planet. Sci.,* 935–949.

[548] Bishop, J. L., C. M. Pieters, T. Hiroi, and J. F. Mustard (1998), Spectroscopic analysis of martian meteorite Allan Hills 84001 powder and applications for spectral identification of minerals and other soil components on Mars, *Meteorit. Planet. Sci., 33,* 699–707.

[549] Borg, L. E., J. N. Connelly, L. E. Nyquist, C.-Y. Shih, H. Wiesmann, and Y. Reese (1999), The age of the carbonates in the martian meteorite ALH 84001, *Science, 286,* 90–94.

[550] Bouvier, A., J. Blichert-Toft, F. Albarede (2009), Martian meteorite chronology and the evolution of the interior of Mars, Earth Planet. *Sci. Lett., 280,* 285–295.

[551] Bradley, J. P., R. P. Harvey, and H. Y. McSween (1996), Magnetite whiskers and platelets in the ALH 84001

martian meteorite: Evidence of vapor phase growth, *Geochim. Cosmochim. Acta, 60,* 5149–5155.

[552] Bradley, J. P., H. Y. McSween, and R. P. Harvey (1998), Epitaxial growth of nanophase magnetite in martian meteorite Allan Hills 84001: Implications for biogenic mineralization, *Meteorit. Planet. Sci., 33,* 765–773.

[553] Brearley, A. J. (2003), Magnetite in ALH 84001: An origin by shock-induced thermal decomposition of iron carbonate, *Meteorit. Planet. Sci., 38,* 849–870.

[554] Buseck, P., et al. (2001), Magnetite morphology and life on Mars, *Proc. Nat. Acad. Sci. USA, 98,* 13490–13495.

[555] Cassata, W. S., D. L. Shuster, P. R. Renne, and B. P. Weiss (2012), Trapped Ar isotopes in meteorite ALH 84001 indicate Mars did not have a thick ancient atmosphere, *Icarus, 221,* 461–465.

[556] Clemett, S. J., M. T. Dulay, J. S. Gillette, X. D. F. Chiller, T. B. Mahajan, and R. N. Zare (1998), Evidence for the extraterrestrial origin of polycyclic aromatic hydrocarbons (PAHs), in the martian meteorite ALH 84001, *Faraday Discuss. R. Soc. Chem., 109,* 417–436, London.

[557] Collinson, D. W. (1997), Magnetic properties of martian meteorites: Implications for an ancient martian magnetic field, *Meteorit. Planet. Sci., 32,* 803–811.

[558] Cooney, T. F., E. R. D. Scott, A. N. Krot, S. K. Sharma, and A. Yamaguchi (1999), Vibrational spectroscopic study of minerals in the martian meteorite ALH 84001. *Am. Mineral., 84,* 1569–1576.

[559] Corrigan, C. M., and R. P. Harvey (2004), Multi-generational carbonate assemblages in martian meteorite Allan Hills 84001: Implications for nucleation, growth and alteration, *Meteorit. Planet. Sci., 39,* 17–30.

[560] Domeneghetti, M. C., A. M. Fioretti, F. Camara, G. Molin, and V. Tazzoli (2008), Thermal history of ALH 84001 meteorite by Fe^{+2}-Mg ordering in orthopyroxene, *Meteorit. Planet. Sci., 42,* 1703–1710.

[561] Eiler, J., J. W. Valley, C. M. Graham, and J. Fournelle (2002), Two populations of carbonate in ALH 84001: Geochemical evidence for discrimination and genesis, *Geochim. Cosmochim. Acta, 66,* 1285–1303.

[562] Eiler, J., N. Kitchen, L. Leshin, and M. Strausberg (2002), Hosts of hydrogen in Allan Hills 84001: Evidence for hydrous martian salts in the oldest martian meteorite, *Meteorit. Planet. Sci., 37,* 395–405.

[563] Eugster, O., A. Weigel, and E. Polnau (1997), Ejection times of martian meteorites, *Geochim. Cosmochim. Acta, 61,* 2749–2757.

[564] Eugster, O., H. Busemann, S. Lorenzetti, and D. Terribilini (2002), Ejection ages from ^{81}Kr-^{83}Kr dating and pre-atmospheric sizes of martian meteorites, *Meteorit. Planet. Sci., 37,* 1345–1360.

[565] Farquhar, J., M. H. Thiemens, and T. Jackson (1998), Atmosphere-surface interactions on Mars: Delta^{17}O measurements of carbonate from ALH 84001, *Science, 280,* 1580–1582.

[566] Fishler, D. K., and R. T. Cygan (1998), Cation diffusion in calcite: Determining closure temperatures and the thermal history for the Allan Hills 84001 meteorite, *Meteorit. Planet. Sci., 33,* 785–789.

[567] Folk, R. L., and L. A. Taylor (2002), Nanobacterial alteration of pyroxene in martian meteorite ALH 84001, *Meteorit. Planet. Sci., 37,* 1057–1069.

[568] Friedmann, E I., J. Wierzchos, C. Ascaso, and M. Winklhofer (2001), Chains of magnetite crystals in the meteorite ALH 84001, *Proc. Nat. Acad. Sci. USA, 98,* 2176–2181.

[569] Garrison, D. H., and D. D. Bogard (1998), Isotopic composition of trapped and cosmogenic noble gases in several martian meteorites, *Meteorit. Planet. Sci., 33,* 721–736.

[570] Gilmour, J. D., J. A. Whitby, and G. Turner (1998), Xenon isotopes in irradiated ALH 84001: Evidence for shock-induced trapping of ancient martian atmosphere, *Geochim. Cosmochim. Acta, 62,* 2555–2571.

[571] Gleason, J. D., D. A. Kring, D. H. Hill, and W. V. Boynton (1997), Petrography and bulk chemistry of martian orthopyroxenite ALH 84001: Implications for the origin of secondary carbonates, *Geochim. Cosmochim. Acta, 61,* 3503–3512.

[572] Golden, D. C., D. W. Ming, C. S. Schwandt, R. V. Morris, S. V. Yang, and G. E. Lofgren (2000), An experimental study of kinetically-driven precipitation of calcium-magnesian-iron carbonates from solution: Implications for the low-temperature formation of carbonates in martian meteorite ALH 84001, *Meteorit. Planet. Sci., 35,* 7457–465.

[573] Golden, D. C., D. W. Ming, C. S. Schwandt, H. V. Lauer, R. A. Socki, R. V. Morris, G. E. Lofgren, and G. A. McKay (2001), A simple inorganic process for the formation of carbonates, magnetite, and sulfides in martian meteorite ALH 84001, *Am. Mineral., 86,* 370–375.

[574] Golden, D. C., and nine authors (2004), Evidence for exclusively inorganic formation of magnetite in martian meteorite ALH 84001, *Am. Mineral., 89,* 681–695.

[575] Goswami, J. N., N. Sinha, S. V. S. Murty, R. K. Mohapatra, and C. J. Clement (1997), Nuclear tracks and light noble gases in Allan Hills 84001: Preatmospheric size, fall characteristics, cosmic-ray exposure duration and formation age, *Meteorit. Planet. Sci., 32,* 91–96.

[576] Grady, M. M., I. P. Wright, and C. T. Pillinger (1998), A nitrogen and argon stable isotope study of Allan Hills 84001: Implications for the evolution of the martian atmosphere, *Meteorit. Planet. Sci., 33,* 795–802.

[577] Greenwood, J. P., L. R. Riciputi, and H. Y. McSween (1997), Sulfide isotopic compositions in shergottites and ALH 84001 and possible implications for life on Mars, *Geochim. Cosmochim. Acta, 61,* 4449–4453.

[578] Greenwood, J. P., S. J. Mojzsis, and C. D. Coath (2000), Sulfur isotopic compositions of individual sulfides in martian meteorites ALH 84001 and Nakhla: Implications for crust-regolith exchange on Mars, *Earth Planet. Sci. Lett., 184,* 23–35.

[579] Greenwood, J. P., and H. Y. McSween (2001), Petrogenesis of Allan Hills 84001: Constraints from impact-melted feldspathic and silica glasses, *Meteorit. Planet. Sci., 36,* 43–61.

[580] Greenwood, J. P., R. E. Blake, and C. D. Coath (2003), Ion microprobe measurements of $^{18}O/^{16}O$ ratios of phosphate minerals in the martian meteorites ALH 84001 and Los Angeles, *Geochim. Cosmochim. Acta, 67*, 2289–2298.

[581] Harvey, R. P., and H. Y. McSween (1996), A possible high-temperature origin for the carbonates in martian meteorite ALH 84001, *Nature, 382*, 49–51.

[582] Jull, A. J. T., C. J. Eastoe , S. Xue, and G. F. Herzog (1995), Isotopic composition of carbonate in the SNC meteorites ALH 84001 and Nakhla, *Meteoritics, 30*, 311–318.

[583] Jull, A. J. T., C. J. Eastoe, and S. Cloudt (1997), Isotopic composition of carbonates in the SNC meteorites, Allan Hills 84001 and Zagami. *J. Geophys. Res., 102*, 1663–1669.

[584] Jull, A. J. T., C. Courtney, D. A. Jeffrey, and J. W. Beck (1998), Isotopic evidence for a terrestrial source of organic compounds found in martian meteorites Allan Hills 84001 and Elephant Moraine 79001, *Science, 279*, 366–369.

[585] Kent, A. J. R., I. D. Hutcheon, F. J. Ryerson, and D. L. Phinney (2001), The temperature of formation of carbonate in martian meteorite ALH 84001: Constraints from cation diffusion, *Geochim. Cosmochim. Acta, 65*, 311–321.

[586] Kirschvink, J. L., A. T. Maine, and H. Vali (1997), Paleomagnetic evidence of a low-temperature origin of carbonate in the martian meteorite ALH 84001, *Science, 275*, 1629–1633.

[587] Kong, P., M. Ebihara, and H. Palme (1999), Siderophile elements in martian meteorites and implications for core formation in Mars, *Geochim. Cosmochim. Acta, 63*, 1865–1875.

[588] Kopp, R. E., and M. Humayun (2003), Kinetic model of carbonate dissolution in martian meteorite ALH 84001, *Geochim. Cosmochim. Acta, 67*, 3247–3256.

[589] Koziol, A. M. (2004), Experimental determination of siderite stability and application to martian meteorite ALH 84001, *Am. Mineral., 89*, 294–300.

[590] Kring, D. A., T. D. Swindle, J. D. Gleason, and J. A. Grier (1998), Formation and relative ages of maskelynite and carbonate in ALH 84001, *Geochim. Cosmochim. Acta, 62*, 2155–2166.

[591] Lapen, T. J., M. Righter, A. D. Brandon, V. Debaille, B. L. Beard, J. T. Shafer, and A. H. Peslier (2010), A younger age for ALH 84001 and its geochemical link to shergottite sources in Mars, *Science, 328*, 347–350.

[592] Lee, D.-C., and A. N. Halliday (1997), Core formation on Mars and differentiated asteroids, *Nature, 388*, 854–857.

[593] Leshin, L. A., S. Epstein, and E. M. Stolper (1996), Hydrogen isotope geochemistry of SNC meteorites, *Geochim. Cosmochim. Acta, 60*, 2635–2650.

[594] Leshin, L. A., K. D. McKeegan, P. K. Carpenter, and R. P. Harvey (1998), Oxygen isotopic constraints on the genesis of carbonates from martian meteorite ALH 84001, *Geochim. Cosmochim. Acta, 62*, 3–13.

[595] Lugmair, G. W., and A. Shukolyukov (1998), Early solar system timescales according to ^{53}Mn-^{53}Cr systematics, *Geochim. Cosmochim. Acta, 62*, 2863–2886.

[596] Marty, B., and K. Marti (2002), Signatures of early differentiation of Mars, Earth Planet. *Sci. Lett., 196*, 251–263.

[597] Mathew, K. J., J. S. Kim, and K. Marti (1999), Martian atmospheric and indigenous components of xenon and nitrogen in the Shergotty, Nakhla and Chassigny group meteorites, *Meteorit. Planet. Sci., 33*, 655–664.

[598] Mathew, K. J., and K. Marti (2001), Early evolution of martian volatiles: Nitrogen and noble gas components in ALH 84001 and Chassigny, *J. Geophys. Res., 106*, 1401–1422.

[599] McKay, D. S., E. K. Gibson, K. L. Thomas-Keprta, H. Vali, C. S. Romanek, S. J. Clemett, X. D. F. Chillier, C. R. Maechling, and R. N. Zare (1996), Search for life on Mars: Possible relic biogenic activity in martian meteorite ALH 84001, *Science, 273*, 924–930.

[600] Mittlefehldt, D. W. (1994), ALH 84001, A cumulate orthopyroxenite member of the martian meteorite clan, *Meteoritics, 29*, 214–221.

[601] Miura, Y. N., K. Nagao, N. Sugiura, H. Sagawa, and K. Matsubara (1995), Orthopyroxenite ALH 84001 and shergottite ALH A77005: Additional evidence for a martian origin from noble gases, *Geochim. Cosmochim. Acta, 59*, 2105–2113.

[602] Miura, Y. N., and N. Sugiura (2000), Martian atmosphere-like nitrogen in the orthopyroxenite ALH 84001, *Geochim. Cosmochim. Acta, 64*, 559–572.

[603] Murty, S. V. S., and R. K. Mohapatra (1997), Nitrogen and heavy noble gasses in ALH 84001: Signatures of ancient martian atmosphere, *Geochim. Cosmochim. Acta, 61*, 5417–5428.

[604] Niles, P. B., L. A. Leshin, and Y. Guan (2005), Microscale carbon isotopic variability in ALH 84001 carbonates and a discussion of possible formation environments, *Geochim. Cosmochim. Acta, 69*, 2931–2944.

[605] Niles, P. B., M. Y. Zolotov, L. A. Leshin (2009), Insights into the formation of Fe- and Mg-rich aqueous solutions on early Mars provided by the ALH 84001 carbonates, Earth Planet. *Sci. Lett., 286*, 122–130.

[606] Pasteris, J. D., and B. Wopenka (2003), Necessary, but not sufficient: Raman identification of disordered carbon as a signature of ancient life, *Astrobiology, 3*, 727–737.

[607] Romanek, C. S., M. M. Grady, I. P. Wright, D. W. Mittlefehldt, R. A. Socki, C. T. Pillinger, and E. K. Gibson (1994), Record of fluid-rock interactions on Mars from the meteorite ALH 84001, *Nature, 372*, 655–657.

[608] Saxton, J. M., I. C. Lyon, and G. Turner (1998), Correlated chemical and isotopic zoning in carbonates in the martian meteorite ALH 84001, Earth Planet. *Sci. Lett., 160*, 811–822.

[609] Scott, E. R. D (1999), Origin of carbonate-magnetite-sulfide assemblages in martian meteorite ALH 84001, *J. Geophys. Res., 104*, 3803–3813.

[610] Scott, E. R. D., A. Yamaguchi, and A. N. Krot (1997), Petrological evidence for shock melting of carbonates in the martian meteorite ALH 84001, *Nature*, *387*, 377–379.

[611] Scott, E. R. D., A. N. Krot, and A. Yamaguchi (1998), Carbonates in fractures of martian meteorite Allan Hills 84001: Petrologic evidence for impact origin, *Meteorit. Planet. Sci.*, *33*, 709–719.

[612] Seitz, H.-M., G. P. Brey, S. Weyer, S. Durali, U. Ott, C. Munker, and K. Mezger (2006), Lithium isotopic compositions of martian and lunar reservoirs, Earth Planet Sci. *Lett.*, *245*, 6–18.

[613] Shearer, C. K., G. D. Layne, J. J. Papike, and M. N. Spilde (1996), Sulfur isotopic systematics in alteration assemblages in martian meteorite ALH 84001, *Geochim. Cosmochim. Acta*, *60*, 2921–2926.

[614] Shearer, C. K., L. A. Leshin, and C. T. Adcock (1999), Olivine in martian meteorite Allan Hills 84001: Evidence for a high-temperature origin and implications for signs of life, *Meteorit. Planet. Sci.*, *34*, 331–339.

[615] Steele, A., D. T. Goddard, I. B. Beech, R. C. Tapper, D. Stapleton, and J. R. Smith (1998), Atomic force microscopy imaging of fragments of martian meteorite ALH 84001, J. *Microscopy*, *189*, 2–7.

[616] Steele, A., D. T. Goddard, D. Stapleton, J. K. W. Toporski, V. Peters, V. Bassinger, G. Sharples, D. D. Wynn-Williams, and D. S. McKay (2000), Investigations into an unknown organism on the martian meteorites Allan Hills 84001, *Meteorit. Planet. Sci.*, *35*, 237–241.

[617] Steele, A., M. D. Fries, H. E. F. Amundsen, B. O. Mysen, M. L. Fogel, M. Schweizer, and N. Z. Boctor (2007), Comprehensive imaging and Raman spectroscopy of carbonate globules from martian meteorite ALH 84001 and a terrestrial analogue from Svalbard, *Meteorit. Planet. Sci.*, *42*, 1549–1566.

[618] Stephan, T., E. K. Jessberger, C. H. Heiss, and D. Rost (2003), TOF—SIMS analysis of polycyclic aromatic hydrocarbons in Allan Hills 84001, *Meteorit. Planet. Sci.*, *38*, 109–116.

[619] Sugiura, N., and H. Hoshino (2000), Hydrogen-isotopic compositions in Allan Hills 84001 and the evolution of the martian atmosphere, *Meteorit. Planet. Sci. 35*, 373–380.

[620] Swindle, T. D., J. A. Grier, and M. K. Burkland (1995), Noble gases in orthopyroxenite ALH 84001: A different kind of martian meteorite with an atmospheric signature, *Geochim. Cosmochim. Acta*, *59*, 793–801.

[621] Terada, K., T. Monde, and Y. Sano (2004), Ion microprobe U-Th-Pb dating of phosphates in martian meteorite ALH 84001, *Meteorit. Planet. Sci.*, *38*, 1697–1703.

[622] Thomas-Keprta, K. L., D. A. Bazylinski, J. L. Kirschvink, S. J. Clemett, D. S. McKay, S. J. Wentworth, H. Vali, E. K. Gibson, and C. S. Romanek (2000), Elongated prismatic magnetite crystals in ALH 84001 carbonate globules: Potential martian magnetofossils, *Geochim. Cosmochim. Acta*, *64*, 4049–4081.

[623] Thomas-Keprta, K. L, S. J. Clemett, D. A. Bazylinski, J. L. Kirschvink, D. S. McKay, S. J. Wentworth, H. Vali, E. K. Gibson, M. F. McKay, and C. S. Romanek (2001), Truncated hexa-octahedral magnetite crystals in ALH 84001: Presumptive biosignatures, *Proc. Nat. Acad. Sci.*, *98*, 2164–2169.

[624] Thomas-Keprta, K. L., S. J. Clemett, D. S. McKay, E. K. Gibson, and S. J. Wentworth (2009), Origins of magnetite nanocrystals in martian meteorite ALH 84001, *Geochim. Cosmochim. Acta*, *73*, 6631–6677.

[625] Treiman, A. H. (1995), A petrographic history of martian meteorite ALH 84001: Two shocks and an ancient age, *Meteoritics*, *30*, 294–302.

[626] Treiman, A. H. (1998), The history of Allan Hills 84001 revised: Multiple shock events, *Meteorit. Planet. Sci.*, *33*, 753–764.

[627] Treiman, A. H. (2003), Submicron magnetite grains and carbon compounds in martian meteorite ALH 84001: Inorganic abiotic formation by shock and thermal metamorphism, *Astrobiology*, *3*, 369–392.

[628] Treiman, A. H., and E. J. Essene (2011), Chemical composition of magnetite in martian meteorite ALH 84001: Revised appraisal from thermochemistry of phases in Fe-Mg-C-O, *Geochim. Cosmochim. Acta*, *75*, 5324–5335.

[629] Treiman, A. H., and C. S. Romanek (1998), Bulk and stable isotopic compositions of carbonate minerals in martian meteorite Allan Hills 84001: No proof of high formation temperature, *Meteorit. Planet. Sci.*, *33*, 737–742.

[630] Turner, G., S. F. Knott, R. D. Ash, and J. D. Gilmour (1997), Ar-Ar chronology of the martian meteorite ALH 84001: Evidence for the timing of the early bombardment of Mars, *Geochim. Cosmochim. Acta*, *61*, 3835–3850.

[631] Valley, J. W., J. M. Eiler, C. M. Graham, E. K. Gibson, C. S. Romanek, and E. M. Stolper (1997), Low-temperature carbonate concretions in the martian meteorite ALH 84001: Evidence from stable isotopes and mineralogy, *Science*, *275*, 1633–1637.

[632] Wadhwa, M., and G. Crozaz (1998), The igneous crystallization history of an ancient martian meteorite from rare earth element microdistributions, *Meteorit. Planet. Sci.*, *33*, 685–692.

[633] Warren, P. H. (1998), Petrologic evidence for low-temperature, possibly flood-evaporitic origin of carbonates in the ALH 84001 meteorite, *J. Geophys. Res.*, *103*, 759–773.

[634] Warren, P. H., and G. W. Kallemeyn (1996), Siderophile trace elements in ALH 84001, other SNC meteorites and eucrites: Evidence of heterogeneity, possibly, time-linked, in the mantle of Mars, *Meteorit. Planet. Sci.*, *31*, 97–105.

[635] Weiss, B. P., J. L. Kirschvink, F. J. Baudenbacher, H. Vali, N. T. Peters, F. A. Macdonald, and J. P. Wikswo (2000), A low temperature transfer of ALH 84001 from Mars to Earth. *Science*, *290*, 791–795.

[636] Weiss, B. P., H. Vali, F. J. Baudenbacher, J. L. Kirschvink, S. T. Stewart, and D. L. Shuster (2002), Records of an ancient martian magnetic field in ALH 84001, Earth Planet. *Sci. Lett.*, *201*, 449–463.

[637] Weiss, B. P., D. L. Shuster, and S. T. Stewart (2002), Temperatures on Mars from $^{40}Ar/^{39}Ar$ thermochronology of ALH 84001, Earth Planet. Sci. Lett., 201, 465–472.

[638] Zolotov, M. Y., and E. L. Shock (2000), An abiotic origin for hydrocarbons in the Allan Hills 84001 martian meteorite through cooling of magmatic and impact-generated gases, Meteorit. Planet. Sci., 35, 629–638.

70. EET A79001—SHERGOTTITE

[639] Becker, L., D. P. Glavin, and J. L. Bada (1997), Polycyclic aromatic hydrocarbons (PAHs) in Antarctic martian meteorites, carbonaceous chondrites and polar ice, Geochim. Cosmochim. Acta, 61, 475–481.

[640] Becker, R. H., and R. O. Pepin (1984), The case for a martian origin of the shergottites: Nitrogen and noble gases in EET A79001, Earth Planet. Sci. Lett., 69, 225–242.

[641] Bogard, D. D., F. Hörz, and P. Johnson (1986), Shock-implanted noble gases: An experimental study with implications for the origin of martian gases in shergottite meteorites, J. Geophys. Res., 91(Suppl.), E99–E114.

[642] Bhandari, N., J. N. Goswami, R. Jha, D. Sen Gupta, and P. N. Shukla (1986), Cosmogenic effects in shergottites, Geochim. Cosmochim. Acta, 50, 1023–1030.

[643] Blichert-Toft, J., J. D. Gleason, P. Telouk, and F. Albarede (1999), The Lu-Hf isotope geochemistry of shergottites and the evolution of the martian mantle-crust system, Earth Planet. Sci. Lett., 173, 25–39.

[644] Bogard, D. D., and P. Johnson (1983), Martian gases in an Antarctic meteorite? Science, 221, 651–654.

[645] Bogard, D. D., L. E. Nyquist, and P. Johnson (1984), Noble gas contents of shergottites and implications for the martian origin of SNC meteorites, Geochim. Cosmochim. Acta, 48, 1723–1739.

[646] Bogard, D. D., and D. H. Garrison (1998), Relative abundances of Ar, Kr and Xe in the martian atmosphere as measured in martian meteorites, Geochim. Cosmochim. Acta, 62, 1829–1835.

[647] Bogard, D. D., and D. H. Garrison (1999), Argon-39-argon-40 "ages" and trapped argon in martian shergottites, Chassigny and Allan Hills 84001, Meteorit. Planet. Sci., 34, 451–473.

[648] Brandon, A. D., R. J. Walker, J. W. Morgan, and G. G. Goles (2000), Re-Os isotopic evidence for early differentiation of the martian mantle, Geochim. Cosmochim. Acta, 64, 4083–4095.

[649] Carr, R. H., I. P. Wright, and C. T. Pillinger (1985), Carbon isotopes in three SNC meteorites. Proc, Lunar Planet. Sci. Conf., 15, J. Geophys. Res., 90(Suppl.), C664–C668.

[650] Carr, R. H., M. M. Grady, I. P. Wright, and C. T. Pillinger (1985), Martian atmospheric carbon dioxide and weathering products in SNC meteorites, Nature, 314, 248–250.

[651] Cisowski, S. M. (1986), Magnetic studies on Shergotty and other SNC meteorites, Geochim. Cosmochim. Acta, 50, 1043–1048.

[652] Clayton, R. N., and T. K. Mayeda (1988), Isotopic composition of carbonate in EET A79001 and its relation to parent body volatiles, Geochim. Cosmochim. Acta, 52, 925–927.

[653] Collinson, D. W. (1986), Magnetic properties of Antarctic shergottite meteorites EET A79001 and ALH A77005: Possible relevance to a martian magnetic field, Earth Planet. Sci. Lett., 77, 159–164.

[654] Collinson, D. W. (1997), Magnetic properties of martian meteorites: Implications for an ancient martian magnetic field, Meteorit. Planet. Sci., 32, 803–811.

[655] Eugster, O., A. Weigel, and E. Polnau (1997), Ejection times of martian meteorites, Geochim. Cosmochim. Acta, 61, 2749–2757.

[656] Farquhar, J., J. Savarino, T. L. Jackson, and M. H. Thiemens (2000), Evidence of atmospheric sulphur in the martian regolith from sulphur isotopes in meteorites. Nature, 404, 50–52.

[657] Garrison, D. H., and D. D. Bogard (1998), Isotopic composition of trapped and cosmogenic noble gases in several martian meteorites, Meteorit. Planet. Sci., 33, 721–736.

[658] Gibson, E. K., C. B. Moore, T. M. Primus, and C. F. Lewis (1985), Sulfur in achondritic meteorites, Meteoritics, 20, 503–511.

[659] Gooding, J. L., and D. W. Muenow (1986), Martian volatiles in shergottite EET A79001: New evidence from oxidized sulfur and sulfur-rich alumino-silicates, Geochim. Cosmochim. Acta, 50, 1049–1059.

[660] Gooding, J. L., S. J. Wentworth, and M. E. Zolensky (1988), Calcium carbonate and sulfate of possible extraterrestrial origin in the EET A79001 meteorite, Geochim. Cosmochim. Acta, 52, 909–915.

[661] Greenwood, J. P., L. R. Riciputi, and H. Y. McSween (1997), Sulfide isotopic compositions in shergottites and ALH 84001 and possible implications for life on Mars, Geochim. Cosmochim. Acta, 61, 4449–4453.

[662] Harper, C. L., L. E. Nyquist, B. Bansal, H. Weismann, and C. Y. Shih (1995), Rapid accretion and early differentiation of Mars indicated by $^{142}Nd/^{144}Nd$ in SNC meteorites, Science, 267, 213–217.

[663] Herd, C. D. K., J. J. Papike, and A. J. Brearley (2001), Oxygen fugacity of martian basalts from electron microprobe oxygen and TEM-EELS analysis of Fe-Ti oxides. Am. Mineral., 86, 1015–1024.

[664] Herd, C. D. K., C. S. Schwandt, J. H. Jones, and J. J. Papike (2002), An experimental and petrographic investigation of Elephant Moraine 79001 lithology A: Implications for its petrogenesis and the partitioning of chromium and vanadium in a martian basalt, Meteorit. Planet. Sci., 37, 987–1000.

[665] Jull, A. J. T., and D. J. Donahue (1988), Terrestrial ^{14}C age of the Antarctic shergottite EET A79001, Geochim. Cosmochim. Acta, 52, 1309–1311.

[666] Jull, A. J. T., C. Courtney, D. A. Jeffrey, and J. W. Beck (1998), Isotopic evidence for a terrestrial source of organic compounds found in martian meteorites Allan

Hills 84001 and Elephant Moraine 79001, *Science, 279,* 366–369.

[667] Kaiden, H., T. Mikouchi, and M. Miyamoto (1998), Cooling rates of olivine xenocrysts in the EET 79001 shergottite, *Antarctic Meteorite Res., 11,* 92–102, NIPR.

[668] Karlsson, H. R., R. N. Clayton, E. K. Gibson Jr., and T. K. Mayeda (1992), Water in SNC meteorites: Evidence for a martian hydrosphere, *Science, 255,* 1409–1411.

[669] Kleine, T., K. Mezger, C. Münker, H. Palme, and A. Bischoff (2004), ^{182}Hf-^{182}W isotope systematics of chondrites, eucrites, martian meteorites: Chronology of core formation and early mantle differentiation in Vesta and Mars, *Geochim. Cosmochim. Acta, 68,* 2935–2946.

[670] Kong, P., M. Ebihara, and H. Palme (1999), Siderophile elements in martian meteorites and implications for core formation in Mars, *Geochim. Cosmochim. Acta, 63,* 1865–1875.

[671] Lee, D.-C., and A. N. Halliday (1997), Core formation on Mars and differentiated asteroids, *Nature, 388,* 854–857.

[672] Lentz, R. C. F., and H. Y. McSween (2000), Crystallization of the basaltic shergottites: Insights from crystal size distribution (CSD), analysis of pyroxenes, *Meteorit. Planet. Sci. 35,* 919–927.

[673] Leshin, L. A., S. Epstein, and E. M. Stolper (1996), Hydrogen isotope geochemistry of SNC meteorites, *Geochim. Cosmochim. Acta, 60,* 2635–2650.

[674] Liu, Y., J. B. Balta, C. A. Goodrich, H. Y. McSween, L. A. Taylor (2013), New constraints on the formation of shergottite Elephant Moraine 79001 lithology A, *Geochim. Cosmochim. Acta, 108,* 1–20.

[675] Lugmair, G. W., and A. Shukolyukov (1998), Early solar system timescales according to ^{53}Mn-^{53}Cr systematics, *Geochim. Cosmochim. Acta, 62,* 2863–2886.

[676] McDonald, G. D., and J. L. Bada (1995), A search for endogenous amino acids in the martian meteorite EET A79001, *Geochim. Cosmochim. Acta, 59,* 1179–1184.

[677] McSween, H. Y., and E. Jarosewich (1983), Petrogenesis of the EET A79001 meteorite: Multiple magma pulses on the shergottite parent body, *Geochim. Cosmochim. Acta, 47,* 1501–1513.

[678] Mikouchi, T., M. Miyamoto, and G. A. McKay (1998), Mineralogy of Antarctic basaltic shergottite Queen Alexandra Range 94201: Similarities to Elephant Moraine A79001 (Lithology B), Martian meteorite, *Meteorit. Planet. Sci., 33,* 181–189.

[679] Mikouchi, T., M. Miyamoto, and G. A. McKay (1999), The role of undercooling in producing igneous zoning trends in pyroxenes and maskelynites among basaltic martian meteorites, *Earth Planet. Sci. Lett., 173,* 235–256.

[680] Mittlefehldt, D. W., D. J. Lindstrom, M. M. Lindstrom, and R. R. Martinez (1999), An impact-melt origin for lithology A of martian meteorite Elephant Moraine A79001, *Meteorit. Planet. Sci., 34,* 357–367.

[681] Niekerk, D., C. A. Goodrich, G. J. Taylor, and K. Keil (2007), Characterization of the lithological contact in the shergottite EET A79001: A record of igneous

differentiation processes on Mars, *Meteorit. Planet. Sci., 42,* 1751–1762.

[682] Nishiizumi, K., J. Klein, R. Middleton, D. Elmore, P. W. Kubik, and J. R. Arnold (1986), Exposure history of shergottites, *Geochim. Cosmochim. Acta, 50,* 1017–1021.

[683] Rajan, R. S., G. W. Lugmair, A. S. Tamhane, and G. Poupeau (1986), Nuclear tracks, Sm isotopes and neutron capture effects in the Elephant Moraine shergottite, *Geochim. Cosmochim. Acta, 50,* 1039–1042.

[684] Reid, A. M., and R. Score (1981), A preliminary report on the achondrite meteorites in the 1979 U.S. Antarctic Meteorite Collection, Proc. 6th Symp. Antarctic Meteorites, in Mem. *Natl. Inst. Polar Res. Spec. Iss., 20,* 33–52.

[685] Sarafin, R., U. Herpers, P. Signer, R. Wieler, G. Bonani, H. J. Hofmann, E. Morenzoni, M. Nessi, M. Suter, and W. Wölfli (1985), ^{10}Be, ^{26}Al, ^{53}Mn and light noble gases in the Antarctic shergottite EET A79001(A), *Earth Planet. Sci. Lett., 75,* 72–76.

[686] Schwenzer, S. P., S. Hermann, R. Mohapatra, and U. Ott (2007), Noble gases in mineral separates from three shergottites: Shergotty, Zagami and EET A79001, *Meteorit. Planet. Sci., 42,* 387–412.

[687] Seitz, H.-M., G. P. Brey, S. Weyer, S. Durali, U. Ott, C. Munker, and K. Mezger (2006), Lithium isotopic compositions of martian and lunar reservoirs, *Earth Planet. Sci. Lett., 245,* 6–18.

[688] Smith, M. R., J. C. Laul, M.-S. Ma, T. Huston, R. M. Verkouteren, M. E. Lipschutz, and R. A. Schmitt (1984), Petrogenesis of the SNC (shergottites, nakhlites, chassignites), meteorites: Implications for their origin from a large, dynamic planet, possibly Mars. *Proc. 14th Lunar Planet. Sci. Conf., in J. Geophys. Res., 89*(Suppl.), B612–B630.

[689] Steele, I. M., and J. V. Smith (1982), Petrography and mineralogy of two basalts and olivine-pyroxene-spinel fragments in achondrite EET A79001, Proc. 13th Lunar Planet. Sci. Conf., in *J. Geophys. Res., 87,* A375–384.

[690] Sunshine, J. M., L. A. McFadden, and C. M. Pieters (1993), Reflectance spectra of the Elephant Moraine A79001 meteorite: Implications for remote sensing of planetary bodies, *Icarus, 105,* 79–91.

[691] Swindle, T. D., M. W. Caffee, and C. M. Hohenberg (1986), Xenon and other noble gases in shergottites, *Geochim. Cosmochim. Acta, 50,* 1001–1015.

[692] Wadhwa, M., H. Y. McSween, and G. Crozaz (1994), Petrogenesis of shergottite meteorites inferred from minor and trace element microdistributions, *Geochim. Cosmochim. Acta, 58,* 4213–4229.

[693] Walton, E. L., P. J. Jugo, C. D. K. Herd, and M. Wilke (2010), Martian regolith in Elephant Moraine 79001 shock melts? Evidence from major element composition and sulfur speciation, *Geochim. Cosmochim. Acta, 74,* 4829–4843.

[694] Walton, E. L. (2013), Shock metamorphism of Elephant Moraine A79001: Implications for olivine-ringwoodite transformation and the complex thermal history of heavily shocked martian meteorites, *Geochim. Cosmochim. Acta, 107,* 299–315.

[695] Wang, A., K. E. Kuebler, B. L. Jolliff, and L. A. Haskin (2004), Raman spectroscopy of Fe-Ti-Cr-oxides, case study: Martian meteorite EET A79001, *Am. Mineral.*, *89*, 665–680.

[696] Warren, P. H., and G. W. Kallemeyn (1997), Yamato-793605, EET 79001 and other presumed martian meteorites: Compositional clues to their origins, *Antarctic Meteorite Research*, *10*, 61–81.

[697] Wiens, R. C. (1988), Noble gases released by vacuum crushing of EET A79001 glass, Earth Planet. *Sci. Lett.*, *91*, 55–65.

[698] Wiens, R. C., R. H. Becker, and R. O. Pepin (1986), The case for a martian origin of the shergottites: II. Trapped and indigenous gas components in EET A79001 glass, *Earth Planet. Sci. Lett.*, *77*, 149–158.

[699] Wiens, R. C., and R. O. Pepin (1988), Laboratory shock emplacement of noble gases, nitrogen, and carbon dioxide into basalt, and implications for trapped gases in shergottite EET A79001, *Geochim. Cosmochim. Acta*, *52*, 295–307.

[700] Wright, I. P., M. M. Grady, and C. T. Pillinger (1988), Carbon, oxygen and nitrogen isotopic compositions of possible martian weathering products in EET A79001, *Geochim. Cosmochim. Acta*, *52*, 917–924.

71. QUE 94201—BASALTIC SHERGOTTITE

[701] Aramovich, C. J., C. D. K. Herd, and J. J. Papike (2002), Symplectites derived from metastable phases in martian basaltic meteorites, *Am. Mineral.*, *87*, 1351–1359.

[702] Blichert-Toft, J., J. D. Gleason, P. Telouk, and F. Albarede (1999), The Lu-Hf isotope geochemistry of shergottites and the evolution of the martian mantle-crust system,. *Earth Planet Sci. Lett.*, *173*, 25–39.

[703] Borg, L. E., L. E. Nyquist, L. A. Taylor, H. Wiesmann, and C.-Y. Shih (1997), Constraints on martian differentiation processes from Rb-Sr and Sm-Nd isotopic analyses of the basaltic shergottite QUE 94201, *Geochim. Cosmochim. Acta*, *61*, 4915–4931.

[704] Eugster, O., A. Weigel, and E. Polnau (1997), Ejection times of martian meteorites, *Geochim. Cosmochim. Acta*, *61*, 2749–2757.

[705] Eugster, O., H. Busemann, S. Lorenzetti, and D. Terribilini (2002), Ejection ages from ^{81}Kr-^{83}Kr dating and pre-atmospheric sizes of martian meteorites, *Meteorit. Planet. Sci.*, *37*, 1345–1360.

[706] Gaffney, A. M., L. E. Borg, and J. N. Connelly (2007), Uranium-lead isotope systematics of Mars inferred from the basaltic shergottite QUE 94201, *Geochim. Cosmochim. Acta*, *71*, 5016–5031.

[707] Garrison, D. H., and D. D. Bogard (1998), Isotopic composition of trapped and cosmogenic noble gases in several martian meteorites, *Meteorit. Planet. Sci.*, *33*, 721–736.

[708] Greenwood, J. P., L. R. Riciputi, and H. Y. McSween (1997), Sulfide isotopic compositions in shergottites and ALH 84001 and possible implications for life on Mars, *Geochim. Cosmochim. Acta*, *61*, 4449–4453.

[709] Karner, J. M., J. J. Papike, S. R. Sutton, C. K. Shearer, G. McKay, L. Le, and P. Burger (2007), Valance state partitioning of Cr between pyroxene-melt: Effects of pyroxene and melt composition and direct determination of Cr valance states by XANES. Application to martian basalt QUE 94201 composition, *Am. Mineral.*, *92*, 2002–2005.

[710] Karner J. M., J. J. Papike, C. K. Shearer, G. McKay, L. Le, and P. Burger (2007), Valance state partitioning of Cr and V between pyroxene-melt: Estimates of oxygen fugacity for martian basalt QUE 94201, *Am. Mineral.*, *92*, 1238–1241.

[711] Karner, J. M., J. J. Papike , S. R. Sutton, C. K. Schearer, P. Burger, G. McKay, and L. Le (2008), Valance state partitioning of V betweeen pyroxene-melt: Effects of pyroxene and melt composition, and direct determination of V valance states by XANES. Application to martian basalt QUE 94201 composition, *Meteor. Planet. Sci.*, *43*, 1275–1285.

[712] Kring, D. A., J. D. Gleason, T. D. Swindle, K. Nishiizumi, M. W. Caffee, D. H. Hill, A. J. T. Jull, and W. V. Boynton (2003), Composition of the first bulk melt sample from a volcanic region on Mars: Queen Alexandra Range 94201, *Meteorit. Planet. Sci.*, *38*, 1833–1848.

[713] Lentz, R. C. F., and H. Y. McSween (2000), Crystallization of the basaltic shergottites: Insights from crystal size distribution (CSD), analysis of pyroxenes, *Meteorit. Planet. Sci.*, *35*, 919–927.

[714] Leshin L. A. (2000), Insight into martian water reservoirs from analyses of martian meteorite QUE 94201, *Geophys. Res. Lett.*, *27*, 2017–2020.

[715] McKay, G., E. Koizumi, T. Mikouchi, L. Le, and C. Schwandt (2002), Crystallization of shergottite QUE 94201: An experimental study, *Lunar Planet. Sci. Conf.*, *33*, abstract 20.

[716] McKay, G., S.-R. Yang, and J. Wagstaff (1996), Complex zoned pyroxenes in shergottite QUE 94201: Evidence for a two-stage crystallization history, *Lunar Planet. Sci. Conf.*, *27*, abstract 851.

[717] McSween, H. Y., D. D. Eisenhour, L. A. Taylor, M. Wadhwa, and G. Crozaz (1996), QUE 94201 shergottite: Crystallization of a martian basaltic magma, *Geochim. Cosmochim. Acta*, *60*, 4563–4569.

[718] Mikouchi, T., M. Miyamoto, and G. A. McKay (1998), Mineralogy of Antarctic basaltic shergottite Queen Alexandra Range 94201: Similarities to Elephant Moraine A79001 (Lithology B), martian meteorite, *Meteorit. Planet. Sci.*, *33*, 181–189.

[719] Mikouchi, T., M. Miyamoto, and G. A. McKay (1999), The role of undercooling in producing igneous zoning trends in pyroxenes and maskelynites among basaltic martian meteorites, Earth Planet. *Sci. Lett.*, *173*, 235–256.

[720] J. J. Papike, J. M. Karner, C. K. Shearer, and P. V Burger (2009), Silicate mineralogy of martian meteorites, *Geochim. Cosmochim. Acta*, *73*, 7443–7485.

[721] Wadhwa, M., G. Crozaz, L. A. Taylor, and H. Y. McSween (1998), Martian basalt (shergottite), Queen Alexandra Range 94201 and lunar basalt 15555, A tale of two pyroxenes, *Meteorit. Planet. Sci.*, *33*, 321–328.

[722] Wang, M.-S., J. A. Mokos, and M. E. Lipschutz (1999), Martian meteorites: Volatile trace elements and cluster analysis, *Meteorit. Planet. Sci.*, *33*, 671–675.

72. ALH A77005—LHERZOLITIC SHERGOTTITE

[723] Berkley, J. L., and K. Keil (1981), Olivine orientation in the ALH A77005 achondrite. *Am. Mineral.*, *66*, 1233–1236.

[724] Bhandari, N., J. N. Goswami, R. Jha, D. Sen Gupta, and P. N. Shukla (1986), Cosmogenic effects in shergottites, *Geochim. Cosmochim. Acta*, *50*, 1023–1030.

[725] Biswas, S., Ngo, H. T., and M. E. Lipschutz (1980), Trace element contents of selected Antarctic meteorites: I. Weathering effects and A77005, A77257, A77278 and A77299, *Zeitschrift für Naturforschung*, *35a*, 191–196.

[726] Blichert-Toft, J., J. D. Gleason, P. Telouk, and F. Albarede (1999), The Lu-Hf isotope geochemistry of shergottites and the evolution of the martian mantle-crust system, *Earth Planet. Sci. Lett.*, *173*, 25–39.

[727] Bogard, D. D., L. E. Nyquist, and P. Johnson (1984), Noble gas contents of shergottites and implications for the martian origin of SNC meteorites, *Geochim. Cosmochim. Acta*, *48*, 1723–1739.

[728] Borg, L. E., L. E. Nyquist, H. Wiesmann, and Y. Reese (2002), Constraints on the petrogenesis of martian meteorites from the Rb-Sr and Sm-Nd isotopic systematics of the lherzolitic shergottites ALH A77005 and LEW 88516, *Geochim. Cosmochim. Acta*, *66*, 2037–2053.

[729] Brandon, A. D., R. J. Walker, J. W. Morgan, and G. G. Goles (2000), Re-Os isotopic evidence for early differentiation of the martian mantle, *Geochim. Cosmochim. Acta*, *64*, 4083–4095.

[730] Burgess, R., I. P. Wright, and C. T. Pillinger (1989), Distribution of sulphides and oxidized sulphur components in SNC meteorites, *Earth Planet. Sci. Lett.*, *93*, 314–320.

[731] Calvin, C., and M. J. Rutherford (2008), The parental melt of lherzolitic shergottite ALH 77005: A study of rehomogenized melt inclusions, *Am. Mineral.*, *93*, 1886–1898.

[732] Collinson, D. W. (1986), Magnetic properties of Antarctic shergottite meteorites EET A79001 and ALH A77005: Possible relevance to a martian magnetic field, *Earth Planet. Sci. Lett.*, *77*, 159–164.

[733] Edmunson, J., L. E. Borg, C. K. Shearer, and J. J. Papike (2005), Defining the mechanisms that disturb the Sm-Nd isotopic systematics of the martian meteorites: Examples from Dar al Gani and Allan Hills 77005, *Meteorit. Planet. Sci.*, *40*, 1159–1174.

[734] Eugster, O., A. Weigel, and E. Polnau (1997), Ejection times of martian meteorites, *Geochim. Cosmochim. Acta*, *61*, 2749–2757.

[735] Garrison, D. H., M. N. Rao, and D. D. Bogard (1995), Solar-proton-produced neon in shergottite meteorites and implications for their origin, *Meteoritics*, *30*, 738–747.

[736] Hamilton, V. E., P. R. Christensen, and H. Y. McSween (1997), Determination of martian meteorite lithologies and mineralogies using vibrational spectroscopy, *J. Geophys. Res.*, *102*, 25593–25603.

[737] Hasan, F. A., M. Haq, and D. W. G. Sears (1986), Thermoluminesence and the reheating history of meteorites: III. The shergottites, *Geochim. Cosmochim. Acta*, *50*, 1031–1038.

[738] Ikeda, Y. (1994), Petrography and petrology of the ALH A77005 shergottite, Proc. NIPR Symp. *Antarctic Meteorites*, *7*, 9–29.

[739] Ikeda, Y. (1998), Petrology of magmatic silicate inclusions in the Allan Hills 77005 lherzolitic shergottite, *Meteorit. Planet. Sci.*, *33*, 803–812.

[740] Ishii, T., H. Takeda, and K. Yanai (1979), Pyroxene geothermometry applied to a three-pyroxene achondrite from Allan Hills, Antarctica and ordinary chondrites. *Miner. J.*, *9*, 460–481.

[741] Jagoutz, E. (1989), Sr and Nd isotopic systematics in ALH A77005: Age of shock metamorphism in shergottites and magmatic differentiation on Mars, *Geochim. Cosmochim. Acta*, *53*, 2429–2441.

[742] Kong, P., M. Ebihara, and H. Palme (1999), Siderophile elements in martian meteorites and implications for core formation in Mars, *Geochim. Cosmochim. Acta*, *63*, 1865–1875.

[743] Kuebler, K. E. (2013), A comparison of the iddingsite alteration products in two terrestrial basalts and the Allan Hills 77005 martian meteorite using Raman spectroscopy and electron microprobe analyses, *J. Geophys. Res.: Planets*, *118*, 803–830.

[744] Laul, J. C. (1987), Rare earth patterns in shergottite phosphates and residues. *Proc. 17th Lunar Planet. Sci. Conf.*, in *J. Geophys. Res.*, *92*(Suppl.), E633–E640.

[745] Lee, D.-C., and A. N. Halliday (1997), Core formation on Mars and differentiated asteroids, *Nature*, *388*, 854–857.

[746] Lundberg, L. L., G. Crozaz, and H. Y. McSween (1990), Rare earth elements in minerals of the ALH A77005 shergottite and implications for its parent magma and crystallization history, *Geochim. Cosmochim. Acta*, *54*, 2535–2547.

[747] Ma, M.-S., J. C. Laul, and R. A. Schmitt (1981), Complementary rare earth element patterns in unique achondrites, such as ALH A77005 and shergottites, and in the Earth, Proc, Lunar Planet. *Sci. Conf.*, *12*, 1349–1358.

[748] McSween, H. Y., L. A. Taylor, and E. M. Stolper (1979), Allan Hills 77005: A new meteorite type found in Antarctica, *Science*, *204*, 1201–1203.

[749] McSween, H. Y., E. M. Stolper, L. A. Taylor, R. A. Muntean, G. D. O'Kelley, J. S. Eldridge, S. Biswas, H. T. Ngo, and M. E. Lipschutz (1979), Petrogenetic relationship between Allan Hills 77005 and other achondrites, *Earth Planet. Sci. Lett. 45*, 275–284.

[750] Mikouchi, T., and M. Miyamoto (2000), Lherzolitic martian meteorites Allan Hills 77005, Lewis Cliff 88516 and Yamoto-793605: Major and minor element zoning in pyroxene and plagioclase glass, *Antarct. Meteorite Res.*, *13*, 256–269.

[751] Miura, Y. N., K. Nagao, N. Sugiura, H. Sagawa, and K. Matsubara (1995), Orthopyroxenite ALH 84001 and shergottite ALH A77005: Additional evidence for a

martian origin from noble gases, *Geochim. Cosmochim. Acta, 59*, 2105–2113.

[752] Nishiizumi, K., J. Klein, R. Middleton, D. Elmore, P. W. Kubik, and Arnold J. R. (1986), Exposure history of shergottites, *Geochim. Cosmochim. Acta, 50*, 1017–1021.

[753] Nishiizumi, K., D. Elmore, and P. W. Kubik (1989), Update on terrestrial ages of Antarctic meteorites, Earth Planet. *Sci. Lett., 93*, 299–313.

[754] Ostertag, R., G. Amthauer, H. Rager, and H. Y. McSween (1984), Fe^{3+} in shocked olivine crystals of the ALH A77005 meteorite, Earth Planet. *Sci. Lett., 67*, 162–166.

[755] Pal, D. K., C. Tuniz, R. K. Moniot, W. Savin, T. Druse, and G. F. Herzog (1986), Beryllium-10 contents of shergottites, nakhlites, and Chassigny, *Geochim. Cosmochim. Acta, 50*, 2405–2409.

[756] Shih, C.-Y., L. E. Nyquist, D. D. Bogard, G. A. McKay, J. L. Wooden, B. M. Bansal, and H. Wiesmann (1982), Chronology and petrogenesis of young achondrites, Shergotty, Zagami, and ALH A77005: Late magmatism on a geologically active planet, *Geochim. Cosmochim. Acta, 46*, 2323–2344.

[757] Smith, J. V., and I. M. Steele (1984), Achondrite ALH A77005: Alteration of chromite and olivine, *Meteoritics, 19*, 121–133.

[758] Smith, M. R., J. C. Laul, M.-S. Ma, T. Huston, R. M. Verkouteren, M. E. Lipschutz, and R. A. Schmitt (1984), Petrogenesis of the SNC (shergottites, nakhlites, chassignites) meteorites: Implications for their origin from a large, dynamic planet, possibly Mars, *Proc. 14th Lunar Planet. Sci. Conf.*, in *J. Geophys. Res., 89*(Suppl.), B612–B630.

[759] Treiman, A. H., G. A. McKay, D. D. Bogard, M.-S. Wang, M. E. Lipschutz, D. W. Mittlefehldt, L. Keller, M. M. Lindstrom, and D. Garrison (1994), Comparison of the LEW 88516 and ALH A77005 martian meteorites: Similar but distinct, *Meteoritics, 29*, 581–592.

[760] Walton, E. L., and C. D. K. Herd (2007), Localized shock melting in lherzolitic shergottite Northwest Africa 1950: Comparison with Allan Hills 77005, *Meteorit. Planet. Sci. 42*, 63–80.

[761] Walton, E. L., and C. D. K. Herd (2007), Dynamic crystallization of shock melts in Allan Hills 77005: Implications for melt pocket formation in martian meteorites, *Geochim. Cosmochim. Acta, 71*, 5267–5285.

[762] Walton, E. L., S. P. Kelley, and C. D. K Herd (2008), Isotopic and petrographic evidence for young martian basalts, *Geochim. Cosmochim. Acta, 72*, 5819–5837.

[763] Wright, I. P., R. H. Carr, and C. T. Pillinger (1986), Carbon abundance and isotopic studies of Shergotty and other shergottite meteorites, *Geochim. Cosmochim. Acta, 50*, 983–991.

73. RBT 04261,262—LHERZOLITIC SHERGOTTITE

[764] Brandon, A. D., I. S. Putchel, R. J. Walker, J. M. D. Day, A. J. Irving, and L. A. Taylor (2012), Evolution of the martian mantle inferred from the ^{187}Re-^{187}Os isotope and high siderophile element systematics of the Shergottite meteorites, *Geochim. Cosmochim. Acta, 76*, 206–235.

[765] Callahan, M. P., A. S. Burton, J. E. Elsila, E. M. Baker, K. E. Smith, D. P. Glavin, and J. P. Dworkin (2013), A search for amino acids and nucleobases in the martian meteorite Roberts Massif 04262 using liquid chromatography-mass spectrometry, *Meteorit. Planet. Sci., 48*, 786–795.

[766] Cartwright, J. A., K. D. Ocker, S. A. Crowther, R. Burgess, and J. D Gilmour (2009), Terrestrial and martian weathering signatures of Xenon components in shergottite mineral separates, *Meteorit. Planet. Sci., 45*, 1359–1378.

[767] Cartwright, J. A., and R. Burgess (2011), Ar-Ar crystallization ages of shergottite RBT 04262 mineral separates (abstract 5465), *Meteorit. Planet. Sci., 46*, A36.

[768] Dalton, H. A., A. H. Peslier, A. D. Brandon, C.-T. A. Lee, and T. J. Lapan (2008), Petrology and mineral chemistry of new olivine-phyric shergottite RBT 04262, *Lunar Planet. Sci., 39*, abstract 2308.

[769] Lapen, T. J., A. D. Brandon, B. L. Beard, A. H. Peslier, C.-T. A. Lee, H. A. Dalton (2008), Lu-Hf age and isotope systematics of the olivine-phyric shergottite RBT 04262 and implications for the sources of enriched shergottites, Lunar Planet. *Sci. Conf., 39*(1391), 2073.

[770] Mikouchi, T., T. Kurihara, and M. Miyamoto (2008), Petrology and mineralogy of RBT 04262: Implications for stratigraphy of the lherzolitic shergottite igneous block, Lunar Planet. *Sci. Conf., 39*(1391), 2403.

[771] Mikouchi, T. (2008), Petrographic and chemical variation of lherzolitic shergottites and implications for the classification of shergottites, *Meteorit. Planet. Sci., 43*, A186.

[772] Mikouchi, T. (2009), Petrogical and mineralogical diversities within the lherzolitic shergottites require a new group name, *Lunar Planet. Sci., 40*, abstract 2272.

[773] Nagao, K., and J. Park (2008), Nobel gases and cosmic-ray exposure ages of two martian shergottites, RBT 04262 and LAR 06319 recovered in Antarctica (abstract 5200), *Meteorit. Planet. Sci., 43*, A107.

[774] Niihara, T. (2011), Uranium-lead age of baddeleyite in shergottite Roberts Massif 04261: Implications for magmatic activity on Mars, *J. Geophys. Res., 116*, E12008, doi: 10.1029/2011JE003802.

[775] Niihara, T. (2012), Correction to "Uranium-lead age of baddeleyite in Shergottite Roberts Massif 04261: Implications for magmatic activity on Mars," *J. Geophys. Res. 117*, E02002, doi:10.1029/2012JE004048.

[776] Nishiizumi, K., and M. W. Caffee (2010), A tale of two shergottites: RBT 04261 and RBT 04262, *Lunar Planet. Sci., 41* abstract 2276.

[777] Satake, W., T. Mikouchi, and M. Miyamoto (2009), Redox state of geochemically-enriched "lherzolitic" shergottites as inferred from Fe Micro-XANES analysis, *Antarctic Meteorites, 32*, 64–65.

[778] Schwenzer, S. P., R. C. Greenwood, S. P. Kelley, U. Ott, A. G. Tindle, R. Haubold, S. Herrmann, J. M. Gibson, M. Anand, S. Hammond, and I. A. Franchi (2013),

Quantifying noble gas contamination during terrestrial alteration in martian meteorites from Antarctica, *Meteorit. Planet. Sci.*, 48, 929–954.

[779] Shearer, C. K., P. V. Burger, J. J. Papike , L. E. Borg, A. J. Irving, and C. Herd (2008), Petrogenetic linkages among martain basalts: Implications based on trace element chemistry of olivine, *Meteorit. Planet. Sci.*, 43, 1241–1258.

[780] Shih, C.-Y., L. A. Nyquist, and Y. Reese (2009), Rb-Sr and Sm-Nd studies of olivine-pyric Shergotties RBT 04262 and LAR 06319: Isotopic evidence for relationship to enriched basaltic shergottites, *Lunar Planet. Sci.*, 40, abstract 1360.

[781] Usui, T., M. E. Sanborn, M. Wadhwa, and H. Y. McSween (2010), Petrology and trace element geochemistry of Robert Massif 04261 and 04262 meteorites, the first examples of geochemically enriched lherzolitic shergottites, *Geochim. Cosmochim. Acta*, 74, 7283–7306.

74. MIL 03346—NAKHLITE

[782] Cartwright, J. A., J. D. Gilmour, and R. Burgess (2013), Martian fluid and martian weathering signatures identified in Nakhla, NWA 998 and MIL 03346 by halogen and noble gas analysis, *Geochim. Cosmochim. Acta*, 105, 255–293.

[783] Day, J. M. D., L. A. Taylor, C. Floss, and H. Y. McSween (2006), Petrology and chemistry of MIL 03346 and its significance in understanding the petrogenesis of nakhlites on Mars, *Meteoritics*, 40, 581–606.

[784] Debaille, V., A. D. Brandon, C. O'Neill, Q.-Z. Yin, and B. Jacobsen (2009), Early martian mantle overturn inferred from isotopic composition of nakhlite meteorites, *Nature Geoscience*, 2, 548–552.

[785] Dyar, M. D., A. H. Treiman, C. M. Pieters, T. Hiroi, M. D. Lane, and V. O'Connor (2005), MIL03346, the most oxidized martian meteorite: A first look at spectroscopy, petrography, and mineral chemistry, *J. Geophys. Res.*, 110, E09005.

[786] Hallis, L. J., and G. J. Taylor (2011), Comparisons of the four Miller Range nakhlites, MIL 03346, 090030, 090032 and 090136: Textural and compositional observations of primary and secondary mineral assemblages, *Meteorit. Planet. Sci.*, 46, 1787–1803.

[787] Hallis, L. J., G. J.Taylor, K. Nagashima, G. R. Huss, A. W. Needham, M. M.Grady, and I. A. Franchi (2012), Hydrogen isotope analyses of alteration phases in the nakhlite martian meteorites, *Geochim. Cosmochim. Acta*, 97, 105–119.

[788] Hallis, L. J. (2013), Alteration assemblages in the Miller Range and Elephant Moraine regions of Antarctica: Comparisons between terrestrial igneous rocks and meteorites, *Meteorit. Planet. Sci.* 48, 165–179.

[789] Hammer, J. E. (2009), Application of a textural geospeedometer to the late-stage magmatic history of MIL 03346, *Meteorit. Planet. Sci.*, 44, 141–154.

[790] Imae, N., and Y. Ikeda (2007), Petrology of the Miller Range 03346 nakhlite in comparison with the Yamoto-000593 nakhlite, *Meteorit. Planet. Sci.*, 42, 171–184.

[791] Imae, N., and Y. Ikeda (2008), Crystallization experiments of intercumulus melts for nakhlites under QFM ±2 at 1 bar, *Meteorit. Planet. Sci.*, 43, 1299–1319.

[792] Kuebler, K. E. (2013), A combined electron microprobe (EMP), and Raman spectroscopic study of the alteration products in martian meteorite MIL 03346, *J. Geophys. Res.*, 118, 347–368.

[793] McCubbin, F. M., N. J. Tosca, A. Smirnov, H. Nekvasil, A. Steele, M. Fries, and D. H. Lindsley (2009), Hydrothermal jarosite and hematite in a pyroxene-hosted melt inclusion in martian meteorite Miller Range (MIL), 03346: Implications for magmatic-hydrothermal fluids on Mars, *Geochim. Cosmochim. Acta*, 73, 4907–4917.

[794] Mikouchi, T., A. Monkawa, E. Koizumi, J. Chokai, and M. Miyamoto (2005), MIL 03346 nakhlite and NWA2737 "Diderot" chassignite: Two new martian cumulate rocks from hot and cold deserts, *Lunar Planet. Sci. Conf. 36*, abstract 1944.

[795] Mikouchi, T., M. Miyamoto, E. Koizumi, J. Makishima, and G. McKay (2006), Relative burial depths of nakhlites: An update, *Lunar Planet. Sci. Conf.*, 37, abstract 1865.

[796] Park, J., D. H. Garrison, and D. D. Bogard (2009), ^{39}Ar-^{40}Ar ages of martian nakhlites, *Geochim. Cosmochim. Acta*, 73, 2177–2189.

[797] Righter, K., and K. M. McBride (2012), The Miller Range nakhlites: A summary of the curatorial subdivision of the main mass in light of newly found paired masses, *Lunar Planet. Sci. Conf.*, 42, LPI Contribution No. 1608, 2161.

[798] Righter, K., H. Yang., G. Costin., and R. T. Downs (2008), Oxygen fugacity in the martian mantle controlled by carbon: New constraints from the nakhlite MIL 03346, *Meteorit. Planet. Sci.*, 43, 1709–1723.

[799] Rutherford, M. J., and J. E. Hammer (2008), Oxidation states in MIL 03346 nakhlite from experiments reproducing phenocryst-melt equilibria as a function of fO_2 and T at 40–150 Mpa, *Lunar Planet. Sci. Conf.*, 39, LPI Contribution No. 1391, 1983.

[800] Rutherford, M. J., C. Calvin, M. Nicholis, and M. C. McCanta (2005), Petrology and melt compositions in nakhlite MIL 00346: Significance of data from natural sample and from experimentally fused groundmass and M.I.'s, *Lunar Planet. Sci. Conf.*, 36, abstract 2233.

[801] Sautter, V., A. Jambon, and O. Boudouma (2006), Cl-amphibole in the nakhlite MIL03346: Evidence for sediment contamination in a martian meteorite, *Earth Planet. Sci. Lett.*, 252, 45–55.

[802] Stopar, J. D., G. J. Taylor, M. A. Velbel, M. D. Norman, E. P. Vicenzi, and L. J. Hallis (2013), Element abundances, patterns, and mobility in Nakhlite Miller Range 03346 and implications for aqueous alteration, *Geochim. Cosmochim. Acta*, 112, 208–225.

[803] Udry, A., H. Y. McSween, Jr., P. Lecumberri-Sanchez, R. J. Bodnar (2012), Paired nakhlites MIL 090030, 090032, 090136, and 03346: Insights into the Miller Range parent meteorite, *Meteorit. Planet. Sci.*, 47, 1575–1589.

75. EET 87500, 501—MESOSIDERITE

[804] Miura, Y. N. (1995), *Studies on differentiated meteorites: Evidence from ^{244}Pu-derived fission Xe, ^{81}Kr, other noble gases and nitrogen, PhD dissertation*, Univ. Tokyo.

[805] Mittlefehldt, D. W. (1990), Petrogenesis of mesosiderites: I. Origin of mafic lithologies and comparison with basaltic achondrites, *Geochim. Cosmochim. Acta, 54*, 1165–1173.

[806] Rubin, A. E., and D. W. Mittlefehldt (1992), Classification of mafic clasts from mesosiderites: Implications for endogenous igneous processes, *Geochim. Cosmochim. Acta, 56*, 827–840.

[807] Eugster, O., G. F. Herzog, K. Marti, and M. W. Caffee (2006), Irradiation records, cosmic-ray exposure ages, and transfer times of meteorites, in *Meteorites and the Early Solar System II*, edited by D. S. Lauretta and H. Y. McSween Jr., pp. 829–851, University of Arizona Press, Tucson.

76. RKP A79015—MESOSIDERITE

[808] Clarke, R. S., and B. Mason (1982), A new metal-rich mesosiderite from Antarctica, RKP A79015, *Mem. Natl. Inst. Polar Res., 25*(Spec. Issue), 78–85.

[809] Hewins, R. H. (1988), Petrology and pairing of mesosiderites from Victoria Land, *Antarctica, Meteoritics, 23*, 123–129.

[810] Hassanzadeh, J., A. E. Rubin, and J. T. Wasson (1990), Compositions of large metal nodules in mesosiderites: Links to iron meteorite group IIIAB and the origin of mesosiderite subgroups, *Geochim. Cosmochim. Acta, 54*, 3197–3208.

[811] Mittlefehldt, D. W. (1990), Petrogenesis of mesosiderites: I. Origin of mafic lithologies and comparison with basaltic achondrites, *Geochim. Cosmochim. Acta, 54*, 1165–1173.

[812] Rubin, A. E., and D. W. Mittlefehldt (1992), Mesosiderites: A chronologic and petrologic synthesis, *Meteoritics, 27*, 282.

[813] Wasson, J. T., B.-G. Choi, E. A. Jerde, and F. Ulff-Møller (1998), Chemical classification of iron meteorites: XII. New members of the magmatic groups, *Geochim. Cosmochim. Acta, 62*, 715–724.

[814] Yang, C.-W., D. B. William, and J. I. Goldstein (1997), Low-temperature phase decomposition in metal from iron, stony-iron, and stony meteorites, *Geochim. Cosmochim. Acta, 61*, 2943–2956.

77. CMS 04069—PALLASITE

[815] Danielson, L. R., K. Righter, and M. Humayun (2009), Trace element chemistry of Cumulus Ridge 04071 pallasite with implications for main group pallasites, *Meteorit. Planet. Sci., 44*, 1019–1032.

78. DRP A78001—IIAB IRON

[816] Clarke, R. S., Jr. (1982), The Derrick Peak, *Antarctica, iron meteorites, Meteoritics, 17*, 129–134.

[817] Wasson, J. T., H. Huber, D. J. Malvin (2007), Formation of IIAB iron meteorites, *Geochim. Cosmochim. Acta, 71*, 760–781.

79. MET 00400—IIIAB IRON

[818] Goldstein, J. I., E. R. D. Scott, and N. L. Chabot (2009), Iron meteorites: Crystallization, thermal history, parent bodies, and origin, *Chemie der Erde, 69*, 293–325.

80. HOW 88403—SULFIDE-RICH IRON

[819] Schrader, D. L., D. S. Lauretta, H. C. Connolly Jr., Y. S. Goreva, D. H. Hill, K. J. Domanik, E. L. Berger, H. Yang, and R. T. Downs (2010), Sulfide-rich metallic impact melts from chondritic parent bodies, *Meteorit. Planet. Sci., 45*, 743–758.

[820] Wasson, J. T. (1990), Ungrouped iron meteorites in Antarctica: Origin of anomalously high abundance, *Science, 249*, 900–902.

[821] Wasson, J. T., X. Ouyang, J. Wang, and E. Jerde (1989), Chemical classification of iron meteorites: XI. Multi-element studies of 38 new irons and the high abundance of ungrouped irons from Antarctica, *Geochimica et Cosmochimica Acta, 53*, 735–744.

4

Primitive Asteroids: Expanding the Range of Known Primitive Materials

Michael K. Weisberg[1,2,3] and Kevin Righter[4]

4.1. INTRODUCTION

Chondrites are ancient rocks derived from primitive asteroids that escaped the processes of melting and differentiation. Their major constituents (chondrules, calcium-aluminum-rich inclusions, metal, and matrix) are interpreted to have formed in the early disk (solar nebula). Thus, chondrites are records of the physical and chemical processes that were active in the early solar system and of the accretion and impact histories of early-formed planetesimals. Chondrites also contain presolar materials that can provide information on stellar and interstellar processes. Additionally, many chondrites record secondary processes of aqueous and thermal alteration that early-formed planetesimals were subjected to.

Most chondrite groups are represented in the Antarctic meteorite collection and some varieties are known only from the Antarctic collection (Figures 4.1a, 4.1b, 4.1c, and 4.1d). The Antarctic meteorites have also provided more pristine varieties of chondrite groups, enabling the study of primary nebular features, thereby better constraining conditions and processes in the early solar nebula. They have given us more heavily altered examples of existing chondrite groups, allowing us to study the range of alteration conditions on primitive asteroids.

The most pristine CR chondrites (QUE 99177 and MET 00426; Plate 26) are Antarctic, as are the most heavily altered CRs (GRO 95577 and MIL 090292; Plate 27).

Nearly all known primitive (unequilibrated) EL chondrites are from Antarctica (Plates 30 and 31) and the CK chondrite group was essentially defined by Antarctic meteorites (Figures 4.1a and 4.1b; Plate 20). An anomalous unequilibrated enstatite chondrite, Lewis Cliff (LEW) 87223, has also been recovered. The first meteorite from the metal-rich CH chondrite group, Allan Hills (ALH) 85085 (Plate 24), is Antarctic and the CB_b chondrite subgroup was not well established until Queen Alexandra Range (QUE 94411; Plate 21) was studied. Grosvenor Mountains (GRO) 95551 (Plate 23) led to the discovery of another type of metal-rich chondrite, not directly related to CH or CB. One of the earliest recognized R3 chondrites is ALH 85151, and one of the two known K chondrites, LEW 87232 (Plate 37), is from Antarctica. The majority of CM (Plates 11 to 14) and a large number of CK chondrites are from Antarctica (Figure 4.1a). The Antarctic meteorite collection has provided ordinary chondrite (impact-) melt rocks, low petrologic types, and rare recrystallized (type 7) examples (Figure 4.1d; Plates 1, 3, 4, and 8). The collection has also provided materials for studies of presolar circumstellar grains and possibly interstellar organic components. With regard to studies of primitive solar system materials, the Antarctic meteorite collection has clearly had a major influence. Here we explore some examples.

4.2. CR CHONDRITES

The CR chondrites are widely considered to be among the most primitive chondrites and, based on Antarctic CRs, exhibit a wide range in degree of alteration from heavily altered (petrologic type 1) to essentially pristine (petrologic type 3). In general, many of the CR chondrites are Antarctic (e.g., EET 87711, EET 96259, GRO 03115, GRO 95577, MIL 07513, MIL 090001, MIL 090292,

[1] Department of Physical Sciences, Kingsborough Community College, City University New York
[2] Earth and Environmental Sciences, Graduate Center, City University New York
[3] Department of Earth Planetary Sciences, American Museum of Natural History, New York
[4] NASA Johnson Space Center

35 Seasons of U.S. Antarctic Meteorites (1976–2010): A Pictorial Guide to the Collection, Special Publication 68, First Edition. Edited by Kevin Righter, Catherine M. Corrigan, Timothy J. McCoy and Ralph P. Harvey.
© 2015 American Geophysical Union. Published 2015 by John Wiley & Sons, Inc.

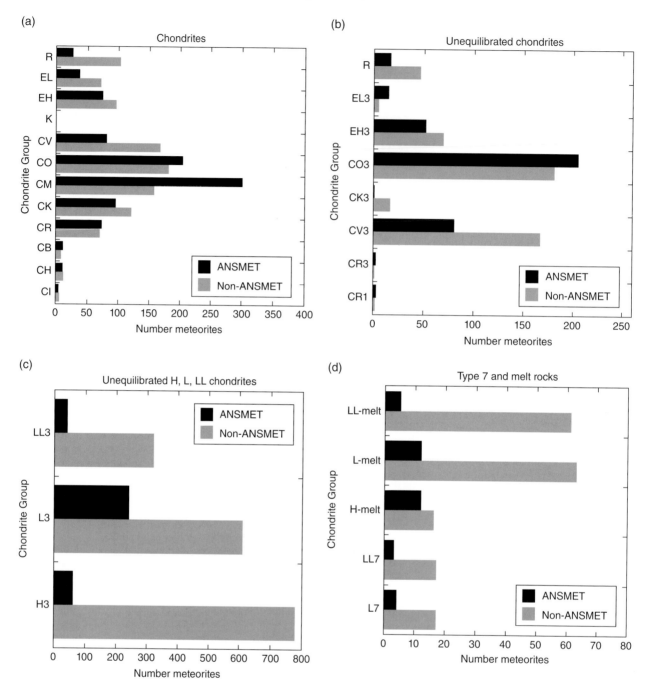

Figure 4.1. (a) Histogram showing the total number of ANSMET vs. non-ANSMET meteorites in the carbonaceous, enstatite, K, and R chondrite classes. (b) Histogram showing unequilibrated Antarctic vs. non-Antarctic meteorites in carbonaceous, enstatite, and R chondrite groups. The only known CR3 and CR1 chondrites are from the ANSMET collection. The majority of EL3 chondrites are also from, and the group is largely defined by, meteorites in the ANSMET collection. (c) Histogram showing ANSMET vs. non-ANSMET type 3 ordinary chondrites. (d) Histogram of ANSMET vs. non-ANSMET ordinary chondrite type 7s and melt rocks. Data were compiled from the Meteoritical Society's Meteoritical Bulletin Database (http://www.lpi.usra.edu/ meteor/metbull.php). Numbers include all Antarctic meteorites and their paired samples. Although all paired samples are included, the ANSMET meteorites are represented in all chondrite groups and in some cases greatly exceed the number of non-ANSMET members, for example, CMs. This is particularly important for a group like the CM that span a wide range of aster-oidal alteration conditions. (Note: the non-ANSMET data include Antarctic meteorites from other collections, specifically from the National Institute of Polar Research (NIPR), Tokyo)

(a)

(b)

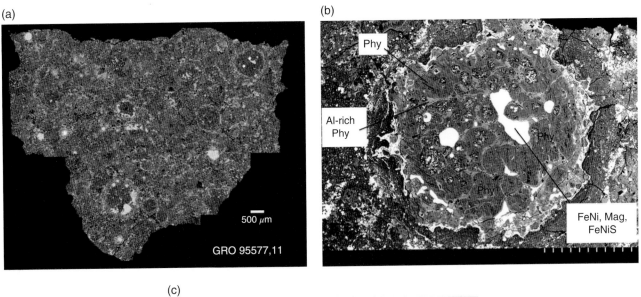

(c)

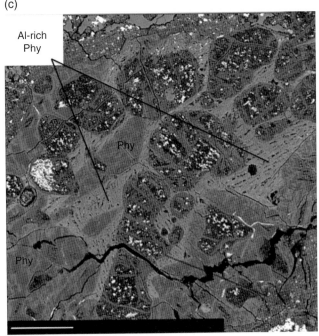

Figure 4.2. (a) Fe Kα X-ray image of a polished thin section of GRO 95577, 11 showing large chondrules up to 1.8 mm in size. The silicates in all the chondrules have been altered to phyllosilicates. The bright white areas are Fe-Ni metal grains, most of which have been altered to magnetite. (b) A chondrule from GRO 95577, 11 (1 mm across) with a porphyritic texture in which the phenocrysts and mesostasis has been completely replaced by phyllosilicates (Phy). The metal (white) has been partially replaced by magnetite (Mag). (c) Enlargement of the interior of the chondrule in (b) showing the phenocrysts replaced by phyllosilicates and the chondrule mesostasis replaced by an Al-rich phyllosilicate. Source: *Weisberg and Huber* [2007].

MIL 090657, QUE 99177 and MET 00426; Plates 25 to 27), and much of our knowledge of the CR group is based on the study of these samples. However, GRO 95577 (Figure 4.2a) is important as the first CR known to be completely hydrated and is considered a CR1 [*Weisberg and Huber*, 2007]. It is a remarkable meteorite in that the chondrules are completely hydrated, consisting almost

entirely of phyllosilicates, magnetite, and sulfides. Although GRO 95577 is completely altered, the initial chondrule textures are preserved, showing perfect pseudomorphic replacement (Figure 4.2b, c) [*Weisberg and Huber*, 2007]. The chondrules are in sharp contact with the matrix, their fine-grained rims are clearly visible, and the boundaries of the dark inclusions can be easily discerned (Figures 4.2b and 4.2c). Many chondrules in GRO 95577 have textures suggestive of the common porphyritic type I (silicates with Mg/(Mg + Fe) > 90) chondrules, observed in CR and other chondrites, but the phenocrysts have undergone pseudomorphic replacement by yellow to brownish serpentine-rich phyllosilicate, with sharp original crystal outlines preserved. The chondrule mesostasis is a green aluminous material, most likely a hydration product of the feldspathic mesostasis commonly found in anhydrous type I chondrules in other CRs. Some chondrules contain magnetite spheres that most likely formed by oxidation of pre-existing metal.

GRO 95577 extends the range of asteroidal alteration recorded in the CRs considerably, from CR chondrites that contain unaltered glassy mesostasis in chondrule interiors [e.g., *Weisberg et al.*, 1993; *Ichikawa and Ikeda*, 1995; *Noguchi*, 1995] and olivine [*Zolensky et al.*, 1993] and amorphous material [*Abreu and Brearley*, 2010] in the matrix to heavily altered, like GRO 95577, in which no anhydrous silicates have survived alteration [*Weisberg and Huber*, 2007]. MIL 090292 is another heavily hydrated CR classified as a CR1 [*Satterwhite and Righter*, 2012].

QUE 99177 and MET 00426 are the first CR chondrites considered to be petrologic type 3 (CR3). They suffered lower degrees of secondary aqueous alteration than all other known CR chondrites [*Abreu and Brearley*, 2010]. Unlike other CR chondrites, which are CR1 or CR2, the matrix material in QUE 99177 and MET 00426 contains little phyllosilicate and is dominated by amorphous silicate, indicating the primitive, unaltered nature of these meteorites [*Abreu and Brearley*, 2010]. Another intriguing characteristic of QUE 99177 and MET 00426 is that they have high presolar silicate and oxide grain abundances, and contain high abundances of phases with anomalous C isotopic compositions [*Floss and Stadermann*, 2011], discussed below. With GRO 95577, the CR group is the only carbonaceous chondrite group that spans the range of asteroidal alteration from completely hydrated (lacking anhydrous silicates) to almost completely anhydrous (lacking hydrous silicates).

4.2.1. Organic Compounds in CR2 Chondrites

Organic compounds in carbonaceous chondrites have been the focus of organic geochemists for many years, and studies of the CM2 Murchison in particular have yielded a diversity of compounds. Antarctic CR2 chondrites, and in particular GRA 95229, LAP 02342, EET 92042, and QUE 99177, have also yielded a great diversity of organic compounds, and quite different from those found in other C chondrites, such as Murchison. For example, the Antarctic CR2s contain several orders of magnitude greater concentrations of amino acids (glycine, isovaline, α-AIB, and alanine), ammonia, amines, and

aldehydes and ketones [*Martins et al.*, 2007; *Pizzarello et al.*, 2008; *Pizzarello and Holmes*, 2009]. Furthermore, the amino acids exhibit chirality that is distinct from that seen in terrestrial amino acids [*Glavin and Dworkin*, 2009]. The presence of N-containing compounds, as well as a great diversity of asymmetric compounds, shows that the more complex molecules (amino acids) could have inherited some of their unique asymmetry from synthesis reactions involving the diversity of chiral prebiotic ingredients such as ketones, aldehydes, and amines.

Insoluble organic matter in CR2 chondrites such as EET 92042 is also known to contain isotopic anomalies in H and N that are most likely acquired from the interstellar medium, although formation in the protoplanetary disk cannot be ruled out [*Busemann et al.*, 2006; *Alexander et al.*, 2007] (Plate 25). The bulk composition of insoluble organic matter in EET 92042, on the other hand, is strikingly similar to the CHON particles found in Comet Halley, for example [*Alexander et al.*, 2007]. The organic matter in CR2 chondrites may provide constraints on the sources for the prebiotic organic matter that helped promote the development of life. The continued discovery of CR2 chondrites by ANSMET teams provides new material with which to test many hypotheses for the origin of organic matter in the early solar system and ultimately better understand the origin of life.

4.3. METAL-RICH CH AND CB CHONDRITES

Ever since the first published studies of ALH 85085 [*Grossman et al.*, 1988; *Scott*, 1988; *Weisberg et al.*, 1988; *Wasson and Kallemeyn*, 1990] (Plate 24), metal-rich, matrix-poor chondrites, such as the CH and CB chondrites, have been among the most perplexing and significant meteorite discoveries. The initial description of ALH 85085 in the 1987 Antarctic Meteorite Newsletter was exciting in that it was unlike any chondrite previously described: a new type of asteroid was sampled. There are now 23 CH chondrites, nine of which are Antarctic from the ANSMET collection.

The CH chondrites are closely related to the CBs and CRs (*Weisberg et al.*, 1995b); they may be among the most primitive, unaltered chondritic materials, and planetesimal collisions may have played major roles in their origin [e.g., *Wasson and Kallemeyn*, 1990; *Krot et al.*, 2005]. They are characterized by small (<100 μm in diameter) chondrules, which are mostly cryptocrystalline in texture. They have high abundances of FeNi-metal (~20 vol%) with a large range in composition, positive Ni versus Co trends, solar Ni/Co ratios, and compositional zoning in many grains. Interchondrule matrix is virtually absent, but hydrated matrix-like clasts are commonly observed. Like the CR and CB chondrites, CH chondrites have ~CI (solar) bulk refractory lithophile element relative abundances but are highly depleted in volatile and moderately volatile lithophile elements. Whole-rock oxygen isotopic compositions of CH chondrites are similar to CR chondrites, suggesting they may have formed from a common oxygen reservoir, and nitrogen isotopic compositions show remarkable positive anomalies in N, with δ^{15}N up to ~800 ‰ [*Krot et al.*, 2002; *Weisberg et al.*, 1989].

The CB (Bencubbin-like) metal-rich chondrites (Plates 21 to 23) also have characteristics that are sharply different from other chondrites [*Weisberg et al.*, 2001]. The Antarctic CB chondrites have given us the opportunity to establish and study this group and its intergroup variations. The characteristics of CB chondrites include (1) high metal abundances (60–80 vol.% metal); (2) most chondrules have cryptocrystalline or barred textures; (3) moderately volatile lithophile elements are highly depleted; and (4) nitrogen is enriched in the heavy isotope. Additionally, most CB chondrites contain areas of impact-produced melt between the silicate and metal consisting of silicate glass with tiny (<1 μm) immiscible blebs of FeNi metal or vice versa. Termed spontaneous fusion texture by *Newsom and Drake* (1979) and interpreted to be melted chondrite matrix by *Meibom et al.* (2005), these regions attest to the major role that impact played in the history of the CB parent asteroid. Similarities in mineral composition, as well as oxygen and nitrogen isotopic compositions of the CB to CR and CH chondrites, are consistent with derivation of these chondrite groups from a common nebular reservoir, hence their grouping in the CR clan [*Weisberg et al.*, 1989, 2001; *Weisberg and Prinz*, 1999; *Krot et al.*, 2002]. A close relationship between CB and CH is strongly supported by the Ischeyevo chondrite, which appears to be a mixture of CH and CB lithologies [*Krot et al.*, 2006; *Ivanova et al.*, 2009].

The CB chondrites have been divided into two subgroups, CB$_a$ and CB$_b$, based on their petrologic characteristics [*Weisberg et al.*, 2001]. The CB$_a$ chondrites contain ~60 vol% metal, chondrules are centimeter-size, FeNi metal ranges from 5% to 8% Ni, and bulk δ^{15}N values are up to ~1000‰, whereas the CB$_b$ contain >70 vol% metal, their chondrules are millimeter-size, their FeNi metal contains 4 to 15% Ni, and δ^{15}N compositions are ~200‰. See *Weisberg et al.* [2001] for detailed description and discussion of the CB chondrite subgroups. The Antarctic meteorite collection has provided samples of both CB$_a$ (MIL 05082) and CB$_b$ (QUE 94411) chondrites (Figure 4.3a; Plates 21 and 22) and thus has helped define the CB group and establish its subgroups.

The origin and relationship of CH and CB chondrites to other chondrites is an open issue. Their chondrules and other components have been interpreted to form in the early solar nebula [e.g., *Newsom and Drake*, 1979; *Weisberg et al.*, 1989; *Weisberg et al.*, 2001; *Meibom et al.*, 1999; *Krot et al.*, 2002; *Campbell et al.*, 2001] or as products of late-stage planetesimal collisions in the early solar system

(a)

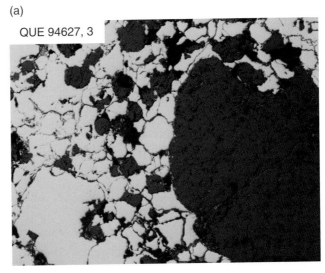

QUE 94627, 3

(b)

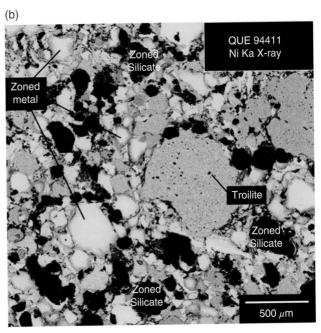

QUE 94411
Ni Ka X-ray

Zoned Silicate

Zoned metal

Troilite

Zoned Silicate

Zoned Silicate

500 μm

Figure 4.3. (a) A reflected light photomicrograph of QUE 94627,3 (paired with QUE 94411), a CB$_b$ chondrite. The section shows a large cryptocrystalline chondrule (dark gray), smaller chondrules (<100 μm in size), and abundant FeNi metal (white). (b) A Ni Kα X-ray image of QUE 94411 showing compositionally zoned metal grains (bright) that decrease in Ni content (brightness) toward the grain edges. Silicates appear black. Source: *Weisberg et al.* [2001].

[e.g., *Wasson and Kallemeyn*, 1990; *Campbell et al.*, 2002; *Amelin and Krot*, 2005; *Krot et al.*, 2005]. The later hypothesis stresses the importance of collision in the early solar system and the possibility of such collisions as a mechanism for chondrule formation. Continued discoveries of new metal-rich chondrites for detailed study could help resolve the issue of their origin.

One of the many important characteristics of the CH and CB chondrites is their zoned metal (Figure 4.3b), initially described in ALH 85085 by *Scott* [1988]. The preservation of zoning in metal grains attests to the primitive character of the CH and CB chondrites; zoning would have been erased under conditions of thermal metamorphism [*Righter et al.*, 2005; *Humayun*, 2012]. The range of compositions of P, Si, Cr, Co, and Ni as well as Ru within individual zoned metal grains in CH and CB_b chondrites is generally consistent with calculated paths for metal condensing from a solar gas [*Meibom et al.*, 1999, 2000, 2001; *Campbell et al.*, 2001, 2005; *Petaev et al.*, 2001]. Analyses of trace siderophile elements of metal in Queen Alexandra Range (QUE) 94411 helped verify that the zoned metal grains in the CB_b chondrites were indeed the products of condensation [*Campbell et al.*, 2001]. However, it remains unclear whether the metal condensed from the solar nebula gas or from a vapor plume produced during a planetary-scale collision.

4.3.1. New Type of Metal-Rich Chondrites

Finding new metal-rich meteorites with textural similarities to CH or CB may provide a better understanding of the origins of these unusual meteorites. GRO 95551 (Plate 23) is one of two meteorites that represent a new type of highly reduced, metal-rich chondrite (Figures 4.4a and 4.4b). The other is Northwest Africa (NWA) 5492. GRO 95551 was initially thought to be a CB chondrite because it contains large metal nodules and large barred olivine chondrules (Figure 4.4b), but it has more reduced mineral compositions and formed from a different oxygen reservoir than the CB and CH chondrites [*Weisberg et al.*, 2001, 2012].

GRO 95551 has similarities to CB chondrites with ~60 vol% metal. Some of the metal and silicate occur as large >500 μm size chondrule-like objects. The silicates include large barred olivine chondrules, like in CB chondrites, but smaller chondrules showing a wider range of textures, including porphyritic, are also present. The compositions of the silicates are highly reduced, with an average olivine composition of $Fa_{1.3}$. Oxygen isotope ratios for chondrules in GRO 95551 and NWA 5492 have unique $\Delta^{17}O$ values that are between values for the ordinary and the enstatite chondrites (Figure 4.4c) [*Weisberg et al.*, 2012] [$\Delta^{17}O = \delta^{17}O - 0.52 (\delta^{18}O)$].

The oxygen isotope ratios coupled with the reduced silicate compositions suggest that GRO 95551 is a new type of chondrite, related to NWA 5492. They may represent a new primitive asteroid with characteristics that may be transitional between ordinary and enstatite chondrites. The discovery of additional chondrites like GRO 95551 will help characterize this new chondrite group and decipher their origin.

4.4. UNEQUILIBRATED ENSTATITE (E3) CHONDRITES (PLATES 28 TO 31)

The enstatite (E) chondrites have important implications for the evolution of the solar system and formation of the inner planets (Mercury through Mars). They have extremely different characteristics than those of other chondrite groups. Their silicate, sulfide, and metal compositions indicate highly reducing conditions [e.g., *Keil*, 1968]. Their major silicate phase is near-pure endmember enstatite (FeO < 1.0 wt.%). They contain higher amounts of FeNi metal than ordinary or carbonaceous chondrites, and their metal is Si-bearing with more than 2 wt.% Si in the EH chondrites. Elements that are strictly lithophile in most other chondrites (e.g., Mg, Ca, Mn, Na, K) in some cases behave as chalcophiles in the E chondrites, occurring in a wider variety of sulfide minerals than in any other chondrite group [e.g., *El Goresy et al.*, 1988]. E chondrites contain the oxi-nitride and nitride phases sinoite (Si_2N_2O) [*Andersen et al.*, 1964; *Keil and Anderson*, 1964; *Mason*, 1966; *Rubin*, 1997], Si_3N_4 [*Alexander et al.*, 1994], and osbornite (TiN) [*Mason*, 1966], which are also indicators of highly reducing conditions. The reduced mineral assemblages of the primitive unequilibrated E3 chondrites formed in a highly reducing solar nebula and/or asteroidal environment, unlike that for other chondrite groups.

Another intriguing aspect of the E chondrites is that they have the oxygen isotopic composition of the Earth and Moon [*Clayton and Mayeda*, 1984; *Javoy*, 1995; *Javoy et al.*, 2010]. The significance of this is unclear, except that they likely formed in the inner solar system and were possibly among the building blocks of the inner planets.

The Antarctic meteorite collection has greatly contributed to the number of known primitive, unequilibrated enstatite (E3) chondrites. The enstatite chondrites are divided into two main groups, EH and EL. MAC 88136 [Figure 4.5a, Plate 31] was the first known primitive (EL3) chondrite [*Lin et al.*, 1991]. Prior to its discovery, only highly metamorphosed examples of EL chondrites were known. Other EL3 chondrites that have since been recognized include ALH 85119, EET 90299, and MAC 02635 and their paired samples.

It has been suggested that asteroidal impact played a major role in the formation of the EL3 chondrites and that they are impact melt breccias that do not preserve their primary textures [*Van Nierkirk and Keil*, 2011]. This interpretation is based on the occurrence of euhedral enstatite in opaque mineral assemblages, the presence of the mineral keilite (Fe, Mg)S, and possible evidence for mobilization of metal. *Rubin* [1997] found euhedral, zoned sinoite grains associated with euhedral enstatite and graphite within impact-melted portions of QUE 94368, the first known EL4 chondrite (Plate 32). He argued that the sinoite crystallized from the impact melt. Arguments

(a)

(b)

(c)

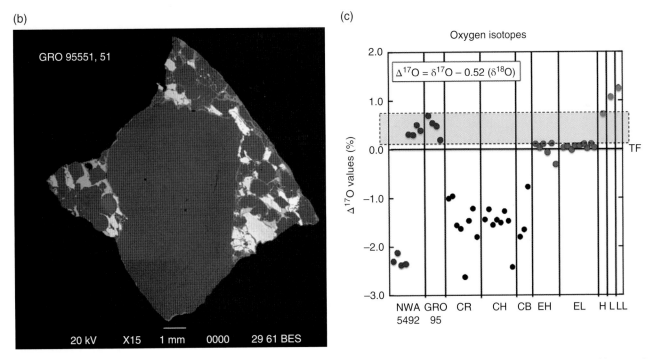

Figure 4.4. (a) A cut surface of GRO 95551 showing large cm-sized silicate chondrules (grey) and metal nodules (white), and numerous smaller metal and silicate chondrules. (b) A backscattered electron image of GRO 95551, 51 showing a large barred olivine (BO) chondrule in the interior. Surrounding the large central BO chondrules are smaller (1 mm) chondrules (grey) and metal grains (white). (c) Diagram showing the $\Delta^{17}O$ values of silicates in GRO 95551 compared with average values for chondrites from other groups. The gray bar shows the range of values for GRO 95551. Data for ordinary chondrites are averages for the H, L, LL. Data are from *Weisberg et al.* [2012] and references therein.

against impact as major process for the history of EL chondrites include the low shock stage of many EL3 chondrites, chondrules and metal with sharp boundaries, the lack of high-pressure minerals, and a wide range of $^{53}Mn/^{55}Mn$ ratios and excess of ^{53}Cr in EL3 sulfides, which should have been equilibrated if they were impact modified [e.g., *Weisberg and Kimura*, 2012; *El Goresy et al.*, 1992].

Notable in many of the EL3 chondrites are metal-rich nodules (Figure 4.5b). Nodules in the MAC 88136 EL3 commonly occur as 400-μm nodules intergrown with lath-shaped or euhedral enstatite and schreibersite, Cr-bearing troilite, daubreelite, and graphite [*Van Niekirk and Keil*, 2011; *Weisberg et al.*, 1994]. These intergrowths of metal and enstatite have been interpreted to be the result of

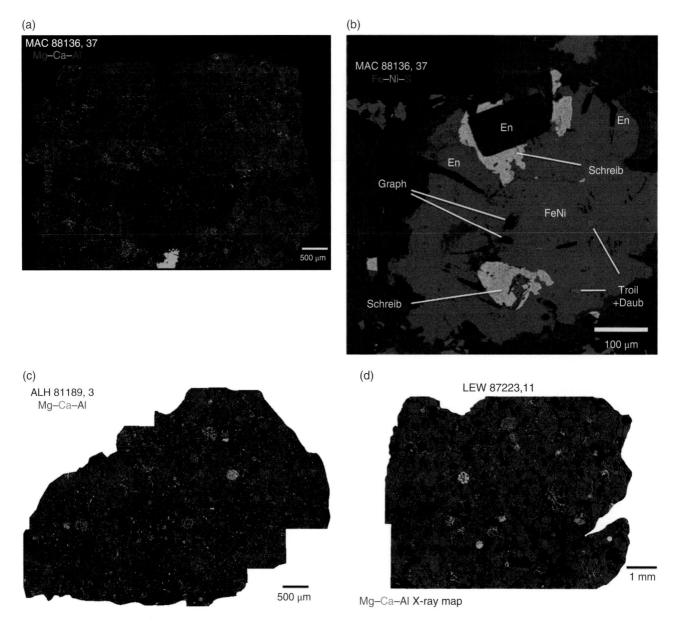

Figure 4.5. (a) A Mg, Ca, Al composite (red, green, blue) X-ray image of the MAC 88136, 37 EL3 chondrite. The section contains chondrules, approximately 500 μm in size, and metal nodules (black), which are slightly smaller, at about 400 μm. The red color of the image illustrates the Mg-rich compositions of the silicates, which are mainly near-endmember enstatite. Small areas of brighter red are less common forsterite grains. (b) A Fe, Ni, S (red, green, blue) composite image of a metal nodule in MAC 88136. The image shows FeNi metal in red, schreibersite in green, and troilite and daubreelite in purple and blue, respectively. Enstatite (black) occurs as euhedral crystals enclosed within and at the edges of the metal. (c) A Mg, Ca, Al (red, green, blue) composite image of ALH 81189, 3. ALH 81189 is a primitive EH3 chondrite with numerous chondrules (up to 500 μm in size) with sharp boundaries. The brightest reds are olivine grains, which are fairly common and suggest that ALH 81189 is a primitive EH3 chondrite. (d) A Mg, Ca, Al (red, green, blue) composite image of LEW 87223, 11 showing enstatite-rich chondrules (red) and several Al-rich chondrules (blue).

impact melting and possibly injection of metal into pore spaces in a loosely consolidated regolith [*Van Nierkirk et al*, 2009; *Van Nierkirk and Keil*, 2011]. Alternatively, they have been interpreted to form from fractional condensation from the early solar gas [*Weisberg et al.*, 1997]. In a reducing solar nebula with a C/O ratio of ≥1, Fe metal

condenses at ~1375 K and olivine, diopside, and enstatite condense at 1100, 1050, and 950 K, respectively [e.g., *Larimer and Bartholomay*, 1979; *Saxena and Eriksson*, 1985; *Ebel*, 2006]. Thus, from 1375 to 1100 K, metal may exist independently of silicates. Aggregates of condensates in this T range would be metal-rich. Metal-rich dust mixing with small amounts of silicate may result in the formation of the metal-enstatite intergrowths observed in MAC 88136 and other EL3 chondrites [*Weisberg et al.*, 1997]. The extent to which EL3 chondrites have been modified by asteroidal impact or preserve their primary structures is an open issue. Continued recovery of E3 chondrites by ANSMET may provide new material with which to test hypotheses for the origin and history of the EL3 chondrites and their parent asteroid.

The Antarctic collection has also greatly increased the number of EH3 chondrites. Among the EH3 chondrites, ALH 81189 (Plate 29) has characteristics (high abundance of Fe-silicates and olivine) that suggest it is one of the most primitive EH3 chondrites (Figure 4.5c) [*Lusby et al.*, 1987; *Hicks et al.*, 2000; *Fagan et al.*, 2010]. Study of the structural ordering of the organic matter in matrix areas also indicate that ALH 84206 (paired with ALH 81189) is among the most primitive EH chondrites [*Quirico et al.*, 2011].

Matrix is an important component of chondrites. In many chondrites, it is a mixture of materials that range from being highly primitive to highly altered. It is the component of chondrites that harbors refractory olivine, primitive amorphous materials, organic matter, and interstellar grains. It reveals various degrees of aqueous alteration, even in cases where chondrules appear to be unaltered. ALH 81189 has been an important chondrite for studies of matrix in E3 chondrites. Estimates of the abundances of matrix material in E3 chondrites are generally lower than in O or C chondrites and as a result E chondrite matrix is not as well characterized as in the other chondrite groups. However, matrix-like material has been identified in some E3 chondrites [e.g., *Kimura*, 1988; *Huss and Lewis*, 1995; *Ebata et al.* 2006; *Rubin et al.*, 2009]. Clastic matrix constitutes 11.7 vol% of the ALH 81189 EH3 chondrite [*Rubin*, 2009], and presolar grains were identified in EH3 clastic matrix [*Ebata et al.*, 2006, 2007; *Ebata and Yurimoto*, 2008].

LEW 87223 is an unusual E3 chondrite that is not EH or EL (Figure 4.5d). The low Si content of its metal, the presence of alabandite (MnS), and its bulk W/Ni ratio are all similar to what is seen in EL chondrites, but its high metal abundance and siderophile element contents are closer to those of EH chondrites [*Grossman et al.*, 1993]. Its oxygen isotope composition is similar to that of equilibrated EH and EL chondrites [*Weisberg et al.*, 1995a]. It has been interpreted to have formed from an EL3 starting material and was modified by impact and brecciation [*Zhang et al.*, 1995]. *Grossman et al.* [1993] suggested it is a new type of chondrite intermediate between E and H chondrite. However, the distribution of the oxygen isotope compositions of its chondrules are more similar to those of E chondrites [*Weisberg et al.*, 2011], and thus it is considered an anomalous E chondrite until

further studies and/or additional examples of this type of E chondrite are found.

4.5. K CHONDRITES

LEW 87232 (Plate 37) is one of two K chondrites. The K-chondrite (Kakangari-like) grouplet consists of two meteorites, Kakangari and LEW 87232, that are chemically, mineralogically, and isotopically distinct from the ordinary, enstatite, and carbonaceous chondrite classes [*Srinivasan and Anders*, 1977; *Davis et al.*, 1977; *Zolensky et al.*, 1989; *Brearley*, 1989; *Weisberg et al.*, 1996]. They have high matrix abundances (70–77 vol%), like some carbonaceous chondrites, metal abundances (6–9 vol%) similar to H chondrites, and their average olivine (Fa$_2$) and enstatite (Fs$_4$) compositions are Fe-poor, indicating an oxidation state intermediate between ordinary and enstatite chondrites. The matrix of K chondrites is mineralogically unique and composed largely of enstatite that is compositionally similar to that in the chondrules. Refractory-lithophile and volatile element abundances of K chondrites are similar to those of the ordinary chondrites. Chalcophile element abundances are between those of H and enstatite chondrites. Their O-isotope compositions, however, are most similar to the CR, CB, and CH chondrites.

4.6. R CHONDRITES (PLATES 35 AND 36)

One of the first identified R chondrites was ALH 85151 [*Weisberg et al.*, 1991]. It is also one of the few unequilibrated R3 chondrites. Like the K chondrites, the R chondrites are not ordinary, carbonaceous, or enstatite chondrites. They are highly oxidized meteorites characterized by NiO-bearing, FeO-rich olivine (Fa$_{37–40}$) and the nearly complete absence of Fe,Ni-metal. They have a very high abundance of matrix (~50 vol%), similar to some carbonaceous chondrites. Most R chondrites are metamorphosed (petrologic type >3.6) and are brecciated. Refractory lithophile abundances of R chondrites are ~0.95 × Cl, intermediate between those in ordinary chondrites (OC) and CI chondrites. Abundances of the volatile elements Se and Zn are greatly enhanced relative to OC, and R chondrites can be distinguished from other chondrite groups on the basis of their Al/Mn and Zn/Mn abundance ratios [*Kallemeyn et al.*, 1996]. The R chondrites have the highest bulk Δ^{17}O values of any other chondrite group [*Weisberg et al.*, 1991; *Schulze et al.*, 1994; *Kallemeyn et al.*, 1996].

The role of water in the evolution of R chondrites and its relationship to their oxidized compositions and high Δ^{17}O values is an open issue. Most R chondrites are completely dry. However, the R6 chondrites LAP 04840 (Plate 36) and MIL 11207 are remarkable in containing abundant amphibole and biotite, water-bearing minerals not previously reported in any other chondrites. The amphibole and biotite are nearly pure hydroxyl species. Additionally, hydrogen in the amphibole is isotopically heavier than any other known H reservoir, with D values of +3660‰. The origin of the water in LAP 04840 is not clear. It may have been derived from internal heating of organics that accreted with the R chondrite matrix, or it was delivered to the parent body from an outside source, such as a comet or water-rich asteroid [*Trieman et al.*, 2007; *McCanta et al.*, 2008].

4.7. PRESOLAR GRAINS

Some primitive Antarctic meteorites have preserved particularly well presolar grains that have contributed to our understanding of nucleosynthetic and early stellar processes. Presolar grains were formed by processes prior to the formation of our own solar system. Presolar silicate and oxide grains fall into four groups: groups 1 and 3 form in red giant stars, and groups 2 and 4 form around low-mass asymptotic giant branch (AGB) stars. Group 4 grains may form in supernovae.

The primitive CO3 ALH 77307 (Plate 15) first showed evidence for presolar diamonds [*Huss et al.*, 2003], and subsequent work has revealed additional material (silicates and oxides) from the same meteorite [*Nguyen et al.*, 2010]. Most of the grains condensed in low-mass red giant and AGB stars, but some are from low-metallicity, low-mass stars, others are from supernova outflows [*Nguyen et al.*, 2010], and a few can offer constraints on galactic chemical evolution.

Presolar grains have also been found in the CRs QUE 99177 and MET 00426. The matrix-normalized presolar SiC abundances in these two CR chondrites is consistent with abundances determined for the more heavily hydrated CR1 and CR2 chondrites, showing that aqueous alteration does not easily destroy presolar SiC. QUE 99177 and MET 00426 contain high abundances (~120 ppm) of carbonaceous matter with anomalous C isotopic compositions that probably formed in cold molecular clouds via ion-molecule reactions [*Floss and Stadermann*, 2009]. C-anomalous material of interstellar origin has previously been found in interplanetary dust particles and in insoluble organic matter from primitive meteorites; however, the amounts observed in QUE 99177 and MET 00426 are significantly higher than in other primitive meteorites and attests to the very primitive nature of the two CR3 chondrites [*Floss and Stadermann*, 2011].

The small EH3 chondrite ALH 81189 has also yielded presolar silicate grains [*Ebata et al.*, 2006] of Group 1 affinity (AGB stars). Finally, presolar grains have been found in the low-grade ordinary chondrites QUE 97008 (Plate 4), WSG 95300 (Plate 1), and MET 00452 and 00526

(L/LL 3.05) [*Nittler et al.*, 2008]. Presolar grains have been found in Antarctic samples of all the major classes of chondrites (O, C, and E), and these samples are highly requested. The search for new types and more extreme examples of existing types of grains can provide important constraints on the stellar and galactic processes operating prior to and during the birth of our solar system.

4.8. CONCLUDING REMARKS

The ANSMET meteorite collection has greatly expanded our knowledge of primitive solar system materials. The examples described in this chapter only scratch the surface. For example, not discussed in detail are the CM chondrites; 413 of the 458 currently known CM chondrites are from Antarctica. Although many are paired, the Antarctic collection clearly provides a major source of material for studying this important chondrite group and understanding the asteroidal process of hydrous alteration they record. Continued recovery of new meteorites by ANSMET teams is essential to furthering our understanding of the primitive materials in the asteroid belt and to deciphering the clues they hold about stellar processes, early solar system evolution, and geologic processes on primitive asteroids.

Acknowledgements. C. Alexander is thanked for doing a thorough reading and review of an earlier version of this chapter.

REFERENCES

Abreu, N. M, and A. J. Brearley (2010), Early solar system processes recorded in the matrices of two highly pristine CR3 carbonaceous chondrites, MET00426 and QUE99177, *Geochim. Cosmochim. Acta, 74*, 1146–1171.

Alexander, C. M. O'D., Swan, P., Prombo, C. A. (1994), Occurrence and implications of silicon nitride in enstatite chondrites, *Meteoritics and Planetary Science, 29*, 79–85.

Alexander, C.M.O'D., M.Fogel, H.Yabuta, and G.D. Cody (2007), The origin and evolution of chondrites recorded in the elemental and isotopic compositions of their macromolecular organic matter, *Geochim. Cosmochim. Acta, 71*, 4380–4403.

Amelin, Y., and A. E. Krot (2005), Young Pb-isotopic ages of chondrules in CB carbonaceous chondrites, *Lunar Planet. Sci. Conf., 36*, abstract #1247.

Andersen, C. A., K. Keil, and B. Mason (1964), Silicon oxynitride: A meteorite mineral, *Science, 146*, 256–257.

Brearley, A. J. (1989), Nature and origin of matrix in the unique type 3 chondrite, Kakangari, *Geochim. Cosmochim. Acta, 53*, 197–214.

Busemann, H., A.F. Young, C.M.O'D. Alexander, P. Hoppe, S. Mukhopadhyay, and L.R. Nittler (2006), Interstellar chemistry recorded in organic matter from primitive meteorites, *Science* 312, 727–730.

Campbell, A. J., M. Humayun, A. Krot, and K. Keil (2001), Origin of zoned metal grains in the QUE 94411 chondrite, *Geochim. Cosmochim. Acta*, *65*, 163–180.

Campbell, A. J., M. Humayun, and M. K. Weisberg (2002), Siderophile element constraints on the formation of metal in the metal-rich chondrites Bencubbin, Gujba and Weatherford, *Geochim. Cosmochim. Acta*, *66*, 647–660.

Campbell, A. J., M. Humayun, and M. K. Weisberg (2005), Compositions of unzoned and zoned metal in the CBb chondrites HH 237 and QUE 94627, *Meteoritics and Planetary Science*, *40*, 1131–1148.

Clayton, R. N., and T. K. Mayeda (1984), Oxygen isotopic compositions of enstatite chondrites and aubrites, *J. Geophys. Res.*, *89*, C245–C249.

Davis, A. M., L. Grossman, and R. Ganapathy (1977), Yes, Kakangari is a unique chondrite, *Nature*, *265*, 230–232.

Ebata, S., and H. Yurimoto (2008), Identification of silicate and carbonaceous presolar grains in the type 3 enstatite chondrites, in *Origin of Matter and Evolution of Galaxies*, edited by T. Suda, T. Nozawa, A. Ohnishi, K. Kato, M. Y. Fujimoto, T. Kajino, and S. Kubono, Melville, New York, American Institute of Physics, pp. 412–414.

Ebata, S., K. Nagashima, S. Itoh, S. Kobayashi, N. Sakamoto, T.J. Fagan, and H. Yurimoto (2006), Presolar silicate grains in enstatite chondrites, *Lunar Planet. Sci. Conf.*, *37*, 1619.

Ebata, S., T. J. Fagan, and H.Yurimoto (2007), Identification of silicate and carbonaceous presolar grains in the type 3 enstatite chondrite ALHA81189 (abstract), *Meteoritics and Planetary Science*, *42*, A38.

Ebel, D.S. (2006), Condensation of rocky material in astrophysical environments, in *Meteorites and the Early Solar System II*, edited by D. Lauretta et al., University of Arizona, Tucson, pp. 253–277.

El Goresy, A., H. Yabuki, K. Ehlers, D. Woolum, and E. Pernicka (1988), Qingzhen and Yamato—691: A tentative alphabet for the EH chondrites, *Proc. NIPR Symp. Antarct. Meteorites*, *1*, 65–101.

El Goresy, A., M. Wadhwa, H.-J. Nagel, E. K. Zinner, J. Janicke, and G. Crozaz (1992), ^{53}Cr-^{53}Mn systematics of Mn-bearing sulfides in four enstatite chondrites, *Lunar Planet. Sci.*, *33*, 331–332.

Fagan, T. J., S. Kataoka, A. Yoshida, and K. Norose (2010), Transition to low oxygen fugacities in the solar nebula recorded by EH3 chondrite ALHA 81189, *Lunar Planet. Sci. Conf.*, *41*, abstract #1534.

Floss, C., and F. J. Stadermann (2009), Auger Nanoprobe analysis of presolar ferromagnesian silicate grains from primitive CR chondrites QUE 99177 and MET 00426, *Geochim. Cosmochim. Acta*, *73*, 2415–2440.

Floss, C., and F. J. Stadermann (2011), High abundance of circumstellar and interstellar C-anomalous phases in the primitive CR3 chondrites QUE 99177 and MET 00426, *Astrophys. J.*, *697*, 1242–1255.

Glavin, D.P., and J.P. Dworkin (2009), Enrichment of the amino acid L-isovaline by aqueous alteration on CI and CM meteorite parent bodies, *PNAS*, *106*, 5487–5492.

Grossman, J. N., A. E. Rubin, and G. J. MacPherson (1988), ALH85085: A unique volatile-poor carbonaceous chondrite with possible implications for nebular fractionation processes, *Earth Planet. Sci. Lett.*, *91*, 33–54.

Grossman, J. N., G. J. MacPherson, and G.Crozaz (1993), LEW 87223: A unique E chondrite with possible links to H chondrites, *Meteoritics*, *28*, 358.

Hicks, T. L., T. J. Fagan, and K. Keil (2000), Metamorphic sequence of unequilibrated EH chondrites using modal olivine and silica abundance, *Lunar Planet. Sci. Conf.*, *31*, abstract #1491.

Humayun, M. (2012), Chondrule cooling rates inferred from diffusive profiles in metal lumps from the Acfer 097 CR2 chondrite, *Meteoritics & Planetary Science*, *47*, 1191–1208. doi: 10.1111/j.1945–5100.2012.01371.x

Huss, G. R., and R. S. Lewis (1995), Presolar diamond, SiC, and graphite in primitive chondrites: Abundances as a function of meteorite class and petrologic type, *Geochim. Cosmochim. Acta*, *59*, 115–160.

Huss, G. R., A. P. Meshik, J.B. Smith, and C.M. Hohenberg (2003), Presolar diamond, silicon carbide, and graphite in carbonaceous chondrites: Implications for thermal processing in the solar nebula, *Geochim. Cosmochim. Acta*, *67*, 4823–4848.

Ichikawa, O., and Y. Ikeda (1995), Petrology of the Yamato-8449 CR chondrite, Proceedings of the Eighth Symposium on Antarctic Meteorites, 63–78.

Ivanova, M. A., N. N. Kononkova, A. N. Krot, R. C. Greenwood, I. A. Franchi, A. B. Verchovsky, M. Trieloff, E. V. Korochantseva, and F. Brandstatter (2009), The Isheyevo meteorite: Mineralogy, petrology, bulk chemistry, oxygen, nitrogen, and carbon isotopic compositions and 40Ar-39 ages, *Meteoritics & Planetary Science*, *43*, 915–940.

Javoy, M. (1995), The integral enstatite chondrite model of the earth, *Geophys. Res. Lett.*, *22*, 2219–2222.

Javoy, M., E. Kaminski, F. Guyot, D. Andrault, C. Sanloup, M. Moreira, S. Labrosse, A. Jambon, P. Agrinier, A. Davaille, and C. Jaupart (2010), The chemical composition of the Earth: Enstatite chondrite models, *Earth and Planetary Science Letters*, *293*, 259–268.

Kallemeyn, G. W., A. E. Rubin, and J. T. Wasson (1996), The compositional classification of chondrites: VII. The R chondrite group, *Geochim. Cosmochim. Acta*, *60*, 2243–2256.

Keil, K. (1968), Mineralogical and chemical relationships among enstatite chondrites, *J. Geophys. Res. 73*, 6945–6976.

Keil, K., and C. A. Andersen (1964), Electron microprobe study of the Jajh deh Kot Lalu enstatite chondrite, *Geochim. Cosmochim. Acta 29*, 621–632.

Kimura, M. (1988), Origin of opaque minerals in an unequilibrated enstatite chondrite, Yamato-691, *Proceedings of the NIPR Symposium on Antarctic Meteorites*, *1*, 51–64.

Krot, A. E., A. Meibom, M. K. Weisberg, and K. Keil (2002), The CR chondrite clan: Implications for early solar system processes, *Meteoritics and Planetary Science*, *37*, 1451–1490.

Krot, A. N., Y. Amelin, P. Cassen, and A. Meibom (2005), Young chondrules in CB chondrites from a giant impact in the early Solar System, *Nature*, *436*, 989–992.

Krot, A. N., M. A. Ivanova, and A. A. Ulyanov (2006), Chondrules in the CB/CH-like carbonaceous chondrite Isheyevo: Evidence for various chondrule-forming mechanisms and multiple chondrule generations, *Chemie der Erde*, 67, 283–300.

Larimer, J. W., and M. Bartholomay (1979), The role of carbon and oxygen in cosmic gases: Some applications to the chemistry and mineralogy of enstatite chondrites, *Geochim. Cosmochim. Acta*, 43,1455–1466.

Lin, Y. T., H. Nagel, L. L. Lundberg, and A. El Goresy (1991), MAC88136: The first EL3 chondrite, *Proc. Lunar Planet. Sci. Conf.*, 22, 811.

Lusby, D., E.R.D. Scott, and K. Keil (1987), Ubiquitous high FeO silicates in enstatite chondrites, *Proc. Lunar Planet. Sci. Conf., 17, J. Geophys. Res. Suppl.*, 92, E679–E695.

Martins, Z., C.M.O'D. Alexander, G. E. Orzechowska, M. L. Fogel, and P. Ehrenfruend (2007), Indigenous amino acids in primitive CR chondrite, *Meteoritics and Planetary Science 42*, 2125–2136.

Mason, B. (1966), The enstatite chondrites, *Geochim. Cosmochim. Acta*, 30, 23–39.

McCanta, M. C., A. H. Treiman, M. D. Dyar, C.M.O'D. Alexander, D. Rumble III, and E. J. Essene (2008), The La Paz 04840 meteorite: Petrology and origin of an amphibole-rich R chondrite, *Geochim. Cosmochim. Acta*, 72, 5757–5780.

Meibom, A., M. I. Petaev, A. K. Krot, J. A. Wood, and K. Keil (1999), Primitive FeNi metal grains in CH carbonaceous chondrites formed by condensation from a gas of solar composition, *J. Geophys. Res.*, 104, 22053–22059.

Meibom, A., S. J. Desch, A. N. Krot, J. N. Cuzzi, M. I. Petaev, L. Wilson, and K. Keil (2000), Large-scale thermal events in the solar nebula: Evidence from Fe,Ni metal grains in primitive meteorites, *Science*, 288, 839–841.

Meibom, A., M. I. Petaev, A. N. Krot, K. Keil, and J. A. Wood (2001), Growth mechanism for Fe,Ni metal condensates in the solar nebula, *J. Geophys. Res.*, 106, 32797–32801.

Meibom, A., K. Righter, N. Chabot, G. Dehn, A. Antignano, T. J. McCoy, A. N. Krot, M. E. Zolensky, M. I. Petaev, and K. Keil (2005), Shock melts in QUE 94411, Hammadah al Hamra 237, and Bencubbin: Remains of the missing matrix? *Meteoritics and Planetary Science*, 40, 1377–1391.

Newsom, H. E., and M. J. Drake, (1979), The origin of metal clasts in the Bencubbin meteoritic breccias, *Geochim. Cosmochim. Acta*, 43, 689–707.

Nguyen, A. N., L. R. Nittler, F. J. Stadermann, R. M. Stroud, C.M.O'D. Alexander, (2010), Coordinated Analysis of presolar grains in the Allan Hills 77307 and Queen Alexandra Range 99177 meteorites, *Astrophys. J.*, 719, 166–189.

Nittler, L. R., C.M.O'D. Alexander, R. M. Gallino, P. Hoppe, A. N. Nguyen, F. J. Stadermann, and E. K. Zinner (2008), Aluminum-, calcium- and titanium-rich oxide stardust in ordinary chondrite meteorites, *Astrophys. J.*, 682,1450–1478.

Noguchi, T. 1995. Petrology and mineralogy of the PCA 91082 chondrite and its comparison with the Yamato-793495 (CR) chondrite, *Proceedings of the Eighth Symposium on Antarctic Meteorites*, 32–62.

Petaev, M. I., A. Meibom, A. N. Krot, J. A. Wood, and K. Keil (2001), The condensation origin of layered metal grains in QUE 94411: Implications for the formation of the Bencubbin-like chondrites, *Meteoritics and Planetary Science*, 36, 93–106.

Pizzarello, S., and W. Holmes (2009), Nitrogen-containing compounds in two CR2 meteorites: [15]N composition, molecular distribution and precursor molecules, *Geochim. Cosmochim. Acta*, 73, 2150–2162.

Pizzarello, S., Y. Huang, and M. R. Alexandre (2008), Molecular asymmetry in extraterrestrial chemistry: Insights from a pristine meteorite, *PNAS*, 105, 3700–3704.

Quirico, E., M. Bourot-denise, C. Robin, G. Montagnac, and P. Beck (2011), A reappraisal of the metamorphic history of EH3 and EL3 enstatite chondrites, *Geochim. Cosmochim. Acta*, 75, 3088–3102.

Righter K., A. J. Campbell, and M. Humayun (2005), Diffusion of trace elements in FeNi metal: Application to zoned metal grains in chondrites, *Geochim. Cosmochim. Acta*, 69, 3145–3158.

Rubin, A. E. (1997), Sinoite (Si2N2O): Crystallization from EL chondrite impact melts, *American Mineralogist*, 82(9-10), 1001–1006.

Rubin, A. E., C. D. Griest, B.-G. Choi, and J. T. Wasson (2009), Clastic matrix in EH3 chondrites, *Meteoritics and Planetary Science*, 44, 589–601.

Satterwhite, C., and K. Righter (2012), *Ant. Met. Newslett.*, 35(2), 19.

Saxena, S. K., and G. Eriksson (1985), Anhydrous phase equilibria in Earth's upper mantle, *J. Petrology*, 26, 1–13.

Schulze, H., A. Bischoff, H. Palme, B. Spettel, G. Dreibus, and J. Otto (1994), Mineralogy and chemistry of Rumuruti: The first meteorite fall of the new R chondrite group, *Meteoritics Planet. Sci.* 29, 275–286.

Scott, E. R. D. (1988), A new kind of primitive chondrite, *Earth Planet. Sci. Lett.*, 91, 1–18.

Srinivasan, B., and E. Anders (1977), Noble gases in the unique chondrite, Kakangari. Meteoritics 12, 417–424.

Treiman, A. H., C. M. Alexander, E. J. Essene, and M. C. McCanta (2007), The amphibole-phlogopite R-chondrite LAP 04840: Hot hydration by heavy H2O, *Lunar and Planetary Institute Science Conference Abstracts*, 38, 1309.

Van Niekirk, D., and K. Keil, (2011), Metal/sulfide–silicate intergrowth textures in EL3 meteorites: Origin by impact melting on the EL parent body, *Meteoritics and Planetary Science*, 46, 1487–1494.

Van Niekirk, D., Humayun, M., and Keil, K. (2009), In situ determination of siderophile trace elements in EL3 meteorites, *Lunar & Planetary Science Conference XL*, The Woodlands, TX, March 23–27.

Wasson, J. T., and G. W. Kallemeyn (1990), Allan Hills 85085: A subchondritic meteorite of mixed nebular and regolithic heritage, *Earth Planet. Sci. Lett.*, 101, 148–161.

Weisberg, M. K., and M. Kimura (2012), The unequilibrated enstatite chondrites, *Chemie der Erde*, 72(2), 101–115.

Weisberg, M. K., M. Prinz, and C. E. Nehru (1988), Petrology of ALH85085: a chondrite with unique characteristics, *Earth Planet. Sci. Lett.*, 91, 19–32.

Weisberg, M. K., M. Prinz, and C.E. Nehru (1989), The Bencubbin breccia and its relationship to CR chondrites and the ALH85085 chondrite, *Meteoritics*, 25, 269–279.

Weisberg, M. K., M. Prinz, H. Kojima, K. Yanai, R. N. Clayton, and T. K. Mayeda (1991), Carlisle Lakes-type chondrites: A new grouplet with high Δ¹⁷O and evidence for nebular oxidation, *Geochim. Cosmochim. Acta*, 55, 2657–2669.

Weisberg, M. K., M. Prinz, R. N. Clayton, and T. K. Mayeda (1993), The CR (Renazzo-type) carbonaceous chondrite group and its implications, *Geochim. Cosmochim. Acta*, 57, 1567–1586.

Weisberg, M. K., J. S. Bosenberg, G. Kozhusko, M. Prinz, R. N. Clayton, and T. K. Mayeda (1995a), EH3 and EL3 chondrites: A petrologic-oxygen isotopic study, *Lunar Planet. Sci. Conf.*, 26, 1481–1482.

Weisberg M. K., M. Prinz, R. N. Clayton, T. K. Mayeda, M. M. Grady, and C. T. Pillinger (1995b), The CR chondrite clan, *Proceedings of the National Institute of Polar Research Symposium on Antarctic Meteorites*, 8, 11–32. Tokyo, Japan.

Weisberg, M. K., M. Prinz, R. N. Clayton, T. K. Mayeda, M. M. Grady, I. Franchi, C. T. Pillinger, and G. W. Kallemeyn (1996), The K (Kakangari) chondrite grouplet, *Geochim. Cosmochim. Acta*, 60, 4253–4263.

Weisberg, M. K., R. A. Fogel, and M. Prinz (1997), Kamacite-enstatite intergrowths in enstatite chondrites, *Lunar Planet. Sci. Conf.*, 28, 1523–1524.

Weisberg, M. K., and M. Prinz (1999), Zoned metal in CR clan chondrites, 24th Symposium on Antarctic Meteorites, National Institute of Polar Research, Tokyo, Japan, 187–189.

Weisberg, M. K., M. Prinz, R. N. Clayton, T. K. Mayeda, N. Sugiura, S. Zashu, and M. Ebihara (2001), A new metal-rich chondrite grouplet, *Meteoritics and Planetary Science*, 36, 401–418.

Weisberg, M. K., and H. Huber (2007), The GRO 95577 CR1 chondrite and hydration of the CR parent body, *Meteoritics and Planetary Science*, 42, 1495–1503.

Weisberg, M. K., D. S. Ebel, H. C. Connolly Jr., N. T. Kita, and T. Ushikubo (2011), Petrology and oxygen isotope compositions of chondrules in E3 chondrites, *Geochim. Cosmochim. Acta*, 75, 6556–6569.

Weisberg, M. K., T. E. Bunch, J. H. Wittke, D. Rumble III, and D. S. Ebel (2012), Petrology and oxygen isotopes of NWA 5492, a new metal-rich chondrite, *Meteoritics and Planetary Science*, 47, 363–373.

Zhang, Y., P. H. Benoit, and D. W. Sears (1995), The classification and complex thermal history of the enstatite chondrites, *J. Geophys. Res.: Planets (1991–2012)*, 100(E5), 9417–9438.

Zolensky, M. E., R. Score, J. W. Schutt, R. N. Clayton, and T. K. Mayeda, (1989), Lea County 001, an H5 chondrite and Lea County 002, an ungrouped chondrite, *Meteoritics*, 24, 227–232.

Zolensky, M. E., R. A. Barrett, and L. Browning (1993), Mineralogy and composition of matrix and chondrule rims in carbonaceous chondrites, *Geochim. Cosmochim. Acta*, 57, 3123–3148.

5

Achondrites and Irons: Products of Magmatism on Strongly Heated Asteroids

David W. Mittlefehldt[1] and Timothy J. McCoy[2]

5.1. INTRODUCTION

The systematic search and recovery of meteorites from Antarctica, precipitated by the serendipitous discovery of nine meteorites on glacial ice near the Yamato Mountains by the 10th Japanese Antarctic Research Expedition (JARE), has been a great boon for understanding the origin of the solar system. This has been especially true for an understanding of strongly heated asteroids. Meteorites from such asteroids have experienced heating to the point where melting has taken place. For some, only incipient melting has occurred, while for others, total melting may have occurred.

In the former case, nascent melts were able to collect and migrate even in the low-gravity environments of asteroids. The results of this process are primitive achondrites that show varying degrees of grain coarsening and mineralogical and elemental fractionation from that of their chondritic precursors. In the latter case, gross-scale chemical segregation occurred due to density-driven separation of immiscible metallic and silicate melts. Some of the melts cooled slowly enough that gravity-induced crystal settling took place, resulting in additional mineralogical and chemical fractionation, in some cases on much smaller scales. Impact-induced fragmentation and debris gardening engendered a heterogeneous debris layer on asteroidal surfaces composed of the disparate products of asteroidal magmatism.

Because of these processes, strongly heated asteroids show extreme heterogeneities on scales of hundreds or tens of kilometers down to meters or decimeters. In order to understand the differentiation and regolith processes,

suites of samples representing different portions of the parent asteroid must be studied. Thus, one key to making progress in understanding magmatic processing on asteroids is having a range of samples available that can be used to make robust models for styles of melting, magma migration, collection and eruption, and crystal fractionation in metallic cores and silicate magma chambers. Delivery of meteorites to Earth is a stochastic process, and for this reason, the likelihood of having especially diagnostic samples for study is directly related to the number of meteorites available.

Antarctica provides an ideal location for collecting large numbers of meteorites. The combination of ice flow, flow barriers, and an ablation mechanism provide locations where meteorites that have fallen over wide areas get concentrated and stranded. The low temperatures and entombment in ice retard terrestrial alteration mechanisms and provide for a long meteorite accumulation period. Together, these environmental factors allow for regions where vast numbers of meteorites can be harvested within relatively short time spans. A typical six-week U.S. Antarctic Search for Meteorites (ANSMET) field season can yield more than 1,000 meteorites from particularly fruitful locations.

The two most well-sampled strongly heated asteroids are the parents of the howardite-eucrite-diogenite (HED) clan (1,135 named meteorites) and the ureilite group (315 named meteorites; census as of November 2012). Prior to the start of systematic meteorite searches in Antarctica (the 15th JARE in 1973–1975 and the first U.S. Antarctic Search for Meteorites in 1976–1977), these two asteroids were represented by 62 and 9 meteorites, respectively. Over the following ~25 years, the numbers of meteorites recovered from all locations in Antarctica from each asteroid reached ~75%–80% of the totals, the proportions only decreasing with the advent of "commercial

[1] Astromaterials Research Office, NASA Johnson Space Center
[2] Department of Mineral Sciences, National Museum of Natural History, Smithsonian Institution

35 Seasons of U.S. Antarctic Meteorites (1976–2010): A Pictorial Guide to the Collection, Special Publication 68,
First Edition. Edited by Kevin Righter, Catherine M. Corrigan, Timothy J. McCoy and Ralph P. Harvey.
© 2015 American Geophysical Union. Published 2015 by John Wiley & Sons, Inc.

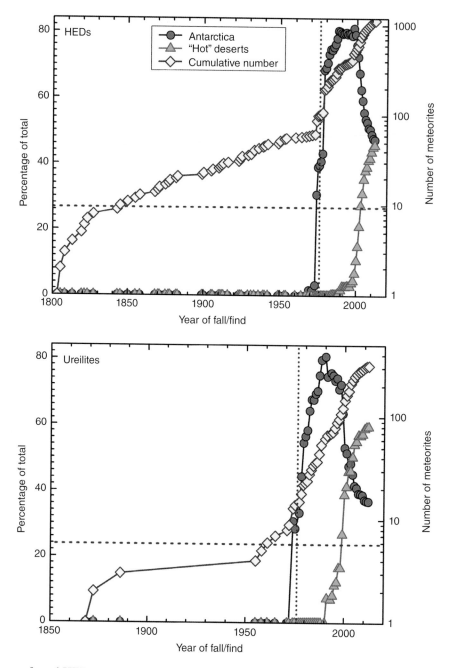

Figure 5.1. Percentages of total HEDs (upper) and ureilites (lower) from Antarctica (all collections) and finds from "hot" deserts (northern Africa, the Arabian Peninsula, and western Australia), and cumulative numbers of named meteorites. Vertical dotted lines mark the start of ANSMET, while the horizontal dashed lines represent the current percentage of each group represented by the ANSMET collection. Statistics as of 1 October 2012 derived from the Meteoritical Bulletin Database (http://www.lpi.usra.edu/meteor/metbull.php).

harvesting" of meteorites from northern Africa and the Arabian Peninsula beginning in the 1990s (Figure 5.1). Presently, ~25% of all HEDs and ureilites are from the U.S. Antarctic collection. Because of the rapid announcement, and efficient request, approval and allocation process, ANSMET HEDs and ureilites (and other meteorites) have had a much greater impact on

cosmochemistry than their numerical proportion would suggest.

The number of types of meteorites from strongly heated asteroids represented in the Antarctic collection is quite large (Plates 38 to 63; 75 to 80), especially if unique meteorites are included. Space limitations preclude discussion of all of them. We have chosen to focus on meteorite types that sample a range of regions in their parent asteroids, that represent a range of processes, that have sufficient numbers to allow for detailed understanding of the processes that occurred on strongly heated asteroids, and for which Antarctic meteorites have been especially important for advancing our understanding. We cover the acapulcoite-lodranite clan, ungrouped iron meteorites, ureilites, brachinites, the howardite-eucrite-diogenite clan, and angrites.

5.2. ULTRAMETAMORPHOSED CHONDRITIC MATERIAL: THE ACAPULCOITE-LODRANITE CLAN OF PRIMITIVE ACHONDRITES

Acapulcoites and lodranites are fine- to medium-grained equigranular rocks composed of orthopyroxene, olivine, chromian diopside, sodic plagioclase, Fe,Ni metal, schreibersite, troilite, whitlockite, chlorapatite, chromite, and graphite. The mineral assemblage is broadly similar to that found in ordinary chondrites, although mineral compositions, abundances, and textures differ from ordinary chondrites. Mineral compositions are substantially more magnesian than observed in ordinary chondrites (olivine mg# 87–97 in aca-lod vs. 70–84 in OCs, where mg# is defined as $100 \times$ molar $MgO/(MgO + FeO)$). Commensurate with the more magnesian mafic mineral compositions, orthopyroxene is more abundant than olivine. Recrystallization is evident from abundant 120° triple junctions.

Acapulcoites are finer-grained and the modal mineralogy is approximately that of ordinary chondrites, containing roughly 10 wt.% plagioclase and 5–6 wt.% troilite. Acapulcoites also contain metal, troilite, and phosphate veins that range in size from a few micrometers in width and hundreds of micrometers in length to several mm in width and cm in length. In contrast, lodranites are coarser-grained and typically exhibit marked depletions in plagioclase and/or troilite relative to ordinary chondrites. A comprehensive description of acapulcoites and lodranites known at that time is given by *Mittlefehldt et al.* [1998] and references therein.

Prior to the collection of Antarctic meteorites, the acapulcoite-lodranite clan was represented by only the type meteorites, Acapulco and Lodran. *Bild and Wasson* [1976] published the first modern, comprehensive study of the petrology of Lodran, concluding that it experienced a multistage formation with a significant role for partial melting and melt removal. Acapulco, which fell in 1976 and was first comprehensively described by *Palme et al.* [1981], was ascribed as representing the early stage of incipient partial melting of a chondritic parent body.

Perhaps no group of meteorites has been more defined (and our understanding of their petrogenesis refined) by

Antarctic meteorites than the acapulcoite-lodranite clan (Plates 39 to 42). Even with the relatively recent influx of hot desert meteorites from Northwest Africa, Antarctic meteorites comprise more than half of the acapulcoite-lodranite clan and those in the U.S. collection more than a quarter of the members. Papers by *Floss* [2000], *McCoy et al.* [1996, 1997a, b], *Mittlefehldt et al.* [1996], *Nagahara* [1992], *Patzer et al.* [2004] and *Takeda et al.* [1994] all centered on Antarctic meteorites. The publications by *McCoy et al.* refined the petrogenesis of acapulcoites-lodranites, arguing that rather than the end members represented by Acapulco and Lodran, these meteorites represented a continuum in partial melting, melt migration, and melt removal. While lodranites are, in general, the residues from more substantial partial melting and melt removal, melt removal was not complete. Elephant Moraine (EET) 84302 experienced removal of troilite without removal of plagioclase. Lewis Cliff (LEW) 86220 samples two distinct lithologies, one of acapulcoite modal mineralogy and grain size and the other of Fe,Ni metal, troilite, and plagioclase-augite (basaltic-gabbroic) mineralogy with a very coarse grain size (Figure 5.2). Plagioclase grains reach almost a centimeter in length in this lithology. This meteorite appears to sample the melt complementary to the residual lodranites, intruding a typical acapulcoite host rock. One of the more interesting members of the clan is Graves Nunataks (GRA) 95209, which samples a physically,

Figure 5.2. Lewis Cliff (LEW) 86220 is an unusual member of the acapulcoite-lodranite clan. While the bulk of the meteorite consists of a finer-grained equigranular host similar to acapulcoites (lower left side), coarse-grained enclaves of plagioclase, calcic pyroxene, metal, and troilite suggest infiltration of a basaltic-Fe,Ni-FeS melt that crystallized in situ (top and right side). The lodranites are residues complementary to that melt. Scale bar is 1 mm.

chemically, and mineralogically complex set of metal-sulfide veins. These veins, particularly when compared to smaller metal particles, record a complex history of melting, melt migration, carbon incorporation, and isotopic homogenization and oxidation-reduction reactions [*McCoy et al.,* 2006]. While recent recoveries from Antarctica continue to provide greater sampling of this group, an obvious gap in our sampling that might be rectified by future recoveries is a regolith breccia that samples the full range of lithologies on the acapulcoite-lodranite parent body.

5.3. ASTEROIDAL CORES: UNGROUPED IRON METEORITES

In general, iron meteorites are severely underrepresented in the Antarctic population relative to either observed falls or non-Antarctic meteorites. Although compiled several years after the start of the Antarctic collection effort, *Graham et al.* [1985] reported 4.6% of observed falls (42 of 905), and 40% of non-Antarctic finds were iron meteorites. The greater resistance to weathering and higher probability of recognition of irons after long residence times on the ground is responsible [*Clarke,* 1986]. In contrast, iron meteorites represent ~0.5% of U.S. Antarctic meteorites (~111/20,000). The reason for this discrepancy is unclear, but Antarctic iron meteorites skew toward lower masses than their non-Antarctic counterparts and differences in transport and exposure of dense, heat-conducting iron meteorites in Antarctica may produce much of the difference.

While Antarctic irons compose a smaller fraction of the total Antarctic meteorite population, ungrouped irons compose a substantially larger fraction of those irons. Iron meteorites are classified based on their structure and chemical composition. Among non-Antarctic iron meteorites, 85% of irons can be classified into 13 well-defined groups. The remaining 15% fall outside these groups and are termed ungrouped. *Clarke* [1986] first recognized that a higher proportion of iron meteorites from Antarctica are ungrouped, an idea confirmed by *Wasson et al.* [1989] and expanded upon by *Wasson* [1990]. In this latter paper, 12 of 31 irons (39%) were recognized as ungrouped.

Among the ungrouped irons are a number of interesting samples. Lewis Cliffs (LEW) 85369 contains Si dissolved in the metal, indicating unusually reducing conditions. It is compositionally similar to the ungrouped iron Horse Creek. The iron sulfide troilite makes up nearly two-thirds the volume of LEW 86211 (Figure 5.3). Troilite is more commonly underrepresented in iron meteorites compared to what is expected based on estimated parent magma compositions. Mount Howe (HOW) 88403 (Plate 80) is a Ni-rich iron whose phosphates

share a common oxygen isotopic signature with group IVA, but whose high-Ni composition and rapid cooling rate argue for formation in a separate parent body [*McCoy et al.,* 2011]. Interestingly, some meteorites originally listed as ungrouped irons (e.g., the sulfide-rich HOW 88403) are more likely segregated metallic impact melts from a chondritic parent body, even though the exact parent body is currently unrecognized [*Schrader et al.,* 2010].

Wasson [1990] argued that smaller fragments of asteroids tended to be ejected at higher velocities than larger fragments, and these smaller fragments experienced more changes in orbital velocity through subsequent collisional or gravitational interactions. The average mass of Antarctic irons is only about 1/100th that of non-Antarctic irons. Thus, these small irons may represent a broader sampling of the asteroid belt than their larger counterparts. *Burbine et al.* [2002] extended this argument, suggesting that our entire meteorite collection might sample ~100 distinct meteorite parent bodies, with ungrouped irons representing perhaps half of these. Thus, small ungrouped irons from Antarctica have a significant potential to reveal the full diversity of materials and conditions present (at least in the cores of differentiated bodies) in the early solar system.

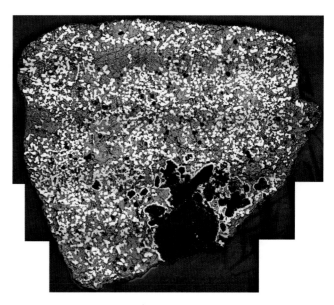

Figure 5.3. Lewis Cliff (LEW) 86211 is an ungrouped iron meteorite with silicate inclusions. Yellowish troilite includes irregular to dendritic blebs of white metal that sometimes rim dark silicate inclusions. Troilite-rich iron meteorites like LEW 86211 are rare and often originate from early cotectic Fe,Ni-FeS partial melts or late cotectic residual melts during asteroidal melting or core solidification. The section is 26 mm across horizontally at the widest point.

5.4. ASTEROIDAL MANTLES: UREILITES

Ureilites are coarse-grained ultramafic achondrites composed of olivine, one or more of the pyroxenes pigeonite, orthopyroxene, and augite, and carbon polymorphs. Their mineralogy, petrology, and compositions are discussed in *Mittlefehldt et al.* [1998] and references therein. The proportion of olivine to pyroxene in ureilites is quite variable, from 0:100 to 100:0. Most ureilites contain pigeonite as the sole pyroxene, but other pyroxene assemblages include orthopyroxene and augite in various combinations. The most common texture is xenomorphic-granular composed of an assemblage of millimeter-sized anhedral olivine and pyroxene grains with curved borders joining in triple junctions. Poikilitic texture in which large olivine grains include small rounded pyroxene grains is common. In these, large low-Ca pyroxene grains may include small olivine and/or augite grains. Elongation of olivine and pyroxene grains in a few ureilites defines foliation and lineation textures. A mosaic texture of small, recrystallized grains characterizes several ureilites. A few ureilites have a bimodal texture in which large, centimeter-sized crystals of low-Ca pyroxene poikilitically enclose regions with the typical xenomorphic-granular texture. The carbon polymorphs (graphite, lonsdaleite, and diamond) occur mostly interstitially along silicate grain boundaries, but they are also present within silicates along fractures or as partial penetrations of silicate grain margins. Olivine grains typically have reduced rims where they are in contact with carbon-rich material.

Most ureilites are unbrecciated (often incorrectly referred to as monomict in the literature) but have been subjected to moderate to high shock stresses. Several ureilites are polymict breccias (Plate 46) containing lithic and mineral clasts of widely differing composition, igneous materials from the ureilite parent asteroid not present in unbrecciated ureilites, and xenolithic material from several chondritic asteroids [e.g., *Cohen et al.*, 2004; *Downes et al.*, 2008; *Goodrich et al.*, 2004]. A few ureilites are dimict breccias composed of two distinct ureilite lithologies [*Goodrich et al.*, 2004].

Within an individual unbrecciated ureilite, olivine core compositions are uniform in mg#. The suite, however, shows a range of olivine core compositions from mg# ~76 to 95. Most ureilites have olivine core compositions of ~77–79, and the more magnesian compositions are less common. Olivine core compositions show a strong correlation between Fe/Mg and Fe/Mn [e.g., *Goodrich et al.*, 2004] and have relative high CaO and Cr_2O_3 contents (0.3–0.4 wt.% and 0.4–0.9 wt.%) [*Berkley et al.*, 1980; *Singletary and Grove*, 2003]. The olivines have very magnesian rims where they are in contact

with carbon phases of up to mg# ~98. These rims are shot through with fine-grain, Ni-poor Fe metal. Low-Ca pyroxenes have a range of Ca contents from ~ Wo_3 to ~ Wo_{11}, but there is no correlation of Ca content with mg#. Augites have Ca contents of Wo_{31-37}, with the lower Ca augites associated with the more ferroan ureilites. There is a strong mg# correlation between olivine and pyroxene.

Compositionally, ureilites are characterized by very low concentrations of incompatible lithophile trace elements, such as the light rare-earth elements (REE), and very low Al contents [*Boynton et al.*, 1976; *Warren et al.*, 2006]. These characteristics are consistent with their ultramafic mineralogy and lack of plagioclase. Siderophile element contents are variable, but some ureilites have relatively high abundances, up to roughly CI levels [*Rankenburg et al.*, 2008; *Warren et al.*, 2006]. The mineralogy and lithophile element compositions bespeak a high temperature origin, yet curiously, ureilites are rich in planetary-type rare gases. These gases are, for the most part, hosted in the diamonds that were produced by shock [*Göbel et al.*, 1978].

Prior to the start of systematic meteorite searches in Antarctica there were only seven widely known ureilites, four of which were falls. (Two other ureilites had been found but were only announced in the Meteoritical Bulletin in 1978 and 1980. They were little known to meteoriticists in 1974.) The baffling mixture of high and low temperature characteristics in these rare rocks made them especially interesting research objects, so much so that the fall of Haverö in 1971 and the discovery of Kenna in 1972 prompted consortia studies of each of them.

The state of knowledge of ureilites based on petrologic studies circa 1980 (prior to significant influence by studies of Antarctic ureilites; Plates 44 to 47) interpreted them as being magmatic adcumulates [*Berkley et al.*, 1976, 1980]. This was based on their ultramafic mineralogy, lack of plagioclase and high-Ca pyroxene, extreme depletion in the most incompatible lithophile elements, and their foliation and lineation textures. Ureilites were divided into three distinct subgroups based on the mg# of olivine cores, 78–79, 84–85, and 91. However, olivine core compositions from the North Haig polymict ureilite showed a continuous range from 76 to 92 [*Berkley et al.*, 1980], presaging the current interpretation as a compositional continuum rather than distinct subgroupings. The petrology-based interpretation was that the graphite was interstitial material trapped between cumulus silicate grains and that lonsdaleite and diamond were formed from it by shock. The graphite was thought to have controlled the oxygen fugacity of the magma through redox with FeO, and that the different mg# groups of ureilites resulted from this process.

Cosmochemists generally had a different interpretation of ureilites and the C-rich materials. They considered ureilites to be restites remaining after partial melting had generated metallic melts that segregated into a core, and silicate melts that erupted as basalts [*Boynton et al.*, 1976; *Wasson et al.*, 1976]. In this scenario, the C-rich material was injected into the silicate host as "veins" by impact processes. The noble gases were trapped in the diamonds produced by the shock that mobilized the C-rich material. In this way, cosmochemists neatly avoided the conundrum of having high noble gas contents in magmatic rocks. Ureilites were found to contain two siderophile element components. One was thought to be a component remaining in the host ureilite after partial melting allowed separation of a metallic core. The C-rich "veins" were also rich in siderophile elements, and the second component was thought to have been injected along with the C-rich material.

Thus, there were two competing models for ureilite genesis circa 1980: (a) fractional crystallization in magma chambers with entrapment of interstitial graphite and adcumulus grain growth to expel interstitial silicate liquid, and (b) partial melting leaving a metal-sulfide and basalt-depleted restite, followed by injection of C-rich materials, including rare gases and siderophile elements, by impact on the parent asteroid. Points of agreement were that the high Ca contents of olivine indicated high equilibration temperatures for the silicates, and the reduced rims on olivine were engendered by an impact that substantially decreased the lithostatic pressure and allowed redox reactions between the C-rich materials and the hot silicates to reduce FeO out of the olivine; CO and/or CO_2 formed by this process escaped the system.

Paired ureilites Allan Hills (ALH) A78019 and ALH A78262, among the earliest Antarctic ureilite recoveries (Plate 44), resolved the issue over internal versus external origin for the C-rich materials (Figure 5.4). These are very low-shock-stage ureilites with well-preserved primary textures of the interstitial graphite [*Berkley and Jones*, 1982]. Graphite occurs as nearly undeformed euhedral grains interstitial to the silicate phases, but intergrown with Fe-Ni metal and sulfide. This clearly established graphite as a primary phase in ureilites, and not material injected by impacts. Thus, the rare gases and siderophile element components associated with the C-rich materials must be an integral part of ureilites. In diamond-free ALH A78019, the noble gases are contained in amorphous C, not the graphite [*Wacker*, 1986].

Perhaps the most astonishing result to come out of ureilite studies that was greatly facilitated by the Antarctic meteorite collections was the recognition that the ureilite group is very heterogeneous in its oxygen isotopic composition, and that the differences are dominated by non-mass-dependent variations. The first three-isotope O isotopic analyses of ureilites included only three non-Antarctic meteorites and showed that they lay along the carbonaceous chondrite anhydrous mineral (CCAM) line [*Clayton et al.*, 1976]. These three meteorites showed variations in $\Delta^{17}O$ (a parameter not yet defined) from −0.61‰ to −1.30‰, but this was not remarked upon. (For comparison, twelve samples of four HEDs in the same study showed variations in $\Delta^{17}O$ only from −0.18‰ to −0.46‰.) A subsequent comprehensive study of the O isotopic composition of ureilites demonstrated that the variation in $\Delta^{17}O$ for the group is from −0.23‰ to −2.53‰ [*Clayton and Mayeda*, 1988; 1996]. The extremes of the range are found in Antarctic ureilites (Figure 5.5). The oxygen isotopic heterogeneity is a characteristic inherited from the proto-ureilite material accreted to form the parent asteroid, and the differentiation process did not homogenize oxygen. The large, non-mass-dependent isotopic variation revealed by Antarctic ureilites effectively precludes cumulate models for the genesis of ureilites, and a restite origin is now considered to be the petrogenetic mechanism.

With the improved sampling of the ureilite parent asteroid represented by the current collection, the variation in mg# of olivine cores has come into better focus. Rather than discrete subgroups, there is a continuum of compositions from mg# 75 to 94, with a peak in the distribution at ~78–79. A most significant finding is that this mg# variation is correlated with the $\Delta^{17}O$ of the meteorites (Figure 5.5). The range in olivine core mg# has been explained by two different mechanisms, and thus the mg#-$\Delta^{17}O$ correlation has two interpretations. One camp considers the olivine core compositions to reflect redox between silicates and graphite during partial melting [*Goodrich et al.*, 2007; *Singletary and Grove*, 2003]. In this scenario, the high mg# ureilites were formed closer to the asteroidal surface such that the lower lithostatic pressure permitted a greater degree of reduction of FeO from the silicates. Material deeper in the proto-ureilite asteroid had the highest $\Delta^{17}O$ and that nearer the surface had the lowest $\Delta^{17}O$. The mg#-$\Delta^{17}O$ correlation then is a juxtaposition of initial heterogeneity and pressure-controlled redox. The second interpretation is that both the mg# and O isotopic heterogeneities reflect those of the proto-ureilite material, with little modification having occurred during magma genesis [*Warren and Huber*, 2006]. The mg#-$\Delta^{17}O$ correlation then is a memory of initial asteroid heterogeneity. The origin of the mg# variations is an unsettled problem in ureilite genesis.

The siderophile element contents of ureilites are still not completely understood, but the improved sampling

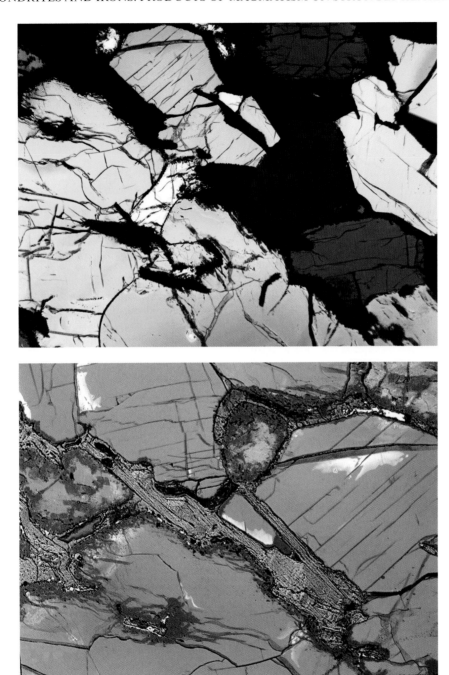

Figure 5.4. Ureilite ALH A78019 is the first low shock-stage ureilite discovered and retains sharp extinction in olivine (upper image) and almost undeformed euhedral graphite (lower image), which demonstrated that graphite was a primary mineral in ureilites. The images are 2.5 mm across (upper) and 1.25 mm across (lower).

of the parent asteroid coupled with modern analytical techniques does support a restite model for ureilite genesis. The highly siderophile element ratios in ureilites can be understood as arising from two components. One is residual metal remaining after extraction of partial melts in the Fe-Ni-S system, while the other may be a total melt of the Fe-Ni-S, a portion of which was trapped in the silicate host, as found by *Rankenburg et al.* [2008]. These authors also considered that the second component could be chondritic material injected during disruption of the ureilite parent asteroid, but this seems less likely considering that the analyzed ureilites are not breccias. This study relied heavily on ANSMET ureilites; 16 of 22 meteorites analyzed were from the US collection.

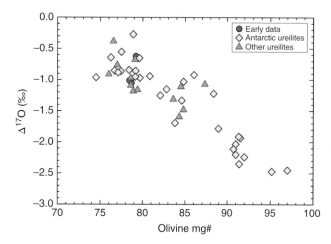

Figure 5.5. Plot of $\Delta^{17}O$ vs. mg# for ureilites. The parameter $\Delta^{17}O$ gives the deviation of the O isotopic composition of a meteorite from the terrestrial fractionation line and is a measure of non-mass-dependent isotopic fractionation in the solar system. Antarctic ureilites significantly extend the compositional ranges of these parameters and provide the clearest evidence for correlated compositional and isotopic heterogeneity.

Figure 5.6. EET 99402 is an ultramafic brachinite with a well-developed xenomorphic-granular texture. Olivines in this rock show preferred orientation, explaining the common low birefringence in the image, and support an igneous-cumulate-origin for it. Image is 2.5 mm across.

5.5. ASTEROIDAL CRUST-MANTLE SUITE: BRACHINITES

The brachinite group membership is poorly defined because there are significant petrologic, compositional, and isotopic variations among bona fide brachinites (Plates 48 and 49), and there are numerous achondrites with generally similar petrologic, compositional, and isotopic characteristics such that the boundaries of the brachinite group are unclear. Indeed, one of the brachinites that is most clearly distinct from the group "type" properties is the eponymous Brachina! The question of how isotopically heterogeneous a single achondrite parent asteroid can be is unresolved [e.g., *Greenwood et al.*, 2008], but regardless, isotopically diverse brachinites are considered to hail from a single asteroid by virtue of petrological similarities [e.g., *Rumble et al.*, 2008].

As of September 2013, the Meteoritical Bulletin Database lists 34 named meteorites as brachinites. There are six achondrites classified as ungrouped that are often considered (with differing degrees of nervousness) as brachinites; Divnoe, GRA 06128, GRA 06129, Northwest Africa (NWA) 1500, NWA 5400, and NWA 6077. Zag (b) is officially classified as a winonaite, but its silicates are far too oxidized to support that classification. It, too, is often considered a potential brachinite. Because of the uncertainties surrounding the definition of brachinite, these additional meteorites will be discussed here along with the bona fide brachinites.

Brachinites are ultramafic and mafic rocks whose origin remains elusive. Petrologically, they are diverse.

The discussion of their mineralogy and petrology is derived from *Day et al.* [2012a], *Floran et al.* [1978], *Goodrich et al.* [2010], *Johnson et al.* [1977], *Mittlefehldt et al.* [2003], *Shearer et al.* [2010], *Swindle et al.* [1998], and *Warren and Kallemeyn* [1989]. Brachina is a fine-grained ultramafic rock dominated by olivine (~80 vol.%). Paired meteorites EET 99402 and EET 99407 (hereafter EET 99402/7) are coarse-grained, moderately shocked olivine-dominated rocks (Figure 5.6), while paired meteorites GRA 06128 and GRA 06129 (hereafter GRA 06128/9) are mafic rocks dominated by plagioclase (~80 vol.%) (Figure 5.7; Plate 49). However, GRA 06128/9 are very much atypical, and most brachinites contain between 79 and 93 vol.% olivine. All brachinites contain high-Ca clinopyroxene, either augite or diopside, at modal abundances of 4–15 vol.%. Low-Ca pyroxene has been reported from only about half of all brachinites, and is typically <2 vol.% when modes are reported. An exception is NWA 595, where orthopyroxene reportedly makes up to ~15 vol.% in some thin sections. Plagioclase is present in about two thirds of all brachinites but is highly variable in abundance. Excluding GRA 06128/9, plagioclase varies from being a trace phase to making up ~10 vol.%. Chromite, sulfide, and metal are common trace to minor phases.

Brachinites commonly have equigranular textures with subhedral to anhedral grains joined at 120° triple junctions (Figure 5.6). In some cases, olivine grains are prismatic with lineations and possible foliations defined by grain alignments. GRA 06128/9 have granoblastic to equigranular textures and include bands of finer-grained granoblastic texture. Grain size is quite variable, from

Figure 5.7. GRA 06128 (shown here) and paired GRA 06129 are unique feldspar-rich brachinites that represent a melt (possibly with some crystal accumulation), rather than cumulates or restites. The image is 2.5 mm across.

0.1–0.3 mm in Brachina up to 2–3 mm in ALH 84025 and GRA 06128/9. High-Ca pyroxene, chromite, and plagioclase are typically interstitial to olivine. Sulfide is anhedral and interstitial, or occurs as veins cross-cutting other phases.

Olivine is homogeneous in composition within each brachinite and ranges from mg# 64 to 72 among the olivine-dominated members. GRA 06128/9 contain olivine grains with mg# of ~40, much more ferroan than in the olivine-rich brachinites. High-Ca pyroxene varies from augite to diopside (42–47 mole% Wo), but augite in Brachina has a lower Ca content, ~39 mole% Wo. The calcic pyroxenes are magnesian with mg# = 79–84 except for GRA 06128/9 whose augites have an mg# of ~65. There are orthopyroxene analyses for only a few brachinites in the literature; they are magnesian (mg# 72–75) with the exception of GRA 06128/9 with mg# of ~54.

Compositional data are available for roughly half the brachinites, and there are substantial differences in their compositions [*Day et al.*, 2009, 2012a; *Goodrich et al.*, 2010; *Johnson et al.*, 1977; *Mittlefehldt et al.*, 2003; *Nehru et al.*, 1983; *Shearer et al.*, 2010, 2011; *Swindle et al.*, 1998; *Warren and Kallemeyn*, 1989]. Brachina is quite unusual compared to the other ultramafic brachinites in that it has approximately chondritic (CI) abundances for most refractory and moderately volatile lithophile elements. Its lithophile element pattern betrays no hint of fractionations imposed by igneous processes. All other ultramafic brachinites show depletions in highly incompatible lithophile

elements relative to less incompatible elements, for example, e.g., La/Yb, Th/Yb <1 × CI. However, lithophile trace element abundances are tricky to interpret in many brachinites because most are finds from "hot deserts" where alteration and contamination by the terrestrial environment can have seriously degraded the samples [*Crozaz et al.*, 2003]. This is not an issue with the Antarctic brachinites. ALH 84025 (Plate 48) has low CI-normalized abundances for Al and Na, consistent with the absence of plagioclase, and is poor in highly incompatible refractory lithophile elements such as La and Sm. Oddly, Eu/Sm is >2 × CI in ALH 84025, suggesting that some plagioclase is present, although it has not been observed. EET 99402/7 have very fractionated lithophile element patterns, with higher abundances of the "plagiophile" elements Na, Al, and Eu and very low abundances of Zr, La, Sm, and Yb, as well as Zr/Yb, La/Yb, and Sm/Yb <1 × CI. With the exceptions of ALH 84025 and EET 99402/7, ultramafic brachinites have approximately chondritic Eu/Sm ratios. The mafic brachinites GRA 06128/9 are rich in most incompatible lithophile elements with abundances between ~3 and 8 × CI. Exceptions are the high field strength elements Ti, Zr, Nb, Hf, and Ta which are at ~1–2 × CI.

Most brachinites have low abundances of the highly siderophile elements with Ru, Pd, Re, Os, Ir, and Pt typically <0.5 × CI, but exceptions are NWA 6077 (0.8–1.1 × CI excluding Pd at ~0.4 × CI) and NWA 5400 (1.7–5.0 × CI excluding Pd at ~0.3 × CI) [*Day et al.*, 2012a].

EET 99402 has the lowest abundances of highly siderophile elements with Ru, Pd, Re, Os, Ir, and Pt in the range of $0.01–0.08 \times$ CI. Cobalt and Ni abundances are less variable (0.3–0.9 and 0.06–0.5) and do not correlate with highly siderophile element abundances.

The oxygen isotopic compositions of brachinites show a substantial range in $\Delta^{17}O$ from ~0‰ to –0.31‰ using modern, high-precision analysis methods [*Day et al.*, 2012a]. This contrasts with the narrow range in $\Delta^{17}O$ for HED meteorites of −0.21‰ to −0.27‰ using these analysis methods [e.g., *Greenwood et al.*, 2005]. Brachina and the possible brachinite Divnoe have the lowest $\Delta^{17}O$ (−0.31‰ and −0.30‰), while the two possible brachinites NWA 5400 and NWA 6077 have the highest (0.00 and −0.02‰). Excluding these four, the brachinite range would be −0.11‰ to −0.24‰.

Brachina was the only brachinite known prior to the onset of systematic collection of meteorites in Antarctica. Because Brachina is a fine-grained ultramafic rock with igneous texture, unfractionated incompatible lithophile trace elements, and relatively high abundances of siderophile elements, it was first thought to be an igneous rock that crystallized from a melt of its own composition [*Johnson et al.*, 1977; *Floran et al.*, 1978], possibly formed by impact melting [*Ryder*, 1982]. In contrast, *Nehru et al.* [1983] included Brachina with their broad category of primitive achondrites, a characterization they found fit three subsequent finds [*Nehru et al.*, 1996]. In their view, brachinites had a formational history somewhat similar to that of the acapulcoite-lodranite clan of meteorites discussed above (section 2), although different in some details. Specifically, *Nehru et al.* [1996] suggest that brachinites evolved from CI-like material that was oxidized during planetary heating, converting orthopyroxene to olivine as FeO increased. Some brachinites were heated to the point of partial melting, allowing removal of metal-sulfide and basaltic melts.

The first Antarctic brachinite find, ALH 84025 (Plate 48), prompted a rethinking of the petrogenesis of the group. *Warren and Kallemeyn* [1989] found that the textures and compositions of ALH 84025 and Brachina are more compatible with an origin as igneous cumulates. *Mittlefehldt et al.* [2003] and *Swindle et al.* [1998] similarly argued for a cumulate origin for EET 99402/7 and Eagles Nest based on mineralogy, texture, and composition. *Mittlefehldt et al.* [2003] argued that the parent asteroid of the brachinites was differentiated and had undergone high degrees of melting. Mafic brachinite GRA 06128/9 is considered to be a melt composition slightly modified by crystal accumulation and possibly metasomatism [*Day et al.*, 2009, 2012a; *Shearer et al.*, 2010]. Recent studies have suggested that many of the ultramafic brachinites may be anatectic residues and not cumulates. There is yet no consensus on the petrogenesis of the brachinite parent asteroid, but it appears that brachinites represent the

products of a range of magmatic processes. Because of the compositional and isotopic variations among brachinites, the membership of the group is not settled, making it difficult to arrive at a consensus regarding the differentiation history of their parent asteroid.

5.6. ASTEROIDAL CRUSTS

5.6.1. The Howardite-Eucrite-Diogenite (HED) Clan

The howardite-eucrite-diogenite clan of meteorites provides us with the most diverse suite of crustal rocks from any asteroid. Though not universally accepted, a very strong case can be made that HEDs are derived from the only intact, differentiated asteroid, 4 Vesta [*Drake*, 2001]. The HED meteorites sample a range of depths of their parent asteroid from the surface down to the lower crust. Geochemical modeling of the differentiation of a Vesta-sized asteroid based on HED data suggests formation at depths of ~20–40 km may be represented by some diogenites [*Righter and Drake*, 1997; *Ruzicka et al.*, 1997]. This is the same range of crustal exposure as seen on Vesta by the Dawn mission where the large Rheasilvia basin floor is found to be ~22 km below the best-fit spheroid [*Schenk et al.*, 2012]. The HED suite thus provides a detailed window into the crustal structure of large, differentiated asteroids. Information derived from HEDs informs interpretation of data returned by the Dawn mission [e.g., *De Sanctis et al.*, 2012; *Prettyman et al.*, 2012], and conversely, Dawn observations influence interpretation of HED data [e.g., *McSween et al.*, 2013]. The discussion here is largely derived from *Mittlefehldt et al.* [1998] and references therein, and updated with recent results.

Eucrites are mafic rocks of rather simple mineralogy. Major phases are pyroxene and plagioclase. Most eucrites are coarse- to fine-grained hypabyssal or volcanic rocks, and in these pyroxene makes up ~51 vol.% and plagioclase ~43 vol.% [*Delaney et al.*, 1984]. Pyroxene originally crystallized as ferroan pigeonite, but metamorphism has rendered it variable mixtures of orthopyroxene, pigeonite, and augite. Some eucrites are plutonic rocks, and these have varied pyroxene:plagioclase ratios from 68:30 to 32:65 [*Delaney et al.*, 1984]. Some of this variation is a reflection of heterogeneity in these coarse-grained rocks, making accurate determinations of modes from thin sections difficult, but there is nevertheless genuine variation in mineral proportions. Minor and accessory phases are silica polymorphs, ilmenite, chromite, apatite, merrillite, zircon, baddeleyite, troilite, and metal. Diogenites are coarse-grained ultramafic rocks composed dominantly of magnesian orthopyroxene with minor chromite and variable amounts of olivine from 0 to ~35 vol.%. One HED meteorite is a dunite that is classified as a diogenite because of similarities in mineral compositions to those in diogenites [*Beck et al.*,

2011]. Most eucrites and diogenites are breccias having been texturally modified by impacts. Howardites are polymict breccias composed mostly of fragments of eucritic and diogenitic debris. HED breccias form a continuum from monomict eucrites through polymict eucrites, howardites, and polymict diogenites to monomict diogenites.

Pyroxenes in HED meteorites vary systematically with petrologic type. Diogenites contain magnesian orthopyroxene. Most fall within a rather narrow range of compositions with mg# ~77–74, $Wo_{~1.5–3.0}$. A few magnesian diogenites have orthopyroxene of mg# 84–79, $Wo_{~1–2}$, and a few ferroan diogenites have orthopyroxene compositions of mg# 73–68, $Wo_{~1–5}$. Yamato Type-B diogenites, a lithology transitional to cumulate eucrites, contain pyroxenes with mg#s ~66.5, $Wo_{~2.7}$. Bulk pyroxene compositions for cumulate and basaltic eucrites are difficult to determine because of subsolidus phase changes during metamorphism that

resulted in coarse exsolution of augite from the primary pigeonite. For cumulate eucrites, the resulting low- and high-Ca pyroxenes range in compositions mg# 65–43, $Wo_{~2–5}$, and mg# 76–56, $Wo_{~42–45}$, while those in basaltic eucrites vary in mg# ~38–32 and $Wo_{~2–6}$, and mg# ~52–38 and $Wo_{~30–42}$. Rarely, original magmatic zoning is preserved in basaltic eucrite pyroxenes; in clasts in several paired Yamato polymict eucrites pyroxene cores have mg#s ~71–72, $Wo_{~4–5}$, and rim subcalcic augite can be mg# ~7, $Wo_{~25}$ [*Takeda et al.*, 1983]. Plagioclases show modest variations in composition of $An_{72–95}$. There is no systematic plagioclase compositional trend with respect to lithology, although those in cumulate eucrites tend to be more calcic than those in basaltic eucrites.

The bulk compositions of diogenites, cumulate eucrites, and basaltic eucrites show a progression from Mg- and Cr-rich to Mg- and Cr-poor compositions (Figure 5.8)

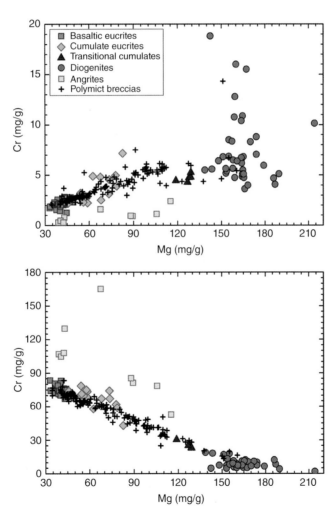

Figure 5.8. Major element compositions of HED igneous lithologies and polymict breccias, with angrites shown for comparison. The Cr-Mg and Ca-Mg correlations demonstrate chemical fractionations that occurred during igneous crystallization to form the crust of the HED parent asteroid. Subsequent impact gardening mixed the igneous progenitors into the polymict breccias. Angrite basalts are compositionally distinct from basaltic eucrites.

reflecting mode of formation from orthopyroxene + chromite ± olivine cumulates, through pigeonite + plagioclase cumulates to ferroan quenched melts. In the same sequence, Na, Al, Ca, and Ti generally increase (Figure 5.8), although some cumulate eucrites have higher Al than basaltic eucrites, a consequence of plagioclase accumulation. Incompatible lithophile trace elements follow the trend of Ti. The polymict breccias form linear trends on element-element diagrams of nonvolatile lithophile elements between diogenite and basaltic eucrite end-members (Figure 5.8). Moderately volatile elements, typified by the alkali elements, and volatile elements, such as Bi, Te, Se, Cd, In, Tl, and halogen elements, are at low abundances in HEDs [*Laul et al.*, 1972; *Paul and Lipschutz*, 1990; *Wolf et al.*, 1983]. Highly siderophile elements are at extremely low abundances in HEDs. In basaltic eucrites the Ir contents are at 10^{-5} to $10^{-6} \times$ CI abundances [*Warren et al.*, 1996]. Diogenites and cumulate eucrites have Ir contents between 10^{-4} to $10^{-6} \times$ CI. The contents of Re and Os are similarly in the range of 10^{-4} to $10^{-6} \times$ CI in igneous lithologies [*Warren et al.*, 2009]. Cobalt, a moderately siderophile element, is at higher abundances of $\sim 10^{-1}$ to $10^{-2} \times$ CI for diogenites and $10^{-2} \times$ CI for basaltic and cumulate eucrites. Howardites contain generally higher siderophile element contents because of the admixture of chondritic debris.

The HED meteorites have been intensively studied since the 1960s and models for HED meteorite genesis were already fairly mature by the time systematic meteorite searches in Antarctica began. There are two competing models for eucrite genesis: (a) they represent primary partial melts of chondritic source regions [*Stolper*, 1977], and (b) they represent residual melts left after crystallization of the global magma ocean [*Ikeda and Takeda*, 1985; *Righter and Drake*, 1997; *Ruzicka et al.*, 1997]. Currently, the consensus favors that latter scenario. In this model, the HED parent asteroid would have a metallic core, a dunitic and harzburgitic mantle which transitions into a harzburgitic/orthopyroxenitic lower crust, topped by a cumulate and basaltic eucrite upper crust. The impact-produced debris layer is represented by howardites, polymict eucrites, and polymict diogenites.

Antarctic HEDs (Plates 55 to 62) have influenced our understanding of HEDs largely through the wider range in lithologic types first recognized among the Antarctic suite, which then influenced interpretations of magmatic and impact processes on their parent asteroid. One example is the identification of some eucrites and diogenites as being polymict breccias. There had been a few HEDs that kept bouncing back and forth between being classified as howardites or eucrites. With the renewed interest in HED petrology prompted by the large number of new Antarctic HEDs, it was recognized that the line between eucrite and howardite (and diogenite and howardite) was indistinct; polymict breccias form a continuum from eucrites to diogenites [*Delaney et al.*, 1983; *Delaney et al.*, 1984; *Miyamoto et al.*, 1978]. Prior to 1976, howardites were considered a distinct rock type; compositional gaps separated them from eucrites and diogenites. They were thought to be regolith breccias by analogy with lunar regolith breccias. With the recognition of the polymict breccia continuum, thinking on the nature of the debris layer on the HED parent asteroid has evolved. It became clear that some polymict breccias are simply the products of individual large impacts mixing different lithologies excavated from the crust, and that regolith gardening was not required [e.g., *Nyquist et al.*, 1986]. Currently, howardites are recognized as of two fundamentally different types; fragmental howardites formed from debris from a limited number of impacts, and regolithic howardites formed from well-mixed regolith [*Warren et al.*, 2009]. Recent work on a suite of howardites and polymict diogenites from the Pecora Escarpment icefield has shown that these breccias came from a single, petrologically heterogeneous meteoroid roughly a meter in size [*Beck et al.*, 2012], giving us a clearer glimpse of the decimeter scale variation of HED polymict breccias. The new view of howardites, in large part due to studies of ANSMET HEDs, is changing our understanding of impact gardening and regolith formation on asteroids, which is different from that on the Moon.

In the mid-1970s, basaltic eucrites were recognized as forming two distinct compositional trends, the Stannern trend with varying incompatible lithophile element contents with nearly uniform mg#, and the Nuevo Laredo trend with concomitantly decreasing mg# and increasing incompatible lithophile element contents [*Stolper*, 1977]. These trends were identified as arising from partial melting to form a suite of primitive melts and fractional crystallization to form a suite of residual melts, respectively. The Stannern trend was sparsely populated; Stannern, Ibitira, Cachari, Haraiya, Juvinas, and Sioux County were its members. The latter four also occupied the primitive end of the Nuevo Laredo trend. With the advent of abundant Antarctic basaltic eucrites, and redefinition of Ibitira as an ungrouped basaltic achondrite not part of the HED suite [*Mittlefehldt*, 2005], the Stannern trend now appears to be subparallel to the Nuevo Laredo–trend eucrites [*McSween et al.*, 2011]. Thus, the partial melting trend as originally defined was a result of too few meteorites and misclassification of one meteorite. The new Stannern trend may be a fractional crystallization trend of parent magmas distinct in composition from those of the Nuevo Laredo trend [*Hewins and Newsom*, 1988; *McSween et al.*, 2011], but that remains to be investigated. An intriguing model is that Stannern-trend eucrites represent basalts that interacted with the earliest basaltic crust to form hybrid magmas

[*Barrat et al.*, 2007]. *Floss et al.* [2000] and *Yamaguchi et al.* [2001] have suggested that metamorphosed eucrite EET 90020 (Plate 55) was partially melted, removing some of its incompatible elements. This rock is then similar to the crustal restite envisioned by *Barrat et al.* [2007] in their model for Stannern-trend eucrite formation. However, EET 90020 cannot be such a restite because its Ar-Ar age indicates that partial melting of it occurred too late in Vestan history [*Yamaguchi et al.*, 2001]. Nevertheless, some metamorphosed eucrites showing light rare-earth-element depletions may represent such restites [*Yamaguchi et al.*, 2009]. The Stannern trend remains a small subset of basaltic eucrites, and half of them are Antarctic. Thus, Antarctic basaltic eucrites played a pivotal role in refining our understanding of the petrogenesis of basalts on the HED parent asteroid.

Originally, diogenites were identified as orthopyroxenites, and olivine was known to be a minor phase in some [*Mason*, 1963]. Olivine-bearing diogenites were recognized as having a distinct major element composition [*Mittlefehldt et al.*, 1979]. The study of U.S. Antarctic diogenites identified a subset of olivine diogenites suggested to represent mantle restites of their parent asteroid [*Sack et al.*, 1991]. One ANSMET diogenite was found to be exceptionally olivine rich [*Bowman et al.*, 1997], and it was suggested that a magnesian harzburgitic component was present in it [*Mittlefehldt*, 2000]. This was inferred from analyses of individual orthopyroxene and olivine grains in the breccia; distinct harzburgite clasts were not identified. Detailed petrologic study of ANSMET diogenites has revealed harzburgite lithologies in some dimict breccias and as individual meteorites, and harzburgitic diogenite is recognized as a subclass [*Beck and McSween*, 2010] (Figure 5.9). Similarly, dunitic diogenite is also a subclass defined based on an ANSMET meteorite [*Beck et al.*, 2011] (Figure 5.10; Plate 62). These harzburgites and dunite are not mantle restites. Minor element contents of the orthopyroxene and mg# trends favor a cumulate origin [*Beck and McSween*, 2010; *Beck et al.*, 2011].

Recent studies in which ANSMET diogenites were critical samples have added complexities to our thinking on HED parent asteroid evolution. Precise oxygen isotope compositions suggest that the diogenite suite may have been derived from at least two parent asteroids [*Day et al.*, 2012b]. If confirmed through further work, this result suggests that the magmatic processes that engendered the HED meteorite suite may have been typical of fully differentiated asteroids. This conclusion has already been suggested by the findings that some basaltic achondrites that are petrologically and compositionally similar to eucrites nevertheless were derived from distinct parent asteroids because of their distinctive oxygen isotopic compositions [cf. *Mittlefehldt*, 2005; *Scott et al.*, 2009].

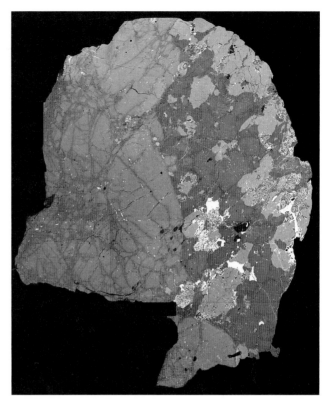

Figure 5.9. Back-scattered electron image mosaic of dimict diogenite LEW 88679. The left side of the image is a more ferroan orthopyroxenite lithology. Dark veins are a cross-cutting network of more magnesian orthopyroxene of uncertain origin. The right side is a more magnesian harzburgite lithology composed of orthopyroxene (dark gray) and olivine (light gray). The mosaic is approximately 1 cm across. Image courtesy of A. W. Beck.

Diogenites contain substantially lower radiogenic ^{26}Mg formed from decay of short-lived ^{26}Al than do basaltic eucrites, and this result demonstrates that they crystallized prior to basaltic eucrite formation [*Schiller et al.*, 2011]. This was expected based on the petrologic and compositional characteristics of the rocks. However, Schiller et al. [2011] further conclude that the time scale for formation of the diogenite and basaltic eucrite suite is too short when considering models for the thermal evolution of Vesta. This conundrum has yet to be explored in detail.

Diogenite petrogenesis has been difficult to fit into the consensus model for Vestan differentiation based on eucrites [*Mittlefehldt*, 1994; *Mittlefehldt et al.*, 2012], and has resulted in various alternate models [*Barrat et al.*, 2010; *Shearer et al.*, 2010; *Yamaguchi et al.*, 2011]. Identification of distinct olivine-rich lithologies and continuing compositional and isotopic investigations of the diogenite suite allow us to more fully explore constraints on the HED parent asteroid geologic evolution, an endeavor that is still evolving.

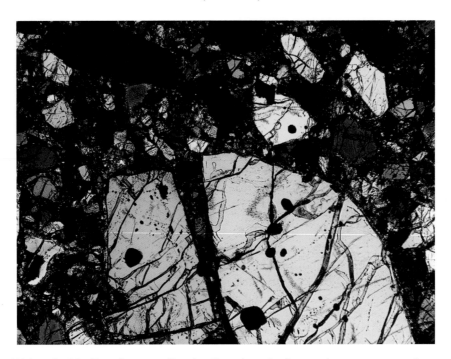

Figure 5.10. MIL 03443 is a dunitic diogenite, extending the diogenite suite from orthopyroxene cumulates to olivine cumulates. Prior to the recovery of this meteorite, dunites were unknown from the HED meteorite suite. Image is 2.5 mm across.

Unbrecciated basaltic eucrite ALH A81001 (Plate 56) has provided evidence that Vesta had a liquid metallic core early in its history that generated a magnetic field, but the evidence is indirect. *Fu et al.* [2012] measured natural remanent magnetization of this meteorite. Their data support a thermoremanent magnetization of the rock that was acquired as a result of cooling in a magnetic field. They also measured an Ar-Ar age of 3.69 Ga, which dates the time the thermoremanent magnetization was acquired. This is far too late in Vestan history for the core to have been still molten, and *Fu et al.* [2012] concluded that the Vestan crust contains remanent magnetization from when the core was producing a magnetic field, and that ALH A81001 acquired its thermoremanent magnetization from the crust.

The HED suite and therefore Vesta has been considered a volatile-poor and essentially anhydrous body. However, petrographic study of a clast in ANSMET howardite EET 92014 showed evidence for interaction with an FeO-rich metasomatizing fluid [*Mittlefehldt and Lindstrom*, 1997]. Additional work has provided more evidence for interactions between eucrites and late-stage aqueous fluids [see *Barrat et al.*, 2011, and references therein]. Analyses of apatite grains in eucrites has shown that while most contain F as the dominant volatile anion species, small amounts of OH are present in some, up to ~0.8 wt.% [*Sarafian et al.*, 2013]. Further, the unusual metamorphosed basaltic eucrite GRA 98098 contains apatite with unusually high Cl contents [*Sarafian et al.*, 2013].

Thus, although Vesta is not awash in volatile elements by any means, recent studies of basaltic eucrites have shown that it is not an anhydrous body as previously thought. Water in even small quantities changes the melting behavior of silicates, and continuing research into the H_2O content of Vesta through studies of HEDs may require refining models of Vestan geologic evolution.

Our understanding of the genesis of HED meteorites and the evolution of their parent asteroid were already very well developed before systematic collection of meteorites in Antarctica began. Nevertheless, studies heavily influenced by Antarctic HEDs have made important contributions to our knowledge of regolith processes, basaltic magmatism, cumulate formation, and the differentiation history of large asteroids. Our understanding of these processes has been used in turn to constrain interpretations of data from the Dawn mission to Vesta.

5.6.2. Angrites

The angrite group is small, and at the start of systematic meteorite searches in Antarctica, only one, Angra dos Reis, was known. This rock is so unusual that a consortium was organized to study it [*Keil*, 1977]. The results of those consortium studies highlighted the importance of Angra dos Reis for understanding asteroidal magmatic processes in the earliest history of the solar system. Yet because it is a one-off meteorite, only limited progress could be made in developing an overall understanding of

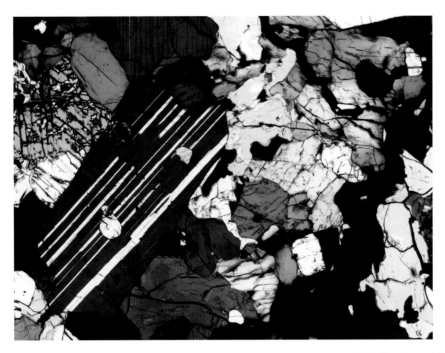

Figure 5.11. Plutonic angrite LEW 86010 was the first member of the group to have a "basaltic" mineralogy (olivine, high-Ca clino-pyroxene, plagioclase), which allowed for combined long-lived and short-lived chronometers to be studied, pinning the absolute age of magmatism to early solar system time-scales of CAI and chondrule formation. Image is 2.5 mm across.

the petrologic evolution of its parent asteroid. The late 1980s saw the recovery of three small angrites from Antarctica, which resulted in a fuller understanding of petrogenesis on their parent asteroid. (The large D'Orbigny stone, found in 1979, only became known to meteoriticists in 2001.) Angrites were recently reviewed, and the synopsis of their petrologic and compositional characteristics given here is derived from *Keil* [2012] and references therein.

The angrites are mafic igneous and metaigneous rocks that are divided petrologically into two types, volcanic and plutonic, based on textures and mineral compositions. Plutonic members are medium- to coarse-grained with hypdiomorphic granular or polygonal-granular textures and equilibrated or modestly zoned mineral compositions. Volcanic angrites have porphyritic to ophitic textures and contain highly zoned minerals. Porphyritic members are olivine-phyric and contain coarse olivine phenocrysts or xenocrysts. Angrites contain the unusual silicate mineral assemblage of Ca-rich olivine, Al-Ti-diopside-hedenbergite (hereafter calcic pyroxene), kirschsteinite, and nearly Na-free anorthite (Figure 5.11). Oxide phases are hercynitic spinel and ulvöspinel. Plutonic angrite Angra dos Reis is an outlier of the group, with a modal mineralogy of ~93 vol.% calcic pyroxene, ~6 vol.% olivine plus kirschsteinite, and is plagioclase-free. (Plagioclase found in mineral separates of Angra dos Reis is likely a laboratory contaminant; see *Mittlefehldt et al.*, 1998.] All other

angrites contain between approximately 20 and 40 vol.% plagioclase. Calcic pyroxene varies from ~24 to ~58 vol.%, while olivine plus kirschsteinite vary from ~13 to ~52 vol.%. Other phases typically are never more than ~1 vol.% each.

In porphyritic volcanic angrites, olivine is highly zoned from core mg#s 85–90 and Ln (larnite) <0.5 mole% for coarse grain cores to mg# <2, Ln 15–20 mole% for grain rims and groundmass grains. The nature of the coarse olivine is uncertain, and the most magnesian cores are thought to be xenocrysts [*Mikouchi et al.*, 1996]. The cores of olivine grains in ophitic angrites are not as magnesian, with mg#s ~65. Core compositions of calcic pyroxene have mg#s of ~60–65 for volcanic angrites and ~75 for plutonic angrites. The TiO_2 and Al_2O_3 contents are high with ~1.5 and ~9.0 wt.% in the cores of volcanic angrite calcic pyroxenes, and ~2.2 and 10.0 wt.% in plutonic angrites. Kirschsteinite is subcalcic in volcanic angrites with Ln of 30–35 mole% and mg# <10, while it is calcic in plutonic angrites with Ln 47–48 mole% and mg# of 53 for Angra dos Reis and 21 for LEW 86010. Plagioclase is nearly pure anorthite, $An_{>99}$. Hercynitic spinel and ulvöspinel compositions require that some of the Fe be ferric in order to satisfy charge balance, indicating a higher oxygen fugacity for their parent magma than typical for basaltic achondrites.

Angrites have unusual major element compositions that are distinct from those of other asteroidal basalts

(Figure 5.8). They are critically silica-undersaturated, and hence with crystallization the silica activity in the magma decreases, resulting in olivine and finally olivine + kirschsteinite crystallizing in the groundmass. This characteristic is unique among basaltic achondrites; all others differentiated towards silica-saturated compositions, crystallizing a silica polymorph in their mesostases. Angrites are highly depleted in some moderately volatile and volatile elements [*Mittlefehldt and Lindstrom*, 1990; *Mittlefehldt et al.*, 1998; *Mittlefehldt et al.*, 2002; *Riches et al.*, 2012]. Sodium abundances are ~0.02–0.04 × CI; hence the nearly pure anorthite plagioclase they contain. The refractory lithophile trace element contents of angrites fall into three groups. Most have CI-normalized abundances of 10–20; Angra dos Reis has abundances of ~30 × CI, while NWA 2999 and NWA 4931 have abundances of ~5 × CI. The latter two are thought to be paired. There is little fractionation of one refractory lithophile trace element from another, and all have relatively flat CI-normalized patterns, with the exception of NWA 4931 which is poor in Nb and Ta compared to other refractory lithophile elements [*Riches et al.*, 2012]. Highly siderophile element contents [Ru, Pd, Re, Os, Ir, and Pt) of angrites are very low in general and highly variable [*Riches et al.*, 2012]. NWA 4931 is exceptional and has the highest abundances at ~0.2–0.3 × CI. At the other extreme, NWA 1296 has CI-normalized abundances of ~10^{-5}. However, most angrites have highly siderophile element abundances of between ~10^{-4} and ~2×10^{-2}. The highly siderophile elements are in roughly chondritic relative ratios, although the most volatile of them, Pd, is at slightly lower abundance in some angrites.

Prior to the first angrite recoveries from Antarctica, our worldview of the differentiation history of the angrite parent body was totally dependent on study of Angra dos Reis, mostly derived from the 1977 consortium investigations. Angra dos Reis is old; the latest Pb isotopic data put its formation at 4557.65 ± 0.13 Ma [*Amelin*, 2008]. The coarse grain size, unusual mineralogy, nearly monomineralic character, and preferred orientation of the groundmass calcic pyroxene led to the model that Angra dos Reis is a cumulate from a highly silica-undersaturated magma [*Prinz et al.*, 1977]. The large ion lithophile element contents of bulk rock Angra dos Reis, at ~20–30 × CI, are high for a cumulate. An acid-washed calcic pyroxene separate was used to construct a trace-element-based model for the petrogenesis of Angra dos Reis: the parent melt of Angra dos Reis was formed by ~1% partial melting, and ~50% of the parent melt had crystallized as calcic pyroxene prior to formation of the Angra dos Reis cumulate [*Ma et al.*, 1977]. To generate a highly silica-undersaturated melt, melting was envisioned to occur at >700 MPa pressure. (For comparison, the central pressure of the largest asteroid, 1 Ceres, is ~140–150 MPa.) An alternative interpretation of the petrology of Angra dos Reis is that it is nearly a melt composition, and that it is thus a metamorphosed volcanic rock [*Treiman*, 1989].

One key aspect of angrites is that because they have not been significantly shock-modified, and many still retain extensive igneous zoning, the parent-daughter relationships of radiogenic isotopic systems have not been disturbed since crystallization for volcanic angrites, or since metamorphism for the plutonic angrites. Because of the very low volatile element contents, the parent asteroid μ ($^{238}U/^{204}Pb$) is very high, and thus very precise ^{207}Pb–^{206}Pb absolute ages can be calculated from measured Pb isotopic compositions. The very low highly siderophile element contents of some indicate efficient separation of metal from silicate during magma genesis, facilitating application of the ^{182}Hf–^{182}W short-lived chronometer. With the recovery of the mineralogically varied angrites from Antarctica, ^{26}Al–^{26}Mg dating became practical because anorthite and olivine provide a long lever arm, while spinel and olivine provide the same for ^{53}Mn–^{53}Cr dating. This combination of very precise absolute Pb–Pb ages with precise dating by several short-lived chronometer systems allowed for the first time pinning absolute ages to different processes that occurred in the earliest history of the solar system. The ^{207}Pb–^{206}Pb, ^{182}Hf–^{182}W, and ^{53}Mn–^{53}Cr systems are concordant [*Kleine et al.*, 2012], while the ^{26}Al–^{26}Mg suggests ages that are ~1 Myr younger for reasons that are not clear [*Spivak-Birndorf et al.*, 2009]. The ^{182}Hf–^{182}W system indicates that core formation on the angrite parent asteroid occurred within ~2 Myr of CAI formation. The oldest angrite, D'Orbigny, crystallized ~4 Myr after CAI formation and has a ^{207}Pb–^{206}Pb age of 4563.4 ± 0.3 Ma. Magmatism persisted for ~7 Myr, until 4556.6 ± 0.2 Ma, the ^{207}Pb–^{206}Pb age of Angra dos Reis.

Starting with the availability of the first two Antarctic angrites, LEW 86010 and LEW 87051 (Plates 53 and 54), more realistic and detailed petrologic models of angrite formation could be constructed. Early results showed that Angra dos Reis is not simply related to LEW 86010 and LEW 87051 in that a distinct source region is required [*Mittlefehldt and Lindstrom*, 1990; *Longhi*, 1999]. However, the compositions of LEW 86010 and LEW 87051 could be simply related via olivine control acting on a common magma composition [*Mittlefehldt and Lindstrom*, 1990]. Further work on additional angrites has shown that olivine control can explain major and minor element compositions of many volcanic angrites; addition of magnesian olivine xenocrysts to a parent melt with a composition like that of D'Orbigny and Sahara 99555 can replicate the composition of olivine-phyric angrites A-881371 and LEW 87051 [*Mikouchi et al.*, 2004]. This model extends to lithophile trace elements and includes NWA 1296, which is compositionally similar to D'Orbigny and Sahara 99555. However, olivine control no longer seems a viable explanation for the compositional differences between LEW 86010 and LEW 87051. Note that as originally promulgated, the olivine-control model considers large magnesian olivine grains in A-881371 and LEW 87051 to be xenocrystic [*Mikouchi et al.*, 2004]. Thus, a mixing process is implied.

Angra dos Reis remains an anomaly. Petrologic considerations show that it was derived from a distinct source [*Longhi*, 1999], and compositionally it is distinct from all other angrites [*Mittlefehldt et al.*, 2002; *Riches et al.*, 2012]. Tungsten isotopic data show that it, along with NWA 2999 and paired NWA 4931, were derived from a source with a much higher Hf/W ratio than were other angrites [*Kleine et al.*, 2012]. The W isotopic data suggest that core formation was not a single, asteroid-wide event but rather was localized. Thus, there is still much that is unknown regarding the differentiation of the angrite parent asteroid.

Perhaps the most significant contribution of the study of angrites to our understanding of the solar system lies in the ability to determine very precise absolute ages for their formation and tie these to several short-lived chronometers. This allows much greater confidence in placing asteroidal magmatism into the context of early solar system processes, first made possible by the recovery of LEW 86010.

5.7. SUMMARY

The collection of meteorites from Antarctica has greatly expanded the numbers and diversities of achondrite and iron meteorites available for study. This has allowed for the development of more robust models for magmatic processes on strongly heated asteroids. In particular, the acapulcoite-lodranite clan could only be defined once numerous examples were available, and their study led directly to development of models for silicate and metallic melt migration on asteroids that were only partially melted. Recovery of angrites has permitted the integration of the very early events in the history of the solar system as defined by nebular materials with the differentiation processes that occurred on some asteroids. The growth of the HED clan and ureilite group through Antarctic meteorite recoveries has helped us refine our knowledge of magmatic processes on asteroids through providing previously unknown or underrepresented lithologies. All of these are ongoing endeavors that will continue to yield new insights into solar system history as new meteorites are returned from Antarctica.

Acknowledgements. The senior author was supported by funding from the NASA Cosmochemistry Program. We thank J.-A. Barrat, whose review was particularly helpful to us while formulating our revisions. An anonymous reviewer also provided a few useful suggestions.

REFERENCES

Amelin, Y. (2008), U-Pb ages of angrites, *Geochim. Cosmochim. Acta, 72*, 221–232.

Barrat, J. A., A. Yamaguchi, R. C. Greenwood, M. Bohn, J. Cotten, M. Benoit, and I. A. Franchi (2007), The Stannern trend eucrites: Contamination of main group eucritic magmas by crustal partial melts, *Geochim. Cosmochim. Acta, 71*, 4108–4124.

Barrat, J. -A., A. Yamaguchi, B. Zanda, C. Bollinger, and M. Bohn (2010), Relative chronology of crust formation on asteroid Vesta: Insights from the geochemistry of diogenites, *Geochim. Cosmochim. Acta, 74*, 6218–6231.

Barrat, J. A., A. Yamaguchi, T. E. Bunch, M. Bohn, C. Bollinger, and G. Ceuleneer (2011), Possible fluid-rock interactions on differentiated asteroids recorded in eucritic meteorites, *Geochim. Cosmochim. Acta, 75*, 3839–3852.

Beck, A. W., and H. Y. McSween Jr (2010), Diogenites as polymict breccias composed of orthopyroxenite and harzburgite, *Meteoritics Planet. Sci., 45*, 850–872.

Beck, A. W., D. W. Mittlefehldt, H. Y. McSween Jr, D. Rumble III, C. -T. A. Lee, R. J. Bodnar (2011), MIL 03443, a dunite from asteroid 4 Vesta: Evidence for its classification and cumulate origin, *Meteoritics Planet. Sci., 46*, 1133–1151.

Beck, A. W., K. C. Welten, H. Y. McSween, C. E. Viviano, and M. W. Caffee (2012), Petrologic and textural diversity among the PCA 02 howardite group, one of the largest pieces of the Vestan surface, *Meteoritics Planet. Sci. 47*, 947–969.

Berkley, J. L., and J. H. Jones (1982), Primary igneous carbon in ureilites: Petrological implications, *J. Geophys. Res., 87*, A353–A364.

Berkley, J. L., H. G. Brown IV, K. Keil, N. L. Carter, J. C. C. Mercier, and G. Huss (1976), The Kenna ureilite: An ultramafic rock with evidence for igneous, metamorphic, and shock origin, *Geochim. Cosmochim. Acta, 40*, 1429–1437.

Berkley, J. L., G. J. Taylor, K. Keil, G. E. Harlow, and M. Prinz (1980), The nature and origin of ureilites, *Geochim. Cosmochim. Acta, 44*, 1579–1597.

Bild, R. W., and J. T. Wasson (1976), The Lodran meteorite and its relationship to the ureilites, *Mineralogical Magazine, 40*, 721–735.

Bowman, L. E., M. N. Spilde, and J. J. Papike (1997), Automated energy dispersive spectrometer modal analysis applied to the diogenites, *Meteoritics Planet. Sci., 32*, 869–875.

Boynton, W. V., P. M. Starzyk, and R. A. Schmitt (1976), Chemical evidence for the genesis of the ureilites, the achondrite Chassigny and the nakhlites, *Geochim. Cosmochim. Acta, 40*, 1439–1447.

Burbine, T. H., T. J. McCoy, A. Meibom, B. Gladman, and K. Keil (2002), Meteoritic parent bodies: Their number and identification, in *Asteroids III*, edited by W. F. J. Bottke, A. Cellino, P. Paolicchi, and R. P. Binzel, University of Arizona Press, Tucson, AZ, pp. 653–667.

Clarke, R. S. J. (1986), Antarctic iron meteorites: An unexpectedly high proportion of falls of unusual interest, International Workshop on Antarctic Meteorites. Lunar and Planetary Institute, Houston, TX, pp. 28–29.

Clayton, R. N., and T. K. Mayeda (1988), Formation of ureilites by nebular processes, *Geochim. Cosmochim. Acta, 52,* 1313–1318.

Clayton, R. N., and T. K. Mayeda (1996), Oxygen isotope studies of achondrites, *Geochim. Cosmochim. Acta, 60,* 1999–2017.

Clayton, R. N., N. Onuma, and T. K. Mayeda (1976), A classification of meteorites based on oxygen isotopes, *Earth and Planetary Science Letters 30,* 10–18.

Cohen, B. A., C. A. Goodrich, and K. Keil (2004), Feldspathic clast populations in polymict ureilites: Stalking the missing basalts from the ureilite parent body, *Geochim. Cosmochim. Acta, 68,* 4249–4266.

Crozaz, G., C. Floss, and M. Wadhwa (2003), Chemical alteration and REE mobilization in meteorites from hot and cold deserts, *Geochim. Cosmochim. Acta, 67,* 4727–4741.

Day, J. M. D., R. D. Ash, Y. Liu, J. J. Bellucci III, W. F. McDonough, R. J. Walker, and L. A. Taylor (2009), Early formation of evolved asteroidal crust, *Nature 457,* 179–182.

Day, J. M. D., R. J. Walker, R. D. Ash, Y. Liu, D. Rumble III, A. J. Irving, C. A. Goodrich, K. Tait, W. F. McDonough, and L. A. Taylor (2012a), Origin of felsic achondrites Graves Nunataks 06128 and 06129, and ultramafic brachinites and brachinite-like achondrites by partial melting of volatile-rich primitive parent bodies, *Geochim. Cosmochim. Acta, 81,* 94–128.

Day, J. M., R. J. Walker, L. Qin, and D. Rumble III (2012b), Late accretion as a natural consequence of planetary growth, *Nature Geoscience 5,* 614–617.

Delaney, J. S., H. Takeda, M. Prinz, C. E. Nehru, and G. E. Harlow (1983), The nomenclature of polymict basaltic achondrites, *Meteoritics, 18,* 103–111.

Delaney, J. S., M. Prinz, and H. Takeda (1984), The polymict eucrites, *J. Geophys. Res., 89,* C251–C288.

De Sanctis, M. C., J.-P. Combe, E. Ammannito, E. Palomba, A. Longobardo, T. B. McCord, S. Marchi, F. Capaccioni, M. T. Capria, D. W. Mittlefehldt, C. M. Pieters, J. Sunshine F., Tosi, F. Zambon, F. Carraro, S. Fonte, A. Frigeri, G. Magni, C. A. Raymond, C. T. Russell, and D. Turrini (2012), Detection of widespread hydratred materials on Vesta by the VIR imaging spectrometer on board the Dawn mission, *Astrophysical Journal Letters, 758,* L36–L40.

Downes, H., D. W. Mittlefehldt, N. T. Kita, and J. W. Valley (2008), Evidence from polymict ureilite meteorites for a disrupted and re-accreted single ureilite parent asteroid gardened by several distinct impactors, *Geochim. Cosmochim. Acta, 72,* 4825–4844.

Drake, M. J. (2001), The eucrite/Vesta story, *Meteoritics Planet. Sci., 36,* 501–514.

Floran, R. J., M. Prinz, P. F. Hlava, K. Keil, C. E. Nehru, and J. R. Hinthorne (1978), The Chassigny meteorite: A cumulate dunite with hydrous amphibole-bearing melt inclusions, *Geochim. Cosmochim. Acta, 42,* 1213–1229.

Floss, C. (2000), Complexities on the acapulcoite-lodranite parent body: Evidence from trace element distributions in silicate minerals, *Meteoritics Planet. Sci., 35,* 1073–1085.

Floss, C., G. Crozaz, A. Yamaguchi, and K. Keil (2000), Trace element constraints on the origins of highly metamorphosed Antarctic eucrites, *Antarctic Meteorite Res, 13,* 222–237.

Fu, R. R., B. P. Weiss, D. L. Shuster, J. Gattacceca, T. L. Grove, C. Suavet, E. A. Lima, L. Li, and A. T. Kuan (2012), An ancient core dynamo in asteroid Vesta. *Science, 338,* 238–241.

Göbel, R., U. Ott, and F. Begemann (1978), On trapped noble gases in ureilites, *J. Geophys. Res., 83,* 855–867.

Goodrich, C. A., E. R. D. Scott, and A. M. Fioretti (2004), Ureilitic breccias: clues to the petrologic structure and impact disruption of the ureilite parent asteroid, *Chemie der Erde, 64,* 283–327.

Goodrich, C. A., J. A. Van Orman, and L. Wilson (2007), Fractional melting and smelting on the ureilite parent body, *Geochim. Cosmochim. Acta, 71,* 2876–2895.

Goodrich, C. A., N. T. Kita, M. J. Spicuzza, J. W. Valley, J. Zipfel, T. Mikouchi, and M. Miyamoto (2010), The Northwest Africa 1500 meteorite: Not a ureilite, maybe a brachinite, *Meteoritics Planet. Sci., 45* 1906–1928.

Graham, A. L., A. W. R. Bevan, and R. Hutchison (1985), *Catalogue of Meteorites,* University of Arizona Press.

Greenwood, R. C., I. A. Franchi, A. Jambon, and P. C. Buchanan (2005), Widespread magma oceans on asteroidal bodies in the early solar system, *Nature, 435,* 916–918.

Greenwood, R. C., I. A. Franchi, and J. M. Gibson (2008), How useful are high-precision delta Δ17O data in defining the asteroidal sources of meteorites?: Evidence from main-group pallasites, primitive and differentiated achondrites, 39th Lunar and Planetary Science Conference. Lunar and Planetary Institute, Houston, TX, Abstract #2445.

Hewins, R., and H. Newsom (1988), Igneous activity in the early solar system, in Meteorites and the early solar system, edited by J. F. Kerridge and M. S. Matthews, University of Arizona Press, Tucson, AZ, pp. 73–101.

Ikeda, Y., and H. Takeda (1985), A model for the origin of basaltic achondrites based on the Yamato 7308 howardite, *J. Geophys. Res.: Solid Earth, 90,* C649–C663.

Johnson, J. E., J. M. Scrymgour, E. Jarosewich, and B. Mason (1977), Brachina meteorite: A chassignite from South Australia, *Records of the South Australian Museum, 17,* 309–319.

Keil, K. (1977), Preface: The Angra dos Reis consortium, *Earth Planet. Sci. Lett, 35,* 271.

Keil, K. (2012), Angrites, a small but diverse suite of ancient, silica-undersaturated volcanic-plutonic mafic meteorites, and the history of their parent asteroid, *Chemie der Erde, 72,* 191–218.

Kleine, T., U. Hans, A. J. Irving, and B. Bourdon (2012), Chronology of the angrite parent body and implications for core formation in protoplanets, *Geochim. Cosmochim. Acta, 84,* 186–203.

Laul, J. C., R. R. Keays, R. Ganapathy, E. Anders, and J. W. Morgan (1972), Chemical fractionations in meteorites: V. Volatile and siderophile elements in achondrites and ocean ridge basalts, *Geochim. Cosmochim. Acta, 36,* 329–345.

Longhi, J. (1999), Phase equilibrium constraints on angrite petrogenesis, *Geochim. Cosmochim. Acta, 63,* 573–585.

Ma, M. S., A. V. Murali, and R. A. Schmitt (1977), Genesis of the Angra dos Reis and other achondritic meteorites, *Earth Planet. Sci. Lett., 35,* 331–346.

Mason, B. (1963), The hypersthene achondrites, *American Museum Novitates, 2155,* 1–13.

McCoy, T. J., K. Keil, R. N. Clayton, T. K. Mayeda, D. D. Bogard, D. H. Garrison, G. R. Huss, I. D. Hutcheon, and R. Wieler (1996), A petrologic, chemical, and isotopic study of Monument Draw and comparison with other acapulcoites: Evidence for formation by incipient partial melting, *Geochim. Cosmochim. Acta, 60,* 2681–2708.

McCoy, T. J., K. Keil, R. N. Clayton, T. K. Mayeda, D. D. Bogard, D. H. Garrison, and R. Wieler (1997), A petrologic and isotopic study of lodranites: Evidence for early formation as partial melt residues from heterogeneous precursors, *Geochim. Cosmochim. Acta, 61,* 623–637.

McCoy, T. J., K. Keil, D. W. Muenow, and L. Wilson (1997), Partial melting and melt migration in the acapulcoite-lodranite parent body, *Geochim. Cosmochim. Acta, 61,* 639–650.

McCoy, T. J., W. D. Carlson, L. R. Nittler, R. M. Stroud, D. D. Bogard, and D. H. Garrison (2006), Graves Nunataks 95209: A snapshot of metal segregation and core formation, *Geochim. Cosmochim. Acta, 70,* 516–531.

McCoy, T. J., R. J. Walker, J. I. Goldstein, J. Yang, W. F. McDonough, D. Rumble, N. L. Chabot, R. D. Ash, C. M. Corrigan, J. R. Michael, and P. G. Kotula (2011), Group IVA irons: New constraints on the crystallization and cooling history of an asteroidal core with a complex history, *Geochim. Cosmochim. Acta, 75,* 6821–6843.

McSween, H. Y., Jr., D. W. Mittlefehldt, A. W. Beck, R. G. Mayne, and T. J. McCoy (2011), HED meteorites and their relationship to the geology of Vesta and the Dawn Mission, *Space Sci. Rev., 163,* 141–174.

McSween, H. Y., Jr., E. Ammannito, V. Reddy, T. H. Prettyman, A. W. Beck, M. Cristina De Sanctis, A. Nathues, L. L. Corre, D. P. O'Brien, N. Yamashita, T. J. McCoy, D. W. Mittlefehldt, M. J. Toplis, P. Schenk, E. Palomba, D. Turrini, F. Tosi, F. Zambon, A. Longobardo, F. Capaccioni, C. A. Raymond, and C. T. Russell (2013), Composition of the Rheasilvia basin, a window into Vesta's interior, *J. Geophys. Res.: Planets, 118,* 335–346.

Mikouchi, T., M. Miyamoto, and G. A. McKay (1996), Mineralogical study of angrite Asuka 881371: Its possible relation to angrite Lew 87051, *Proceedings of the Symposium on Antarctic Meteorites, 9th,* 174–188.

Mikouchi, T., G. A. McKay, and J. H. Jones (2004), Sahara 99555 and D'Orbigny: Possible pristine parent magma of quenched angrites, 35th Lunar and Planetary Science Conference. Lunar and Planetary Institute, Houston, TX, Abstract #1504.

Mittlefehldt, D. W. (1994), The genesis of diogenites and HED parent body petrogenesis, *Geochim. Cosmochim. Acta, 58,* 1537–1552.

Mittlefehldt, D. W. (2000), Petrology and geochemistry of the Elephant Moraine A79002 diogenite: A genomict breccia containing a magnesian harzburgite component, *Meteoritics Planet. Sci., 35,* 901–912.

Mittlefehldt, D. W. (2005), Ibitira: A basaltic achondrite from a distinct parent asteroid and implications for the Dawn mission, *Meteoritics Planet. Sci., 40,* 665–677.

Mittlefehldt, D. W., and M. M. Lindstrom (1990), Geochemistry and genesis of the angrites, *Geochim. Cosmochim. Acta, 54,* 3209–3218.

Mittlefehldt, D. W., and M. M. Lindstrom (1997), Magnesian basalt clasts from the EET 92014 and Kapoeta howardites and a discussion of alleged primary magnesian HED basalts, *Geochim. Cosmochim. Acta, 61,* 453–462.

Mittlefehldt, D. W., C.-L. Chou, and J. T. Wasson (1979), Mesosiderites and howardites: igneous formation and possible genetic relationships, *Geochim. Cosmochim. Acta, 43,* 673–688.

Mittlefehldt, D. W., M. M. Lindstrom, D. D. Bogard, D. H. Garrison, and S. W. Field (1996), Acapulco- and Lodran-like achondrites: Petrology, geochemistry, chronology, and origin, *Geochim. Cosmochim. Acta, 60,* 867–882.

Mittlefehldt, D. W., T. J. McCoy, C. A. Goodrich, and A. Kracher (1998), Non-chondritic meteorites from asteroidal bodies, in J. J. Papike (Ed.), Planetary Materials. Rev. Material. Geochem, 36, Mineralogical Society of America, Washington, DC, pp. 4.1–4.195.

Mittlefehldt, D. W., M. Killgore, and M. T. Lee (2002), Petrology and geochemistry of D'Orbigny, geochemistry of Sahara 99555, and the origin of angrites, *Meteoritics Planet. Sci., 37,* 345–369.

Mittlefehldt, D. W., D. D. Bogard, J. L. Berkley, and D. H. Garrison (2003), Brachinites: Igneous rocks from a differentiated asteroid, *Meteoritics Planet. Sci., 38,* 1601–1625.

Mittlefehldt, D. W., A. W. Beck, C.-T. A. Lee, H. Y. McSween, and P. C. Buchanan (2012), Compositional constraints on the genesis of diogenites, *Meteoritics Planet. Sci., 47,* 72–98.

Miyamoto, M., H. Takeda, and K. Yanai (1978), Yamato achondrite polymict breccias, *Memoirs of the National Institute of Polar Research, Special Issue, 8,* 185-197.

Nagahara, H. (1992), Yamato-8002: Partial melting residue on the "unique" chondrite parent body, *Proceedings of the NIPR Symposium on Antarctic Meteorites, 5,* 191–223.

Nehru, C. E., M. Prinz, J. S. Delaney, G. Dreibus, H. Palme, B. Spettel, and H. Wänke (1983), Brachina: A new type of meteorite, not a chassignite, *J. Geophys. Res., 88,* B237–B244.

Nehru, C. E., M. Prinz, M. K. Weisberg, M. E. Ebihara, R. N. Clayton, and T. K. Mayeda (1996), A New brachinite and petrogenesis of the group, 27th Lunar and Planetary Science Conference. Lunar and Planetary Institute, Houston, TX, Abstract #1472.

Nyquist, L. E., H. Takeda, B. M. Bansal, C.-Y. Shih, H. Wiesmann, and J. L. Wooden (1986), Rb-Sr and Sm-Nd internal isochron ages of a subophitic basalt clast and a matrix sample of the Y75011 eucrite, *J. Geophys. Res., 91,* 8137–8150.

Palme, H., L. Schultz, B. Spettel, H. W. Weber, H. Wänke, M. C. Michel-Levy, and J. C. Lorin (1981), The Acapulco meteorite: Chemistry, mineralogy and irradiation effects, *Geochim. Cosmochim. Acta, 45,* 727–752.

Patzer, A., D. H. Hill, and W. V. Boynton (2004), Evolution and classification of acapulcoites and lodranites from a chemical point of view, *Meteoritics Planet. Sci., 39,* 61–85.

Paul, R. L., and M. E. Lipschutz (1990), Chemical studies of differentiated meteorites: I. Labile trace elements in Antarctic and non-Antarctic eucrites. *Geochim. Cosmochim. Acta,* 3185–3196.

Prettyman, T. H., D. W. Mittlefehldt, N. Yamashita, D. J. Lawrence, A. W. Beck, W. C. Feldman, T. J. McCoy, H. Y. McSween, M. J. Toplis, T. N. Titus, P. Tricarico, R. C. Reedy, J. S. Hendricks, O. Forni, L. Le Corre, J.-Y. Li, H. Mizzon, V. Reddy, C. A. Raymond, and C. T. Russell (2012), Elemental mapping by Dawn reveals exogenic H in Vesta's regolith, *Science, 338,* 242–246.

Prinz, M., K. Keil, P. F. Hlava, J. L. Berkley, C. B. Gomes, and W. S. Curvello (1977), Studies of Brazilian meteorites: III. Origin and history of the Angra dos Reis achondrite, *Earth Planet. Sci. Lett., 35,* 317–330.

Rankenburg, K., M. Humayun, A. D. Brandon, and J. S. Herrin (2008), Highly siderophile elements in ureilites, *Geochim. Cosmochim. Acta, 72,* 4642–4659.

Riches, A. J. V., J. M. D. Day, R. J. Walker, A. Simonetti, Y. Liu, C. R. Neal, and L. A. Taylor (2012), Rhenium-osmium isotope and highly-siderophile-element abundance systematics of angrite meteorites, *Earth Planet. Sci. Lett., 353–354,* 208–218.

Righter, K., and M. J. Drake (1997), A magma ocean on Vesta: Core formation and petrogenesis of eucrites and diogenites, *Meteoritics Planet. Sci., 32,* 929–944.

Rumble, D. I., A. J. Irving, T. E. Bunch, J. H. Wittke, and S. M. Kuehner (2008), *Oxygen isotopic and petrological diversity among brachinites NWA 4872, NWA 4874, NWA 4882 and NWA 4969: How many ancient parent bodies?* 39th Lunar and Planetary Science Conference. Lunar and Planetary Institute, Houston, TX, Abstract #1974.

Ruzicka, A., G. A. Snyder, and L. A. Taylor (1997), Vesta as the howardite, eucrite and diogenite parent body: Implications for the size of a core and for large-scale differentiation, *Meteoritics Planet. Sci., 32,* 825–840.

Ryder, G. (1982), Siderophiles in the Brachina meteorite: impact melting? *Nature, 299,* 805–807.

Sack, R. O., W. J. Azeredo, and M. E. Lipschutz (1991), Olivine diogenites: The mantle of the eucrite parent body, *Geochim. Cosmochim. Acta, 55,* 1111–1120.

Sarafian, A. R., M. F. Roden, and A. E. Patiño-Douce (2013), The volatile content of Vesta: Clues from apatite in eucrites, *Meteoritics & Planetary Science, 48,* 2135–2154.

Schenk, P., D. P. O'Brien, S. Marchi, R. Gaskell, F. Preusker, T. Roatsch, R. Jaumann, D. Buczkowski, T. McCord, H. Y. McSween, D. Williams, A. Yingst, C. Raymond, and C. Russell (2012), The geologically recent giant impact basins at Vesta's south pole, *Science, 336,* 694–697.

Schiller, M., J. Baker, J. Creech, C. Paton, M. A. Millet, A. Irving, & M. Bizzarro (2011), Rapid timescales for magma ocean crystallization on the howardite-eucrite-diogenite parent body, *Astrophys. J. Lett, 740,* L22.

Schrader, D. L., D. S. Lauretta, H. C. Connolly Jr, Y. S. Goreva, D. H. Hill, K. J. Domanik, E. L. Berger, H. Yang, and R. T. Downs (2010), Sulfide-rich metallic impact melts from chondritic parent bodies, *Meteoritics Planet. Sci., 45,* 743–758.

Scott, E. R. D., R. C. Greenwood, I. A. Franchi, and I. S. Sanders (2009), Oxygen isotopic constraints on the origin and parent bodies of eucrites, diogenites, and howardites, *Geochim. Cosmochim. Acta, 73,* 5835–5853.

Shearer, C. K., P. Burger, and J. J. Papike (2010), Petrogenetic relationships between diogenites and olivine diogenites: Implications for magmatism on the HED parent body, *Geochim. Cosmochim. Acta, 74,* 4865–4880.

Shearer, C. K. Burger, P. V. Papike, J. J. Sharp, Z. D. McKeegan, K. D. (2011), Fluids on differentiated asteroids: Evidence from phosphates in differentiated meteorites GRA 06128 and GRA 06129, *Meteoritics Planet. Sci., 46,* 1345–1362.

Singletary, S. J., and T. L. Grove (2003), Early petrologic processes on the ureilite parent body, *Meteoritics Planet. Sci., 38,* 95–108.

Spivak-Birndorf, L., M. Wadhwa, and P. Janney (2009), 26Al-26Mg systematics in D'Orbigny and Sahara 99555 angrites: Implications for high-resolution chronology using extinct chronometers, *Geochim. Cosmochim. Acta, 73,* 5202–5211.

Stolper, E. (1977), Experimental petrology of eucritic meteorites, *Geochim. Cosmochim. Acta, 41,* 587–611.

Swindle, T. D., D. A. Kring, M. K. Burkland, D. H. Hill, and W. V. Boynton (1998), Noble gases, bulk chemistry, and petrography of olivine-rich achondrites Eagles Nest and Lewis Cliff 88763: Comparison to brachinites, *Meteoritics Planet. Sci., 33,* 31–48.

Takeda, H. Wooden, J. L. Mori, H. Delaney, J. S. Prinz, M. Nyquist, L. E. (1983), Comparison of Yamato and Victoria Land polymict eucrites: A view from mineralogical and isotopic studies. *J. Geophys. Res. 88,* B245–B256.

Takeda, H, H. Mori, T. Hiroi, and J. Saito (1994), Mineralogy of new Antarctic achondrites with affinity to Lodran and a model of their evolution in an asteroid, *Meteoritics, 29,* 830–842.

Treiman, A. H. (1989), An alternate hypothesis for the origin of Angra dos Reis: Porphyry, not cumulate, *Proceedings of the Lunar and Planetary Science Conference, 19,* 443–450.

Wacker, J. F. (1986), Noble gases in the diamond-free ureilite, ALHA 78019: The roles of shock and nebular processes, *Geochim. Cosmochim. Acta, 50,* 633–642.

Warren, P. H., and H. Huber (2006), Ureilite petrogenesis: A limited role for smelting during anatexis and catastrophic disruption, *Meteoritics Planet. Sci., 41,* 835–849.

Warren, P. H., and G. W. Kallemeyn (1989), Allan Hills 84025: The second brachinite, far more differentiated than Brachina, and an ultramafic achondritic clast from L chondrite Yamato-75097. *Proceedings of the Lunar and Planetary Science Conference, 19,* 475–486.

Warren, H. P., W. G. Kallemeyn, T. Arai, & K. Kaneda (1996), Compositional-petrologic investigation of eucrites and the QUE94201 Shergottite, *Antarctic Meteorites XXI,* 195–197.

Warren, P. H., F. Ulff-Møller, H. Huber, and G. W. Kallemeyn (2006), Siderophile geochemistry of ureilites: A record of early stages of planetesimal core formation, *Geochim. Cosmochim. Acta, 70,* 2104–2126.

Warren, P. H., G. W. Kallemeyn, H. Huber, F. Ulff-Møller, and W. Choe (2009), Siderophile and other geochemical constraints on mixing relationships among HED-meteoritic breccias, *Geochim. Cosmochim. Acta, 73,* 5918–5943.

Wasson, J. T. (1990), Ungrouped iron meteorites in Antarctica: Origin of anomalously high abundance, *Science, 249,* 900–902.

Wasson, J. T., C.-L. Chou, R. W. Bild, and P. A. Baedecker (1976), Classification of and elemental fractionation among ureilites, *Geochim. Cosmochim. Acta, 40,* 1449–1458.

Wasson, J. T., X. Ouyang, J. Wang, and J. Eric (1989), Chemical classification of iron meteorites: XI. Multi-element studies of 38 new irons and the high abundance of ungrouped irons from Antarctica, *Geochim. Cosmochim. Acta, 53,* 735–744.

Wolf, R., M. Ebihara, G. R. Richter, and E. Anders (1983), Aubrites and diogenites: Trace element clues to their origin, *Geochim. Cosmochim. Acta, 47,* 2257–2270.

Yamaguchi, A., G. J. Taylor, K. Keil, C. Floss, G. Crozaz, L. E. Nyquist, D. D. Bogard, D. H. Garrison, Y. D. Reese, H. Wiesmann, and C.-Y. Shih (2001), Post-crystallization reheating and partial melting of eucrite EET90020 by impact into the hot crust of asteroid 4Vesta ~4.50 Ga ago, *Geochim. Cosmochim. Acta, 65,* 3577–3599.

Yamaguchi, A., J. A. Barrat, R. C. Greenwood, N. Shirai, C. Okamoto, T. Setoyanagi, M. Ebihara, I. A. Franchi, and M. Bohn (2009), Crustal partial melting on Vesta: Evidence from highly metamorphosed eucrites, *Geochim. Cosmochim. Acta, 73,* 7162–7182.

Yamaguchi, A., J.-A. Barrat, M. Ito, and M. Bohn (2011), Posteucritic magmatism on Vesta: Evidence from the petrology and thermal history of diogenites, *J. Geophys. Res.: Planets, 116,* E08009.

6

ANSMET Meteorites from the Moon

Randy L. Korotev[1] and Ryan A. Zeigler[2]

"The occurrence of secondary craters in the rays extending as much as 500 km from some large craters on the moon shows that fragments of considerable size are ejected at speeds nearly half the escape velocity from the moon (2.4 km/sec). At least a small amount of material from the lunar surface and perhaps as much or more than the impacting mass is probably ejected at speeds exceeding the escape velocity by impacting objects moving in asteroidal orbits. Some small part of this material may follow direct trajectories to the earth, some will go into orbit around the earth, and the rest will go into independent orbit around the sun. Much of it is probably ultimately swept up by earth."

Shoemaker et al. [1963]

6.1. INTRODUCTION

Despite the prediction of *Shoemaker et al.* [1963], it took another 19 years before the first lunar meteorite was recognized, Allan Hills 81005, a 31-g stone discovered during an ANSMET search in January of 1982 (Table 6.1) [*Marvin et al.*, 1983; *Cassidy*, 2003]. As of 2013, ANSMET has collected 24 lunar meteorite stones representing, when pairings are taken into account, 14 different meteorites (Table 6.1). In total, 22% of the known lunar meteorites have been found in Antarctica,17% by ANSMET and 7% by searches done by the NIPR (National Institute of Polar Research) of Japan; the rest have been found in hot deserts (Australia: 2%, Africa 46%, and Oman: 29%).

Unlike the asteroids, the Moon did not accrete directly from dust and planetesimals. Instead, the Moon is the likely product of a giant impact between two differentiated bodies [*Hartmann and Davis*, 1975; *Cameron and Ward*, 1976; *Canup and Asphaug*, 2001]. It subsequently experienced further differentiation and a more prolonged period of igneous geologic activity than any asteroid.

As a result, lunar meteorites are more diverse, particularly in terms of chemical composition, than meteorites from any other solar system body. The Moon is differentiated both vertically and laterally. Vertically, it has a small, most likely metallic core, a thick mantle (which itself may be layered) consisting mainly of olivine and pyroxene, and a crust that is rich in the anorthite extreme of plagioclase feldspar [*Wieczorek et al.*, 2006]. Laterally, many of the huge impact basins that formed early in lunar history are filled with basalt from volcanism: melting of the mantle from the heat of radioactive decay with the magmas rising to fill the low spots in the crust. This lateral heterogeneity is very evident, even from Earth, with the light-colored feldspathic "terra" or "highlands" pockmarked by dark circular "maria" (seas; singular: "mare;" Figure 6.1). Chemically, the maria are rich in iron, magnesium, and in some places titanium (olivine, pyroxene, ilmenite), whereas the highlands are rich in aluminum and calcium (anorthite). Although the mare-highlands dichotomy has been recognized for at least hundreds of years, it was not until samples were collected during the *Apollo* missions (1969–1972) and the composition of the lunar surface was systematically mapped by the Lunar Prospector mission (1998–1999) that a second lateral heterogeneity was discovered. The geochemically incompatible elements such as K, REE (rare earth elements), P, Zr, Th, and U, which do not partition into the major minerals of the lunar crust (plagioclase, pyroxene, olivine, and ilmenite), are concentrated in the northwest quadrant of the nearside in a region known as the Procellarum KREEP Terrane (Figure 6.2). The last basin-forming impact to occur on the nearside of the Moon, Imbrium, occurred in the Procellarum KREEP Terrane (Figure 6.2), spreading KREEP-rich rocks (i.e., rocks with high concentrations of Th and other incompatible elements) over the surface of the Moon [*Haskin*, 1998]. The lunar meteorites reflect the

[1] *Department of Earth and Planetary Sciences and McDonnell Center for the Space Sciences, Washington University*

[2] *Astromaterials Research and Exploration Science Directorate, Acquisition and Curation, NASA Johnson Space Center*

35 Seasons of U.S. Antarctic Meteorites (1976–2010): A Pictorial Guide to the Collection, Special Publication 68,
First Edition. Edited by Kevin Righter, Catherine M. Corrigan, Timothy J. McCoy and Ralph P. Harvey.
© 2015 American Geophysical Union. Published 2015 by John Wiley & Sons, Inc.

Table 6.1. ANSMET lunar meteorites.

Plot symbol	Stone name	Paired with	New	Mass (g)	Norm. plag. (%)	Lithology	Find date	Likely or possible launch pairs
A	ALH A81005		1	31.4	71.5	FRegBrx	1982-01-18	
E	EET 87521	EET 96008	1	30.7	38.4	BRegBrx	1987-12-20	Yam 793274/981031, QUE 94281, NWA 4884 (YQEN)
e	EET 96008	EET 87521		52.97	38.4	BRegBrx	1996-12-14	
G	GRA 06157		1	0.788	79.9	FRegBrx	2007-01-08	
1	LAP 02205	other 5 LAPs	1	1226.3	28.1	MB	2002-12-18	NWA 4734, NWA 032/479 (NNL)
2	LAP 02224	other 5 LAPs		252.5	28.1	MB	2002-12-12	
3	LAP 02226	other 5 LAPs		244.1	28.1	MB	2002-12-13	
4	LAP 02436	other 5 LAPs		58.97	28.1	MB	2002-12-12	
5	LAP 03632	other 5 LAPs		92.566	28.1	MB	2004-01-01	
6	LAP 04841	other 5 LAPs		55.992	28.1	MB	2005-01-03	
L	LAR 06638		1	5.293	82.9	FRegBrx	2007-01-03	
m	MAC 88104	MAC 88105	1	61.2	79.5	FRegBrx	1989-01-13	Dhofar 1428
M	MAC 88105	MAC 88104		662.5	79.5	FRegBrx	1989-01-13	
T	MET 01210		1	22.83	46.4	BRegBrx	2001-12-23	Yam 793274, Asuka 88175 MIL 05035 (YAMM)
B	MIL 05035		1	142.216	26.7	MB	2005-12-11	Yam 793274, Asuka 88175 MET 01210 (YAMM)
I	MIL 07006		1	1.368	75.6	FRegBrx	2007-12-31	Yam 791197, Dhofar 1436/1443
£	MIL 090034	MIL 090070 & 090075	1	195.565	86.0	FRegBrx	2009-12-18	
¥	MIL 090036		1	244.830	76.1	FRegBrx	2009-12-22	
$	MIL 090070	MIL 090034 & 090075		137.461	86.8	FRegBrx	2009-12-20	
¢	MIL 090075	MIL 090034 & 090070		143.523	82.7	FRegBrx	2009-12-20	
P	PCA 02007		1	22.372	72.8	FRegBrx	2003-01-05	
Q	QUE 93069	QUE 94269	1	21.4	80.1	FRegBrx	1993-12-11	
q	QUE 94269	QUE 93069		3.2	80.1	FRegBrx	1994-12-10	
U	QUE 94281		1	23.4	44.9	BRegBrx	1994-12-12	Yam 793274/981031, EET 87/96, NWA 4884 (YQEN)
Total	24		14	3733.4				

Note: New = first or only stone of a pair group; Norm. plag. = normative plagioclase abundance based on data of Table 6.2; BRegBrx = basaltic regolith breccias; FRegBrx = feldspathic regolith breccias; MB = mare basalt.

Moon's lateral variability. Among the 14 ANSMET lunar meteorites, 9 are feldspathic regolith (soil) breccias from the lunar highlands, 2 are basalts from the maria, and 3 are regolith breccias dominated by mare material but containing some feldspathic material (Table 6.1). Among the ANSMET lunar meteorites, concentrations of incompatible elements vary by a factor of 10 (Figures 6.3 and 6.4).

In this chapter we review the characteristics of the ANSMET lunar meteorites and compare them with *Apollo* samples and lunar meteorites from elsewhere.

6.2. EXPERIMENTAL METHODS

We present here new compositional data obtained by INAA (instrumental neutron activation analysis) and EPMA (electron probe microanalysis) of fused powders for some ANSMET lunar meteorites. The analytical procedures are described in detail in *Korotev et al.* [2009b]. Briefly, we analyzed multiple (2–16, typically 6–10) sub-samples (typically 25–35 mg in mass) of each stone. Mass-weighted mean concentration values and total mass analyzed are presented in Table 6.2 and data for individual

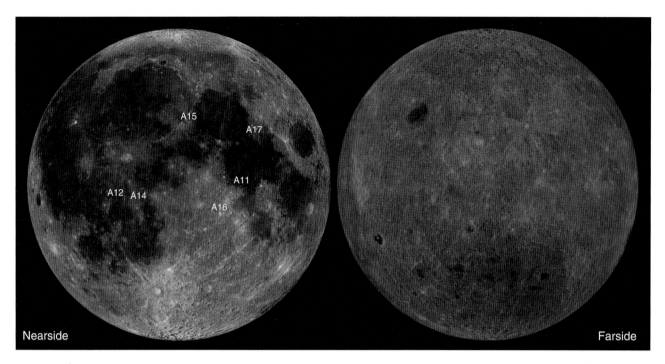

Figure 6.1. The nearside and farside of the Moon as captured by the wide angle camera on the *Lunar Reconnaissance Orbiter* mission. Compare with the terranes map of Figure 6.2. The dark, circular maria are prominent on the nearside, where the locations of the six *Apollo* landing sites are indicated. Images courtesy of NASA/GSFC/Arizona State University.

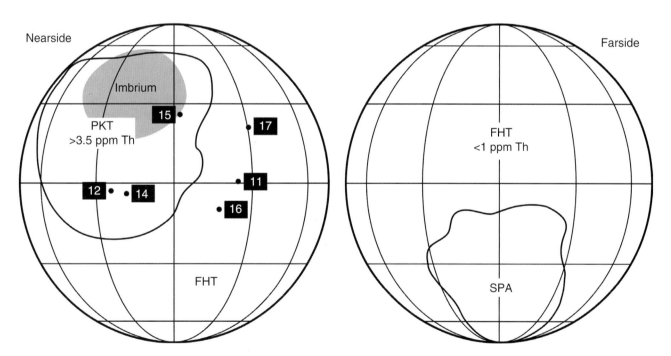

Figure 6.2. Schematic terrane map of the Moon after *Jolliff et al.* [2000]. The six *Apollo* landing sites were all near the center of the nearside. The distance between *Apollo* 12 (west) and *Apollo* 17 (east) is 16% of the lunar circumference. Three of the missions (12, 14, and 15) landed in the geochemically anomalous Procellarum KREEP Terrane (PKT), a region with high concentrations of incompatible elements such as Th [*Lawrence et al.*, 2000; *Jolliff et al.*, 2000]. All pixels with >3.5 ppm Th in the Th map of *Lawrence et al.* [2000] lie within the PKT boundary of *Jolliff et al.* [2000], as depicted here. The *Apollo* 11, 12, 15, and 17 lunar modules landed in areas resurfaced by mare basalt. Only *Apollo* 16 landed in the Feldspathic Highlands Terrane (FHT) distant from a mare. The South Pole-Aitken Terrane encompasses the giant and ancient South Pole-Aitken (SPA) basin on the farside.

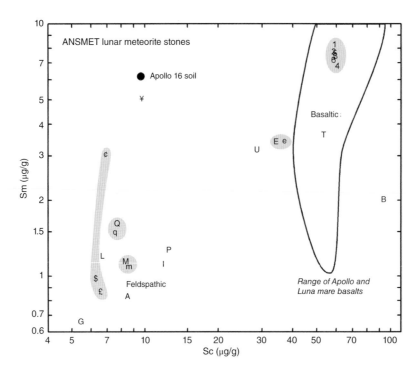

Figure 6.3. ANSMET lunar meteorites in Sc-Sm space; note logarithmic axes. Each point represents the mean composition of a named stone; see Table 6.1 for symbol key. Gray fields encompass paired stones. For reference, the mean composition of typical mature regolith from *Apollo* 16 is represented by the filled circle [*Korotev*, 1997]. Sc increases with increasing pyroxene abundance. For the nonbasaltic meteorites, Sm increases with increasing abundance of KREEP components.

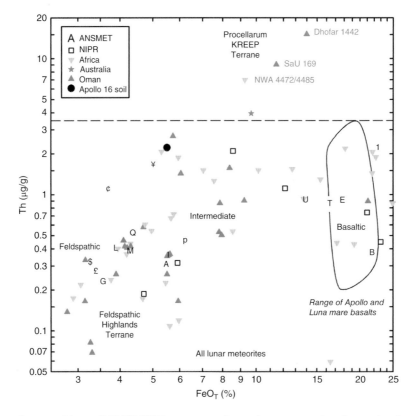

Figure 6.4. Comparison of compositions of ANSMET lunar meteorites to lunar meteorites from other locations in FeO-Th space (total Fe as FeO_T); note logarithmic axes. Each point represents the mean composition of a meteorite. For paired ANSMET meteorites, the symbol is that for the largest stone (see legend in Table 6.1). The two "Africa" points that overlap with the LAP 02205 point (symbol "1" representing the mean of all six LAP stones) are NWA 032 and NWA 4734. Meteorites with 7%–12% FeO are absent in the ANSMET collection as are Th-rich meteorites. The horizontal dashed line represents the Procellarum KREEP Terrane boundary of Figure 6.2.

Table 6.2. Mean compositions of ANSMET lunar meteorites.

	Unit	ALH A81005	EET 87/96	GRA 06157	LAP N=6	LAR 06638	MAC 88104/5	MET 01210	MIL 05035	MIL 07006	MIL 090034	MIL 090036	MIL 090070	MIL 090075	PCA 02007	QUE 93/94	QUE 94281	%±*
SiO$_2$	%	44.9	46.9	44.1	45.3	44.6	45.1	44.6	47.1	45.1	44.3	45.4	44.5	45.2	44.2	44.7	47.3	0.4
TiO$_2$	%	0.27	0.73	0.17	3.23	0.16	0.24	1.53	1.45	0.27	0.15	0.57	0.14	0.21	0.28	0.25	0.69	6
Al$_2$O$_3$	%	25.8	13.6	28.8	9.9	29.9	28.7	16.6	9.45	27.2	31.0	27.1	31.3	29.8	26.2	28.9	16.0	1.9
Cr$_2$O$_3$	%	0.13	0.27	0.08	0.30	0.08	0.09	0.22	0.31	0.12	0.07	0.10	0.06	0.08	0.14	0.08	0.26	1.0
FeO$_T$	%	5.47	17.75	3.55	22.5	3.90	4.29	16.36	21.7	5.54	3.40	5.01	3.26	3.70	6.24	4.37	13.93	1.0
MnO	%	0.07	0.23	0.05	0.29	0.09	0.06	0.22	0.32	0.08	0.05	0.07	0.05	0.05	0.09	0.07	0.19	12
MgO	%	8.20	8.35	6.81	6.45	4.52	4.11	6.31	7.3	5.33	2.84	5.18	2.75	3.85	6.84	4.58	8.52	5
CaO	%	14.9	11.4	15.5	11.1	16.4	16.9	13.1	11.8	15.8	17.1	15.5	17.0	16.6	15.3	16.4	12.4	1.2
Na$_2$O	%	0.30	0.37	0.36	0.35	0.34	0.33	0.30	0.26	0.35	0.37	0.62	0.37	0.39	0.34	0.34	0.36	1.0
K$_2$O	%	0.02	0.05	0.03	0.09	0.02	0.03	0.05	0.03	0.02	0.03	0.10	0.04	0.06	0.03	0.03	0.07	11
P$_2$O$_5$	%	0.02	0.05	0.02	0.12	0.05	0.03	0.11	0.04	0.02	0.02	0.10	0.02	0.06	0.02	0.03	0.06	22
Σ	%	100.1	99.7	99.5	99.6	100.1	99.9	99.4	99.8	99.8	99.4	99.7	99.5	100.0	99.6	99.8	99.8	—
Mg'	%	72.8	45.6	77±4	33.9	67.4	63.0	40.7	37.	63.2	59.8	64.8	60.1	65.0	66.1	65.1	52.2	—
Sc	μg/g	9.07	41.5	5.45	58.2	6.69	8.60	54.9	95.3	12.0	6.62	9.77	6.29	6.90	12.70	7.50	34.2	1
Co	μg/g	21.0	47.9	13.4	37.0	21.1	15.2	29.3	24.0	17.6	8.1	17.8	8.1	11.1	28.1	22.7	44.2	1
Ni	μg/g	201	61	96	<150	280	152	196	<200	174	45	218	43	83	339	308	213	2
Sr	μg/g	135	107	159	129	153	155	155	117	147	159	198	161	162	141	152	106	2–15
Zr	μg/g	31	110	15	193	30	36	96	42	30	24	153	27	94	42	43	113	5–50
Ba	μg/g	30	73	24	149	34	32	80	33	31	26	131	27	78	46	43	72	2–16
La	μg/g	2.18	7.31	1.37	12.7	2.63	2.57	6.04	2.03	2.21	1.80	11.11	2.15	6.78	2.52	3.38	6.21	1
Ce	μg/g	5.69	19.5	3.57	34	6.51	6.54	14.2	5.67	5.72	4.72	29.0	5.38	17.2	6.72	8.54	15.6	2
Nd	μg/g	3.38	11.9	2.05	23	3.97	4.14	10.3	4.86	3.54	2.99	17.2	3.31	10.4	4.38	5.01	9.7	10
Sm	μg/g	0.99	3.52	0.66	7.45	1.20	1.20	3.59	2.01	1.12	0.86	5.04	0.98	3.02	1.35	1.54	3.09	1
Eu	μg/g	0.687	0.870	0.845	1.25	0.825	0.804	1.07	0.836	0.801	0.839	1.38	0.860	0.941	0.804	0.817	0.813	1
Tb	μg/g	0.21	0.75	0.14	1.82	0.25	0.25	0.88	0.56	0.25	0.18	0.98	0.20	0.61	0.31	0.32	0.65	1
Yb	μg/g	0.85	2.73	0.58	6.43	0.95	1.01	3.45	2.90	0.99	0.70	3.25	0.78	2.22	1.10	1.21	2.41	1
Lu	μg/g	0.128	0.39	0.081	0.92	0.134	0.145	0.50	0.42	0.140	0.099	0.45	0.110	0.30	0.157	0.173	0.35	1
Hf	μg/g	0.76	2.58	0.45	5.2	0.86	0.89	2.51	1.34	0.84	0.62	3.86	0.66	2.31	0.95	1.16	2.27	1

(*Continued*)

105

Table 6.2. (Continued)

	Unit	ALH A81005	EET 87/96	GRA 06157	LAP N=6	LAR 06638	MAC 88104/5	MET 01210	MIL 05035	MIL 07006	MIL 090034	MIL 090036	MIL 090070	MIL 090075	PCA 02007	QUE 93/94	QUE 94281	%±*
Ta	µg/g	0.09	0.32	0.07	0.73	0.11	0.11	0.23	0.13	0.10	0.08	0.44	0.08	0.26	0.17	0.15	0.29	1–10
Ir	ng/g	6.7	0.6	3.2	<8	12.3	6.8	7.1	<8	6.6	1.9	8.7	2.4	3.0	16.8	15.4	6.6	3
Au	ng/g	2.2	0.4	0.8	<8	3.7	2.8	1.9	<5	1.6	0.2	1.8	0.0	0.2	6.7	4.8	1.7	25
Th	µg/g	0.31	0.92	0.23	2.17	0.41	0.39	0.85	0.37	0.36	0.28	1.65	0.32	1.11	0.44	0.52	0.91	2
U	µg/g	0.103	0.26	0.069	0.54	0.22	0.102	0.28	0.11	0.098	0.073	0.46	0.126	0.33	0.126	0.14	0.23	4–50
INAA	mg			57	1271				423	153	481	556	318	326				
sources of data		5, 14, 18, 24, 29, 34, 37	8, 15, 20, 31, 35, 36	1	1, 2, 6, 11	1	9, 13, 16, 25, 30, 37, 37	4, 7, 13, 22	1, 12, 27	1	1	1	1	1	7, 13, 21	1, 17, 19, 23, 26, 28, 32, 33, 38	3, 7, 10, 15, 20	

Note: FeO_T = Total Fe as FeO; Mg' = Mole % $MgO/(MgO+FeO_T)$;

*Analytical uncertainties for data of this lab: 95% confidence limits, in percentage of mean value, based on counting statistics and number of subsamples analyzed (INAA data) or EPMA spots analyzed (SiO_2, TiO_2, Al_2O_3, CaO, K_2O, and P_2O_5). Uncertainties associated with sampling are considerably greater (e.g., *Korotev et al.,* 2009; *Korotev*, 2012).

Sources of data: 1. This work. 2. *Anand et al.* [2006]. 3. *Arai and Warren* [1999]. 4. *Arai et al.* [2010]. 5. *Boynton and Hill* [1983]. 6. *Day et al.* [2006a]. 7. *Day et al.* [2006b]. 8. *Dreibus et al.* [1996]. 9. *Jolliff et al.* [1991]. 10. *Jolliff et al.* [1998]. 11. *Joy et al.* [2008]. 12. *Joy et al.* [2010a]. 13. *Joy et al.* [2006b]. 14. *Kallemeyn and Warren* [1983]. 15. *Karouji et al.* [2002]. 16. *Koeberl et al.* [1996]. 17. *Koeberl et al.* [1991a]. 18. *Korotev et al.* [1983]. 19. *Korotev et al.* [1996]. 20. *Korotev et al.* [2003b]. 21. *Korotev et al.* [2006]. 22. *Korotev et al.* [2009b]. 23. *Kring et al.* [1995]. 24. *Laul et al.* [1983]. 25. *Lindstrom et al.* [1991a]. 26. *Lindstrom et al.* [1995]; 27. *Liu et al.* [2009]; 28. *Nishiizumi et al.* [1996]. 29. *Palme et al.* [1983]. 30. *Palme et al.* [1991]. 31. *Snyder et al.* [1999]. 32. *Spettel et al.* [1995]. 33. *Thalmann et al.* [1996]. 34. *Verkouteren et al.* [1983]. 35. *Warren and Kallemeyn* [1989]. 36. *Warren and Kallemeyn* [1991a]. 37. *Warren and Kallemeyn* [1991b]. 38. *Warren et al.* [1995].

subsamples are presented in some figures. After INAA, we pulverized two or three representative INAA subsamples in an agate mortar and pestle, fused the resultant powders into glass with a Mo strip heater, and analyzed the glasses by EPMA to determine concentrations of major elements [*Korotev et al.*, 2009b]. For those meteorites for which literature data are available, means calculated from all data are presented in Table 6.2.

Also presented here are petrographic descriptions of ANSMET lunar meteorites. For "historical" lunar meteorites (pre-2001), the descriptions are culled and summarized from literature sources. For more recently recovered meteorites, particularly for meteorites found since 2006, more detailed petrography is provided from our own research on the meteorites. The petrographic descriptions are based on optical, electron, and x-ray imaging of polished thin sections of the meteorites, coupled with quantitative mineral analyses obtained by wavelength-dispersive EPMA [*Zeigler et al.*, 2005].

6.3. LUNAR FRAGMENTAL AND REGOLITH BRECCIAS

The lunar crust has been battered by countless impacts of very large to very small meteorites since its formation 4.5 billion years ago. The Moon has virtually no atmosphere, so asteroidal materials impact the Moon at very high velocities, tens of kilometers per second. The shock and heat of meteorite impacts both break large rocks into smaller rocks and fuse small rocks into larger rocks. As a consequence, nearly all rocks from the lunar highlands are breccias: rocks consisting of fragments of older rocks. Most brecciated lunar meteorites are fragmental and regolith breccias.

Fragmental breccias consist of material of the megaregolith (upper few kilometers of unconsolidated material) that has been lithified by shock compression and sintering during meteorite impacts. Regolith breccias are fragmental breccias consisting of fine-grained material ("soil") of the upper few meters of the Moon that has been mixed by many small impacts. Regolith breccias are recognized by the presence of lithologies that formed at (e.g., agglutinates, small glassy breccias formed by impacts of micrometeorites) or above (volcanic and impact glass spherules) the lunar surface [*Stöffler et al.*, 1980; *Stöffler and Grieve*, 2007]. Most brecciated lunar meteorites are regolith breccias derived from material that was within about 3 m of the lunar surface [*Warren*, 1994]. The relative abundance of regolith breccias among the lunar meteorites is greater than that among *Apollo* samples. There are at least three possible explanations for this discrepancy: (1) astronaut selection bias, (2) differences in coherency between rocks of the lunar regolith and the subset that can survive being launched from the Moon and landing on Earth,

and (3) the possibility that some meteorites were lithified by the impact that accelerated them to lunar escape velocity [*Warren*, 2001].

For no lunar meteorite has the source crater been identified, although from the first lunar meteorite to be recognized it has been fashionable, if not mandatory, to speculate about possible craters or general source locations on the basis of mineralogy and composition [e.g., *Palme et al.*, 1983; *Ryder and Ostertag*; 1983; *Pieters et al.*, 1983; *Anand et al.*, 2006; *Joy et al.*, 2008; *Arai et al.*, 2010]. Most lunar meteorites were launched by impacts making craters less than a kilometer in diameter [*Warren*, 1994; *Head*, 2001; *Basilevsky et al.*, 2010]; thus, locating the source crater of a lunar meteorite is a challenging assignment. Perhaps the best constrained are SaU (Sayh al Uhaymir) 169 from Oman [*Gnos et al.*, 2004], NWA (Northwest Africa) 4472/4485 [*Joy et al.*, 2011], and Dhofar 1442 from Oman [*Korotev*, 2012]. Because of their high concentrations of Th (Figure 6.4), the source craters of these meteorites are limited to only a few regions within the Procellarum KREEP Terrane (Figure 6.2) [*Lawrence et al.*, 2000].

In at least two respects, the fact that most lunar meteorites are regolith breccias makes them especially valuable. First, regolith breccias consist of well-mixed, near-surface materials. We can reasonably assume, on the basis of our experience with *Apollo* regolith samples, that their compositions represent the surface of the Moon where they originated on the scale of a few kilometers at a minimum, for example, the *Apollo* 15 or 17 sites, and more realistically on a scale of a few tens of kilometers, for example, the *Apollo* 11, 12, or 16 sites [*Warren*, 1994; *Korotev et al.*, 2003a]. Second, although we do not know the precise point of origin for any lunar meteorite, we do know, to first order, that lunar meteorites come from randomly distributed locations on the lunar surface [*Warren*, 1994, *Korotev et al.*, 2009b]. Effects of spatially nonuniform cratering rates on the Moon are not large [*Warren*, 1994; *Gallant et al.*, 2009; *Le Feuvre and Wieczorek*, 2008]. These two features have led to use of the mean composition of the feldspathic lunar meteorites as an estimate of the typical composition of the lunar highlands surface and, by inference, the upper crust [*Palme et al.*, 1991; *Warren*, 1994; *Korotev et al.*, 1996, 2003a; *Warren*, 2005] and to providing "ground truth" for orbital remote sensing, particularly for Lunar Prospector [*Korotev et al.*, 2003a; *Gillis et al.*, 2004; *Korotev*, 2005; *Warren*, 2005].

Regolith exposed at the surface of the Moon undergoes a variety of space weathering effects and "matures" with time [e.g., *McKay et al.*, 1991; *Lucey et al.*, 2006]. Consequences of space weathering include decrease in mean grain size, increase in the relative abundance of agglutinates, increase in the relative abundance of solar wind–implanted gases such as ^{36}Ar, increase in the

Figure 6.5. The thick, vesicular fusion crust on regolith breccia QUE 93069. The cube is 1 cm in dimension. NASA photo.

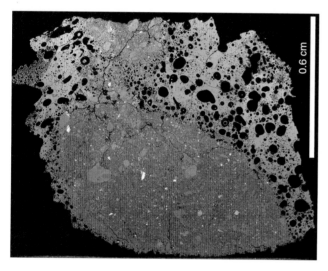

Figure 6.6. BSE (backscattered electron) image of a thin section of PCA 02007, a regolith breccia. At the top of the image is the thick, vesicular fusion crust caused largely by release, during atmospheric entry, of gases implanted by the solar wind while the material of the meteorite was fine-grained regolith on the lunar surface. The lower part of the image shows typical regolith-breccia texture. In BSE images, brightness increases with mean atomic mass. The darkest areas are rich in plagioclase (Al) and the lightest areas are rich in mafic (Fe-bearing) minerals.

concentration of siderophile elements like Ni and Ir, which are mainly derived from impacts of micrometeorites, and increase in the magnitude of the ferromagnetic resonance parameter I_S/FeO [*McKay et al.*, 1974, 1986; *Morris*, 1978; *Korotev et al.*, 2006]. Most samples of regolith fines collected on the *Apollo* missions are submature ($30 < I_S/FeO < 59$) or mature ($I_S/FeO \geq 60$) [*Morris*, 1978]. Most regolith breccias from *Apollo* 15 and 16, however, are immature ($I_S/FeO < 29$) [*McKay et al.*, 1986]. Unfortunately, I_S/FeO has been measured on only three lunar meteorites. ALH A81005 ($I_S/FeO = 5$) [*Morris*, 1983] and MAC 88105 ($I_S/FeO = 0.8$) [*Lindstrom et al.*,

1995] are highly immature. QUE 93069 was formed from submature regolith ($I_S/FeO = 34$) [*Lindstrom et al.*, 1995].

Fusion crusts of nearly all stony meteorites are vesicular from release of gases at the surface of the meteorite during atmospheric entry [*Genge and Grady*, 1999]. A prominent characteristic of several ANSMET lunar regolith breccias is the presence of thick and highly vesicular fusion crusts. These vesicular crusts are most apparent on those meteorites that have high abundances of gaseous solar wind–implanted elements (e.g., H, He, N, Ne, Ar), most notably ALH A81005, QUE 93069/94269 (Figure 6.5), QUE 94281, and PCA 02007 (Figure 6.6). (Vesicular fusion crusts are seldom seen on hot-desert meteorites because the crusts on those meteorites have largely been eroded away by dust-bearing wind.) As *Thaisen and Taylor* [2009] warn, however, unbrecciated, basaltic lunar meteorites LAP 02005 and MIL 05035 also have vesicular fusion crusts, and these are not likely due to solar wind–implanted gases. Nevertheless, it is among the regolith-breccia meteorites where thick, vesicular fusion crusts are most evident.

6.4. COMPOSITIONAL SYSTEMATICS

To a good first approximation, compositions of polymict lunar samples (breccias and regolith) can be modeled as mixtures of three classes of material [*Korotev*, 2005; *Lucey et al.*, 2006]: rocks of the Feldspathic Highlands Terrane [*Jolliff et al.*, 2000], rocks of the Procellarum KREEP Terrane [*Jolliff et al.*, 2000], and rocks (and pyroclastic glass beads) from the basaltic maria, which occur in both terranes. Rocks of the Feldspathic Highlands Terrane (Figure 6.2) are mainly anorthosites, noritic anorthosites, troctolitic anorthosites, and their brecciated derivatives. They are characterized by low concentrations of elements associated with mafic minerals (e.g., Fe and Sc; Figures 6.3 and 6.4), high concentrations of Al_2O_3 and CaO (Figure 6.7), and low concentrations of incompatible elements (e.g., Sm and Th; Figures 6.3 and 6.4). Materials of the Procellarum KREEP Terrane are typically noritic in composition, with intermediate concentrations of FeO and Al_2O_3 but high concentrations of incompatible elements. Rocks and glass from the basaltic maria are rich in FeO and Sc but poor in Al_2O_3, with low to intermediate concentrations of incompatible elements. This *Apollo* view of lunar meteorites may be oversimplified in that some lunar meteorites of intermediate FeO concentration (Figure 6.4) may not be mixtures of feldspathic material and mare basalt. Rather, they may represent mixtures that include moderately mafic material of the Feldspathic Highlands Terrane unlike that sampled by the *Apollo* missions [*Korotev et al.*, 2009b].

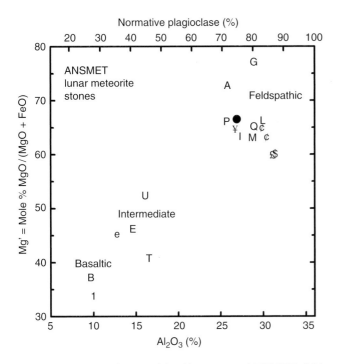

Figure 6.7. *Mg'* varies considerably among ANSMET feldspathic lunar meteorites. Each point represents a named stone; see Table 6.1 for symbol key. For reference, the mean composition of typical mature regolith from *Apollo* 16 is represented by the filled circle [*Korotev*, 1997]. The normative plagioclase axis assumes an An₉₆ composition for the plagioclase.

6.5. ROSTER OF ANSMET LUNAR METEORITES

Below, we discuss the ANSMET lunar meteorites in order of discovery. Few data are available for the most recently discovered meteorites. In contrast, we are aware of 38 peer-reviewed papers reporting data for Allan Hills 81005.

6.5.1. Allan Hills A81005

ALH A81005, the first meteorite recognized to be from the Moon [*Marvin*, 1983; *Cassidy*, 2003], was collected nine years after the last *Apollo* mission (Plate 64). The rock is a regolith breccia from the lunar highlands. Like most feldspathic regolith breccias, lithic clasts are mainly other types of breccias: granulitic breccias, impact-melt breccias, and monomict breccias all derived from anorthosites, noritic anorthosites, troctolitic anorthosites, and their more mafic variants. The meteorite also contains rare clasts of mare basalts and impact-glass spherules all set in a dark matrix of minerals clasts and impact-derived vesicular glass [*Korotev et al.*, 1983; *Kurat and Brandstätter*, 1983; *Marvin*, 1983; *Ryder and Ostertag*, 1983; *Simon et al.*, 1983; *Treiman and Drake*, 1983; *Warren et al.*, 1983; *Goodrich et al.*, 1984; *Gross and Treiman*, 2010].

Petrographically the most unusual feature is the wide range of magnesium numbers (*Mg'*=mole% MgO/(MgO+FeO)) in the mafic minerals associated with the feldspathic lithologies, 36–87 [*Goodrich et al.*, 1984; *Gross and Treiman*, 2010; *Gross et al.*, 2014].

Early compositional studies [*Boynton and Hill*, 1983; *Kallemeyn and Warren*, 1983; *Korotev et al.*, 1983; *Laul et al.*, 1983; *Palme et al.*, 1983] noted that ALH A81005 had low concentrations of elements associated with KREEP compared to polymict materials (regoliths and breccias) from the *Apollo* 16 site (Figure 6.3), the only *Apollo* mission to have landed in the feldspathic highlands at a point distant from basalt-filled impact basins (Figure 6.1). As more feldspathic lunar meteorites were found and data from the Lunar Prospector mission [*Lawrence et al.*, 2000] were assimilated, it became apparent that the meteorites are, in fact, typical of the feldspathic highlands, and it is the *Apollo* 16 site that is anomalous as a result of its proximity to the Procellarum KREEP Terrane (Figure 6.2) and the presence of high-Th Imbrium ejecta at the site [*Korotev et al.*, 2003a] (Figure 6.4). The meteorite remains among the most magnesian (high *Mg'*) of the feldspathic lunar meteorites (Figure 6.7) [*Korotev*, 2012; *Gross et al.*, 2014].

6.5.2. Elephant Moraine 87521 and 96008

EET 87521 was originally classified as a brecciated eucrite but was later recognized to be a lunar meteorite composed mainly of brecciated VLT (very low Ti, <1% TiO₂) mare basalt [*Delaney*, 1989; *Warren and Kallemeyn*, 1989]. A paired stone, EET 96008, was found nine years later. Prior to EET 87/96, VLT basalt was known only from small fragments in the *Apollo* 17 and *Luna* 24 regoliths [*Vaniman and Papike*, 1977; *Ma et al.*, 1978]. The VLT basalt of EET 87/96 has a crystallization age of about 3.5 Ga (Figure 6.8; Plate 65).

Early works classified the meteorite as a fragmental breccia, but the presence of rare glass spherules and agglutinates [*Mikouchi*, 1999] confirm that it is a highly immature regolith breccia. The meteorite is characterized by coarse pyroxenes with exsolution lamellae and intergrowths of silica, fayalite, and hedenbergite; both features are rare in *Apollo* basalts [*Anand et al.*, 2003; *Warren and Kallemeyn*, 1989; *Takeda et al.*, 1992]. The meteorite is also unusual in that there are no breccias in the *Apollo* collection that, like EET 87/96, consist mainly of mineral grains from coarse-grained volcanic rocks and have such little glass [*Korotev et al.*, 2003b].

Although petrographic evidence for feldspathic clasts is rare [*Warren and Kallemeyn*, 1989; *Anand et al.*, 2003; *Takeda et al.*, 1992; *Warren and Ulff-Møller*, 1999], the FeO concentration of the meteorite

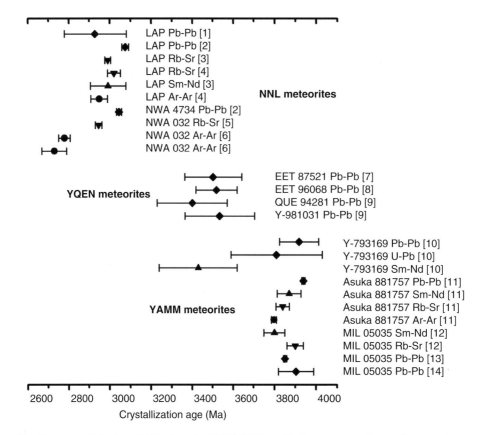

Figure 6.8. Crystallization ages of the basalt lithology in ANSMET basaltic lunar meteorites and comparison to lunar meteorites from elsewhere that are likely or possible launch pairs. At least three different age groups are indicated. Symbol shapes represent different isotopic techniques. Sources of data: [1] *Anand et al.*, 2006; [2] *Wang and Hsu*, 2010; [3] *Rankenburg et al.*, 2007, [4] *Nyquist et al.*, 2005; [5] *Borg et al.*, 2009; [6] *Fernandes et al.*, 2003; [7] *Terada et al.*, 2005; [8] *Anand et al.*, 2003; [9] *Terada et al.*, 2006; [10] *Torigoye-Kita et al.*, 1995; [11] *Misawa et al.*, 1993; [12] *Nyquist et al.*, 2007; [13] *Zhang et al.*, 2010; and [14] *Terada et al.*, 2007). The YQEN also group includes NWA 4884 [*Korotev et al.*, 2009b], for which there are no isotopic data.

is at the low end of the range for mare basalts from the *Apollo* and *Luna* missions (Figure 6.4), suggesting that the meteorite may contain a non-negligible proportion of nonmare material. *Arai and Warren* [1999] suggest 5%–10%. If the proportion is even that great, then the nonmare material is petrographically cryptic, presumably because it is finer grained than the mare material. Compositionally, the meteorite is heterogeneous and consists of two mafic components, (1) a primitive component with 12%–15% FeO, high *Mg'*, and low concentrations of incompatible elements; and (2) an evolved component with >24% FeO, low *Mg'*, and moderate concentrations of incompatible elements [*Lindstrom et al.*, 1999; *Korotev et al.*, 2003b]. Curiously, concentrations of solar wind–implanted gases [*Eugster et al.*, 1996, 2000] are two to three times greater in EET 87/96 than in MAC 88105 [*Eugster et al.*, 1991; *Palme et al.*, 1991], which is more obviously a regolith breccia. EET 87/96 does, however, have the lowest concentration of Ir of the brecciated ANSMET meteorites (Figure 6.9), indicating little input of asteroidal meteorite material.

6.5.3. MacAlpine Hills 88104 and 88105

The two MAC 88104/05 stones (Plate 68) were found close together in the field and are indistinguishable compositionally and petrographically [*Jolliff et al.*, 1991; *Lindstrom et al.*, 1991a,b; *Neal et al.*, 1991]. The meteorite is a feldspathic regolith breccia (Figure 6.10). The bulk composition is typical of that for a feldspathic lunar meteorite (Figs. 6.3 and 6.4), but *Mg'* is at the ferroan end of the range (Figure 6.7). Lithic clasts are dominated by granulitic and impact-melt breccias, but the meteorite shows a wide variety of relict igneous clasts nearly all associated with the ferroan-anorthositic suite of lunar plutonic rocks [*Warren*, 1990]. *Cohen et al.* [2005] have dated (^{40}Ar–^{39}Ar) nine clasts of feldspathic and K-poor impact-melt breccia in MAC 88105. Ages ranged from 2.5 ± 1.5 Ga to 3.9 ± 0.1 Ga, all similar to or younger than the ubiquitous 3.8–3.9 Ga obtained for the K-rich impact-melt breccias in the *Apollo* collection [*Haskin et al.*, 1998]. The meteorite also contains impact glasses mainly of nonmare origin and rare fragments of mare basalt [*Delano*, 1991;

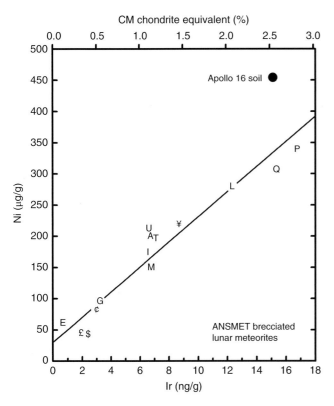

Figure 6.9. Concentrations of Ir and Ni in ANSMET brecciated lunar meteorites and comparison to mature *Apollo* 16 regolith [*Korotev*, 1997]. Igneous and plutonic lunar rocks have effectively zero ng/g Ir; all of the Ir and most of the Ni in lunar regoliths and breccias derive from asteroidal meteorites that have impacted to Moon. Much of the Ir in the lunar regolith derives from micrometeorites of CM chondrite composition [*Wasson et al.*, 1975]. At *Apollo* 16, a significant fraction also derives from an iron meteorite component with a non-chondritic Ni/Ir ratio [*Korotev*, 1997]. For reference, the diagonal line is a mixing line between an assumed lunar component with 0 ng/g Ir and 30 μg/g Ni and CM chondrites [*Wasson and Kallemeyn*, 1988]. Data are from Table 6.2. Symbol key is in Table 6.1.

Figure 6.10. Sawn faces of two regolith breccias from the lunar highlands, lunar meteorite MAC 88105 (top) and *Apollo* 16 sample 60019 (bottom). Note that the clasts show no size sorting or preferred orientation, they are only moderately rounded, and their aspect ratios are rarely greater than 3:1. Both rocks are highly coherent. Clasts and matrix have equal strength as fractures pass from matrix through clasts without deviation. (Many *Apollo* regolith breccias are highly friable, however, and clasts can easily be separated from the matrix.) The clasts are mainly impact-melt breccias and granulitic breccias. Although the clasts are lighter colored than the matrix, they are not more feldspathic. The matrix is dark because it is fine grained and space weathered. NASA photos.

Jolliff et al., 1991; *Koeberl*, 1991a; *Lindstrom et al.*, 1991b; *Neal et al.*, 1991; *Palme et al.*, 1991; *Takeda et al.*, 1991; *G.J. Taylor*, 1991; *Warren and Kallemeyn*, 1991a; *Robinson and Treiman*, 2010]. *Joy et al.* [2010b] report the occurrence of a rare magnesian troctolitic clast in MAC 88105. Dhofar 1428 from Oman is similar in composition to MAC 88104/5 and may be a launch pair [*Korotev*, 2012], that is, both rocks may have been ejected from the Moon by a single impact but fell to Earth at different times and places. MAC 88105 has the lowest concentrations of solar wind implanted gases among the five ANSMET meteorites for which gas concentrations have been measured. CRE (cosmic-ray exposure) data indicate a lunar ejection age of 275 ± 15 ka and a long terrestrial age of 230 ± 70 ka [*Nishiizumi et al.*, 1996; Figure 6.11].

6.5.4. Queen Alexandra Range 93069 and 94269

QUE 03069/94269 is a feldspathic, glass-rich regolith breccia. The two stones, found a year apart, are identical in composition [*Warren et al.*, 2005] and have the same CRE history [*Nishiizumi et al.*, 1996; *Polnau and Eugster*, 1998]. The bulk composition is typical of that for feldspathic lunar meteorites (Figure 6.3). The rock is dominated by lithic and mineral clasts from feldspathic rocks; clasts of mafic nonmare rocks and mare basalt are rare. *Cohen et al.* [2005] dated five clasts of feldspathic and K-poor impact-melt breccia in QUE 93069. Similar to MAC 88105, all are younger than 3.7 ± 0.5 Ga.

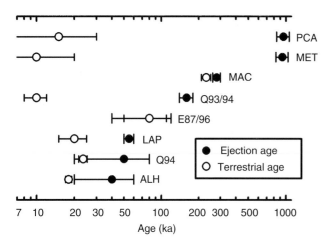

Figure 6.11. Ejection ages and terrestrial ages for ANSMET lunar meteorites ALH A81005, QUE 94281, LAP 02205 and pairs, EET 87521/96008, QUE 93069/94269, MAC 88015, MET 01210, and PCA 02207 [after *Nishiizumi et al.*, 1996; *Nishiizumi*, 2003]. Data are from *Jull* [2006], *Nishiizumi* [2003], *Nishiizumi and Caffee* [1996], *Nishiizumi et al.* [1991, 1996, 1999, 2006], and *Thalmann et al.* [1996].

The meteorite is rich in impact-glass spherules and fragments. Most glasses have feldspathic compositions, but a small fraction have mafic composition, some corresponding to medium-Ti mare basalt. Glass concentrations have a wide range of *Mg'* (55–96) and some have high-aluminum, silica-poor [*Naney et al.*, 1976] compositions. Agglutinate clasts are present and grains of FeNi metal are abundant [*Bischoff*, 1996; *Grier et al.*, 1995; *Koeberl et al.*, 1996, *Korotev et al.*, 1996; *Robinson and Treiman*, 2010; *Spettel et al.*, 1995; *Warren and Kallemeyn*, 1995; *Warren et al.*, 2005].

The most unusual aspect of the meteorite is that the regolith of which it is composed is more mature than that of most meteoritic regolith breccias. As a consequence of the high surface exposure, concentrations of solar wind–implanted noble gases [*Thalmann et al.*, 1996], siderophile elements (Figure 6.9) [*Korotev et al.*, 2006], and impact-glass spherules and fragments are all at the high end of the ranges observed among lunar meteorites. The meteorite also has a strikingly vesicular fusion crust from release of the solar wind–implanted gases from the surface melt during atmospheric entry (Figure 6.5).

6.5.5. Queen Alexandra Range 94281

QUE 94281 is a clast-rich regolith breccia consisting of both basaltic and feldspathic material. "The matrix is very heterogeneous. In most areas, it is a fine-grained, glass-poor fragmental material, but several large regions—up to 5 mm across—have extremely fine-grained (glassy aphanitic) dark-brown matrix along with relatively low fragment/matrix ratio" [*Arai and Warren*, 1999].

The meteorite contains agglutinates and abundant glass spherules. Lithic clasts are predominantly VLT basalt or gabbro, but feldspathic clasts are also present. Pyroxene grains in the basalt are coarse grained and finely exsolved, suggesting a hypabyssal setting [summary from *Jolliff et al.*, 1998; *Arai and Warren*, 1999; and *Korotev et al.*, 2003b].

There is no evidence to suspect that QUE 94281 is paired with QUE 93069/94269 (Figures 6.3, 6.7, and 6.11). There is, however, strong compositional, mineralogical, textural, and CRE evidence that QUE 94281 is launch paired with paired stones Yamato 793274 and Yamato 981031 [*Dreibus et al.*, 1996; *Jolliff et al.*, 1998; *Kring et al.*, 1996; *Polnau and Eugster*, 1998]. *Arai and Warren* [1999] make the strongest argument by showing that the same two distinctive types of mare volcanic glass dominate the glass populations in each of the two meteorites. QUE 94281 and Yamato 79/98 are similar in composition (Figure 6.12). There is also strong evidence that the "YQ" meteorites [*Arai and Warren*, 1999] are launch paired with EET 87/96. Both compositionally and texturally, the coarse-grained basalt or gabbro component of the YQ meteorites is similar to that of EET 87/96 [*Kring et al.*, 1996; *Korotev et al.*, 2003b] (Figure 6.12). Crystallization ages of the basalt lithologies are the same, within uncertainty (Figure 6.8). Finally, NWA 4884 from the Sahara Desert is identical in composition to QUE 94281 for lithophile elements [*Korotev et al.*, 2009b] (Figure 6.12). The preliminary petrographic description of NWA 4884 is consistent with that of QUE 94281 [*A. Irving and S. Kuehner*, in *Connolly et al.*, 2007].

On any of a number of two-element plots, the "YQEN" meteorites (Yamato 79/98, QUE 94281, EET 87/96, and NWA 4884) plot along a mixing line between VLT basalt and a nonmare component that is richer in incompatible elements, Na_2O, and TiO_2 and is more magnesian than most feldspathic lunar meteorites (Figures 6.12 and 6.13) [*Korotev et al.*, 2003b]. On the basis of compositional mass balance, *Jolliff et al.* [1998] estimated 46% nonmare material in QUE 94281, whereas *Arai and Warren* [1999] estimated ~33%; the difference reflects different assumptions about the compositions of the mare and nonmare components.

EET 87/96 is richer in the mare component than the other YQEN meteorites. As with the YAMM meteorites (next section), the YQEN meteorites appear to originate from a place where feldspathic ejecta overlies a deposit of mare basalt. EET 87/96 represents deeper, immature, basaltic regolith, whereas QUE 94281, Yamato 793274/981031, and NWA 4884 represent a shallower, more mature regolith consisting of basaltic material, feldspathic material, and perhaps some KREEP-bearing material.

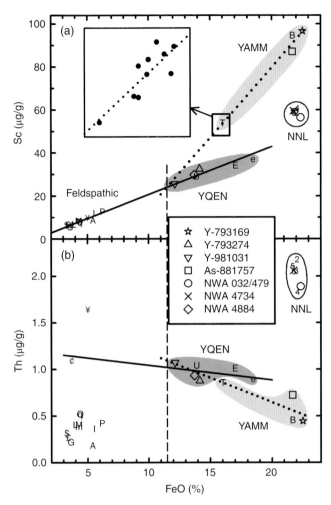

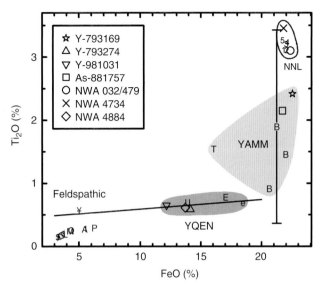

Figure 6.13. Comparison of TiO_2 concentrations in ANSMET (Table 6.1) basaltic meteorites (>12% FeO) to likely launch pairs (meteorites in legend). The diagonal line is a least-squares fit to the YQEN data showing that, with <1% TiO_2, the basaltic component is a VLT (very-low-Ti) basalt and that the feldspathic component is somewhat richer in Ti than most feldspathic lunar meteorites. For MIL 05035, three points are plotted (B), representing the data (low to high TiO_2) of *Joy et al.* [2008], *Liu et al.* [2009], and this work. The error bar for our data represents the 95% confidence interval based on analysis of 11 subsamples. The meteorite is very coarse grained.

6.5.6. Meteorite Hills 01210

MET 01210 is a glassy-matrix regolith breccia dominated by low-Ti or VLT mare basalt. *Huber and Warren* [2005] note that "roughly half of the pyroxenes display exsolution to extents that vastly exceed the norm (<<0.1 µm typical lamella width) for *Apollo* mare basalts" and that "MET 01210 thus joins lunaites Y-793274, QU94281, EET87521, and (to a lesser degree) As-881757, in having undergone a remarkable extent of slow cooling or annealing, in comparison to *Apollo* mare basalts." There is no evidence that MET 01210 is related to the YQEN clan, however. CRE data for MET 01210 are distinct from that of QUE 94281 and EET 87/96 (Figure 6.11), and the compositions are different (Figures 6.12 and 6.13). The meteorite contains some feldspathic material in the form of impact-melt and granulitic breccias, symplectites from breakdown of pyroxferroite, and rare agglutinates and spherules [*Arai et al.*, 2005; *Huber and Warren*, 2005; *Zeigler et al.*, 2005; *Day et al.*, 2006b; *Joy et al.*, 2006a; 2010a].

Arai et al. [2005] note that a basalt clast in MET 01210 is very similar to that of gabbroic lunar meteorite Asuka 881757, collected in Antarctica 2850 km away, and that the two meteorites might be "petrogenetically related."

Figure 6.12. (a) Sc/FeO is nearly constant among feldspathic and very-low-Ti dominated, basaltic lunar meteorites (YQEN field). The diagonal solid line is a least-squares fit to the "feldspathic" and seven YQEN points. The dotted line is a least squares fit to the four YAMM meteorite points. If MET 01210 (T) is a simple binary mixture of mare basalt such as that of the high-FeO YAMM meteorites and some low-FeO, nonmare component, then the low-FeO component has a composition that is not highly feldspathic, having a composition at the intersection of the diagonal lines at 11.5% FeO. The inset shows data for 10 subsamples of MET 01210 [*Korotev et al.*, 2009b]. (b) The implied low-FeO component of MET 01210 (intersection of diagonal dotted line and dashed vertical lines) has about 1.1 µg/g Th, greater than most feldspathic lunar meteorites. The fact that the implied nonmare component plots near the field of the YQEN meteorites is a coincidence in that two launch-pair groups have different CRE exposure histories. The solid diagonal line is a least-squares fit to the seven YQEN points. If these breccias represent simple binary mixtures, then the feldspathic component is also richer in Th (~1.1 µg/g) than most feldspathic lunar meteorites. For the ANSMET meteorites, plot symbols are in Table 6.1. Data for the meteorites in the legend are from *Barrat et al.* [2005], *Fagan et al.* [2002], *Fukuoka* [1990], *Jolliff et al.* [1993], *Koeberl et al.* [1991b, 1993], *Korotev et al.*, [1993, 2009b], *Lindstrom et al.* [1991c], *Warren and Kallemeyn* [1991b, 1993], *Yanai and Kojima* [1991], *Zeigler et al.* [2005], and unpublished data (this lab).

The bulk composition of MET 01210 corresponds to a mixture of Asuka 881757 (also MIL 05035, below) and feldspathic material (Figure 6.12), suggesting that the meteorites are launch paired [*Zeigler et al.*, 2007]. Asuka 881757 is believed to be launch paired with Yamato 793169, collected about 370 km away, on the basis of similar CRE histories [*Nishiizumi et al.*, 1992; *Thalmann et al.*, 1996], chemical composition [*Jolliff et al.*, 1993; *Warren and Kallemeyn*, 1993], and crystallization ages [*Fernandes et al.*, 2009] (Figure 6.8). Crystallization ages of the MET 01210 basalt are consistent with the pairing hypothesis (Figure 6.8) as are preliminary CRE data [*Nishiizumi et al.*, 2006]. *Arai et al.* [2010] review the evidence for the launch pairing of the "YAMM" meteorites (Yamato 793169, Asuka 881757, MET 01210, and MIL 05035) and suggest that MET 01210 derives from a cryptomare, a place where mare volcanic deposits are overlain by feldspathic ejecta from an impact into the highlands [*Head and Wilson*, 1992] and where subsequent smaller impacts mixed the mare and highlands materials.

Arai et al. [2010] estimate that MET 01210 contains 32% nonmare material, whereas *Zeigler et al.* [2007] estimate 35%. Both estimates assume that the nonmare material is feldspathic. However, if MET 01210 is a simple binary mixture, then the low-FeO component is unlikely to have less than about 11.5% FeO (Figure 6.12), an anorthositic norite composition. A scenario that accounts for the data is that MET 01210 is a three-component mixture of (1) a high-Sc basalt like Asuka 881757 and a regolith with 11.5% FeO, which itself is a mixture of (2) feldspathic material and (3) a low-Sc (likely VLT) basalt. The fact that the MET 01210 subsample data do not scatter in a triangular pattern but trend along the YAMM mixing line (Figure 6.12, inset) argues that the feldspathic and low-Sc mare component are well mixed, that is, a regolith. Such regolith-rock mixing relationships are seen in other sample sets [*Jolliff et al.*, 1996; *Korotev*, 1997].

6.5.7. Pecora Escarpment 02007

PCA 02007 is a feldspathic regolith breccia with clasts set in a matrix that is partially glassy and partially fragmental. Most of the clasts are feldspathic. Our own interpretation is that "the lithic-clast population of PCA 02007 … consists entirely of clasts that have been modified by impacts, that is, there are no igneous lithic clasts, only clasts that have been extensively metamorphosed. The 100 clasts of this study fall into the following textural categories: glassy impact-melt breccias ($N = 84$), quenched mafic breccias (3), crystalline impact-melt breccias (8), regolith breccia (5), and agglutinates (3 or more)" [*Korotev et al.*, 2006]. Others, however, have interpreted the textures of some clasts in terms of plutonic highlands rocks [*L. A. Taylor et al.*, 2004; *Day et al.*, 2006b; *Joy et al.*,

2006a] and have observed clasts of mare basalt [*L. A. Taylor et al.*, 2004; *Joy et al.*, 2010a, *Vaughan et al.*, 2011]. *Day et al.* [2006b] found an unusual lithic clast from a chondrite.

Perhaps because of the mare component (basalt clasts, impact glass, pyroclastic glass), PCA 02007 is compositionally at the mafic end of the range of feldspathic lunar meteorites (Figs. 6.3 and 6.4). Among ANSMET lunar meteorites it is the richest in siderophile elements (Figure 6.9), probably because the regolith from which it formed is moderately mature [*Korotev et al.*, 2006; *Joy et al.*, 2010a]. The meteorite has a thick, vesicular fusion crust (Figure 6.6). There are no data for any of the usual regolith-maturity parameters, however.

PCA 02007 and MET 01210 have the same ejection age (Figure 6.11), suggesting that PCA 02007 may be another member of the YAMM launch-pair group and that PCA-like feldspathic material might the feldspathic component of MET 01210. As noted above, however, the implied feldspathic component of MET 01210 is considerably richer in incompatible elements (~1.5 µg/g Th, Figure 6.12) than PCA 02007 (0.44 µg/g), so we suspect that the meteorites are unrelated.

6.5.8. LaPaz Icefield 02205, 02224, 02226, 02436, 03632, and 04841

The six stones of the LaPaz Icefield lunar meteorite were found over three field seasons (Table 6.1; Plate 67). At 1.226 kg, LAP 02205 is the largest ANSMET lunar meteorite stone. The six stones are essentially identical in composition (e.g., Figures 6.3 and 6.14), texture, and mineralogy.

The meteorite is an unbrecciated low-Ti mare basalt. There are numerous petrographic descriptions [*Righter et al.*, 2005; *Zeigler et al.*, 2005; *Anand et al.*, 2006; *Joy et al.*, 2006b; *Day et al.*, 2006a; *Hill et al.*, 2009]. Modally, the meteorite is 50.2% pyroxene (with a ~3:1 augite to pigeonite ratio), 35.1% plagioclase, 3.1% olivine, 3.7% ilmenite, 3.5% fayalite, 2.1% silica (likely cristobalite), 0.3% chromite, ~1% shock-induced melt veins, and ~1% mesostasis glass [*Zeigler et al.*, 2005; see also *Day et al.*, 2006a]. Minor and trace phases include troilite, ulvöspinel, phosphates (RE-merrillite, fluorapatite), and FeNi metal of nonmeteoritic origin. Mesostasis areas (symplectites) contain pyroxferroite, fayalite, silica, baddelyite, barian K-feldspar, K-Si-rich glasses, and Si-rich glasses. Although unbrecciated, the meteorite is fractured, contains glassy melt veins, and some of the plagioclase is isotropic (maskelynite).

The LAP basalt is similar in composition to the low-Ti basalts of *Apollo* 12 and 15, but in detail it is not identical to any *Apollo* basalt. The crystallization age, ~3.0 Ga (Figure 6.8), is younger than that of the *Apollo*

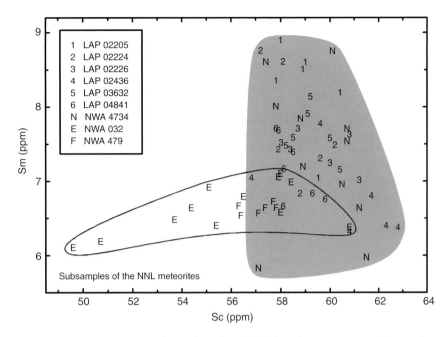

Figure 6.14. Each LAP point represents a 35-mg subsample and each NWA point represents a 25-mg subsample, on average. The six LAP basalts are indistinguishable in composition from each other and from NWA 4734 on any two-element plot such as this. NWA 032 and NWA 479 are paired. NWA 032/479 is richer in olivine than LAP and, consequently, poorer in Sc and richer in Co [*Zeigler et al.*, 2005]. All data are from the Washington University INAA lab [*Fagan et al.*, 2002; *Zeigler et al.*, 2005].

12 and 15 basalts (3.1–3.4 Ga) [Table 6 of *S. R. Taylor*, 1982]. The LAP basalt is all but identical, however, in composition and texture (Figure 6.14) [*Korotev et al.*, 2009a; *Fernandes et al.*, 2009] as well as crystallization age [*Korotev et al.*, 2009a; *Fernandes et al.*, 2009; *Wang and Hsu*, 2010; *Elardo et al.*, 2012, 2013] to 1.37-kg NWA 4734 from Morocco. These data strongly suggest that the LAP basalts and NWA 4734 are launch paired [*Korotev et al.*, 2009a].

Paired stones NWA 032 and NWA 479 have several chemical (Figures 6.12, 6.13, and 6.14) and petrographical features in common with the LAP stones, suggesting that the two meteorites are launch paired [*Jolliff et al.*, 2004; *Zeigler et al.*, 2005]. The crystallization ages are very similar (Figure 6.8) as are the ejection ages [*Nishiizumi et al.*, 2006]. However, the two meteorites have very different values of ε_{Nd}, making any simple genetic relationship unlikely [*Elardo et al.*, 2012, 2013]. Thus, the similarities between LAP and NWA 032/479 may be coincidental.

6.5.9. Miller Range 05035

MIL 05035 is an unbrecciated low-Ti mare basalt with a coarse-grained, gabbroic texture (Plate 66). Pyroxenes range from 2 to 10 mm in size [*Arai et al.*, 2010]. Modally, the meteorite is 54%–62% pyroxene, 25%–36% plagioclase, 6%–7% symplectites, 0.6%–3.5% fayalite, 0.8%–1.0% silica, 0.5%–1.0% ilmenite, 0.5%–0.6% glass, and 0.2%–1.5%

ulvöspinel, with trace amounts of silica, troilite, baddelyite, K- and Ba-rich alkali feldspar, and phosphates [*Joy et al.*, 2008; *Liu et al.*, 2009]. Pyroxenes are strongly shocked and all of the plagioclase has been converted to maskelynite with some recrystallization.

The whole rock TiO_2 concentration of MIL 05035 is not well constrained. For this work we obtained a mean of 1.9% ± 1.5% (95% confidence limits) on a 0.42-g sample (mean of 11 subsamples, Figure 6.13), *Joy et al.* [2008] obtained 0.9% on a 0.31-g sample, and *Liu et al.* [2009] obtained 1.44% on a 1.9-g sample. The mass-weighted mean, 1.45%, is listed in Table 6.2. From modal recombination of the data of *Joy et al.* [2008] and *Liu et al.* [2009] we estimate 1.6% TiO_2. Similarly, MgO concentrations range from 6.1% (this work) to 7.8% [*Joy et al.*, 2008] among the three laboratories. For comparison, *Thaisen and Taylor* [2009], who note that the composition of the fusion crust of MIL 05035 is highly variable from place to place, obtained means and standard deviations of 2.0% ± 0.9% TiO_2 and 5.8% ± 2.4% MgO from 170 analyses of the fusion crust.

It was recognized in the earliest studies of MIL 05035 that it was very similar in texture, mineralogy, and composition to Asuka 881757, collected 2500 km away in Antarctica, and that the two meteorites were thus likely launch pairs [*Arai et al.*, 2007; *Joy et al.*, 2007; *Liu et al.*, 2007; *Korotev and Zeigler*, 2007; *Zeigler et al.*, 2007]. Subsequent studies have shown that the crystallization ages also match (Figure 6.8). Although, as noted above,

Yamato 793169 and MET 01210 are also likely launch pairs to Asuka 881757, neither is as similar to Asuka 881757 as is MIL 05035.

6.5.10. Graves Nunataks 06157

At 0.79 g, GRA 06157 is the smallest lunar meteorite so far recovered. It is a glassy-matrix regolith breccia containing abundant lithic and mineral clasts and a few small glassy spherules (Figure 6.15). All of the lithic clasts identified are granulite breccias. The mineral clast population is dominated by plagioclase, pyroxene, and olivine, with trace amounts of FeTiCr oxides, FeNi metal, and FeS. Our thin section (Figure 6.15) [also *Zeigler et al.*, 2012b] is roughly hemisphere shaped with a 5-mm diameter. It has a discontinuous vesicular fusion crust on the curved side (the exterior surface of the meteorite).

The most abundant type of lithic clasts are highly feldspathic (>90% plagioclase) granulite breccias consisting of large plagioclase grains (An_{94-97}) up to 1 mm in their longest dimension that contain tiny rounded inclusions of pyroxene almost always less than 2 μm wide. Rarely there are strings of larger pyroxene grains (up to 20 μm wide) present along the borders of some plagioclase grains. Augite is the dominant pyroxene ($En_{44}Wo_{42}Fs_{14}$), with rare occurrences of hypersthene ($En_{64}Wo_3Fs_{33}$); there is no compositional difference among the different-sized pyroxenes. Minor olivine grains also occur, with a relatively wide range in compositions (Fo_{60-78}). There are also troctolitic granulite clasts, including the largest lithic clast in the section (~1.5 mm on a side). These clasts contain up to 50% magnesian olivine (Fo_{85-90}), minor (<5%) amounts of magnesian augite ($En_{53}Wo_{37}Fs_9$), nearly end-member magnesian spinel, and trace amounts of bronzite ($En_{77}Wo_2Fs_{20}$). A few small (~150 μm) mafic granulite clasts (plagioclase is <20% by mode) are compositionally similar to the more feldspathic granulite clasts, albeit with slightly more ferroan pyroxene compositions ($En_{39}Wo_{42}Fs_{19}$; $En_{55}Wo_2Fs_{43}$).

Plagioclase compositions fall in a narrow range (An_{94-98}). In contrast, the olivine clasts analyzed span almost the

Figure 6.15. Back-scattered electron image (Figure 6.6) and an RGB elemental x-ray map of a thin section of lunar meteorite GRA 06157. In the x-ray map, areas rich in Al (e.g., plagioclase) are bright red, areas rich in Mg are bright green (e.g., olivine), and areas rich in Fe (e.g., pyroxene or FeNi metal) are bright blue. The scale bar applies only to the BSE image.

entire range of olivine compositions (Fo$_{7-93}$); most clasts are at the Mg-rich end of the range, however. Most pyroxene mineral clasts are exsolved, with augite and orthopyroxene lamellae ranging from submicron all the way up to nearly 100 μm thickness. A single large (200 × 150 μm) basaltic spherule is present. It is partially devitrified with magnesian olivine (Fo$_{78}$) quenching out of the glass. It has a bulk composition very similar to *Apollo* 15 yellow pyroclastic glass [*Shearer and Papike*, 1993].

Our bulk sample is only 57 mg in mass, from which we made two INAA subsamples, both of which we subsequently analyzed for major elements. Despite being a regolith breccia, GRA 06157 has low concentrations of siderophile elements compared to most regolith breccias [*Korotev et al.*, 2009a] (Figure 6.9). GRA 06157 is one of the most feldspathic meteorites studied here (29 wt.% Al$_2$O$_3$). Our sample is distinct in being more magnesian (*Mg'* = 77 ± 4) than both any other lunar meteorite from Antarctica (Table 6.1) and any lunar meteorite composed of feldspathic regolith breccia of which we are aware (Figure 6.7). The magnesian nature is likely caused by the presence of the highly magnesian olivine present in the troctolitic granulite clasts and olivine mineral clasts.

6.5.11. Larkman Nunatak 06638

LAR 06638 is another small (5.3 g) meteorite. The rock is a dark-matrix regolith breccia with millimeter-size light-colored clasts and one large (~1 cm), light-colored breccia clast that represents a significant proportion of the total mass of the meteorite [*Satterwhite and Righter*, 2007]. The large white clast is a fragmental breccia (there is no glassy matrix). The thin section in this study (LAR 06638,13) has both prominent lithologies present: 60% dark-matrix material and 40% light-colored clast material (Figure 6.16) [*Zeigler and Korotev*, 2013]. The dark-matrix area consists of granulite, impact-melt breccia, mineral, and glass clasts set in a glassy matrix, whereas the light-colored clast area consists of granulite and mineral clasts set in a fragmental matrix. The lithic clast population of the dark-matrix area is dominated by feldspathic granulite clasts, with a few feldspathic impact-melt breccia clasts also present. Both the granulite and impact-melt clasts are dominated by calcic plagioclase (An$_{95-99}$; almost always >90% by mode). The granulite clasts contain minor amounts of pyroxene and olivine that range widely in size (1–50 μm; typically <10 μm).

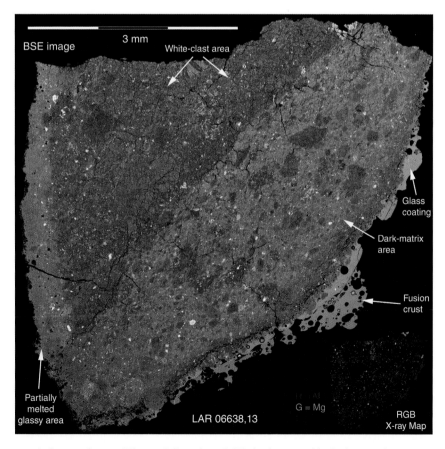

Figure 6.16. Back-scattered electron image (Figure 6.6) and an RGB (red, green, blue) elemental x-ray map of a thin section of lunar meteorite LAR 06638. The scale bar applies only to the BSE image.

Pyroxene compositions are typically high in Ca with a relatively constant Fe:Mg ratio ($En_{45-67}Wo_{9-35}$); minor amounts of low-Ca pyroxene are also observed ($En_{70-77}Wo_{2-4}$). Olivine grains have a restricted compositional range (Fo_{63-76}, most between Fo_{64-68}). Clasts of feldspathic impact-melt breccia typically have subparallel plagioclase laths with very thin "lamellae" (1–2 μm) of mafic glass separating them. The textures and mineral compositions observed in granulite clasts in the light-colored clast area are identical to those described in the dark-matrix area.

Mineral clast populations in both lithologies are dominated by plagioclase (An_{95-98}), with minor amounts of pyroxene ($En_{40-50}Wo_{18-40}$; $En_{60-70}Wo_{2-4}$) and olivine (Fo_{65-76}), and with trace amounts of FeTiCr oxides, FeNi metal, and Fe sulfide. Glass clasts are found only in the dark-matrix area and range in shape from irregular fragments to spherules. They are typically feldspathic (<6 wt.% FeO) and incompatible-element poor (e.g., <0.05 wt.% K_2O; <0.3 wt.% TiO_2). A few glass clasts are moderately mafic (9–12 wt.% FeO) and richer in incompatible elements (0.2 wt.% K_2O; 2.5 wt.% TiO_2). The spherules are typically small (<50 μm) and have feldspathic compositions; the exception is a single large (400 micron) devitrified feldspathic spherule.

LAR 06638 has three distinct glass coatings (Figure 6.16). The inner, moderately mafic glass coating (16 wt.% FeO + MgO) is slightly vesicular and contains schlieren and abundant nanophase Fe globules. This inner glass layer is overlain by a more feldspathic fusion crust (10.4 wt.% FeO + MgO) that is highly vesicular and free of schlieren and nanophase Fe. Finally, a partially melted glassy area occurs, where many of the mafic silicate grains that occur in the adjacent textures are absent and presumably incorporated into the moderately mafic glass. The presence of multiple generations of glass coatings on LAR 06638 is, to our knowledge, unique among lunar meteorites. The more mafic, schlieren, and nano-phase-Fe-bearing glass is similar in morphology to the South Ray Crater glass coatings at the *Apollo* 16 site [see *McKay et al.*, 1986] and likely has a similar origin. The outer, more feldspathic glass has a morphology typical of fusion crust observed on other feldspathic lunar meteorites. It is unclear whether the partially melted glass area represents a partially formed fusion crust or incipient melting due to heating on the lunar surface, likely from an overlying (and possibly ablated) glass splash coating.

Overall, LAR 06638 is a highly feldspathic (30 wt% Al_2O_3), incompatible-element poor (0.4 μg/g Th) breccia. We obtained INAA data for both the dark-mafic and light-clastic lithologies. Although the compositions are similar, the dark-matrix material is slightly less mafic (0.8 × Sc, 0.95 × FeO) and slightly richer in both Na_2O (1.06×) and siderophile elements (1.2× for Ni and Ir) than the large light-colored clast. The small differences in composition

are likely explained by the incorporation of small amounts of more diverse material into the regolith breccia, for example, KREEPy glass clasts to account for the higher siderophile and incompatible-element concentrations and anorthosite to account for the lower concentrations of mafic elements and increased Na concentrations. Among lunar meteorites from Antarctica, LAR 06638 most closely resembles MAC 88104/5 in composition, although it is slightly more feldspathic (Figures 6.3 and 6.4) and 1.8 times richer in siderophile elements (Figure 6.9). Compositionally, however, it is more similar to Dhofar 490/1084 from Oman [*Korotev*, 2012].

6.5.12. Miller Range 07006

MIL 07006 is a basalt-bearing, feldspathic regolith breccia with a dark, aphanitic-to-glassy matrix. The lithic clasts are predominantly granulitic and feldspathic impact-melt breccias [*Zeigler et al.*, 2012a]; however, clasts of VLT basalt have also been observed [*Joy et al.*, 2010a; *Liu et al.*, 2009]. The clasts have sharp boundaries with the surrounding glassy matrix, in contrast to the other regolith breccia meteorites from the MIL site (below). Plagioclase compositions in MIL 07006 are largely invariant (An_{94-98}). There is a dichotomy in the compositions of pyroxene (En_{48-78}) and olivine (Fo_{40-80}) in the lithic clasts when compared to mineral clasts of pyroxene (En_{10-52}) and olivine (Fo_{6-61}).

MIL 07006 is among the most Sc- and Fe-rich of the ANSMET feldspathic lunar meteorites (Figures 6.3 and 6.4) as a result of the presence of basaltic lithic and mineral clasts. Compositionally, MIL 07006 is similar to PCA 02007 and Yamato 791197 from Antarctica, but it is much more similar to Dhofar 1436/1443 from Oman [*Korotev*, 2012].

6.5.13. Miller Range 090034, 090070, and 090075

MIL 090034, MIL 090070, and MIL 090075 are glassy-matrix, clast-rich, immature, feldspathic regolith breccias. Centimeter-sized lithic clasts are common in all three stones. The clasts are dominantly feldspathic impact-melt breccias (particularly the largest clasts), with lesser amounts of granulitic breccias and rare norites, troctolites, and gabbros [*Liu et al.*, 2011; *Korotev et al.*, 2011; *Zeigler et al.*, 2012a].

MIL 090070 and MIL 090075 were found close together in the field and there is no compositional [*Korotev et al.*, 2011; *Shirai et al.*, 2012] or petrographic [*Zeigler et al.*, 2012a] evidence to suspect that they are not paired. MIL 090034 was found 15 km away from the MIL 090070/75 stones. MIL 090034 and MIL 090070/75 appear different from each other in hand specimen [R. Harrington, in *Satterwhite and Righter*, 2010]. We were allocated three

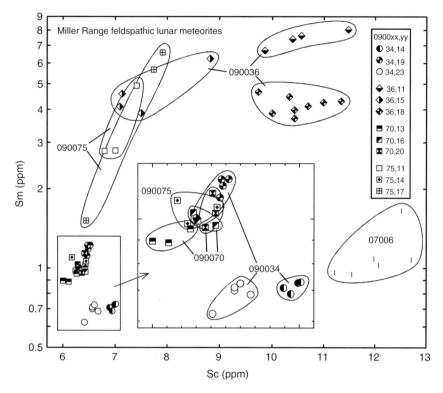

Figure 6.17. Subsamples (mean mass: 34 mg each; range: 20–41 mg) of the Miller Range feldspathic lunar meteorites in Sc-Sm space (compare with Figure 6.3). For the MIL 0900xx stones, we were allocated three different samples and analyzed four or more subsamples of each.

samples of each stone and analyzed 3–5 subsamples of each by INAA. For the most part, the stones are highly feldspathic (low Sc) and poor in incompatible elements (Figure 6.3). Two of the MIL 090075 samples, however, contain some Sm-rich lithology, but the third overlaps in composition with the three MIL 090070 samples and one of the MIL 090034 samples (Figure 6.17). Despite the differences, preliminary data of *Nishiizumi and Caffee* [2013] show that MIL090034, MIL 090070, and MIL 090075 all have similar concentrations of cosmogenic radionuclides. Thus, we assume that MIL090034 is paired with MIL 090070 and MIL 090075 and that the meteorite is compositionally heterogeneous at the scale of our sampling.

6.5.14. *Miller Range 090036*

MIL 090036 is also a glassy-matrix, clast-rich, immature, feldspathic regolith breccia. Like MIL 090036/70/75, it is dominated by clasts of impact-melt breccias. Those in MIL 090036 are compositionally distinct, having more sodic plagioclase and more magnesian mafic silicates than the melt-breccia clasts in the other MIL 0900xx stones. As with MIL 090036/70/75, we analyzed multiple subsamples of three MIL 090036 samples. The three subsamples do not overlap in composition, again indicating

that the meteorite is heterogeneous at the scale of our sampling (Figure 6.17). In detail, MIL 090036 shows no compositional overlap with the other three MIL 0900xx stones. It is significantly richer in incompatible elements Ti, Na, Eu, and siderophile elements (Table 6.2; Figures 6.3, 6.9, and 6.18) [*Korotev et al.*, 2011]. MIL 090036 is compositionally and petrographically similar to NWA 7022 [*Kuehner et al.*, 2012] (Figure 6.18), but the preliminary CRE data do not substantiate launch pairing [*Nishiizumi and Caffee*, 2013].

There is no petrographic or compositional evidence to suspect that any of the MIL 0900xx stones is paired with either MIL 05035 (basalt) or MIL 07006 (basalt-bearing feldspathic regolith breccia). None of the MIL 0900xx stones contain basaltic clasts [*Liu et al.*, 2011] and the texture of MIL 07006 is very different from those of any of the MIL 0900xx stones. The preliminary CRE data [*Nishiizumi and Caffee*, 2013] also argue that four distinct lunar meteorites have been found at the Miller Range site.

6.6. DISCUSSION AND SUMMARY

Lunar meteorites collected on ANSMET expeditions cover much of the compositional range observed among all lunar meteorites (Figure 6.4). Absent, however, are meteorites with 7% to 12% FeO and those with very high

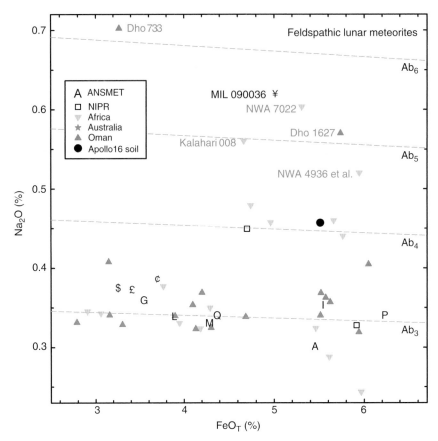

Figure 6.18. Whole-rock Na_2O concentrations in feldspathic lunar meteorites reflect the average albite (Ab) content of the plagioclase. Most feldspathic lunar meteorites have Na_2O concentrations corresponding to plagioclase of about Ab_{97} composition, but some are more albitic. MIL 090036 is substantially richer in Na_2O than all other ANSMET lunar meteorites.

concentrations of incompatible elements. Curiously, although a number of the feldspathic lunar meteorites from hot deserts have been identified as impact-melt or granulitic breccias [e.g., *Daubar et al.*, 2002; *Hudgins et al.*, 2011; *Kuehner et al.*, 2010; *Korotev*, 2012], all of the brecciated ANSMET lunar meteorites are regolith breccias.

Nearly two thirds (64%) of the ANSMET lunar meteorites are breccias from the feldspathic highlands. Until lunar meteorites were discovered, most of our knowledge of the composition, mineralogy, and petrography of the feldspathic highlands was based on samples from the *Apollo* 16 landing site. It was recognized at the time of the mission (April 1972) that the site geology had been influenced by the impact that formed the giant Imbrium basin (Figure 6.2), centered 1650 km to the northwest [*Muehlberger et al.*, 1980]. During the decade following the mission, however, the extent to which the Imbrium impact had affected the chemical composition of materials from the *Apollo* 16 landing site was largely unappreciated. Mafic (noritic and troctolitic), KREEP-bearing impact-melt breccias were common at the site, but because such breccias were also found at the *Apollo* 12,

14, 15, and 17 sites, it was assumed that such breccias were characteristic of the highlands [*Ryder and Wood*, 1977; *Spudis*, 1984]. Results from the *Apollo* orbiting gamma-ray spectrometers, however, showed that the Imbrium-Procellarum area was rich in Th (3–5 µg/g), whereas vast regions of the farside highlands were in the 0.3–0.5 µg/g range [*Metzger et al.*, 1977], much lower than the 2–2.5 µg/g characteristic of the *Apollo* 16 site (Figure 6.19). Thus, the low concentration of Th in the first lunar meteorite (0.3 µg/g, ALH A81005, Figure 6.19) led to immediate speculation that its point of origin was distant from the Imbrium-Procellarum area [*Kallemeyn and Warren*, 1983; *Korotev et al.*, 1983]. Most of the feldspathic lunar meteorites found after ALH A81005 have lower concentrations of Th and other incompatible elements than does the *Apollo* 16 regolith (Figures 6.4 and 6.19).

Among ANSMET meteorites, only MIL 090036 is comparable to the *Apollo* 16 regolith in concentrations of incompatible elements; thus, it is likely to originate from some place in the feldspathic highlands near the Procellarum KREEP Terrane. By the time of MAC 88104/5, the fourth feldspathic meteorite, it was clear that

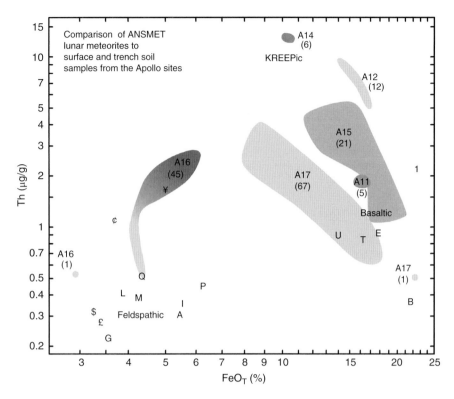

Figure 6.19. Like Figure 6.4, but comparison of ANSMET lunar meteorites, most of which are regolith breccias, to regolith (soil) samples from the *Apollo* missions (gray fields). Most feldspathic lunar meteorites have lower concentrations of incompatible elements (Th) than the *Apollo* 16 regolith. Most *Apollo* 16 regolith samples plot on the dark (high-Th) part of the field. The number in parentheses is the number of soil samples that define the field. The anomalous *Apollo* 16 sample is highly immature 67711 and the anomalous *Apollo* 17 sample is 74220, the "orange glass" soil. Symbol key is in Table 6.1.

the feldspathic lunar meteorites provided a more typical sampling of the upper crust of the feldspathic highlands, particularly its composition, than did the *Apollo* 16 samples [*Palme et al.*, 1991]. Another difference between the *Apollo* 16 collection and the feldspathic lunar meteorites involves anorthosite. A common lithology of the *Apollo* 16 regolith is shocked, but unbrecciated, highly feldspathic (>95% plagioclase) anorthosite of plutonic origin [e.g., *Warren*, 1990, 1993]. The ubiquity of "ferroan anorthosite" (*Mg'*: 50–70) at the *Apollo* 16 site supports the magma ocean model in which the Moon was mainly molten early in its history and the anorthosite was a flotation cumulate [*Warren*, 1985; *Wieczorek et al.*, 2006]. Recently, large exposures of anorthosite consisting of >99% plagioclase have been identified from orbit at numerous locations on the lunar surface [*Ohtake et al.*, 2009]. Thus, it is curious that lithic clasts of ferroan anorthosite are rare in feldspathic lunar meteorites [*Korotev et al.*, 2003a, 2009b]. Most lithic clasts in the feldspathic meteorites from ANSMET and hot deserts are granulitic and impact-melt breccias that are more mafic than the *Apollo* 16 ferroan anorthosites. The reason for this discrepancy is not yet understood. Perhaps the most unexpected observation about the

crust to be revealed by the feldspathic lunar meteorites is that despite having a narrow range of normative plagioclase abundances (72–87%; Figure 6.7), the range in *Mg'* is large, 59–77. This observation provides a challenge for models of crust formation [*Korotev et al.*, 2003a; *Korotev*, 2005; *Arai et al.*, 2008; *Treiman et al.*, 2010; *Gross et al.*, 2014].

The remaining 36% of the ANSMET lunar meteorites, two basalts and three breccias, are of mare affinity. They probably originate from three source craters on the Moon, craters represented by the YQEN, YAMM, and NNL launch-pair groups (Table 1; Figures 6.8, 6.12, and 6.13). Compositionally and texturally, the NNL group (NWA 032/479, NWA 4734, and the LAP stones) resembles basalts of the *Apollo* 12 and 15 collection but are a bit younger. The basalts and basalt clasts of the YQEN (Yamato 793274/981031, QUE 94281, EET 87521/96008, and NWA 4884) and YAMM (Yamato 793169, Asuka 881757, MET 01210, and MIL 05035) meteorites are distinct in several ways from any *Apollo* mare basalt, most notably their compositions and evidence for slow cooling. In studies in which new analytical techniques are applied to mare basalts for the purpose of addressing planetary issues such as formation of the Moon, water on the

Moon, or basalt petrogenesis, it has become standard to include samples of MIL 05035 or a LAP basalt along with *Apollo* samples to increase the sample diversity [e.g., *Rankenburg et al.*, 2006, 2007; *Day et al.*, 2007; *Spicuzza et al.*, 2007; *Zhang et al.*, 2010; *Elardo et al.*, 2012; *Paniello et al.*, 2012; *Wang et al.*, 2012].

The four YAMM stones are thought to have been launched from one crater on the Moon. The find sites for MET 01210 and MIL 05035 are separated by 400 km, those for Yamato 793169 and Asuka 881757 are separated by 380 km, and the ANSMET sites are >2500 km from the NIPR sites. These distances are all too great for the meteorites to be terrestrially paired. Thus, it is noteworthy that no prospective launch pairs of the four YAMM meteorites have been found outside of Antarctica.

For the five ANSMET lunar meteorites for which the parameters have been measured, ejection ages from the Moon range from 40 to 275 thousand years and terrestrial residence times range from 10 to 230 thousand years (Figure 6.11). On the basis of lunar meteorite finds from Oman, an average of 1.0 kg per year of lunar rocks in the 1-g to 10-kg mass range have reached the surface of the Earth in modern times [*Korotev*, 2012]. This calculation

has not been done with the ANSMET collection, and such a comparison might be informative.

Among the ~19,000 named meteorite stones collected by ANSMET (through the 2011–2012 season), 0.13% are lunar. GRA 06157, the smallest lunar meteorite stone, was collected at a site where only 352 meteorites have been collected. Thirteen of the named ANSMET sites have produced 352 or more meteorites. Ten of those have produced at least one lunar meteorite. The Queen Alexandra Range site (3444 named stones) has yielded two unpaired lunar meteorites and the Miller Range site (2038) has yielded four.

The fact that launch-paired meteorites EET 87/98 (immature regolith breccia), QUE 94281 (heterogeneous glassy- and fragmental-matrix regolith breccia), and Yamato 79/98 (either glassy-matrix regolith breccia [*Arai et al.*, 2002; *Koeberl et al.*, 1991b] or fragmental-matrix regolith breccia [*Sugihara et al.*, 2004]) are texturally and compositionally distinct from one another argues that a single impact on the Moon can, in fact, launch different kinds of rocks. The YAMM launch pairs are even more diverse in that MET 01210 is a breccia and MIL 05035, Asuka 881757, and Yamato 793169 are unbrecciated basalts.

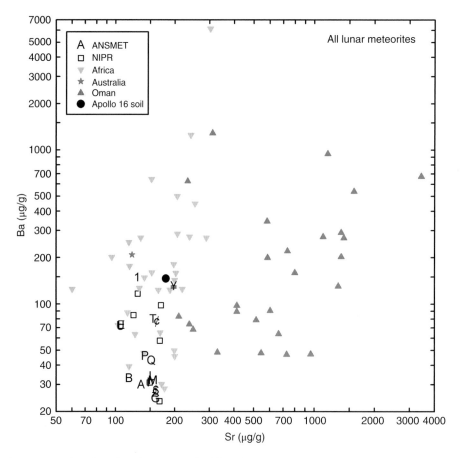

Figure 6.20. Many lunar meteorites from hot deserts are significantly contaminated with Sr, Ba, and other elements as a result of terrestrial alteration. Meteorites from Antarctica are not contaminated with these elements.

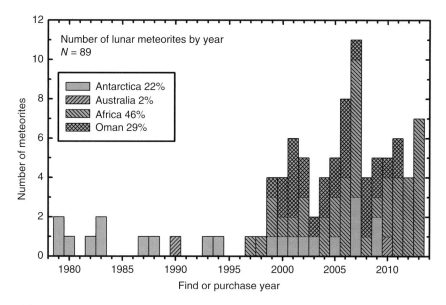

Figure 6.21. Number of lunar meteorites by find or purchase year. For paired stones, the find year of the first found stone of the pair group is represented. The rate of lunar finds from Antarctica (ANSMET and NIPR) has remained rather constant over the 35-year time period.

Although the number of lunar meteorites from hot deserts (about 78%) substantially exceeds those from Antarctica, most meteorites from hot deserts are contaminated, to varying degrees, with products of terrestrial alteration, whereas there is little evidence for even minor postfall changes in bulk composition among the lunar meteorites in the ANSMET and NIPR collection. Most hot-desert meteorites are seriously contaminated with Sr and Ba (Figure 6.20), but enrichments in Na, K, Ca, P, Zn, As, Se, Br, Sb, Au, U, carbonates, and sulfates have also been recorded [e.g., *Zeigler et al.*, 2006; *Korotev*, 2012].

Perhaps because of the systematic way in which ANSMET and NIPR searches are done, the find rate of lunar meteorites in Antarctica has been constant over the past 35 years (Figure 6.21). This constancy suggests that continued searching will yield more lunar meteorites from Antarctica.

Acknowledgments This work was funded by NASA grants NAG5-8609, NAG5-4172, NNG04GG10G, and NNX07AI44G. We thank the Antarctic Search for Meteorites (U.S.), Antarctic Meteorite Research Center of the National Institute of Polar Research (Japan), Cecilia Satterwhite, Kathleen McBride, and John Schutt for finding and providing these wonderful rocks and information about them. Many thanks to Mark Robinson for calling the *Shoemaker et al.* [1963] work to our attention and Brad Jolliff for constructing Figure 6.1. We appreciate the helpful reviews of Barb Cohen, Paul Warren, and editor Cari Corrigan.

REFERENCES

Anand, M., L. A. Taylor, C. R. Neal, G. A. Snyder, A. Patchen, Y. Sano, and K. Terada (2003), Petrogenesis of lunar meteorite EET 96008, *Geochim. Cosmochim. Acta*, *67*, 3499–3518.

Anand, M., L. A. Taylor, C. Floss, C. R. Neal, K. Terada, and S. Tanikawa (2006), Petrology and geochemistry of LaPaz Icefield 02205: A new unique low-Ti mare-basalt meteorite, *Geochim. Cosmochim. Acta*, *70*, 246–264.

Arai, T., and P. H. Warren (1999), Lunar meteorite Queen Alexandra Range 94281: Glass compositions and other evidence for launch pairing with Yamato 793274, *Meteorit. Planet. Sci.*, *34*, 209–234.

Arai, T., T. Ishi, and M. Otsuki (2002), Mineralogical study of new lunar meteorite Yamato 981031, *Lunar Planet. Sci.*, *XXXIII*, abstract 2064.

Arai, T., K. Misawa, and H. Kojima (2005), A new lunar meteorite MET 01210: Mare breccia with a low-Ti ferrobasalt, *Lunar Planet. Sci.*, *XXXVI*, abstract 2361.

Arai, T., K. Misawa, and H. Kojima (2007), Lunar meteorite MIL 05035: mare basalt paired with Asuka-881757, *Lunar Planet. Sci.*, *XXXVIII*, abstract 1582.

Arai, T., H. Takeda, A. Yamaguchi, and M. Ohtake (2008), A new model of lunar crust: asymmetry in crustal composition and evolution, *Earth, Planets and Space*, *60*, 433–444.

Arai, T., B. R. Hawke, T. A. Giguere, K. Misawa, M. Miyamoto, and H. Kojima (2010), Antarctic lunar meteorites Yamato-793169, Asuka-881757, MIL 05035, and MET 01210 (YAMM): Launch pairing and possible cryptomare origin, *Geochim. Cosmochim. Acta*, *74*, 2231–2248.

Barrat, J. A., M. Chaussidon, M. Bohn, Ph. Gillet, C. Göpel, and M. Lesourd (2005), Lithium behavior during cooling of a dry basalt: An ion-microprobe study of the lunar meteorite Northwest Africa 479 (NWA 479), *Geochim. Cosmochim. Acta*, *69*, 5597–5609.

Basilevsky, A. T., G. Neukum, and L. Nyquist (2010), The spatial and temporal distribution of lunar mare basalts as deduced from analysis of data for lunar meteorites, *Planetary and Space Science*, *58*, 1900–1905.

Bischoff, A. (1996), Lunar meteorite Queen Alexandra Range 93069: A lunar highland regolith breccia with very low abundances of mafic components, *Meteorit. Planet. Sci.*, *31*, 849–855.

Borg, L. E., A. M. Gaffney, C. K. Shearer, D. J. DePaolo, I. D. Hutcheon, T. L. Owens, E. Ramon, and G. Brennecka (2009), Mechanisms for incompatible-element enrichment on the Moon deduced from the lunar basaltic meteorite Northwest Africa 032, *Geochim. Cosmochim. Acta*, *73*, 3963–3980.

Boynton, W. V., and D. H. Hill (1983), Composition of bulk fragments and a possible pristine clast from Allan Hills 81005, *Geophys. Res. Lett.*, *10*, 837–840.

Cameron, A. G. W., and W. R. Ward (1976), *The origin of the Moon.* Lunar Planet. Sci., VII, 120–122.

Canup, R. M., and Asphaug E. (2001), Origin of the Moon in a giant impact near the end of the Earth's formation, *Nature*, *412*, 708–712.

Cassidy, W. A. (2003), *Meteorites, Ice, and Antarctica*, Cambridge Univ. Press, Cambridge, UK, 349 pp.

Cohen, B. A., T. D. Swindle, and D. A. Kring (2005), Geochemistry and 40Ar-39Ar geochronology of impact-melt clasts in feldspathic lunar meteorites: Implications for lunar bombardment history, *Meteorit. Planet. Sci.*, *40*, 755–777.

Connolly, H. C., Jr., J. Zipfel, L. Folco, C. Smith, R. H. Jones, G. Benedix, K. Righter, A. Yamaguchi, H. Chennaoui Aoudjehane, and J. N. Grossman (2007), The Meteoritical Bulletin, 91, 2007 March, *Meteorit. Planet. Sci.*, *42*, 413–466.

Daubar, I. J., D. A. Kring, T. D. Swindle, and A. J. T. Jull (2002), Northwest Africa 482: A crystalline impact-melt breccia from the lunar highlands, *Meteorit. Planet. Sci.*, *37*, 1797–1814.

Day, J. M. D., L. A. Taylor, C. Floss, A. D. Patchen, D. W. Schnare, and D. G. Pearson (2006a), Comparative petrology, geochemistry and petrogenesis of evolved, low-Ti lunar mare basalt meteorites from the La Paz Icefield, Antarctica, *Geochim. Cosmochim. Acta*, *70*, 1581–1600.

Day, J. M. D., C. Floss, L. A. Taylor, M. Anand, and A. D. Patchen (2006b), Evolved mare basalt magmatism, high Mg/Fe feldspathic crust, chondritic impactors, and the petrogenesis of Antarctic lunar breccia meteorites Meteorite Hills 01210 and Pecora Escarpment 02007, *Geochim. Cosmochim. Acta*, *70*, 5957–5989.

Day, J. M. D., D. G. Pearson, and L. A. Taylor (2007), Highly siderophile element constraints on accretion and differentiation of the Earth-Moon system, *Science*, *315*, 217–219.

Delaney, J. S. (1989), Lunar basalt breccia identified among Antarctic meteorites, *Nature*, *342*, 889–890.

Delano, J. W. (1991), Geochemical comparison of impact glasses from lunar meteorites ALH81005 and MAC88105 and Apollo 16 regolith 64001, *Geochim. Cosmochim. Acta*, *55*, 3019–3029.

Dreibus, G., B. Spettel, F. Wlotzka, K. P. Jochum, L. Schultz, H. W. Weber, and H. Wänke (1996), Chemistry, petrology, and noble gases of basaltic lunar meteorite QUE 94281, *Meteorit. Planet. Sci.*, *31*, A38–A39.

Elardo, S. M., C. K. Shearer Jr., A. L. Fagan, C. R. Neal, P. V. Burger, and L. E. Borg (2012), Diversity in low-Ti mare magmatism and mantle sources: A perspective from lunar meteorites NWA 4734, NWA 032, and LAP 02205, *Lunar Planet. Sci.*, *XLIII*, abstract 2648.

Elardo, S. M., C. K. Shearer, A. L. Fagan, L. E. Borg, A. M. Gaffney, P. V. Burger, C. R Neal., and F. M. McCubbin (2013), The origin of young mare basalts inferred from lunar meteorites NWA 4734, NWA 032, and LAP 02205, *Lunar Planet. Sci.*, *XLIV*, abstract 2762.

Eugster, O., M. Burger, U. Krähenbühl, Th. Michel, J. Beer, H. J. Hofmann, H. A. Synal, W. Woelfli, and R. C. Finkel (1991), History of the paired lunar meteorites MAC88104 and MAC88105 derived from noble gas isotopes, radionuclides, and some chemical abundances, *Geochim. Cosmochim. Acta*, *55*, 3139–3148.

Eugster, O., Ch. Thalmann, A. Albrecht, G. F. Herzog, J. S. Delaney, J. Klein, and R. Middleton (1996), Exposure history of glass and breccia phases of lunar meteorite EET87521, *Meteorit. Planet. Sci.*, *31*, 299–304.

Eugster, O., E. Polnau, E. Salerno, and D. Terribilini (2000), Lunar surface exposure models for meteorites Elephant Moraine 96008 and Dar al Gani 262 from the Moon, *Meteorit. Planet. Sci.*, *35*, 1177–1181.

Fagan, T. J., G. J. Taylor, K. Keil, T. E Bunch., J. H Wittke., R. L. Korotev, B. L. Jolliff, J. J. Gillis, L. A. Haskin, E. Jarosewich, R. N. Clayton, T. K. Mayeda, V. A. Fernandes, R. Burgess, G. Turner, O. Eugster, and S. Lorenzetti (2002), Northwest Africa 032: Product of lunar volcanism, *Meteorit. Planet. Sci.*, *37*, 371–394.

Fernandes, V. A., R. Burgess, and G. Turner (2003), $^{40}Ar–^{39}Ar$ chronology of lunar meteorites Northwest Africa 032 and 773, *Meteorit. Planet. Sci.*, *38*, 555–564.

Fernandes, V. A., R. Burgess, and A. Morris (2009), $^{40}Ar–^{39}Ar$ age determinations of lunar basalt meteorites Asuka 881757, Yamato 793169, Miller Range 05035, LaPaz Icefield 02205, Northwest Africa 479, and basaltic breccia Elephant Moraine 96008, *Meteorit. Planet. Sci.*, *44*, 805–821.

Fukuoka, T. (1990), Chemistry of Yamato-793274 lunar meteorite, *Papers Presented to the Fifteenth Symposium on Antarctic Meteorites*, 122–123. Nat. Inst. Polar Res., Tokyo.

Gallant, B., M. Gladman, and M. Ćuk (2009), Current bombardment of the Earth–Moon system: Emphasis on cratering asymmetries, *Icarus*, doi:10.1016/j.icarus.2009.03.025.

Genge, M. J., and Grady M. M. (1999), The fusion crusts of stony meteorites: Implications for the atmospheric reprocessing of extraterrestrial materials, *Meteorit. Planet. Sci.*, *34*, 341–356.

Gillis, J. J., B. L. Jolliff, and R. L. Korotev (2004), Lunar surface geochemistry: Global concentrations of Th, K, and FeO as derived from Lunar Prospector and Clementine data, *Geochim. Cosmochim. Acta*, *68*, 3791–3805; Erratum: *69*, 5147–5148.

Gnos, E., B. A. Hofmann, A. Al-Kathiri, S. Lorenzetti, O. Eugster, M. J. Whitehouse, I. Villa, A. J. T. Jull, J. Eikenberg, B. Spettel, U. Krähenbühl, I. A. Franchi, and G. C. Greenwood (2004), Pinpointing the source of a lunar meteorite: Implications for the evolution of the Moon, *Science, 305*, 657–659.

Goodrich, C.A., G. J. Taylor, K. Keil, W. V. Boynton, and D. H. Hill (1984), Petrology and chemistry of hyperferroan anorthosites and other clasts from lunar meteorite ALH81005, *Proc. 15th Lunar Planet. Sci. Conf.*, in *J. Geophys. Res., 89*, C87–C94.

Grier, J. A., D. A. Kring, and T. D. Swindle (1995), Impact melts and anorthositic clasts in lunar meteorites QUE93069 and MAC88105, *Lunar Planet. Sci., XXVI*, 513–514.

Gross, J., and A. H. Treiman (2010), "Massif" anorthosites in Allan Hills 81005: Possible origin from a diapir? *73th Annual Meeting of the Meteoritical Society*, abstract 5435.

Gross, J., A. H. Treiman, and C. M. Mercer (2014), Lunar feldspathic meteorites: Constraints on the geology of the lunar highlands, and the origin of the lunar crust, *Earth Planet. Sci. Lett., 388*, 318–328.

Hartmann, W. K., and D. R. Davis (1975), Satellite-sized planetesimals and lunar origin. *Icarus, 24*, 504–515.

Haskin, L. A. (1998), The Imbrium impact event and the thorium distribution at the lunar highlands surface. *J. Geophys. Res., 103*, 1679–1689.

Haskin, L. A., R. L. Korotev, K. M. Rockow, and B. L. Jolliff (1998), The case for an Imbrium origin of the Apollo Th-rich impact-melt breccias. *Meteoritics & Planetary Science, 33*, 959–975.

Head, J. N. (2001), Lunar meteorite source crater size: Constraints from impact simulations, *Lunar Planet. Sci., XXXII*, abstract 1768.

Head, J. W., and L. Wilson (1992), Lunar mare volcanism: Stratigraphy, eruption conditions, and evolution of secondary crusts, *Geochim. Cosmochim. Acta, 56*, 2155–2175.

Hill, E., L. A. Taylor, C. Floss, and Y. Liu (2009), Lunar meteorite LaPaz Icefield 04841: Petrology, texture, and impact-shock effects of a low-Ti mare basalt, *Meteorit. Planet. Sci., 44*, 87–94.

Huber, H., and P. H. Warren (2005), MET01210: Another lunar mare meteorite (regolith breccia), with extensive pyroxene exsolution, and not part of the YQ launch pair, *Lunar Planet. Sci., XXXVI*, abstract 2401.

Hudgins, J. A., S. P. Kelley, R. L. Korotev, and J. G. Spray (2011), Mineralogy, geochemistry, and ^{40}Ar–^{39}Ar geochronology of lunar granulitic breccia Northwest Africa 3163 and paired stones: Comparisons with Apollo samples, *Geochim. Cosmochim. Acta, 75*, 2865–2881.

Jolliff, B. L., R. L. Korotev, and L. A. Haskin (1991), A ferroan region of the lunar highlands as recorded in meteorites MAC88104 and MAC88105, *Geochim. Cosmochim. Acta, 55*, 3051–3071.

Jolliff, B. L., R. L. Korotev, and L. A Haskin (1993), Lunar basaltic meteorites Yamato-793169 and Asuka-881757: Samples of the same low-Ti mare-lava? in *Papers Presented to the Eighteenth Symposium on Antarctic Meteorites*, pp. 214–217, National Institute of Polar Research, Tokyo.

Jolliff, B. L., K. M. Rockow, R. L. Korotev, and L. A. Haskin (1996), Lithologic distribution and geologic history of the Apollo 17 site: The record in soils and small rock particles from the highlands massifs, *Meteorit. Planet. Sci., 31*, 116–145.

Jolliff, B. L., K. M. Rockow, and R. L. Korotev (1998), Geochemistry and petrology of lunar meteorite Queen Alexandra Range 94281, a mixed mare and highland regolith breccia, with special emphasis on very-low-Ti mafic components, *Meteorit. Planet. Sci., 33*, 581–601.

Jolliff, B. L., J. J. Gillis, L. A. Haskin, R. L. Korotev, and M. A. Wieczorek (2000), Major lunar crustal terranes: Surface expressions and crust-mantle origins, *J. Geophys. Res., 105*, 4197–4416.

Jolliff, B. L., R. A. Zeigler, and R. L. Korotev (2004), Petrography of lunar meteorite LAP 02205, a new low-Ti basalt possibly launch paired with NWA 032, *Lunar Planet. Sci., XXXV*, abstract 1438.

Joy, K. H., I. A. Crawford, S. S. Russell, B. Swinyard, B. Kellett, and M. Grande (2006a), Lunar regolith breccias MET 01210, PCA 02007 and DAG 400: Their importance in understanding the lunar surface and implications for the scientific analysis of D-CIXS data, *Lunar Planet. Sci., XXXVII*, abstract 1274.

Joy, K. H., I. A. Crawford, H. Downes, S. S. Russell, and A. T. Kearsley (2006b), A petrological, mineralogical, and chemical analysis of lunar mare basalt meteorite LaPaz Icefield 02205, 02224, and 02226, *Meteorit. Planet. Sci., 41*, 1003–1025.

Joy, K. H., R. Burgess, R. Hinton, V. A. Fernandes, Crawford I. A., A. T. Kearsley, and A. J. Irving (2011), Petrogenesis and chronology of lunar meteorite Northwest Africa 4472: A KREEPy regolith breccia from the Moon, *Geochim. Cosmochim. Acta, 75*, 2420–2452.

Joy, K. H., M. Anand, I. A. Crawford, and S. S. Russell (2007), Petrography and bulk composition of Miller Range 05035: A new lunar VLT gabbro, *Lunar Planet. Sci., XXXVIII*, abstract 1867.

Joy, K. H., I. A. Crawford, M. Anand, R. C. Greenwood, , I. A. Franchi, S. S. Russell (2008), The petrology and geochemistry of Miller Range 05035: A new lunar gabbroic meteorite, *Geochim. Cosmochim. Acta, 72*, 3822–3844.

Joy, K. H., I. A. Crawford, S. S. Russell, and A. T. Kearsley (2010a), Lunar meteorite regolith breccias: An in situ study of impact melt composition using LA-ICP-MS with implications for the composition of the lunar crust, *Meteorit. Planet. Sci., 45*, 917–946.

Joy, K. H., G. J. Taylor, G. R. Huss, K. Nagashima, and I. A. Crawford (2010b), An unusual magnesian troctolitic gabbro in lunar meteorite MAC 88105: An example of new rock types found in lunar meteorites, *73th Annual Meeting of the Meteoritical Society*, abstract 5426.

Jull, A. J. T. (2006), Terrestrial ages of meteorites, in *Meteorites and the Early Solar System* II, edited by D. S. Lauretta and H. Y. McSween, pp. 889–905, University of Arizona Press, Tucson.

Kallemeyn, G. W., and P. H. Warren (1983), Compositional implications regarding the lunar origin of the ALH81005 meteorite, *Geophys. Res. Lett., 10*, 833–836.

Karouji, Y., Y. Oura, and M. Ebihara (2002), Chemical composition of lunar meteorites including Yamato 981031, *Antarctic Meteorites*, *XXVII*, 52–54, National Institute of Polar Research, Tokyo.

Koeberl, C., G. Kurat, and F. Brandstätter (1991a), MAC88105—A regolith breccia from the lunar highlands: Mineralogical, petrological, and geochemical studies, *Geochim. Cosmochim. Acta*, *55*, 3073–3087.

Koeberl, C., G. Kurat, and F. Brandstätter (1991b), Lunar meteorites Yamato 793274: Mixture of mare and highland components, and barringerite from the Moon, *Proceedings of the NIPR Symposium on Antarctic Meteorites*, *4*, 33–55. Nat. Inst. Polar Res., Tokyo.

Koeberl, C., G. Kurat, and F. Brandstätter (1993), Gabbroic lunar mare meteorites Asuka-881757 (Asuka-31), and Yamato 793169: Geochemical and mineralogical study. *Proceedings of the NIPR Symposium on Antarctic Meteorites*, *6*, 14–34. Nat. Inst. Polar Res., Tokyo.

Koeberl, C., G. Kurat, and F. Brandstätter (1996), Mineralogy and geochemistry of lunar meteorite Queen Alexandra Range 93069, *Meteorit. Planet. Sci.*, *31*, 897–908.

Korotev, R. L. (1997), Some things we can infer about the Moon from the composition of the Apollo 16 regolith, *Meteorit. Planet. Sci.*, *32*, 447–478.

Korotev, R. L. (2005), Lunar geochemistry as told by lunar meteorites. *Chemie der Erde, 65*, 297–346.

Korotev, R. L. (2012), Lunar meteorites from Oman, *Meteorit. Planet. Sci.*, *47*, 1365–1402.

Korotev, R. L., and R. A. Zeigler (2007), Keeping up with the lunar meteorites, *Lunar Planet. Sci.*, *XXXVIII*, abstract 1340.

Korotev, R. L., M. M Lindstrom., D. J. Lindstrom, and L. A. Haskin (1983), Antarctic meteorite ALH81005—Not just another lunar anorthositic norite, *Geophys. Res. Lett.*, *10*, 829–832.

Korotev, R. L., B. L. Jolliff, and K. M. Rockow (1996), Lunar meteorite Queen Alexandra Range 93069 and the iron concentration of the lunar highlands surface, *Meteorit. Planet. Sci.*, *31*, 909–924.

Korotev, R. L., B. L. Jolliff, R. A Zeigler., J. J. Gillis, and L. A. Haskin (2003a), Feldspathic lunar meteorites and their implications for compositional remote sensing of the lunar surface and the composition of the lunar crust, *Geochim. Cosmochim. Acta*, *67*, 4895–4923.

Korotev, R. L., B. L. Jolliff, R. A. Zeigler, and L. A. Haskin (2003b), Compositional constraints on the launch pairing of three brecciated lunar meteorites of basaltic composition, *Antarctic Meteorite Research*, *16*, 152–175.

Korotev, R. L., R. A. Zeigler, and B. L. Jolliff (2006), Feldspathic lunar meteorites Pecora Escarpment 02007 and Dhofar 489: Contamination of the surface of the lunar highlands by post-basin impacts, *Geochim. Cosmochim. Acta*, *70*, 5935–5956.

Korotev, R. L., R. A. Zeigler, A. J. Irving, and T. E. Bunch (2009a), Keeping up with the lunar meteorites—2009, *Lunar Planet. Sci.*, *XL*, abstract 1137.

Korotev, R. L., R. A. Zeigler, B. L. Jolliff, A. J. Irving, and T. E. Bunch (2009b), Compositional and lithological diversity among brecciated lunar meteorites of intermediate iron composition, *Meteorit. Planet. Sci.*, *44*, 1287–1322.

Korotev, R. L., B. L Jolliff., and P. K. Carpenter (2011), Miller Range feldspathic lunar meteorites, *Lunar Planet. Sci.*, *XLII*, abstract 1999.

Kring, D. A., D. H. Hill, and W. V. Boynton (1995), The geochemistry of a new lunar meteorite, QUE93069, a breccia with highland affinities, *Lunar Planet. Sci.*, *XXVI*, 801–802.

Kring, D. A., D. H. Hill, and W. V. Boynton (1996), A glass-rich view of QUE94281, a new meteoritic sample from a mare region of the Moon, *Lunar Planet. Sci.*, *XXVII*, 707–708.

Kuehner, S. M., A. J. Irving, M. Gellissen, and R. L. Korotev (2010), Petrology and composition of lunar troctolitic granulite Northwest Africa 5744: A unique recrystallized, magnesian crustal sample, *Lunar Planet. Sci.*, *XLI*, abstract 1552.

Kuehner, S. M., A. J. Irving, and R. L. Korotev (2012), Petrology and composition of lunar meteorite Northwest Africa 7022: An unusually sodic anorthositic gabbroic impact melt breccia with compositional similarities to Miller Range 090036, *Lunar Planet. Sci.*, *XLIII*, abstract 1524.

Kurat, G., and F. Brandstätter (1983), Meteorite ALH81005: Petrology of a new lunar highland sample, *Geophys. Res. Lett.*, *10*, 795–798.

Laul, J. C., M. R. Smith, and R. A. Schmitt (1983), Allan Hills 81005 meteorite: Chemical evidence for lunar highland origin, *Geophys. Res. Lett.*, *10*, 825–828.

Lawrence, D. J., W. C. Feldman, B. L. Barraclough, A. B. Binder, R. C. Elphic, S. Maurice, M. C. Miller, and T. H. Prettyman (2000), Thorium abundances on the lunar surface, *J. Geophys. Res.*, *105*, 20,307–20,331.

Le Feuvre, M., and M. A. Wieczorek (2008), Nonuniform cratering of the terrestrial planets, *Icarus*, *197*, 291–306.

Lindstrom, M. M., C. Schwarz, R. Score, and B. Mason (1991a), MacAlpine Hills 88104 and 88105 lunar highland meteorites: General description and consortium overview, *Geochim. Cosmochim. Acta*, *55*, 2999–3007.

Lindstrom, M. M., S. J. Wentworth, R. R. Martinez, D. W. Mittlefehldt, D. S. McKay, M.-S. Wang, and M. J. Lipschutz (1991b), Geochemistry and petrography of the MacAlpine Hills lunar meteorites, *Geochim. Cosmochim. Acta*, *55*, 3089–3103.

Lindstrom, M. M., D. W. Mittlefehldt, R. R. Martinez, M. J. Lipschutz, and M.-S. Wang (1991c), Geochemistry of Yamato-82192, −86032 and −793274 lunar meteorites, *Proceedings of the NIPR Symposium on Antarctic Meteorites*, *4*, 12–32. Nat. Inst. Polar Res., Tokyo.

Lindstrom, M. M., D. W. Mittlefehldt, R. V. Morris, R. R. Martinez, and S. J. Wentworth (1995), QUE93069, a more mature regolith breccia for the Apollo 25th anniversary, *Lunar Planet. Sci.*, *XXVI*, 849–850.

Lindstrom, M. M., D. W. Mittlefehldt, and R. R. Martinez (1999), Basaltic lunar meteorite EET96008 and evidence for pairing with EET87521, *Lunar Planet. Sci.*, *XXX*, abstract 1921.

Liu, Y., E. Hill, A. Patchen, and L. A. Taylor (2007), New lunar meteorite MIL 05035: Petrography and mineralogy, *Lunar Planet. Sci.*, *XXXVIII*, abstract 2103.

Liu, Y., A. Zhang, K. G. Thaisen, M. Anand, and L. A. Taylor (2009), Mineralogy and petrography of a lunar highland breccia meteorite, MIL 07006, *Lunar Planet. Sci.*, *XL*, abstract 2105.

Liu, Y., A. Patchen, and L. A. Taylor (2011), Lunar highland breccias MIL 090034/36/70/75: A significant KREEP component, *Lunar Planet. Sci., XLII*, abstract 1261.

Lucey, P., R. L. Korotev, J. J. Gillis, L. A. Taylor, D. Lawrence, R. Elphic, B. Feldman, L. L. Hood, D. Hunten, M. Mendillo, S. Noble, J. J. Papike, and R. C. Reedy (2006), Understanding the lunar surface and space-moon interactions, in *New Views of the Moon*, pp. 83–219. Published in *Reviews in Mineralogy and Geochemistry, 60,* edited by B. L. Jolliff, M. A. Wieczorek, C. K. Shearer, and C. R. Neal, Mineralogical Society of America, Washington, D. C.

Ma, M.-S., R. A. Schmitt, G. J. Taylor, R. D. Warner, D. E. Lange, and K. Keil (1978), Chemistry and petrology of Luna 24 lithic fragments and <250 μm soils: Constraints on the origin of VLT mare basalts, in *Mare Crisium: The View from Luna 24*, edited by R. B. Merrill and J. J. Papike, pp. 569–592, Pergamon Press, New York.

Marvin, U. B. (1983), The discovery and initial characterization of Allan Hills 81005: The first lunar meteorite, *Geophys. Res. Lett., 10*, 775–778.

McKay, D. S., R. M. Fruland, and G. H. Heiken (1974), Grain size and the evolution of lunar soil, *Proc. 5th Lunar Sci. Conf.*, 887–906.

McKay, D. S., D. D. Bogard, R. V. Morris, R. L Korotev., P. Johnson, and S. J. Wentworth (1986), Apollo 16 regolith breccias: Characterization and evidence for early formation in the mega-regolith, Proc. 16th Lunar Planet. Sci. Conf., *J. Geophys. Res., 91*, D277–D303.

McKay, D. S., G. Heiken, A. Basu, G. Blanford, S. Simon, R. Reedy, B. M. French, and J. Papike (1991), The lunar regolith, in *Lunar Sourcebook*, edited by G. Heiken et al., pp. 285–356, Cambridge Univ. Press, Cambridge, UK.

Metzger, A. E., E. L. Haines, R. E. Parker, and R. G. Radocinski (1977), Thorium concentrations in the lunar surface: I. Regional values and crustal content, *Proc. 8th Lunar. Sci. Conf.*, 949–999.

Mikouchi, T. (1999), Mineralogy and petrology of a new lunar meteorite EET96008: Lunar basaltic breccia similar to Y-793274, QUE94281 and EET87521, *Lunar Planet. Sci., XXX*, abstract 1558.

Misawa, K., M. Tatsumoto, G. B. Dalrymple, and K. Yanai (1993), An extremely low U/Pb source in the Moon: U-Th-Pb, Sm-Nd, Rb-Sr, and ⁴⁰Ar/³⁹Ar isotopic systematics and age of lunar meteorite Asuka 881757, *Geochim. Cosmochim. Acta, 57*, 4687–4702.

Morris, R. V. (1978), The surface exposure (maturity), of lunar soils: Some concepts and I$_s$/FeO compilation, *Proc. 9th Lunar Planet. Sci. Conf.*, 2287–2297.

Morris, R. V. (1983), Ferromagnetic resonance and magnetic properties of ALH81005, *Geophys. Res. Lett., 10*, 807–808.

Muehlberger, W. R., F. Hörz, J. R. Sevier, and G. E. Ulrich (1980), Mission objectives for geological exploration of the Apollo 16 landing site, in *Proc. Conf. Lunar Highlands Crust*, edited by J. J. Papike and R. B. Merrill, pp. 1–49, Pergamon Press, New York.

Naney, M. T., D. M. Crowl, and J. J. Papike (1976), The Apollo 16 drill core: Statistical analysis of glass chemistry and the characterization of a high alumina–silica poor (HASP) glass, *Proc. 7th Lunar Sci. Conf.*, 155–184.

Neal, C. R., L. A. Taylor, Y. Liu, and R. A. Schmitt (1991), Paired lunar meteorites MAC88104 and MAC88105: A new "FAN" of lunar petrology, *Geochim. Cosmochim. Acta, 55*, 3037–3049.

Nishiizumi, K. (2003), Exposure histories of lunar meteorites, in *Evolution of Solar System Materials: A New Perspective from Antarctic Meteorites*, National Institute of Polar Research, Tokyo, p. 104.

Nishiizumi, K., and M. W. Caffee (1996), Exposure histories of lunar meteorites Queen Alexandra Range 94281 and 94269, *Lunar Planet. Sci., XXVII*, 959–960.

Nishiizumi, K., and M. W. Caffee (2013), Relationships among six lunar meteorites from Miller Range, Antarctica based on cosmogenic radionuclides, *Lunar Planet. Sci., XLIV*, abstract 2715.

Nishiizumi, K., J. R. Arnold, J. Klein, D. Fink, R. Middleton, P. W. Kubik, P. Sharma, D. Elmore, and R. C. Reedy (1991), Exposure histories of lunar meteorites: ALH81005, MAC81004, MAC81005, and Y791197, *Geochim. Cosmochim. Acta, 55*, 3149–3155.

Nishiizumi, K., J. R. Arnold, M. W. Caffee, R. C. Finkel, J. Southon, and R. C. Reedy (1992), Cosmic ray exposure histories of lunar meteorites Asuka 881757, Yamato 793169, and Calcalong Creek, in *Papers Presented to the 17th Symposium on Antarctic Meteorites*, p. 129–132, August 19–21, Tokyo, Natl. Inst. Polar Res.

Nishiizumi, K., M. W. Caffee, A. J. T. Jull, and R. C. Reedy (1996), Exposure history of lunar meteorite Queen Alexandra Range 93069 and 94269, *Meteorit. Planet. Sci., 31*, 893–896.

Nishiizumi, K., J. Masarik, M. W. Caffee, and A. J. T. Jull (1999), Exposure histories of pair lunar meteorites EET 96008 and EET 87521, *Lunar Planet. Sci., XXX*, abstract 1980.

Nishiizumi, K., D. J. Hillegonds, and K. C. Welten (2006), Exposure and terrestrial histories of lunar meteorites LAP 02205/02224/02226/02436, MET 01210, and PCA 02007, *Lunar Planet. Sci., XXXVII*, abstract 2369.

Nyquist, L. E., C.-Y. Shih, Y. Reese, and D. D. Bogard (2005), Age of lunar meteorite LAP02205 and implications for impact-sampling of planetary surfaces, *Lunar Planet. Sci., XXXVI*, abstract 1374.

Nyquist, L. E., C.-Y. Shih, and Y. D. Reese (2007), Sm-Nd and Rb-Sr ages for MIL 05035: Implications for surface and mantle sources, *Lunar Planet. Sci., XXXVIII*, abstract 1702.

Ohtake, M., T. Matsunaga, J. Haruyama, Y. Yokota, T. Morota, C. Honda, Y. Ogawa, M. Torii, H. Miyamoto, T. Arai, N. Hirata, A. Iwasaki, R. Nakamura, T. Hiroi, T. Sugihara, H. Takeda, H. Otake, C. M. Pieters, K. Saiki, K. Kitazato, M Abe., N. Asada, H. Demure, Y. Yamaguchi, S. Sasaki, S. Kodama, J. Terazono, M. Shirao, A Yamaji., S. Minami, H. Akiyama, and J.-L. Josset (2009), The global distribution of pure anorthosite on the Moon, *Nature, 461*, doi:10.1038/nature08317.

Palme, H., B. Spettel., G. Weckwerth, and H. Wänke (1983), Antarctic meteorite Allan Hills 81005, a piece from the ancient lunar crust, *Geophys. Res. Lett., 10*, 817–820.

Palme, H., B. Spettel, K. P. Jochum, G. Dreibus, H. Weber, G. Weckwerth, H. Wänke, A Bischoff., and D. Stöffler (1991),

Lunar highland meteorites and the composition of the lunar crust, *Geochim. Cosmochim. Acta, 55*, 3105–3122.

Paniello, R. C., J. M. D. Day, and, F. Moynier (2012), Zinc isotopic evidence for the origin of the Moon, *Nature, 490*, 376–379.

Pieters, C. M., B. R. Hawke, M. Gaffey, and L. A. McFadden (1983), Possible lunar source areas of meteorite ALH81005: Geochemical remote sensing information, *Geophys. Res. Lett., 10*, 813–816.

Polnau, E., and Eugster O. (1998), Cosmic-ray produced, radiogenic, and solar noble gases in lunar meteorites Queen Alexandra Range 94269 and 94281, *Meteorit. Planet. Sci., 33*, 313–319.

Rankenburg, K., A. D. Brandon, and C. R. Neal (2006), Neodymium isotope evidence for a chondritic composition of the Moon, *Science, 312*, 1369–1372.

Rankenburg, K., A. D. Brandon, and M. D. Norman (2007), A RbSr and SmNd isotope geochronology and trace element study of lunar meteorite LaPaz Icefield 02205, *Geochim. Cosmochim. Acta, 71*, 2120–2135.

Righter, K., S. J. Collins, and A. D. Brandon (2005), Mineralogy and petrology of the LaPaz Icefield lunar mare basaltic meteorites, *Meteorit. Planet. Sci., 40*, 1703–1722.

Robinson, K. L., and A. H. Treiman (2010), Mare basalt fragments in lunar highlands meteorites: Connecting measured Ti abundances with orbital remote sensing, *Lunar Planet. Sci., XLI*, abstract 1788.

Ryder, G., and R. Ostertag (1983), Allan Hills 81005: Moon, Mars, petrography, and Giordano Bruno, *Geophys. Res. Lett., 10*, 791–794.

Ryder, G., and J. A. Wood (1977), Serenitatis and Imbrium impact melts: Implication for large-scale layering in the lunar crust, *Proc. 8th Lunar Sci. Conf.*, 655–668.

Satterwhite, C., and Righter K., editors (2007), *Antarctic Meteorite Newsletter, 30*(2). Houston, TX, NASA Johnson Space Center.

Satterwhite, C., and Righter K., editors (2010), *Antarctic Meteorite Newsletter, 33*(2). Houston, TX: NASA Johnson Space Center.

Shearer, C. K., and J. J. Papike (1993), Basaltic magmatism on the Moon: A perspective from volcanic picritic glass beads, *Geochim. Cosmochim. Acta, 57*, 4785–4812.

Shirai, N., M. Ebihara, S. Sekimoto, A Yamaguchi., L. Nyquist, C.-Y. Shih, J. Park, and K. Nagao (2012), Geochemistry of lunar highland meteorites MIL 090034, 090036 and 090070, *Lunar Planet. Sci., XLIII*, abstract 2003.

Shoemaker, E. M., R. J. Hackman, and R. E. Eggleton (1963), Interplanetary correlation of geologic time. *Advances in Astronautical Sciences, 8*, 70–89.

Simon, S. B., J. J. Papike, and C. K. Shearer (1983), Petrology of ALH81005, the first lunar meteorite, *Geophys. Res. Lett., 10*, 787–790.

Snyder, G. A., C. R. Neal, A. M. Ruzicka, and L. A. Taylor (1999), Lunar meteorite EET 96008, Part II. Whole-rock trace-element and PGE chemistry, and pairing with EET 87521, *Lunar Planet. Sci., XXX*, abstract 1705.

Spettel, B., G. Dreibus, A. Burghele, K. P Jochum., L. Schultz, H. W. Weber, F. Wlotzka, and H. Wänke (1995), Chemistry, petrology, and noble gases of lunar highland meteorite Queen Alexandra Range 93069, *Meteoritics, 30*, 581–582.

Spicuzza, M. J., J. M. D. Day, L. A. Taylor, and J. W. Valley (2007), Oxygen isotope constraints on the origin and differentiation of the Moon, *Earth Planet. Sci. Lett., 253*, 254–265.

Spudis, P. D. (1984), Apollo 16 site geology and impact melts: Implications for the geologic history of the lunar highlands, *Proc. 15th Lunar Planet. Sci. Conf.*, in *J. Geophys. Res., 89*, C95–C107.

Stöffler, D., H.-D. Knöll, U. B. Marvin, C. H. Simonds, and P. H. Warren (1980), Recommended classification and nomenclature of lunar highlands rocks: A committee report, in *Proc. Conf. Lunar Highlands Crust*, edited by J. J. Papike and R. B. Merrill, pp. 51–70, Pergamon Press, New York.

Stöffler, D. and R. A. F. Grieve (2007), Impactites, Chapter 2.11, in *Metamorphic Rocks: A Classification and Glossary of Terms*, edited by D. Fettes and J. Desmons, pp. 82–92, 111–125, and 126–242, Cambridge Univ. Press, Cambridge, UK.

Sugihara, T., M Ohtake., A. Owada, T. Ishii, M. Otsuki, and H. Takeda (2004), Petrology and reflectance spectroscopy of lunar meteorite Yamato 981031: Implications for the source region of the meteorite and remote-sensing spectroscopy, *Antarctic Meteorite Research 17*, 209–230.

Takeda, H., H. Mori, J. Saito, and M. Miyamoto (1991), Mineral-chemical comparisons of MAC88105 with Yamato lunar meteorites, *Geochim. Cosmochim. Acta, 55*, 3009–3018.

Takeda, H., H. Mori, J. Saito, and M. Miyamoto (1992), Mineralogical studies of lunar mare meteorites EET87521 and Y793274, *Proc. Lunar Planet. Sci., Vol. 22*, 275–301.

Taylor, G. J. (1991), Impact melts in the MAC88105 lunar meteorite: Inferences for the lunar magma ocean hypothesis and the diversity of basaltic impact melts, *Geochim. Cosmochim. Acta, 55*, 3031–3036.

Taylor, L. A., M. Anand, C. Neal, A Patchen., and G. Kramer (2004), Lunar meteorite PCA 02007: A feldspathic regolith breccia with mixed mare/highland components, *Lunar Planet. Sci., XXXV*, abstract 1755.

Taylor, S. R. (1982), *Planetary Science: A Lunar Perspective*, Lunar and Planetary Institute, Houston, Tex.

Terada, K., T. Saiki, Y. Oka, Y. Hayasaka, and Y. Sano (2005), Ion microprobe U-Pb dating of phosphates in lunar basaltic breccia, Elephant Moraine 87521, *Geophys. Res. Lett., 32*, L20202, doi 10.1029/2005GL023909.

Terada, K., Y. Sasaki, and Y. Sano (2006), Ion microprobe U-Pb dating of phosphates in very-low-Ti basaltic breccia, *Meteorit. Planet. Sci., 41*, abstract 5129.

Terada, K., Y. Sasaki, M. Anand, K. H. Joy, and Sano Y. (2007), Uranium-lead systematics of phosphates in lunar basaltic regolith breccia, Meteorite Hills 01210, *Earth Planet. Sci. Lett.*, doi:10.1016/j.epsl.2007.04.029.

Thaisen, K. G., and L. A. Taylor (2009), Meteorite fusion crust variability, *Meteorit. Planet. Sci., 44*, 871–878.

Thalmann, C., O. Eugster, G. F. Herzog, J. Klein, U. Krähenbühl, S. Vogt, and S. Xue (1996), History of lunar meteorites Queen

Alexandra Range 93069, Asuka 881757, and Yamato 793169 based on noble gas isotopic abundances, radionuclide concentrations, and chemical composition, *Meteorit. Planet. Sci.*, *31*, 857–868.

Torigoye-Kita, N., K. Misawa, G. B. Dalrymple, and M. Tatsumoto (1995), Further evidence for a low U/Pb source in the Moon: U-Th-Pb, Sm-Nd, and Ar-Ar isotopic systematics of lunar meteorite Yamato-793169, *Geochim. Cosmochim. Acta*, *59*, 2621–2632.

Treiman, A. H., and M. J. Drake (1983), Origin of lunar meteorite ALH81005: Clues from the presence of terrae clasts and a very low-titanium mare basalt clast, *Geophys. Res. Lett.*, *10*, 783–786.

Treiman, A. H., A. K. Maloy, C. K. Shearer Jr., and J. Gross (2010), Magnesian anorthositic granulites in lunar meteorites Allan Hills 81005 and Dhofar 309: Geochemistry and global significance, *Meteorit. Planet. Sci.*, *45*, 163–180.

Vaniman, D. T., and J. J. Papike (1977), Very low Ti /VLT/ basalts: A new mare rock type from the Apollo 17 drill core, *Proc. 8th Lunar Planet. Sci.*, 1443–1471.

Vaughan, W. M., A. Wittmann, K. H. Joy, T. J. Lapen, and D. A. Kring (2011), Provenance of impact melt and granulite clasts in lunar meteorite PCA 02007, *Lunar Planet. Sci.*, *XLII*, abstract 1247.

Verkouteren, R. M., J. E. Dennison, and M. E. Lipschutz (1983), Siderophile, lithophile and mobile trace elements in the lunar meteorite Allan Hills 81005, *Geophys. Res. Lett.*, *10*, 821–824.

Wang, Y., and W. Hsu (2010), SIMS Pb/Pb dating of Zr-rich minerals from NWA 4734 and LAP 02205/02224: Evidence for the same crater on the Moon, *73rd Annual Meeting of the Meteoritical Society*, abstract 5024.

Wang, Y., Y. Guan, W. Hsu, and J. M. Eiler (2012), Water content, chlorine and hydrogen isotope compositions of lunar apatite, *75th Annual Meeting of the Meteoritical Society*, abstract 5170.

Warren, P. H. (1985), The magma ocean concept and lunar evolution, *Ann. Rev. Earth Planet. Sci. 1985*, *13*, 201–240.

Warren, P. H. (1990), Lunar anorthosites and the magma-ocean plagioclase-flotation hypothesis: Importance of FeO enrichment in the parent magma, *Am. Mineralogist*, *75*, 46–58.

Warren, P. H. (1993), A concise compilation of petrologic information on possibly pristine nonmare Moon rocks, *Am. Mineralogist*, *78*, 360–376.

Warren, P. H. (1994), Lunar and martian meteorite delivery systems, *Icarus*, *111*, 338–363.

Warren, P. H. (2001), Porosities of lunar meteorites: Strength, porosity, and petrologic screening during the meteorite delivery process, *J. Geophys. Res.*, *106*, 10,101–10,111.

Warren, P. H. (2005), "New" lunar meteorites: Implications for composition of the global lunar surface, lunar crust, and the bulk Moon. *Meteorit. Planet. Sci.*, *40*, 477–506.

Warren, P. H., and G. W. Kallemeyn (1989), Elephant Moraine 87521: The first lunar meteorite composed of predominantly mare material, *Geochim. Cosmochim. Acta*, *53*, 3323–3300.

Warren, P. H., and G. W. Kallemeyn (1991a), The MacAlpine Hills lunar meteorite and implications of the lunar meteorites collectively for the composition and origin of the Moon, *Geochim. Cosmochim. Acta*, *55*, 3123–3138.

Warren, P. H., and G. W. Kallemeyn (1991b), Geochemical investigations of five lunar meteorites: Implications for the composition, origin and evolution of the lunar crust, *Proceedings of the NIPR Symposium on Antarctic Meteorites*, *4*, 91–117. Nat. Inst. Polar Res., Tokyo.

Warren, P. H., and G. W. Kallemeyn (1993), Geochemical investigations of two lunar mare meteorites: Yamato-793169 and Asuka-881757, *Proceedings of the NIPR Symposium on Antarctic Meteorites*, *6*, 35–57. Nat. Inst. Polar Res., Tokyo.

Warren, P. H., and G. W. Kallemeyn (1995), QUE93069: a lunar meteorite rich in HASP glasses, *Lunar Planet. Sci.*, *XXVI*, 1465–1466.

Warren, P. H. and F. Ulff-Møller (1999), Lunar meteorite EET96008: Paired with EET87521, but rich in diverse clasts, *Lunar Planet. Sci.*, *XXXI*, abstract 1450.

Warren, P. H., G. J. Taylor, and K. Keil (1983), Regolith breccia Allan Hills 81005: Evidence of lunar origin and petrography of pristine and nonpristine clasts, *Geophys. Res. Lett.*, *10*, 779–782.

Warren, P. H., F. Ulff-Møller, and G. W. Kallemeyn (2005b), "New" lunar meteorites: Impact melt and regolith breccias and large-scale heterogeneities of the upper lunar crust, *Meteorit. Planet. Sci.*, *40*, 989–1014.

Wasson, J. T., and G. W. Kallemeyn (1988), Compositions of chondrites, *Phil. Trans. R. Soc. Lond., A 325*, 535–544.

Wasson, J. T., W. V. Boynton, C.-L Chou., and P. A. Baedecker (1975), Compositional evidence regarding the influx of interplanetary materials onto the lunar surface, *The Moon*, *13*, 121–141.

Wieczorek, M. A., B. L. Jolliff, A. Khan, M. E. Pritchard, B. P. Weiss, J. G. Williams, L. L. Hood, K. Righter, C. R. Neal, C. K. Shearer, I. S. McCallum, S. Tompkins, B. R. Hawke, C. Patterson, J. J. Gillis, and B. Bussey (2006), Chapter 3. The constitution and structure of the lunar interior, in *New Views of the Moon*, pp. 221–364. *Reviews in Mineralogy and Geochemistry*, *60*, edited by Jolliff, B. L. M. A. Wieczorek, C. K. Shearer, and C. R. Neal. Mineralogical Society of America, Washington, D. C.

Yanai, K., and H. Kojima (1991), Varieties of lunar meteorites recovered from Antarctica, *Proceedings of the NIPR Symposium on Antarctic Meteorites*, *4*, 70–90.

Zeigler, R. A., and R. L. Korotev (2013), Petrography and geochemistry of feldspathic lunar meteorite Larkman Nunatak 06638, *Lunar Planet. Sci.*, *XLIV*, abstract 1767.

Zeigler, R. A., R. L. Korotev, B. L. Jolliff, and L. A. Haskin (2005), Petrology and geochemistry of the LaPaz icefield basaltic lunar meteorite and source-crater pairing with Northwest Africa 032, *Meteorit. Planet. Sci.*, *40*, 1073–1102.

Zeigler, R. A., R. L. Korotev, and B. L. Jolliff (2006), Geochemical signature of terrestrial weathering in hot-desert lunar meteorites, *Meteorit. Planet. Sci.*, *41*, A215.

Zeigler, R. A., R. L. Korotev, and B. L. Jolliff (2007), Miller Range 05035 and Meteorite Hills 01210: Two basaltic lunar meteorites, both likely source-crater paired with Asuka 881757 and Yamato 793169, *Lunar Planet. Sci.*, *XXXVIII*, abstract 2110.

Zeigler, R. A., R. L. Korotev, and B. L. Jolliff (2012a), Pairing relationships among feldspathic lunar meteorites from Miller Range, Antarctica, *Lunar Planet. Sci.*, *Lunar Planet. Sci.*, *XLIII*, abstract 2377.

Zeigler, R. A., R. L. Korotev, and B. J. Jolliff (2012b), Feldspathic lunar meteorite Graves Nunataks 06157, a magnesian piece of the lunar highlands crust, *Second Conference on the Lunar Highlands Crust, abstract 9033*, Lunar and Planetary Institute, Houston.

Zhang, A., W. Hsu, Q. Li, Y. Liu, Y. Jiang, and G. Tang (2010), SIMS Pb/Pb dating of Zr-rich minerals in lunar meteorites Miller Range 05035 and LaPaz Icefield 02224: Implications for the petrogenesis of mare basalt, *Science China Earth Sciences*, *53*, 327–334.

7

Meteorites from Mars, via Antarctica

Harry Y. McSween, Jr.[1], Ralph P. Harvey[2], and Catherine M. Corrigan[3]

7.1. INTRODUCTION

Among the most exciting and certainly most valuable samples in the U.S. Antarctic meteorite collection are igneous rocks from Mars. Martian meteorites include basalts and cumulate gabbros ("shergottites"), ultramafic cumulates of clinopyroxene ("nakhlites"), olivine ("chassignites"), and orthopyroxene (ALH 84001), and a complex regolith breccia of basaltic composition (NWA 7034 and its pairs). The shergottites are divided into "basaltic," "olivine-phyric," and "lherzolitic" types. Martian meteorites are often collectively referred to as "SNC" (shergottite, nakhlite, chassignite) meteorites, an abbreviation begun before ALH 84001 was discovered (adding A to make "SNAC" might produce a more inclusive term, but it is unlikely to catch on; the addition of NWA 7034 exacerbates the problem). Of the seven martian lithologies so far represented as meteorites, nearly half were first discovered by ANSMET teams: ALH A77005 was the first known lherzolitic shergottite, EET A79001 was the first olivine-phyric shergottite, and ALH 84001 was the first (and only) orthopyroxenite.

The initial proposal that SNC meteorites are from Mars was made in 1979, the same year that the EET A79001 shergottite was collected in Antarctica. This meteorite had nothing to do with the original idea that shergottites were martian rocks (EET A79001 would not be described until two years later), although ironically it eventually would provide the clinching argument for a Mars origin. The radiometric crystallization ages of the few shergottites and nakhlites that had been analyzed by 1979 were ~180 million and ~1.3 billion years, respectively, much younger than the ~4.5 billion year ages of other achondrites. It is difficult to understand how such relatively recent magmatic activity could have occurred on asteroids, whose much smaller volumes could not retain radioactively generated heat for long periods of time. Based on that logic, *McSween et al.* [1979a] and *Walker et al.* [1979] suggested that the shergottites and nakhlites might be from a large planet, specifically Mars. This novel hypothesis garnered much attention without broad acceptance; at the time, transfer of materials between planetary bodies was considered effectively impossible [*Wetherill*, 1976; *Vickery and Melosh*, 1983].

Two events would quickly revive the martian origin hypothesis. First, ANSMET recovered ALHA 81005, a lunar meteorite serving as *prima facie* evidence that impact events transfer fragments of the Moon to Earth [e.g., *Marvin*, 1983]. Second, analyses of impact melt pockets in EET A79001, originally aimed at determining the timing of ejection from this rock's parent body, discovered trapped gases whose relative elemental and isotopic abundances (Figures 7.1a and 7.1b) exactly matched those determined for the martian atmosphere by *Viking* landers [*Bogard and Johnson*, 1983; *Bogard et al.*, 1984; *Pepin*, 1985]. Given the unique composition of the martian atmosphere and the extraordinary precision of the match with trapped gases in EET A79001, it was difficult to see how the gases could originate anywhere *except* on the planet's surface. This new evidence provided the impetus for rapid development of new models for the transfer of material between planetary bodies [*Melosh*, 1984]. Thirty years of continued recovery of martian specimens support models suggesting that even modest impacts should be able to eject low-shock, near-surface spalls from Mars into Earth-crossing orbits [*Head et al.*, 2002].

[1] *Department of Earth and Planetary Sciences, University of Tennessee*

[2] *Department of Earth, Environmental, and Planetary Sciences, Case Western Reserve University*

[3] *Department of Mineral Sciences, National Museum of Natural History, Smithsonian Institution*

35 Seasons of U.S. Antarctic Meteorites (1976–2010): A Pictorial Guide to the Collection, Special Publication 68,
First Edition. Edited by Kevin Righter, Catherine M. Corrigan, Timothy J. McCoy and Ralph P. Harvey.
© 2015 American Geophysical Union. Published 2015 by John Wiley & Sons, Inc.

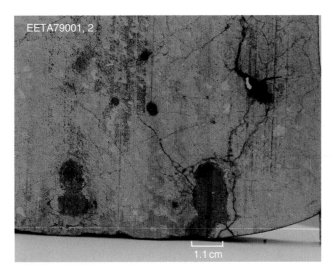

Figure 7.1a. Dark gray pockets of impact melted glass in the EETA 79001 shergottite.

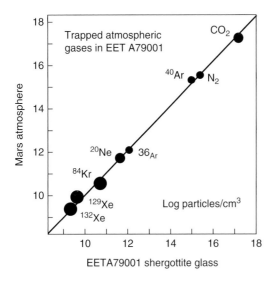

Figure 7.1b. Trapped gas composition in impact glass, compared to martian atmosphere analyzed by the *Viking* lander. This correspondence provides the best evidence for martian origin of SNC meteorites.

In this chapter, we will focus on a subset of martian meteorites in the U.S. Antarctic collection chosen because they have significantly influenced our understanding of the formation, differentiation, and magmatic evolution of Mars. The meteorites will be discussed in chronological order of their recovery. Photographs of these meteorites are presented in the book plates, and additional technical information such as subsampling strategies is given in chapter 3.

7.2. BACKGROUND AND GEOLOGIC CONTEXT

Mars has been the focus of NASA's planetary exploration program for several decades, and staggering amounts of information have been collected by orbiting spacecraft, landers, and rovers (a comprehensive book edited by *Bell* [2008] provides thorough reviews). The planet's surface can be divided into ancient, heavily cratered highlands in the southern hemisphere and younger northern lowlands composed of layered sediments and volcanics. Four magmatic centers host enormous volcanoes with eruptions apparently extending to recent times. Mars lacks any conclusive evidence of plate tectonics; rather the volcanoes are products of plume activity. Martian stratigraphy is defined within three time periods. The oldest (Noachian) is characterized by possibly episodic warmer, wetter environments that produced sediments, as well as pervasive impacts. This period was followed by the drier Hesperian and subsequently by the youngest units (the Amazonian); by Amazonian time, Mars had become an arctic desert.

All of the known martian meteorites except NWA 7034 are igneous rocks, and the consensus among the majority of scientists is that, with two exceptions, they are relatively young, with crystallization ages within the last third of martian history. Radiometrically determined crystallization ages range from 165 to 575 million years for the shergottites, 1260 to 1380 million years for the nakhlites and chassignites (ages and references are summarized by *McSween* [2008]), and 4.09 billion years for ALH 84001 [*Lapen et al.*, 2010]. The newly discovered NWA 7034 breccia has an age of 4.4 billion years [*Agee et al.*, 2013], which the authors interpret as a crystallization age. This bias toward young and relatively strong lithologies is thought to be a selection effect favoring those rocks that can remain intact during ejection by impact and disallowing typical sedimentary rocks or weak impact breccias [*Head et al.*, 2002]. ALH 84001 has a Noachian age, but it is an exceptionally strong breccia with clasts cemented by later minerals, and thus may be an exception to the trend.

By its nature, the impact mechanism that delivers martian rocks to Earth removes their geographic context; the locations on Mars from which the meteorites were derived are therefore unknown. The most promising launch sites are probably recent rayed craters identified in thermal infrared images [*Tornabene et al.*, 2006]. These impacts occurred on young volcanic surfaces and offer direct evidence of large-scale transport of ejecta, but with few constraints on lithology, individual craters cannot yet be correlated with specific meteorites. The launch ages for martian meteorites can be determined from measurement of cosmogenic nuclides, produced through bombardment by cosmic rays when the meteorites were unshielded in space. These ages (ages and references are summarized by *McSween*, [2008]) occur in clusters (Figure 7.2), with

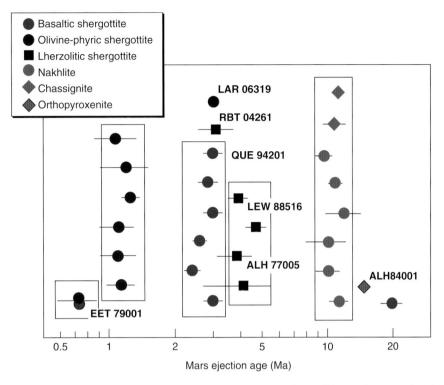

Figure 7.2. Mars ejection ages for SNC meteorites, calculated from cosmogenic nuclides. Only meteorites in the U. S. Antarctic collection are labeled.

most groupings being composed of a single type of SNC meteorite. These data suggest that as many as eight different launch sites on Mars have provided the meteorites.

Many attempts have been made to restore some of the geological context. Source craters within the young basaltic Tharsis terrain were identified as possible shergottite sources by *Mouginis-Mark et al.* [1992]. Later work specifically suggested the crater Zunil as a potential source for some shergottites based on a similar match between crater dynamics and source materials [*Swindle et al.*, 2004]. Craters in the ancient (Noachian age) southern highlands of Mars have been also been suggested as possible sources for ALH 84001 [*Barlow*, 1997]. Spectral mapping from Mars orbit has proven to be of relatively limited use in identifying SNC source regions, primarily because their igneous mineral assemblages (particularly the basaltic lithologies) are hardly unique within the setting of planetary crusts, and globally-distributed eolian dust commonly obscures bedrock mineralogy. Yet some potential SNC launch sites have been identified. Similarities between the relatively unique mineralogy of the nakhlites (Ca-rich pyroxenes and Fe-rich olivines) and the Syrtis Major volcanic complex (as determined from orbit) has been used to suggest the latter as the source for both the nakhlites and chassignites [*Harvey and Hamilton*, 2006]. *Lang et al.* [2009] found a rayed crater on a volcanic flow in Tharsis that had similar spectra to shergottites.

Despite the biases inherent in sampling by impacts and the lack of geographical context, the martian meteorites are an extremely important source of information that complements what has been learned from spacecraft exploration. Remote sensing spectroscopy can identify only a few minerals, whereas complete mineral assemblages can be determined from meteorites. These assemblages constrain the conditions of magmatic crystallization and define subsequent alteration processes and shock history. Although relative chronology of planetary surfaces can be determined from crater counting and stratigraphic mapping, absolute ages can only be measured using radiogenic isotopes. Radiogenic and stable isotopes, as well as most of the elements in trace abundances, in rocks cannot presently be analyzed by remote sensing techniques. These data can be obtained from laboratory measurements on samples, however, and they carry important information about the planet's composition, origin, and differentiation, as well as alteration processes and the cycling of volatile elements between the atmosphere and lithosphere. Analyses of the isotopes of oxygen, the most abundant element in SNCs, also provide a means of identifying these meteorites as SNCs and linking the different SNC types to the same parent planet. The ratios of $^{17}O/^{16}O$ and $^{18}O/^{16}O$ (Figure 7.3) define straight lines with slopes of ~0.5 due to mass fractionation of the different isotopes. Martian meteorites define the mass fractionation trend

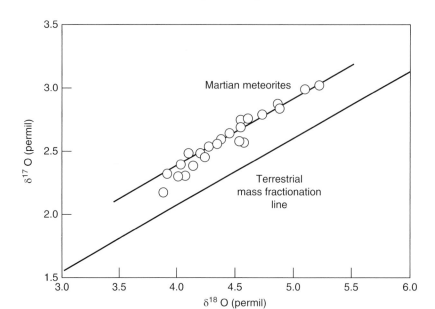

Figure 7.3. Oxygen isotope compositions of SNC meteorites define a mass fractionation line for Mars. The δ notation refers to the ratio of ^{17}O or ^{18}O to ^{16}O, compared to a terrestrial standard.

for Mars, which is distinct from those of the Earth and other bodies. Reviews of the properties and compositions of martian meteorites and what has been learned from them have been published by *McSween* [1994, 2002, 2008], *McSween and Treiman* [1998], *Treiman* [2005], and *Bridges and Warren* [2006].

With the exception of ALH 84001 and NWA 7034, martian meteorites represent Amazonian-age volcanism, and their compositions can be compared with those of older martian igneous rocks analyzed by remote sensing to explore magmatic evolution through time. For example, the Spirit rover APXS analyzed hundreds of Noachian igneous rocks and soils. Their compositions are compared with shergottites and nakhlites in Figure 7.4 [*McSween et al.*, 2009]. Also included in this figure is a box enclosing the compositions of large areas of the martian surface analyzed by a gamma-ray spectrometer on an orbiting spacecraft [*McSween et al.*, 2009]. This plot demonstrates that these martian meteorites are not compositionally representative of the whole martian crust or the mantle source regions that partially melted to produce magmas. However, the composition of the older NWA 7034 breccia [*Agee et al.*, 2013] is similar to the rocks analyzed by Spirit (Figure 7.4). There is also at least one rock on Mars that is similar to SNCs. The Opportunity rover analyzed Bounce Rock, an ejecta block lofted onto Meridiani Planum from elsewhere. This rock was found to be mineralogically and chemically similar (Figure 7.4) to basaltic shergottites [*Zipfel et al.*, 2011], further supporting their martian origin. Similarly, the Spirit rover

identified an outcrop in the Columbia Hills as mineralogically and chemically similar to ALH 84001, including significant carbonates [*Morris et al.*, 2010].

We now consider specific martian meteorites in the U.S. Antarctic collection that have been especially influential in understanding Mars and its rocks.

7.3. LHERZOLITIC SHERGOTTITE: ALH A77005

The importance of Allan Hills A77005 (480 g) is as the first known sample of its kind, and as the first "new" kind of SNC discovered since the group was defined in the mid-seventies (Plate 72). ALH A77005, a cumulate rock obviously related to the shergottites but formed from a less fractionated magma, became the first recognized lherzolitic shergottite, although that classification term was not applied in the original descriptions [*McSween et al.*, 1979a, 1979b; *Ishii et al.*, 1979]. The descriptor "lherzolitic" is not ideal, as it implies modal mineral abundances seen only in some portions of large specimens; ALH A77005 is more properly characterized as a heterogeneous olivine gabbro, containing ultramafic segregations of olivine and pyroxene. The meteorite was initially described at the same time that shergottites were first postulated to be martian rocks, and its unique properties expanded the known geologic complexity of the shergottite parent planet.

Mineralogically ALH A77005 consists of cumulus olivine and chromite, with low- and high-calcium pyroxenes, maskelynite, ilmenite, troilite, and whitlockite. The olivines show preferred orientation [*Berkley*

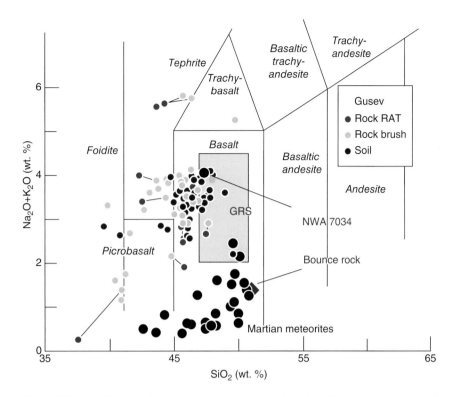

Figure 7.4. Geochemical classification diagram for volcanic rocks, comparing the different compositions for martian meteorites and Noachian-age rocks and soils in Gusev crater analyzed by the Spirit rover on Mars. A box encloses analyses of large areas of the martian surface analyzed by an orbiting GRS.

and Keil, 1981] and unusual brown coloration due to oxidation of iron during shock [*Ostertag et al.*, 1984]. Olivines and pyroxenes are more magnesian than in other shergottites, as appropriate for a cumulate rock. The compositions of spinels in ALH A77005 were reported by *Goodrich et al.* [2003], who inferred an oxygen state, defined in terms of oxygen fugacity, equivalent to QFM - 2.6 (where QFM is the quartz-fayalite-magnetite buffer) during crystallization. ALH A77005 is heavily shocked, accounting for the transformation of plagioclase to maskelynite and the occurrence of pockets of impact melt (now glass with skeletal olivine crystals). *Smith and Steele* [1984] and *Ikeda* [1994] published additional petrographic descriptions of the meteorite.

The mineralogy of ALH A77005 is consistent with early accumulation of crystals from a magma having the crystallization sequence of the parent magmas of olivine-phyric shergottites [*McSween et al.*, 1979a; *Goodrich et al.*, 2003]. The crystallization history was as follows:
• Crystallization and accumulation of olivine and chromite
• Crystallization of pyroxenes, in some cases enclosing olivine and chromite
• Crystallization of plagioclase and other minerals from intercumulus melt

• Equilibration of previously formed phases, including pyroxene exsolution, formation of iron-rich olivine, reaction of chromite to form ulvöspinel
• Shock metamorphism, producing maskelynite and impact melt pockets

Trace element abundances in ALH A77005 minerals [*Lundberg et al.*, 1990] confirm the crystallization sequence inferred from mineral compositions and rock texture.

The meteorite is coarse grained and heterogenous, showing variation in texture and modal abundances on a centimeter scale. Local areas can vary from cumulate olivine-phyric gabbro to lherzolite composed of olivine poikilitically enclosed by pyroxene. Consequently, small samples of lherzolitic shergottites may not be representative, and bulk chemical analyses should be viewed with caution. Bulk analyses for major, minor, and trace elements in ALH A77005, with original sources, were summarized by *Lodders* [1998]. *Shih et al.* [1982], *Jagoutz* [1989], and *Nyquist et al.* [2001] reported its crystallization age as approximately 179 Ma. The name *lherzolitic shergottite* (first applied by *Jagoutz* [1989]) is not ideal, as it implies modal abundances seen only in some portions of large specimens. Harzburgite is a more appropriate term in some regions, and other specimens might be more properly termed gabbroic shergottites.

Additional lherzolitic shergottites occur in the U.S. Antarctic collection: LEW 88516 [*Harvey et al.*, 1993; *Treiman et al.*, 1994] and the paired meteorites RBT 04261 and 04262 [*Usui et al.*, 2010]. All the lherzolitic shergottites are similar in petrology and chronology, with similar Mars launch ages (Figure 7.2), with the possible exception of RBT 04261/2, suggesting that they were derived from the same unit and sampled by the same impact. The unique characteristics of RBT 04261/2 (Plate 73) are discussed by *Usui et al.* [2010].

7.4. OLIVINE-PHYRIC SHERGOTTITE: EET A79001

The case can be made that Elephant Moraine A79001 (Plate 70) is the most important of the SNC meteorites, at least in an historical sense; this was the rock first identified as martian based on trapped gases, as described in the introduction. But EET A79001 is an extremely important meteorite for other reasons as well. A large (7.94 kg) sample, it received considerable attention because it is the first meteorite known to contain two different lithologies (termed A and B) joined along an apparently igneous contact (Figure 7.5) [*Steele and Smith*, 1982; *McSween and Jarosewich*, 1983; *Mikouchi et al.*, 1999; *Goodrich*, 2003]. Lithology A is an olivine-phyric shergottite, the first known sample of that lithology. It consists of megacrysts of olivine, orthopyroxene, and chromite, set in a

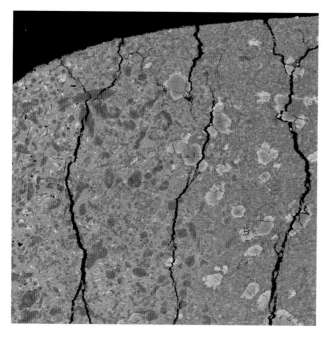

Figure 7.5. Backscatter electron image of EET A79001, showing textural and mineralogical difference between lithology A (on right, with large, zoned olivine megacrysts) and lithology B (on left). Image is ~4 cm across.

matrix of pigeonite, augite, and maskelynite with minor phosphates, ilmenite, pyrrhotite, and mesostasis. This lithology comprises most of the meteorite, and being less fractionated than basaltic shergottites, more closely approximates a primitive martian magma. In contrast, lithology B is a basaltic shergottite consisting of the same phases as in the matrix of lithology A but is finer grained and without the olivine and pyroxene megacrysts.

The megacrysts of olivine and pyroxene in lithology A are not in equilibrium with the groundmass, and they appear to be xenocrysts that reacted with and were partly resorbed by the enclosing magma. They have been suggested to be disaggregated xenoliths of harzburgite, similar to ALH A77005 [*Steele and Smith*, 1982]. *McSween and Jarosewich* [1983] explored the possibility that assimilation of megacrysts by a melt having the composition of lithology B could have produced the groundmass composition of lithology A. However, *Wadhwa et al.* [1994] calculated that the energy required to assimilate sufficient harzburgite was more than could be provided by the latent heat of crystallization. *Mittlefehldt et al.* [1999] suggested that the energy required for assimilation could be satisfied by impact melting. Another possibility is that lithology A is a hybrid magma, formed by mixing of two dissimilar magmas [*Wadhwa et al.*, 1994]. Neither model satisfies all the geochemical constraints, although the idea that the megacrysts are xenolithic is generally accepted. *Goodrich* [2003] argued that the proportion of actual xenocrysts is small, however, and that a highly fractionated late-stage liquid was lost from the rock. *Herd et al.* [2002] provided further data supporting the idea that only the cores of megacrysts in EET A79001 are xenocrystic. The crystallization histories of both lithologies were studied using crystal size distribution (CSD) analysis by *Lentz and McSween* [2000]. The crystallization age of EET A79001 is approximately 180 Ma [*Nyquist et al.*, 2001].

Bulk analyses of major, minor, and trace elements in lithology A and B were summarized by *Lodders* [1998]. The abundances of trace elements in minerals were reported by *Wadhwa et al.* [1994].

The shergottites have provided important insights into the oxidation state of the martian mantle source. Three different oxybarometers have been developed. One [*Wadhwa*, 2001] uses the partitioning of europium in pyroxenes [*Wadhwa*, 2001], another [*Herd et al.*, 2001] is based on the composition of spinel, and a third [*Herd et al.*, 2002] utilizes the behavior of chromium and vanadium in silicates [*Herd et al.*, 2002]. All three methods give similar results of approximately QFM - 1.8 for the two EET A79001 lithologies.

EET A79001 is highly shock metamorphosed. In the curatorial description and following papers, the veins and pockets of impact melt found throughout the rock were designated as lithology C to distinguish this material from the other lithologies. All plagioclase has been converted to maskelynite, and other high-pressure phases (ringwoodite and majorite) have been found in veins.

One other olivine-phyric shergottite, LAR 06319, has been recovered by ANSMET and has been described by *Basu Sarbadhikari et al.* [2009] and *Peslier et al.* [2010]. Although texturally and mineralogically similar to the other olivine-phyric shergottites, and thus considered close to its parental melt, LAR 06319 is unique in several ways. The minerals in LAR 06319 show a continuous oxidation trend with crystallization over two full log units, from the reduced conditions (QFM - 2.0) for early crystallizing phases to oxidizing conditions (QFM + 0.3) for mesostasis phases [*Peslier et al.*, 2010]. LAR 06319 is also a member of the "enriched" shergottites, according to Lu-Hf, Sm-Nd isotopes and trace element data [*Shafer et al.*, 2010]. Thus, LAR 06319 has the dubious distinction of being a "primitive" melt derived from an "enriched" mantle source or from assimilated crust [*Usui et al.*, 2012]. LAR 06319 also has a distinct Mars launch age (Figure 7.2) and is presumably from a different location than EET A79001.

7.5. ORTHOPYROXENITE: ALH 84001

Gram for gram, Allan Hills 84001 may be the most intensely studied rock in history. As the oldest of the martian meteorites, it represents the ancient Noachian crust. It also contains the highest abundance of preterrestrial secondary minerals found among the martian meteorites and thus has witnessed the vast majority of that planet's history. It is also perhaps the most publicly recognizable of the martian meteorites, given claims that its secondary minerals reveal a history that includes biological activity.

Allan Hills 84001 is an orthopyroxene cumulate (1.93 kg) containing minor olivine, chromite, carbonate, sulfide, phosphate, augite, and glass (Plate 69). Originally misclassified as a diogenite, it was not recognized as a martian sample until a decade after its recovery [*Mittlefehldt*, 1994], based primarily on its oxygen isotopic composition. Geochemical analyses of ALH 84001 [*Warren and Kallemeyn*, 1996; *Wadhwa and Crozaz*, 1998] are consistent with origin as a magmatic cumulate. The crystallization age of 4.09 Ga [*Lapen et al.*, 2010] identifies it as the only available meteoritic sample, other than NWA 7034, of the ancient martian crust. The age of secondary carbonate mineralization has been dated at 3.90 Ga [*Borg et al.*, 1999]. The meteorite was heavily shocked, forming brecciated granular bands and feldspathic and silica glasses. Multiple shock events bracket the precipitation of carbonate globules and pancakes (Figure 7.6) within the

Figure 7.6. Carbonate globules in ALH 84001. Orange interiors are siderite, white rims are magnesite and dolomite, and black bands contain magnetite. Globules are several hundred μm across. Figure credit: Monica Grady.

breccia zones [*Treiman*, 1998; *Greenwood and McSween*, 2001; *Corrigan and Harvey*, 2004].

McKay et al. [1996] presented the intriguing proposal that ALH 84001 contains biochemical markers, biogenic minerals, and microfossils (Figure 7.7), all supporting extraterrestrial life. Summaries of arguments in support of [e.g., *Gibson et al.*, 2001; *Thomas-Keprta et al.*, 2009] and against [e.g., *McSween*, 1997; *Treiman*, 2003] this idea have been published in the scientific and popular literature. Polycyclic aromatic hydrocarbons (PAHs) found in ALH 84001 do not require a biologic source and play no significant role in biochemistry [*Zolotov and Shock*, 1999] but might have formed through diagenesis of decomposed biologic matter [*Clement and Zare*, 1996]. The PAHs in ALH 84001 represent only a tiny fraction of the organic matter in this meteorite, whose bulk carbon isotopic composition indicates terrestrial contamination [*Bada et al.*, 1998; *Jull*, 1998; *Becker et al.*, 1999]. Nanophase magnetite grains found in the carbonates were described as morphologically similar to those produced by terrestrial magnetotactic bacteria [*McKay et al.*, 1996; *Thomas-Keprta et al.*, 2009]. However, magnetite whiskers and platelets [*Bradley et al.*, 1996] and epitaxial growth of magnetites on carbonate [*Bradley et al.*, 1998] are inconsistent with a biologic origin, and their occurrence with periclase in carbonate voids suggests formation by carbonate decomposition during impact heating [*Barber and Scott*, 2002; *Brearley*, 2003]. Elongated, segmented forms in carbonates were suggested to be microfossils [*McKay et al.*, 1996]; this claim was later partially retracted when similar features were shown to be artifacts of sample preparation [Bradley et al., 1997; McKay et al., 1997]. The biological community was very unsupportive, given that these forms are smaller than most viruses, with internal volumes too tiny to hold enough

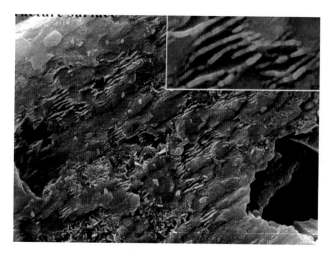

Figure 7.7. Secondary electron image of elongated forms in ALH 84001 cited as evidence for microfossils. Large image is 6 µm across, and higher resolution image is ~900 nm across.

Figure 7.8. Backscatter electron map of QUE 94201, showing zoned pyroxenes (light gray) and maskelynite (dark). Image is 1.3 cm across.

organic molecules to carry out metabolic and genetic functions [*Morowitz*, 1996].

All of the cited evidence occurs associated with the carbonate globules, so their environment (especially temperature) of formation is critical to the argument for life. The globules are chemically zoned, with calcite-ankerite cores and magnesite-siderite rims. Based on oxygen isotopes, *Romanek et al.* [1994] suggested that the carbonate formed by reaction with aqueous fluid at low temperature. *Harvey and McSween* [1996] noted that the absence of phyllosilicates in ALH 84001 was inconsistent with hydrothermal reactions and favored formation at high temperature. Subsequent stable isotope [*Valley et al.*, 1997; *Leshin*, 1998; *Eiler et al.*, 2002] and paleomagnetic [*Kirschvink et al.*, 1997] measurements have established a low-temperature origin, either by precipitation from hydrothermal fluids [*Valley et al.*, 1997; *Treiman et al.*, 2002] or evaporation of brines [*McSween and Harvey*, 1998; *Warren*, 1998]. Following the precipitation of carbonates, likely during multiple events [*Eiler et al.*, 2002; *Corrigan and Harvey*, 2004], they experienced thermal decomposition and sometimes melting by shock [*Scott et al.*, 1997; *Golden et al.*, 2001].

Abiotic explanations for all the purported evidence for life in ALH 84001 are now well established and accepted, and the argument has mostly subsided. However, the proposal of *McKay et al.* [1996] was influential in fueling a vigorous, international Mars exploration program focused on the search for water and life, and the efforts to assess the purported evidence in this meteorite greatly increased our understanding of the environmental constraints on living organisms and of the methods necessary to explore for extraterrestrial life.

7.6. BASALTIC SHERGOTTITE: QUE 94201

Queen Alexandra Range 94201 was the first basaltic shergottite added to the U.S. Antarctic collection (excepting lithology B of EET A79001, which was not an individual sample). Although small (12 g), its composition and properties identify it as an end member to the basaltic shergottite suite, making it unique among this group [*McSween et al.*, 1996; *Mikouchi et al.*, 1998] (Plate 71). Pyroxenes are complexly zoned (sector zoned in many cases) with cores of magnesian pigeonite mantled by magnesian augite, followed by rims of pigeonite that extend to pyroxferroite (Figure 7.8). Other basaltic shergottites are thought to have a two-stage cooling history, with cumulus magnesian pyroxene cores formed at depth, and iron-rich rims formed during ascent or eruption. In contrast, QUE 94201 pyroxene is thought to have continuously crystallized without liquid removal or assimilation, thus representing a liquid composition. This interpretation is bolstered by crystal size distribution (CSD) analysis of pyroxenes [*Lentz and McSween*, 2000] and the high abundance of maskelynite relative to other shergottites; plagioclase crystallization was late in most shergottites but synchronous with pyroxene in QUE 94201. The melt was highly fractionated, producing iron-rich minerals and a high proportion of phosphates.

The bulk composition of QUE 94201 was given by *Lodders* [1998]. It is depleted in incompatible trace elements. Trace element patterns in minerals [*Wadhwa et al.*, 1998] support continuous, closed-system fractional crystallization of the magma. Crystallization occurred under slightly more reducing conditions than for other shergottites [*Herd et al.*, 2001; *Wadhwa*, 2001].

There has been considerable controversy about whether shergottite magmas were dry or hydrous [e.g., *McSween et al.*, 2001, and references therein; *Usui et al.*, 2012]. Analyses of apatites in QUE 94201 clearly demonstrate that the magma that formed this particular meteorite was hydrous, containing 730 to 2130 ppm H_2O [*McCubbin et al.*, 2012].

The crystallization age of QUE 94201 is 327 Ma [*Nyquist et al.*, 2001], older than most other basaltic shergottites. However, its Mars ejection age groups it with other basaltic shergottites (Figure 7.2), suggesting derivation from the same location.

7.7. NAKHLITE: MIL 03346

Like QUE 94201, the importance of the Miller Range 03346 nakhlite lies in its place within the range of known lithologies. The nakhlites preserve a record of martian magmatic processes that are akin to terrestrial komatiites: among the earliest lavas formed when the Earth's interior was hotter than it is today. Their best terrestrial analogs are voluminous ultramafic flows, deep enough and fluid enough to allow significant mineral accumulation. Variation among the nakhlites thus mirrors that seen within a vertical sequence through such a flow, providing a look at the fractionation, crystallization sequence, and equilibration that would have occurred [*Lentz et al.*, 1999].

MIL 03346 is the largest sample among a set of four nakhlites all found in close proximity (Plate 74). MIL 03346 was described by *Day et al.* [2006], and *Udry et al.* [2012] showed that MIL 090030, 090032, and 090136 are petrologically identical to MIL 03346. These paired meteorites are the only nakhlites in the U.S. Antarctic collection. Within error, MIL 03346 shares the same ~1.3 Ga crystallization and 11 Ma ejection age as the other nakhlites and the chassignites (Figure 7.2), suggesting all originate from the same differentiated flow on Mars [*Murty et al.*, 2005; *Shih et al.*, 2006].

Like other nakhlites, MIL 03346 is composed mostly of cumulus augite, with minor olivine and about 20% vitrophyric intercumulus material, or mesostasis. This mesostasis is both more abundant and less crystalline than in other nakhlites. The pyroxenes have homogeneous cores but strongly zoned rims. The vitrophyre is composed of glass with fayalite laths and skeletal grains of iron-titanium oxide and sulfide, and lacks the feathery feldspars found in other nakhlites. This meteorite is among the most highly oxidized nakhlites [*Dyar et al.*, 2005; *Righter et al.*, 2008]. MIL 03346 is less equilibrated than most other nakhlites, presumably because of faster cooling. This, along with the glassy mesostasis, suggests fast cooling near the top of the thick parent lava flow or near-surface sill, with other nakhlites originating further down in the sequence and experiencing slower cooling and

increased equilibration [*Harvey and McSween*, 1992; *Mikouchi et al.*, 2003].

Minor secondary minerals, formed by hydrothermal alteration, occur within MIL 03346 [*Hallis and Taylor*, 2011]. Like the other nakhlites, most of the alteration assemblage consists of "iddingsite": a mixture of Fe-oxides, sulfate, carbonates, and clays found primarily as alteration veins in olivine phenocrysts. In MIL 03346 the absence of siderite and halite suggests a weakly acidic alteration brine relatively less concentrated and depleted in HCO_3^- as compared to those projected for some other nakhlites [*Hallis and Taylor*, 2011]. MIL 03346 and other nakhlites serve as an intermediate data point within the range of martian water-rock interactions recorded in martian meteorites, between the ancient, relatively wet conditions recorded by ALH 84001 and the young and nearly secondary-mineral-free basaltic shergottite QUE 94201.

Because it is a cumulate rock, the bulk chemical analysis of MIL 03346 [*Day et al.*, 2006] does not represent a liquid composition. The meteorite, like other nakhlites, is highly enriched in incompatible trace elements. The history of this meteorite is summarized as follows [*Day et al.*, 2006]:
• Partial melting and ascent of magma
• Slow crystallization of augite and olivine in a shallow magma chamber
• Eruption or emplacement in a sill of the crystal-laden magma
• Crystal settling, rapid cooling, and formation of overgrowths on pyroxenes and skeletal crystals in intercumulus melt
• Secondary alteration by hydrothermal fluid.

7.8. SUMMARY

The U.S. Antarctic meteorite collection has been invaluable in providing new martian meteorites that have had a profound effect on our understanding of the Red Planet. Of the six types of martian meteorites that have been recognized, three were first recovered by ANSMET.

We have described five of these meteorites, including
• Allan Hills A77005, the first known lherzolitic shergottite, which expanded the varieties of martian cumulate rocks
• Elephant Moraine A79001, the first recognized olivine-phyric shergottite, which provided the critical evidence that these meteorites were from Mars
• Allan Hills 84001, the only known orthopyroxenite, which generated a Mars exploration program focused on the search for life
• Queen Alexandra Range 94201, a basaltic shergottite that defines the range of magmatic fractionation in these lavas
• Miller Range 03346, the only nakhlite (with four paired meteorites) in the collection, which samples the top of the nakhlite cumulus pile.

These meteorites have provided a record of martian magmatic activity extending from Noachian time to nearly the present. In addition, their secondary minerals serve as physical witnesses to martian history and provide critical insight into the dramatic climate changes Mars has experienced. The ability to analyze their geochemistry, radiometric ages, and petrology complements and extends what has been learned from spacecraft exploration, and helps to make Mars the most completely studied extraterrestrial body.

REFERENCES

Agee, C. B., N. V. Wilson, F. M. McCubbin, K. Ziegler, V. J. Polyak, Z. D. Sharp, Y. Asmerom, M. H. Nunn, R. Shaheen, M. H. Thiemens, A. Steele, M. L. Fogel, R. Bowden, M. Glamoclija, Z. Zhang, and S. M. Elardo (2013), Unique meteorite from Early Amazonian Mars: Water-rich basaltic breccia Northwest Africa 7034, *Science*, *339*, 780–785.

Bada, J. L., D. P. Glavin, G. D. McDonald, and L. Becker (1998), A search for endogenous amino acids in martian meteorite ALH84001, *Science*, *279*, 362–365.

Barber, D. J., and E. R. D. Scott (2002), Origin of supposedly biogenic magnetite in the martian meteorite Allan Hills 84001, *Proc. Nat. Acad. Sci.*, *99*, 6556–6561.

Barlow, N. (1997), The search for possible source craters for martian meteorite ALH84001 (abstract), *Lunar Planet. Sci.*, *XXVIII*, 65–66.

Basu Sarbadhikari, A., J. M. D. Day, Y. Liu, D. Rumble, and L. A. Taylor (2009), Petrogenesis of olivine-phyric shergottite Larkman Nunatak 06319: Implications for enriched components in martian basalts, *Geochim. Cosmochim. Acta*, *73*, 2190–2214.

Becker, L., B. Popp, T. Rust, and J. L. Bada (1999), The origin of organic matter in the martian meteorite ALH84001, *Earth Planet. Sci. Lett.*, *167*, 71–79.

Bell, J. F., ed. (2008), *The Martian Surface: Composition, Mineralogy, and Physical Properties*, Cambridge University Press, Cambridge, UK.

Berkley, J. L., and K. Keil (1981), Olivine orientation in the ALHA77005 chondrite, *Amer. Mineral.*, *66*, 1233–1236.

Bogard, D. D., and P. Johnson (1983), Martian gases in an Antarctic meteorite, *Science*, *221*, 651–654.

Bogard, D. D., L. E. Nyquist, and P. Johnson (1984), Noble gas contents of shergottites and implications for the Martian origin of SNC meteorites, *Geochim. Cosmochim. Acta*, *48*, 1723–1739.

Borg, L. E., J. N. Connelly, L. E. Nyquist, C.-Y. Shih, H. Wiesmann, and Y. Reese (1999), The age of the carbonates in martian meteorite ALH84001, *Science*, *286*, 90–94.

Bradley, J. P., R. P. Harvey, and H. Y. McSween Jr. (1996), Magnetite whiskers and platelets in the ALH84001 martian meteorite: Evidence of vapor phase growth, *Geochim. Cosmochim. Acta*, *60*, 5149–5155.

Bradley, J. P., R. P. Harvey, and H. Y. McSween Jr. (1997), No "nanofossils" in martian meteorite, *Nature*, *390*, 454.

Bradley, J. P., H. Y. McSween Jr., and R. P. Harvey (1998), Epitaxial growth of nanophase magnetite in martian meteorite Allan Hills 84001: Implications for biogenic mineralization, *Meteorit. Planet. Sci.*, *33*, 765–773.

Brearley, A. (2003), Magnetite in ALH 84001: An origin by shock-induced thermal decomposition of iron carbonate, *Meteorit. Planet. Sci.*, *38*, 849–870.

Bridges, J. C., and P. H. Warren (2006), The SNC meteorites: Basaltic igneous processes on Mars, *J. Geol. Soc. London*, *163*, 229–251.

Clement, S. J., and R. N. Zare (1996), Response to technical comment, *Science*, *274*, 2122–2123.

Corrigan, C. M., and R. P. Harvey (2004), Multi-generational carbonate assemblages in martian meteorite Allan Hills 84001: Implications for nucleation, growth, and alteration, *Meteorit. Planet. Sci.*, *39*, 17–30.

Day, J. M. D., L. A. Taylor, C. Floss, and H. Y. McSween Jr. (2006), Petrology and chemistry of MIL 03346 and its significance in understanding the petrogenesis of nakhlites on Mars, *Meteorit. Planet. Sci.*, *41*, 581–606.

Dyar, M. D., A. H. Treiman, C. M. Pieters, T. Hiroi, M. D. Lane, and V. O'Connor (2005), MIL 03346, the most oxidized martian meteorite: A first look at spectroscopy, petrography, and mineral chemistry, *J. Geophys. Res.*, *110*, E09005, 2005JE00246.

Eiler, J. M., J. W. Valley, C. M. Graham, and J. Fournelle (2002), Two populations of carbonate in ALH 84001: Geochemical evidence for discrimination and genesis, *Geochim. Cosmochim. Acta*, *66*, 1285–1303.

Gibson, E. K., D. S. McKay, K. L. Thomas-Keprta, S. J. Wentworth, F. Westall, A. Steele, C. S. Romanek, M. S. Bell, and J. Toporski (2001), Life on Mars: Evaluation of the evidence within martian meteorites ALH84001, *Nakhla, and Shergotty, Precambrian Res.*, *106*, 15–34.

Golden, D. C., H. V. Lauer, G. E. Lofgren, G. A. McKay, D. W. Ming, R. V. Morris, C. S. Schwandt, and R. A. Socki (2001), A simple inorganic process for formation of carbonates, magnetite, and sulfides in martian meteorite ALH 84001, *Amer. Mineral.*, *86*, 370–375.

Goodrich, C. A. (2003), Petrogenesis of olivine-phyric shergottites Sayh al Uhaymir 005 and Elephant Moraine A79001 lithology A, *Geochim. Cosmochim. Acta*, *67*, 3735–3771.

Goodrich, C. A., C. D. K. Herd, and L. A. Taylor (2003), Spinels and oxygen fugacity in olivine-phyric and lherzolitic shergottites, *Meteorit. Planet. Sci.*, *38*, 1773–1792.

Greenwood, J. P., and H. Y. McSween Jr. (2001), Petrogenesis of Allan Hills 84001: Constraints from impact-melted feldspathic and silica glasses, *Meteorit. Planet. Sci.*, *36*, 43–61.

Hallis, L. J., and G. J. Taylor (2011), Comparisons of the four Miller Range nakhlites, MIL 03346, 090030, 090032 and 090136: Textural and compositional observations of primary and secondary mineral assemblages, *Meteorit. Planet. Sci.*, *46*, 1787–1803.

Harvey, R. P., and V. E. Hamilton (2006), Syrtis Major as the source region of the nakhlite/Chassigny martian meteorites (abstract), *Meteorit. Planet. Sci.*, *40*, A65.

Harvey, R. P., and H. Y. McSween Jr. (1992), Petrogenesis of the nakhlite meteorites: Evidence from cumulate mineral zoning, *Geochim. Cosmochim. Acta*, 56, 1655–1663.

Harvey, R. P., and H. Y. McSween Jr. (1996), A possible high-temperature origin for the carbonates in the martian meteorite ALH84001, *Nature*, 382, 49–51.

Harvey, R. P., M. Wadhwa, H. Y. McSween Jr., and G. Crozaz (1993), Petrography, mineral chemistry and petrogenesis of Antarctic shergottite LEW88516, *Geochim. Cosmochim. Acta*, 57, 4769–4783.

Head, J. N., H. J. Melosh, and B. A. Ivanov (2002), Martian meteorite launch: High-speed ejecta from small craters, *Science*, 298, 1752–1756.

Herd, C. D. K., J. J. Papike, and A. J. Brearley (2001), Oxygen fugacity of martian basalts from electron microprobe oxygen and TEM-EELS analyses of Fe-Ti oxides, *Amer. Mineral.*, 86, 1015–1024.

Herd, C. D. K., C. S. Schwandt, J. H. Jones, and J. J. Papike (2002), An experimental and petrographic investigation of Elephant Moraine 79001 lithology A: Implications for its petrogenesis and the partitioning of chromium and vanadium in a martian basalt, *Meteorit. Planet. Sci.*, 37, 987–1000.

Ikeda, Y. (1994), Petrography and petrology of the ALH 77005 shergottite, *Proc. NIPR Symp. Antarc. Met.*, 7, 9–29.

Ishii, T., H. Takeda, and K. Yanai (1979), Pyroxene geothermometry applied to a three-pyroxene achondrite from Allan Hills, Antarctica, and ordinary chondrites, *Mineral. J.*, 9, 460–481.

Jagoutz, E. (1989), Sr and Nd systematics in ALHA 77005: Age of shock metamorphism in shergottites and magmatic differentiation on Mars, *Geochim. Cosmochim. Acta*, 53, 2429–2441.

Jull, A. J. T., C. Courtney, D. A. Jeffrey, and J. W. Beck (1998), Isotopic evidence for a terrestrial source of organic compounds found in martian meteorites Allan Hills 84001 and Elephant Moraine 79001, *Science*, 279, 366–369.

Kirschvink, J. L., A. T. Maine, and H. Vali (1997), Paleomagnetic evidence of a low-temperature origin of carbonate in the martian meteorite ALH84001, *Science*, 275, 1629–1632.

Lang, N. P., L. L. Tornabene, H. Y. McSween, Jr., and P. R. Christensen (2009), Tharsis-sourced relatively dust-free lavas and their possible relationship to martian meteorites, *J. Volcan. Geotherm. Res.*, 185, 103–115.

Lapen, T. J., M. Righter, A. D. Brandon, V. Debaille, B. L. Beard, J. T. Shafer, and A. H. Peslier (2010), A younger age for ALH84001 and its geochemical link to shergottite sources in Mars, *Science*, 328, 347–351.

Lentz, R. C. F., and H. Y. McSween, Jr. (2000), Crystallization of the basaltic shergottites: Insights from crystal size distribution (CSD), analysis of pyroxenes, *Meteorit. Planet. Sci.*, 35, 919–927.

Lentz, R. C. F., G. J. Taylor, and A. H. Treiman (1999), Formation of a martian pyroxenite: A comparative study of the nakhlite meteorites and Theo's flow, *Meteorit. Planet. Sci.*, 34, 919–932.

Leshin, L. A., K. D. McKeegan, P. K. Carpenter, and R. P. Harvey (1998), Oxygen isotopic constraints on the genesis of carbonates from martian meteorite ALH84001, *Geochim. Cosmochim. Acta*, 62, 3–13.

Lodders, K. (1998), A survey of shergottite, nakhlite and chassigny meteorites whole-rock compositions, *Meteorit. Planet. Sci.*, 33, A183–A190.

Lundberg, L. L., G. Crozaz, and H. Y. McSween Jr. (1990), Rare earth elements in minerals of the ALHA77005 shergottite and implications for its parent magma and crystallization history. *Geochim. Cosmochim. Acta*, 54, 2535–2547.

Marvin, U. B. (1983), The discovery and initial characterization of Allan Hills 81005: The first lunar meteorite, *Geophys. Res. Lett.*, 10, 775–778.

McCubbin, F. M., E. H. Hauri, S. M. Elardo, K. E. Vander Kaaden, J. Wang, and C. K. Shearer (2012), Hydrous melting of the martian mantle produced both depleted and enriched shergottites, *Geology*, 40, 683–686.

McKay, D. S., E. K. Gibson, K. L. Thomas-Keprta, H. Vali, C. S. Romanek, S. J. Clemett, X. D. F. Chillier, C. R. Maechling, and R. N. Zare (1996), Search for past life on Mars: Possible relic biogenic activity in martian meteorite ALH84001, *Science*, 273, 924–930.

McKay, D. S., E. K. Gibson, K. L. Thomas-Keprta, and H. Vali, (1997), *Reply, Nature*, 390, 455.

McSween, H. Y., Jr. (1994), What we have learned about Mars from SNC meteorites, *Meteoritics*, 29, 757–779.

McSween, H. Y., Jr. (1997), Evidence for life in a martian meteorite? *GSA Today*, 7, 1–7.

McSween, H. Y., Jr. (2002), The rocks of Mars, from far and near, *Meteorit. Planet. Sci.*, 37, 7–25.

McSween, H. Y., Jr. (2008), Martian meteorites as crustal samples, in *The martian Surface: Composition, Mineralogy, and Physical Properties*, pp. 383–396, edited by J. F. Bell III, Cambridge University Press, Cambridge, UK.

McSween, H. Y., Jr., and R. P. Harvey (1998), An evaporation model for formation of carbonates in the ALH84001 martian meteorite, *Intern. Geol. Rev.*, 40, 774–783.

McSween, H. Y., Jr., and E. Jarosewich (1983), Petrogenesis of the Elephant Moraine A79001 meteorite: Multiple magma pulses on the shergottite parent body, *Geochim. Cosmochim. Acta*, 47, 1501–1513.

McSween, H. Y., Jr., and A. H. Treiman (1998), Martian meteorites, in *Planetary Materials*, edited by J. J. Papike (pp. 6-1–6-53), *1 36*, Mineral. Soc. Amer., Washington, D. C.

McSween, H. Y., Jr., E. M. Stolper, L. A. Taylor, R. A. Munteen, G. D. O'Kelley, J. S. Eldridge, S. Biswas, H. T. Ngo, and M. E. Lipschutz (1979a), Petrogenetic relationship between Allan Hills 77005 and other achondrites, *Earth Planet. Sci. Lett*, 45, 275–284.

McSween, H. Y., Jr., L. A. Taylor, and E. M. Stolper (1979b), Allan Hills 77005: A new meteorite type found in Antarctica, *Science*, 204, 1201–1203.

McSween, H. Y., Jr., D. D. Eisenhour, L. A. Taylor, M. Wadhwa, and G. Crozaz (1996), QUE94201 shergottite: Crystallization of a martian basaltic magma, *Geochim. Cosmochim. Acta*, 60, 4563–4569.

McSween, H. Y., Jr., T. L. Grove, R. C. F. Lentz, J. C. Dann, A. H. Holzheid, L. R. Riciputi, and J. G. Ryan (2001), Geochemical evidence for magmatic water within Mars from pyroxenes in the Shergotty meteorite, *Nature*, 409, 487–490.

McSween, H. Y., Jr., G. J. Taylor, and M. B. Wyatt (2009), Elemental composition of the martian crust, *Science*, *324*, 736–739.

Melosh, H. J. (1984), Impact ejection, spallation, and the origin of meteorites, *Icarus*, *59*, 234–260.

Mikouchi, T., M. Miyamoto, and G. A. McKay (1998), Mineralogy of Antarctic basaltic shergottite Queen Alexandra Range 94201: Mineralogical similarities to Elephant Moraine A79001 (lithology B), martian meteorite, *Meteorit. Planet. Sci.*, *33*, 181–189.

Mikouchi, T., M. Miyamoto, and G. A. McKay (1999), The role of undercooling in producing igneous zoning trends in pyroxenes and maskelynites among basaltic martian meteorites, *Earth Planet. Sci. Lett.*, *173*, 235–256.

Mikouchi, T., E. Koizumi, A. Monkawa, Y. Ueda, and M. Miyamoto (2003), Mineralogy and petrology of Yamato-000593: Comparison with other martian nakhlite meteorites, *Antarct. Met. Res.*, *16*, 34–57.

Mittlefehldt, D. W. (1994), ALH84001, a cumulate orthopyroxenite member of the SNC meteorite group, *Meteoritics*, *29*, 214–221.

Mittlefehldt, D. W., D. J. Lindstrom, M. M. Lindstrom, and R. R. Martinez (1999), An impact-melt origin for lithology A of martian meteorite Elephant Moraine A79001, *Meteorit. Planet. Sci.*, *34*, 357–367.

Morowitz, H. J. (1996), Technical comment, *Science*, *273*, 1639–1640.

Morris, R. V., S. W. Ruff, R. Gellert, D. W. Ming, R. E. Arvidson, R. C. Clark, D. C. Golden, K. Siebach, G. Klingelhoefer, C. Schroeder, I. Fleischer, A. S. Yen, and S. W. Squyres (2010), Identification of carbonate-rich outcrops on Mars by the Spirit Rover, *Science*, *329*, 421–424.

Mouginis-Mark, P. J., T. J. McCoy, G. J. Taylor, and K. Keil (1992), Martian parent craters for the SNC meteorites, *J. Geophys. Res.*, *97*, 10,213–10,225.

Murty, S. V. S, R. R. Mahajan, J. N. Goswami, and N. Sinha (2005), Noble gases and nuclear tracks in the nakhlite MIL 03346 (abstract), *LPSC, XXXVI*, #1280 (CD-ROM).

Nyquist, L. E., D. D. Bogard, C.-Y. Shih, A. Greshake, D. Stoffler, and O. Eugster (2001), Ages and geologic histories of martian meteorites, *Space Sci. Rev.*, *96*, 105–164.

Ostertag, R., G. Amthauer, H. Rager, and H. Y. McSween (1984), Fe3+ in shocked olivine crystals of the ALHA 77005 meteorite, *Earth Planet. Sci. Lett.*, *67*, 162–166.

Pepin, R. O. (1985), Evidence of martian origins, *Nature*, *317*, 473–475.

Peslier, A. H., D. Hnatyshin, C. D. K. Herd, E. L. Walton, A. D. Brandon, T. J. Lapen, and J. T. Shafer (2010), Crystallization, melt inclusion, and redox history of a martian meteorite: Olivine-phyric shergottite Larkman Nunatak 06319, *Geochim. Cosmochim. Acta*, *74*, 4543–4576.

Righter, K., H. Yang, G. Costin, and R. T. Downs (2008), Oxygen fugacity in the martian mantle controlled by carbon: New constraints from the nakhlite MIL 03346, *Meteorit. Planet. Sci.*, *43*, 1709–1723.

Romanek, C. S., M. M. Grady, I. P. Wright, D. W. Mittlefheldt, R. A. Socki, C. T. Pillinger, and E. K. Gibson (1994), Record of fluid-rock interactions on Mars from the meteorite ALH84001, *Nature*, *372*, 655–657.

Scott, E. R. D., A. Yamaguchi, and A. N. Krot (1997), Petrological evidence for shock melting of carbonates in the martian meteorite ALH84001, *Science*, *387*, 377–379.

Shafer, J. T., A. D. Brandon, T. J. Lapen, M. Righter, B. L. Beard, A. H. Peslier, and B. L. Beard (2010), Trace element systematics and ^{147}Sm–^{143}Nd and ^{176}Lu–^{176}Hf ages of Larkman Nunatak 06319: Closed-system fractional crystallization of an enriched shergottite magma, *Geochim. Cosmochim. Acta*, *74*, 7307–7328.

Shih, C.-Y., L. E. Nyquist, D. D. Bogard, G. A. McKay, J. L. Wooden, B. M. Bansal, and H. Wiesmann (1982), Chronology and petrogenesis of young achondrites, Shergotty, Zagami, and ALHA77005: Late magmatism on a geologically active planet, *Geochim. Cosmochim. Acta*, *46*, 2323–2344.

Shih, C.-Y., L. E. Nyquist, and Y. Reese (2006), Rb-Sr and Sm-Nd isotopic studies of Antarctic nakhlite MIL 03346 (abstract), *LPSC, XXXVII*, #1701 (CD-ROM).

Smith, J. V., and I. M. Steele (1984), Achondrite ALHA77005: Alteration of chromite and olivine, *Meteoritics*, *19*, 121–133.

Steele, I. M., and J. V. Smith (1982), Petrography and mineralogy of two basalts and olivine-pyroxene-spinel fragments in achondrite EETA79001, *J. Geophys. Res.*, *87*(Suppl.), A375–S384.

Swindle, T. D., J. B. Plescia, and A. S. McEwen (2004), Exploring a possible shergottite source crater and calibrating martian cratering chronology, *Meteorit. Planet. Sci.*, *39*(Suppl.), abstract #5149.

Thomas-Keprta, K. L., S. J. Clemett, D. S. McKay, E. K. Gibson, and S. J. Wentworth (2009), Origins of magnetite crystals in martian meteorite ALH84001, *Geochim. Cosmochim. Acta*, *73*, 6631–6637.

Tornabene, L. L., J. E. Moersch, H. Y. McSween, Jr., J. L. Piatek, K. A Milam, A. S. McEwen, and P. R. Christensen (2006), Identification of large (2–10 km), rayed craters on Mars in THEMIS thermal infrared images: Implications for possible martian meteorite source regions, *J. Geophys. Res.*, *111*, E10006, doi:10.1029/2005JE002600.

Treiman, A. H. (1998), The history of Allan Hills 84001 revised: Multiple shock events, *Meteorit. Planet. Sci.*, *33*, 753–764.

Treiman, A. H. (2003), Submicron magnetite grains and carbon compounds in martian meteorite ALH 84001: Inorganic, abiotic formation by shock and thermal metamorphism, *Astrobiol.*, *3*, 369–392.

Treiman, A. H. (2005), The nakhlite meteorites: Augite-rich igneous rocks from Mars, *Chem. Erde*, *65*, 203–270.

Treiman, A. H., G. A. McKay, D. D. Bogard, D. W. Mittlefehldt, M. S. Wang, L. Keller, M. E. Lipschutz , M. M. Lindstrom, and D. Garrison (1994), Comparison of the LEW 88516 and ALH77005 martian meteorites: Similar but distinct, *Meteoritics*, *29*, 581–592.

Treiman, A. H., H. E. F. Amundsen, D. F. Blake, and T. Bunch (2002), Hydrothermal origin for carbonate globules in martian meteorite ALH84001: A terrestrial analogue from Spitsbergen (Norway), *Earth Planet. Sci. Lett.*, *204*, 323–332.

Udry, A., H. Y. McSween Jr., P. Lecumberri-Sanchez, and R. J. Bodnar (2012), Paired nakhlites MIL 090030, 090032, 090136, and 03346: Insights into the Miller Range parent meteorite, *Meteoritics & Planetary Science*, *47*, 1575–1589.

Usui, T., M. Sanborn, M. Wadhwa, and H. Y. McSween Jr. (2010), Petrology and trace element geochemistry of RBT 04261 and RBT 04262 meteorites, the first examples of geochemically enriched lherzolitic shergottites, *Geochim. Cosmochim. Acta*, *74*, 7283–7306, doi:10.1016/j.gca.2010.09.010.

Usui, T., C. M. Alexander, J. Wang, J. I. Simon, and J. H. Jones (2012), Origin of water and mantle-crust interactions on Mars inferred from hydrogen isotopes and volatile element abundances of olivine-hosted melt inclusions of primitive shergottites, *Earth Planet. Sci. Lett.*, *357–358*, 119–129.

Valley, J. W., J. M. Eiler, C. M. Graham, E. K. Gibson, C. S. Romanek, and E. M. Stolper (1997), Low-temperature carbonate concretions in the martian meteorite ALH84001: Evidence from stable isotopes and mineralogy, *Science*, *275*, 1633–1638.

Vickery, A. M., and H. J. Melosh, (1983), The origin of the SNC meteorites: An alternative to Mars, *Icarus*, *56*, 299–318.

Wadhwa, M. (2001), Redox state of Mars' upper mantle and crust from Eu anomalies in shergottite pyroxenes, *Science*, *291*, 15,271–15,530.

Wadhwa, M., and G. Crozaz (1998), The igneous crystallization history of an ancient martian meteorite from rare earth element microdistributions, *Meteorit. Planet. Sci.*, *33*, 685–692.

Wadhwa, M., H. Y. McSween Jr., and G. Crozaz (1994), Petrogenesis of shergottite meteorites inferred from minor and trace element microdistributions, *Geochim. Cosmochim. Acta*, *58*, 4213–4229.

Wadhwa, M., G. Crozaz, L. A. Taylor, and H. Y. McSween Jr. (1998), martian basalt (shergottite), QUE94201 and lunar basalt 15555: A tale of two pyroxenes, *Meteor. Planet. Sci.*, *33*, 321–328.

Walker, D., E. M. Stolper, and J. F. Hays (1979), Basaltic volcanism: The importance of planet size, *Proc. Lunar Planet Sci. Conf.*, *10*, 1995–2015.

Warren, P. H. (1998), Petrologic evidence for low-temperature, possibly flood evaporatic origin of carbonates in the ALH84001 meteorite, *J. Geophys. Res.*, *103*, 16,759–16,773.

Warren, P. H., and G. W. Kallemeyn (1996), Siderophile trace elements in ALH84001, other SNC meteorites and eucrites: Evidence of heterogeneity, possibly time-linked, in the mantle of Mars, *Meteorit. Planet. Sci.*, *31*, 97–105.

Wetherill, G.W. (1976), Where do the meteorites come from? A re-evaluation of the Earth-crossing Apollo objects as sources of chondritic meteorites. *Geochim. Cosmochim. Acta* *40*, 1297–1317.

Zipfel, J., C. Schroder, B. L. Jolliff, R. Gellert, K. E. Herkenhoff, R. Rieder, R. Anderson, J. F. Bell, Brückner J., J. A. Crisp, P. R. Christensen, B. C. Clark, P. A. de Souza, G. Dreibus, C. d'Uston, T. Economou, S. P. Gorevan, B. C. Hahn, G. Klingelhöfer, T. J. McCoy, H. Y. McSween Jr., D. W. Ming, R. V. Morris, D. S. Rodionov, S. W. Squyres, H. Wänke, S. P. Wright, M. B. Wyatt, and A. S. Yen (2011), Bounce Rock: a shergottite-like basalt encountered at Meridiani Planum, Mars, *Meteor. Planet. Sci.*, *46*, 1–20, doi:10.1111/j.1945-5100.2010.01127.x.

Zolotov, M., and E. Shock (1999), Abiotic synthesis of polycyclic aromatic hydrocarbons on Mars, *J. Geophys. Res.*, *104*, E6, 14,033–14,049.

8

Meteorite Misfits: Fuzzy Clues to Solar System Processes

Timothy J. McCoy

8.1. INTRODUCTION

Since its inception in 1976, meteorites from the U.S. Antarctic collection have revolutionized our understanding of the processes that occurred during the birth and evolution of the solar system. Many of these advancements are discussed in the preceding chapters of this book, ranging from understanding the full array of materials present in the solar nebula to gaining insights about habitable environments on early Mars. In spite of these advancements, our knowledge of the history of our solar system remains woefully incomplete.

In this chapter, I examine meteorites that may offer clues to these processes, but those clues remain unclear. A partial list of misfit meteorites from the U.S. Antarctic meteorite program is given in Table 8.1, along with brief notes as to the nature or relationships of these meteorites. The uncertain origin of these meteorites results from a wide range of factors. For a few of the meteorites, no detailed follow-up studies have ever been conducted, probably as a result of the subsequent recovery of meteorites deemed more interesting. For others, detailed initial studies lacked the analytical tools or paradigms that exist today and prompt us to revisit earlier-recovered older meteorites. As I show, some of these enigmatic meteorites await recovery of related meteorites for the emergence of a complete understanding of the chemical and physical processes that formed and altered asteroids. The origin of some meteorites, although discussed by one or more authors, is often unclear even in the processes

that formed them (e.g., impact melting vs. asteroid differentiation). In other cases, these meteorites can only be fully understood in the context of their parent asteroids, and spacecraft missions are posing new challenges and opportunities.

8.2. MATERIALS AND PROCESSES IN THE SOLAR NEBULA

Chondritic meteorites are essentially sedimentary rocks composed of materials formed prior to their lithification on an asteroid. The components themselves formed either on the parent asteroid prior to final lithification or, as is widely argued at least for chondrules and CAIs, in the solar nebula. As such, the full range of materials found in chondrites provides important clues to the range of materials present in the solar nebula. Much of the research in this field has centered on either the abundant ferromagnesian chondrules dominated by the mafic silicates olivine and pyroxene or on the ubiquitous, but volumetrically minor (<10 vol.%), calcium-aluminum-rich inclusions (CAIs) whose bulk composition, mineralogy, and isotopic signatures suggest they were among the earliest formed phases in the solar nebula. Far less attention has been paid to the aluminum-rich chondrules [*Bischoff and Keil*, 1983]. The compositions of these objects suggest they are intermediate between ferromagnesian chondrules and CAIs, with bulk compositions controlled either by nebular condensation or through remelting of mixed components [*MacPherson and Huss*, 2005].

Within the Antarctic meteorite collection, two occurrences of Al-rich chondrules are of note. *Grossman et al.* [1995] noted that Lewis Cliff (LEW) 87234 (paired with

Department of Mineral Sciences, National Museum of Natural History, Smithsonian Institution

35 Seasons of U.S. Antarctic Meteorites (1976–2010): A Pictorial Guide to the Collection, Special Publication 68, First Edition. Edited by Kevin Righter, Catherine M. Corrigan, Timothy J. McCoy and Ralph P. Harvey.

Table 8.1. List of ungrouped or anomalous meteorites.

ALH A80104	Post-eutectic crystallization iron meteorite?
EET 83230	High-Ni iron related in process to IVA irons
EET 96031 pairing group	Low-FeO chondrites
GRA 06128/06129	Albite-rich achondrites related to brachinites
GRO 95555	Unbrecciated, equigranular diogenite
ILD 83500	High-Ni iron related to South Byron trio and, possibly, the Milton pallasite
LAP 04757/04773	Low-FeO chondrites similar in bulk composition to H chondrites
LAP 03719	Unbrecciated, olivine-bearing aubrite
LAP 04840/10031/10033	Hornblende-phlogopite-bearing R6 chondrite
LAR 04315	Metal-sulfide-veined ureilite
LAR 06298/06299	Kg-sized LL chondrite impact melt breccias
LEW 85332	Low petrologic type, ungrouped carbonaceous chondrite
LEW 85369	Silicon-bearing iron meteorite
LEW 86211	Troilite-rich ungrouped iron meteorite
LEW 87051	Quenched angrite; main mass of 0.611 g
LEW 87232	Kakangari-like chondrite
LEW 87234 pairing group	Ungrouped type 3 enstatite chondrites rich in Ca-Al-rich chondrules
LEW 88055	Silicate-bearing iron meteorite related to aubrites
LEW 88631	Silicide-bearing iron meteorite
LEW 88774	Ureilite exhibiting coarse exsolution of augite in low-Ca pyroxene
MAC 87300/87301	C2 chondrite related to CM and CO groups
MAC 88107	C2 chondrite related to CM and CO groups
MET 01085	Olivine-free ureilite
MIL 03443	Dunite related to HED meteorites
MIL 090405 pairing group	Brachinite-like achondrites
MIL 090807/090978	Enstatite chondrite impact melts similar to Happy Canyon
MIL 11207	Hornblende-bearing R6 chondrite
QUE 93148	Coarse-grained olivine-pyroxene-metal; pyroxene pallasite?
QUE 94204 pairing group	Coarse-grained enstatite meteorites; impact melts or primitive achondrites?
QUE 94570	Low-FeO chondrite similar in bulk composition to L chondrites
RBT 04239/04255	Ungrouped olivine-rich achondrites similar to Divnoe
RKP A79015	Anomalous mesosiderite

LEW 87057/87220/87223) contained an unusually high abundance of Ca- and Al-rich objects relative to other enstatite chondrites, including chondrules that contain primary (igneous) calcic plagioclase. This meteorite also contains a high abundance of FeO-bearing pyroxene for an enstatite chondrite [*Weisberg et al.*, 2011]. *Weisberg and Righter* [2014 (this volume)] have noted the anomalous nature of this meteorite grouping, suggesting they don't fall within either the EH or EL group. *Grossman et al.* [1995] reported that three analyses revealed no resolvable excesses of ^{26}Mg indicative of the former presence of ^{26}Al. These results are supportive of the hypothesis that Al-rich chondrules formed after CAIs. Far more commonly, a single interesting chondrule will be noted in the initial description of a meteorite, without subsequent study. A relevant example of this is found in the LL6 chondrite Miller Range (MIL) 07065. The initial description indicated the presence of a semicircular, 1-mm diameter skeletal spinel. Subsequent examination by the author suggested that a circular area including this grain is enriched in feldspar. Despite extensive

metamorphism, this suggests the possibility of an Al-rich chondrule in which spinel was the liquidus phase and rapid cooling or few nuclei produced a skeletal morphology. Given the potential of these objects as intermediate between chondrules and CAIs and significant improvements in our ability to constrain relative ages in the earliest history of the solar system, revisiting these and other examples of Al-rich chondrules in the Antarctic collection might prove fruitful.

8.3. NEBULAR OR PARENT BODY?

Given the complexity of meteorite petrogenesis, it is of little surprise that the properties of some meteorites cannot be confidently ascribed to a particular setting. Attributing the properties of chondritic meteorites to nebular or parent body settings are among the most difficult connections to establish. Two groupings of meteorites that illustrate this difficulty are Queen Alexandra Range (QUE) 94570 and the LaPaz Icefield (LAP) 04773/04757 pairing groups (Plate 10). QUE 94570 was originally classified as an unusual carbonaceous chondrite similar to Coolidge, with mafic silicate compositions that are more magnesian than typical for ordinary chondrites (olivine of Fa_{10-13} vs. Fa_{16-30} for ordinary chondrites). Subsequent bulk chemical analyses by *Kallemeyn et al.* [1998] revealed that QUE 94570 is chemically similar to L chondrites, although with much more magnesian silicate compositions, similar to the non-Antarctic meteorites Moorabie and Suwahib Buwah.

In contrast, the paired meteorites LAP 04757/04773 are compositionally and isotopically indistinguishable from H chondrites but have mafic silicate compositions more reduced than H chondrites (olivine of Fa_{13} vs. Fa_{16-20} for H chondrites) [*Troiano et al.*, 2011]. *Troiano et al.* [2011] review models for formation of chondrites similar to ordinary chondrites but with lower FeO concentrations in the mafic silicates. These authors favor a parent body model in which less accretion of reduced components (e.g., metal) occurred relative to oxidized components (e.g., ice) within the H chondrite parent body, producing more mafic silicates. Alternatively, these meteorites could sample materials from the solar nebula that were more reduced initially. While these theories may well be correct, they have greater difficulty accounting for the formation of L chondrites with FeO-poor mafic silicate compositions. Further, they should predict a continuum of compositions within the H chondrite population, from that of typical H chondrites down to that of the low-FeO chondrites. A detailed examination of H chondrite mafic silicate compositions might further our understanding of the number of ordinary chondrite parent bodies and their formation.

8.4. PARENT BODY AQUEOUS ALTERATION

The phyllosilicate-bearing matrix of CI, CM, and CR and some low petrologic type ordinary chondrites have long testified to the importance of water in parent body alteration. However, recent studies have revealed an increasingly diverse set of aqueous alteration products beyond the hydrated carbonaceous chondrites. Rare salt crystals, some of which contain fluid inclusions, have been identified in two ordinary chondrite breccias [*Zolensky et al.*, 1999], although none to date in Antarctic meteorites. An occurrence of hydrated phases first identified in the Antarctic collection was in LAP 04840 (Plate 36), where biotite and amphibole were observed [*McCanta et al.*, 2008]. Containing ~13% ferri-magnesio-hornblende and ~0.4% phlogopite by volume, *McCanta et al.* [2008] suggest that LAP 04840 formed when a hot, dry R chondrite was juxtaposed to a cooler, hydrogen-bearing rock. While the hydrogen could have been liberated from insoluble organic matter, explaining its

remarkably deuterium-rich nature, the proximity of hot and cold materials may have required a role for impact, either through crater formation on the parent asteroid or disruption and reassembly. Although further insights into understanding these scenarios could come from additional studies of LAP 04840, this meteorite illustrates the value of additional sample recovery. Subsequent field parties have recovered a number of similar meteorites from the LaPaz Icefield. In this case, these are thought to be fragments of the same meteorite which either broke apart during atmospheric entry, creating a shower of stones that fell, or during transport in the ice. These fragments, essentially identical to LAP 04840, offer few new insights. In contrast, a recent recovery from a geographically (and glacially) distinct region, Miller Range (MIL) 11207 is very different in degree of metamorphism and is certainly not paired with LAP 04840 [*Gross et al.*, 2013]. Future studies of this meteorite will undoubtedly further elucidate our understanding of the role of water in the alteration of chondritic bodies.

8.5. IMPACT PRODUCTS

Study of Antarctic meteorites has made significant contributions to understanding the impact history of our solar system. *Cohen et al.* [2000] reported $^{40}Ar-^{39}Ar$ ages for impact melt clasts from lunar meteorites, including MacAlpine Hills (MAC) 88105 (Plate 68) and Queen Alexandra Range (QUE) 93069. These authors found no clasts older than 4.0 billion years, supporting the idea that an increase in the impact flux produced most of the large basins on the Moon, the so-called lunar cataclysm. *Bogard and Garrison* [2003] reported $^{39}Ar-^{40}Ar$ ages for eucrites that suggest the same cataclysm might have occurred on the parent body of these meteorites, thought to be asteroid 4 Vesta. Few detailed studies of other types of breccias from the Antarctic collection have been conducted. In particular, substantial numbers of large Antarctic ordinary chondrites have never been adequately examined for evidence of brecciation [*Corrigan et al.*, 2012]. *Welzenbach et al.* [2005] report a variety of clasts in MAC 87302 (Plate 5) and *Corrigan et al.* [2012] in a variety of ordinary chondrite breccias, including impact melt clasts that could be dated to further elucidate whether the impact events recorded on the Moon and 4 Vesta are a solar system–wide phenomenon. Coupled with other interesting phases recorded in ordinary chondrite breccias (e.g., halides), this largely unstudied population deserves closer scrutiny.

Some of the more enigmatic and poorly studied groups of Antarctic meteorites are those that exhibit igneous textures but mineralogies that are essentially chondritic. Some of these meteorites, such as the acapulcoites

Figure 8.1. Photomosaic of Queen Alexandra Range 94204 in cross-polarized light. Rounded to irregular polysynthetically twinned enstatite crystals are set in a matrix and include Fe,Ni metal, troilite, and plagioclase. Scale bar is 1 mm. Despite numerous petrographic descriptions, debate continues as to whether this meteorite formed by low degrees of incipient partial melting or complete impact melting of an enstatite chondrite.

and lodranites, are discussed by *Mittlefehldt and McCoy* [2014 (this volume)]. Another example is the enigmatic enstatite meteorite QUE 94204 (Figure 8.1), which is the first recovered sample of a significant pairing group (Plate 34). Consisting of rounded to ovoid millimeter-sized enstatite grains with included and interstitial metal, sulfide, and feldspar, the mineralogy is similar to enstatite chondrites, but the texture is distinctly achondritic. *Burbine et al.* [2000] argued that the presence of zoned feldspar and polysynthetic twinning in enstatite indicates rapid cooling, suggestive of impact melting as the mechanism to retain the bulk chondritic composition. In contrast, *Izawa et al.* [2011] suggest that QUE 94204 formed through partial melting of an enstatite chondrite-like protolith, analogous to the formation of acapulcoites and lodranites, although these authors could not unambiguously rule out impact melting. Recently, *van Niekerk and Keil* [2012] compared the two competing hypotheses for the formation of the QUE 94204 pairing group and favored an impact melt origin. Despite the extensive petrologic studies referenced above, the petrogenesis of these meteorites is still unclear, yet the issue of impact versus indigenous melting of asteroids remains one of the central questions in understanding the melting of asteroids in the early history of the solar system.

8.6. CHONDRITE-ACHONDRITE LINKS

It is generally accepted that differentiated asteroids and planets began, at some point in their history, as chondritic agglomerates that were subsequently processed. With more than a dozen known types of chondrites and a comparable or larger number of achondrites (including the ungrouped irons; see *Mittlefehldt and McCoy* [2014 (this volume)], one of the more compelling questions in meteoritics is whether we have any differentiated bodies that formed by melting of known chondrites. Arguments have been presented to link silicate-bearing IIE irons with H chondrites and silicate-bearing IVA irons with LL chondrites, but the study of Antarctic meteorites has done little to augment these arguments. In contrast, one of the most hotly debated chondrite-achondrite links is that between the differentiated aubrites and the similarly highly reduced enstatite chondrites. It is particularly noteworthy that this topic continues to garner fierce debate given that it dates to at least the earliest days of the U.S. Antarctic meteorite program [e.g, *Watters and Prinz*, 1979].

Enstatite chondrites and aubrites share a wide range of similar properties, including major mineralogy, mineral chemistry, and oxygen isotopic composition. Both formed under highly reducing conditions in which FeO is essentially absent from the mafic silicates and elements normally found in silicates combine to form exotic sulfides, metal, and nitrides. Despite these similarities, several differences exist between aubrites and enstatite chondrites, with the former containing higher abundances of diopside and olivine, along with relatively Ti-rich troilite. A variety of solutions have been offered to explain how these differences could have arisen during igneous differentiation, most involving either crystallization or oxidation-reduction reactions. In this context, two Antarctic meteorites are particularly relevant. Elephant Moraine (EET) 90102 was described as the first diopside-bearing equilibrated enstatite chondrite [*Fogel*, 1997]. Genesis of aubrites from such a protolith could help explain the increased abundances of diopside in these meteorites. A second meteorite relevant to this debate is LAP 03719 (Plate 51). Most aubrites are heavily brecciated, presenting a significant challenge in understanding the lithologies present and relationships between them. In contrast, LAP 03719 (Figure 8.2) is unbrecciated and contains ~15 vol.% olivine. *McCoy et al.* [2005] noted that one of the theories for olivine crystallization in aubrites, that it was stabilized by the crystallization of roedderite in a peralkaline melt, is inconsistent with the absence of roedderite in LAP 03719. No further studies have been conducted on this meteorite, and the question of whether enstatite chondrites and aubrites are, in fact, directly related to one another remains unresolved.

Figure 8.2. The unbrecciated, olivine-bearing aubrite LaPaz Ice Field 03719 has been little studied but could provide insights into the link between differentiated meteorites and their primitive, chondritic precursors. An important distinction between aubrites and the related enstatite chondrites is the abundance of olivine. Scale bar is 1 mm.

8.7. IGNEOUS DIFFERENTIATION

Meteorites hold particular significance in our understanding of how primitive, chondritic bodies were transformed into layered, differentiated worlds like the Earth. Unchanged by later plate tectonics, volcanism, or erosion, primitive differentiated bodies provide some of our best clues to the processes and timescales of differentiation early in the history of the solar system.

An interesting example of an understudied meteorite is Lewis Cliff (LEW) 87051 (Plate 54). LEW 87051 is an angrite, a group of meteorites whose FeO-rich mineralogy, unusual phase assemblage (dominated by Ti-,Al-rich calcic pyroxene), and quench textures suggest that they formed as near-surface igneous units as a result of melting under conditions that are highly oxidized compared to most meteorites. Prior to 1999, only four angrites were known. The type meteorite, Angra dos Reis, was observed to fall in 1869. The other three were recovered in Antarctica, two by the U.S. program. Each exhibits a distinct petrology, with LEW 87051 showing a quenched texture. Since 1999, numerous angrites have been discovered, most from Northwest Africa, and many approach or surpass a kilogram in mass. This has renewed interest in angrites and spurred reexamination of the original four. Unfortunately, the cautionary tale of LEW 87051 is that it was found as a mass of only 0.6 g, of which only

two thin sections with minute amounts of material (areas of ~1 mm) exist. Thus, LEW 87051 records an important part of the angrite petrogenetic story but one that is extremely limited by our lack of ability to conduct further analyses.

Other small meteorites present challenges in establishing linkages to known groups. QUE 93148 (Plate 63) was collected as a mass of 1.09 g and originally classified as a lodranite. It is dominated by coarse-grained, magnesian olivine (Fa$_{15}$) and pyroxene (Fs$_{13}$), along with high concentrations of metal. *Goodrich and Righter* [2000] convincingly demonstrated that QUE 93148 is not related to the lodranites, although its linkages to other groups remains uncertain. These authors suggested that it may be a pyroxene pallasite, a relatively rare group of metal-rich olivine-pyroxene rocks that are isotopically distinct from the main group pallasites. Alternatively, they speculated that QUE 93148 could be derived from the mantle of the pallasite parent body. At that time, a genetic link between pallasites and the basaltic to orthopyroxenitic cumulate howardite-eucrite-diogenite (HED) series of rocks was widely accepted. Reexamination of QUE 93148 has occurred in the last few years as a result of both new technological developments and a changing paradigm established by new meteorites. Laser ablation inductively coupled plasma mass spectrometry now permits analysis of siderophile element concentrations in metal at the

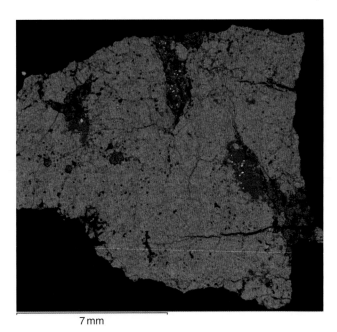

7 mm

Figure 8.3. Mineral map derived from composite elemental maps of the Graves Nunataks 06128 meteorite. The high abundance of sodic feldspar (green) and clastlike appearance of olivine and pyroxene (brown), phosphates (purple), and sulfides (yellow) suggest the formation of a lithology significantly enriched in feldspar. Whether such a lithology derived by partial melting or required plagioclase fractionation akin to the lunar magma ocean is unresolved.

scale of tens of micrometers. *Righter et al.* [2008] applied this technique to QUE 93148 but were limited by the availability of a sample with minimal metal. Further, other olivine-bearing rocks from the HED parent body, namely the harzburgitic diogenites, are now known [*Beck and McSween*, 2010]. A search for metal within the harzburgitic diogenites and a subsequent comparison of compositions to metal in QUE 93148 may yield new insights into the origin of that meteorite.

In other cases, some meteorites remain anomalous not because of the availability of material but because ancillary studies cannot fully distinguish between two competing models. Graves Nunataks (GRA) 06128 and 06129 (Figure 8.3) are two paired, relatively large (~0.65 kg total) meteorites that are exceptional in containing a highly oxidized mafic assemblage with ~85 vol.% sodic plagioclase (Plate 49). A relationship to the brachinites seems certain, although brachinites are olivine-dominated meteorites. In most respects, GRA 06128/129 appear to represent the complementary partial melts removed during relatively low degrees of partial melting, leaving the brachinites as residues. Complementary basaltic partial melts are known from a number of other meteorite groups, including silicate-bearing IAB irons, aubrites, ureilites and acapulcoites-lodranites. What is far

less certain is whether GRA 06128/129 was derived solely from crystallization of a low-degree partial melt, as has been argued by, for example, *Day et al.* [2012]. Alternatively, mineral accumulation akin to the plagioclase flotation that formed the lunar highlands crust or crystal segregation in a magma chamber may be responsible for the extraordinarily high plagioclase concentration, as argued by *Shearer et al.* [2010]. If plagioclase accumulation is required, GRA 06128/129 samples a process that was previously only known from larger planetary bodies. It seems clear that existing experimental studies of partial melting of oxidized chondritic meteorites are insufficient in number or range of conditions to fully decide between these two possible models.

8.8. SPACECRAFT MISSIONS

Among the rarest opportunities afforded to a planetary scientist is that of visiting, for the first time, a planetary body from which we have meteorites samples. These encounters, even if only brief flybys, allow us to test paradigms established in the laboratory and, in most cases, realize that our understanding is both shallow and incomplete. Such an opportunity has recently been afforded by the visit of the Dawn spacecraft to asteroid 4 Vesta. A variety of arguments have been presented that the HED meteorites originate from this asteroid [see *Mittlefehldt and McCoy*, 2014 (this volume)]. As such, we have a chance to evaluate our models for the formation of the most abundant group of igneous meteorites, the HEDs. One of the major surprises of Dawn's encounter with 4 Vesta was the presence of not just the single southern polar basin imaged by Hubble [*Thomas et al.*, 1997] but two major basins, the formation of which appear to have dominated the geological history of the asteroid since its crystallization and cooling [*Jaumann et al.*, 2012]. This finding presents an opportunity to revisit two of the more interesting Antarctic meteorites from the HED group. Grosvenor Mountains (GRO) 95555 (Figure 8.4; Plate 59) is an unbrecciated diogenite with a polygonal-granular texture and an orthopyroxene grain size of ~1 mm [*Papike et al.*, 2000]. MIL 07001, a harzburgitic diogenite, exhibits a similarly equigranular texture, but with ~10 vol.% small olivines interspersed [*Beck et al.* 2012]. *Papike et al.* [2000] argued that GRO 95555 formed at sufficient depth to escape brecciation where high-grade thermal metamorphism might have occurred. In addition, a significant percentage of eucrites within the Antarctic population appear to contain disparate textures or lithologies without sharp delineation of boundaries between these areas, suggestive of metamorphism after brecciation. Our new understanding of the geology of Vesta challenges us to place these metamorphosed meteorites in context. Did the metamorphism result from impact? Were the

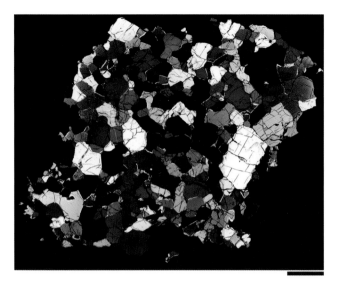

Figure 8.4. The unbrecciated, metamorphosed diogenite Grosvenor Mountains 95555 formed during high-grade thermal metamorphism at sufficient depth to escape brecciation. The Dawn mission has revealed the importance of impact basins in the geologic evolution of 4 Vesta, which is thought to be the source asteroid for the howardite-eucrite-diogenite clan of meteorites. GRO 95555 might provide important clues to both the nature and timing of metamorphism on 4 Vesta. Scale bar is 1 mm.

basin-forming events a major heat source? Alternatively, was metamorphism unrelated to impact, instead resulting from global burial metamorphism [e.g., *Yamaguchi et al.*, 1997]? In any of these models, metamorphosed HEDs might provide important clues to both the nature and timing of metamorphism on 4 Vesta.

8.9. CONCLUSIONS

The Antarctic meteorite program has revolutionized our understanding of the evolution of our solar system in large part because of the steady production of new, interesting, and sometimes unexpected meteorites. These meteorites challenge our concepts and fill in gaps in our knowledge. It is important to recognize, however, that the missing pieces of the puzzle often already exist in our collection, awaiting rediscovery through future recovery, analytical breakthrough, or changing paradigm that will allow us to fully reveal the picture of the early solar system.

Acknowledgments. This chapter would not have been possible without the dedicated efforts of all involved in the U.S. Antarctic Meteorite Program and the passion of the meteoritics community in driving our scientific exploration of the early solar system. My colleagues Ralph Harvey (Case), Kevin Righter (JSC), Cari Corrigan, Emma Bullock, Glenn MacPherson, Yulia Goreva, Andrew Beck, Linda Welzenbach, and Kat Gardner-Vandy (SI) all contributed ideas and commented on earlier versions of this manuscript, which significantly improved its clarity and utility. I am particularly grateful to the Smithsonian Institution for supporting my efforts in studying these weird and wonderful samples for the past 17 years.

REFERENCES

Beck, A. W., and H. Y. McSween Jr. (2010), Diogenites as polymict breccias composed of orthopyroxenite and harzburgite, *Meteoritics Planet. Sci., 45*, 850–872.

Beck, A. W., J. M. Sunshine, T. J. McCoy, and T. Hiroi (2012), Challenges to finding olivine on the surface of 4Vesta (abstract), *Lunar Planet. Sci., 43*, CD #2218.

Bischoff, A., and K. Keil (1983), Ca-Al-rich chondrules and inclusions in ordinary chondrites, *Nature, 303*, 588–592.

Bogard, D. D., and D. H. Garrison (2003), ^{39}Ar–^{40}Ar ages of eucrites and thermal history of asteroid 4 Vesta, *Meteoritics Planet. Sci., 38*, 669–710.

Burbine, T. H., T. J. McCoy, and T. L. Dickinson (2000), Origin of plagioclase-"enriched," igneous, enstatite meteorites (abstract), 63rd Meteoritical Society Meeting, #5158.

Cohen, B. A., T. D. Swindle, and D. A. Kring (2000), Support for the lunar cataclysm hypothesis from lunar meteorite impact melt ages. *Science, 290*, 1754–1756.

Corrigan, C. M., B. A. Cohen, K. Hodges, N. Lunning, and E. Bullock (2012), 3. 9 billion years ago and the asteroid belt: Impact melts in ordinary chondrites, *Lunar Planet. Sci., 43*, #1577.

Day, J. M. D., R. J. Walker, R. D. Ash, Y. Liu, D. Rumble III, A. J. Irving, C. A. Goodrich, K. Tait, W. F. McDonough, and L. A. Taylor (2012), Origin of felsic achondrites Graves Nunataks 06128 and 06129, and ultramafic brachinites and brachinite-like achondrites by partial melting of volatile-rich primitive parent bodies, *Geochim. Cosmochim. Acta, 81*, 94–128.

Fogel, R. A. (1997), On the significance of diopside and oldhamite in enstatite chondrites and aubrites, *Meteoritics Planet. Sci., 32*, 577–591.

Goodrich, C. A., and K. Righter (2000), Petrology of unique achondrite Queen Alexandra Range 93148: A piece of the pallasite (howardite-eucrite-diogenite?) parent body? *Meteoritics Planet. Sci., 35*, 521–535.

Gross, J., A. H. Treiman, and H. C. Connolly Jr. (2013), A new subgroup of amphibole-bearing R-chondrites: Evidence from the new R chondrite MIL 11207, 44th Lunar Planet. Sci. Conf., #2212.

Grossman, J. N., G. J. MacPherson, W. Hsu, and E. K. Zinner (1995), Plagioclase-rich objects in the ungrouped E3 chondrite Lewis Cliff 87234: Petrology and aluminum-magnesium isotopic data (abstract), *Meteoritics, 30*, 514.

Izawa, M. R. M., R. L. Flemming, N. R. Banerjee, and S. Mateev (2011), QUE 94204: A primitive enstatite achondrite produced by the partial melting of an E chondrite-like protolith, *Meteoritics Planet. Sci., 46*, 1742–1753.

Jaumann, R., D. A. Williams, D. L. Buczkowski, R. A. Yingst, F. Preusker, H. Hiesinger, N.Schmedemann, T. Kneissl, J. B. Vincent, D. T. Blewett, B. J. Buratti, U. Carsenty, B. W.Denevi, M. C. De Sanctis, W. B. Garry, H. U. Keller, E. Kernten, K. Krohn, J.-Y. Li, S.Marchi, K. D. Matz, T. B. McCord, H. Y. McSween, S. C. Mest, D. W. Mittlefehldt, S.Mottola, A. Nathues, G. Neukum, D. P. O'Brien, C. M. Pieters, T. H. Prettyman, C.Raymond, T. Roatsch, C. Russell, P. Schenk, B. E. Schmidt, F. Scholten, L. Stephan,M. V. Sykes, P. Tricarico, R. Wagner, M. T. Zuber, and H. Sierks (2012), Vesta's shape and morphology, *Science, 336*, 687–694.

Kallemeyn, G. W., M. Ebihara, and S. Latif (1998), Prompt-gamma analysis and instrumental neutron activation analysis studies of a new reduced L chondrite (abstract), *Meteoritics Planet. Sci., 33*, A81.

Macpherson, G. J., and G. R. Huss (2005), Petrogenesis of Al-rich chondrules: Evidence from bulk compositions and phase equilibria, *Geochim. Cosmochim. Acta, 69*, 3099–3127.

McCanta, M. C., A. H. Treiman, M. D. Dyar, C. M. O'D. Alexander, D. Rumble III, and E. J. Essene (2008), The LaPaz Icefield 04840 meteorite: Mineralogy, metamorphism, and origin of an amphibole- and biotite-bearing R chondrite, *Geochim. Cosmochim. Acta, 72*, 5757–5780.

McCoy, T. J., A. Gale, and T. L. Dickinson (2005), The early crystallization history of the aubrite parent body (abstract), 68th Meteoritical Society Meeting, #5146.

Mittlefehldt, D. W., and T. J. McCoy (2014), Achondrites and irons: Products of magmatism on strongly heated asteroids, in *35 Seasons of U.S. Antarctic Meteorites (1976–2010): A Pictorial Guide to the Collection, Special Publication 68*, edited by K. Righter, C. M. Corrigan, R. P. Harvey, and T. J. McCoy, American Geophysical Union/John Wiley & Sons, Washington, D. C.

Papike, J. J., C. K. Shearer, M. N. Spilde, and J. M. Karner (2000), Metamorphic diogenite Grosvenor Mountains 95555: Mineral chemistry of orthopyroxene and spinel and comparisons to the diogenite suite, *Meteoritics Planet. Sci., 35*, 875–879.

Righter, M., T. Lapen, and K. Righter (2008), Relationships between HEDs, mesosiderites, and ungrouped achondrites: Trace element analyses of mesosiderite RKP A79015 and ungrouped achondrite QUE 93148 (abstract), 39th Lunar Planet. Sci. Conf. #2468.

Shearer, C. K., P. V. Burger, C. Neal, Z. Sharp, L. Spivak-Birndorf, L. Borg, V. A. Fernandes, J. J. Papike, J. Karner, M. Wadhwa, A. Gaffney, J. Shafer, J. Geissman, N.-V. Atudorei, C. Herd, B. P. Weiss, P. L. King, S. A. Crowther, and J. D. Gilmour (2010), Non-basaltic asteroidal magmatism during the earliest stages of solar system evolution: A view from Antarctic achondrites Graves Nunatak 06128 and 06129, *Geochim. Cosmochim. Acta, 74*, 1172–1199.

Thomas, P. C., R. P. Binzel, M. J. Gaffey, A. D. Storrs, E. N. Wells, and B. H. Zellner (1997), Impact excavation on asteroid 4Vesta: Hubble Space Telescope results, *Science, 277*, 1492–1495.

Troiano, J., D. Rumble III, M. L. Rivers, and J. M. Friedrich (2011), Compositions of three low-FeO ordinary chondrites: Indications of a common origin with the H chondrites, *Geochim. Cosmochim. Acta, 75*, 6511–6519.

van Niekerk, D., and K. Keil (2012), Anomalous enstatite meteorites Queen Alexandra Range 94204 and pairs: The perplexing question of impact melts or partial melt residues, either way, unrelated to Yamato 793225 (abstract), 43rd Lunar Planet. Sci. Conf. #2644.

Watters, T. R., and M. Prinz (1979), Aubrites: Their origin and relationship to enstatite chondrites, *Proc. Lunar Planet. Sci. Conf., 10*, 1073–1093.

Weisberg, M. K., D. S. Ebel, H. C. Connolly Jr., N. T. Kita, and T. Ushikubo (2011), Petrology and oxygen isotope compositions of chondrules in E3 chondrites, *Geochim. Cosmochim. Acta, 75*, 6556–6569.

Weisberg, M. K., and K. Righter, (2014), Primitive asteroids: Expanding the range of known primitive materials, in *35 Seasons of U.S. Antarctic Meteorites (1976–2010): A Pictorial Guide to the Collection, Special Publication 68*, edited by K. Righter, C. M. Corrigan, R. P. Harvey, and T. J. McCoy, American Geophysical Union/John Wiley & Sons, Washington, D. C.

Welzenbach, L. C., T. J. McCoy, A. Grimberg, and R. Wieler (2005), Petrology and noble gases of the regolithic breccias MAC 87302 and implications for the classification of Antarctic meteorites (abstract), 36th Lunar Planet. Sci. Conf., #1425.

Yamaguchi, A., G. J. Taylor, and K. Keil (1997), Metamorphic history of eucritic crust of 4Vesta, *J. Geophys. Res. Planets, 102*, 13381–13386.

Zolensky, M. E., R. J. Bodnar, E. K. Gibson, L. E. Nyquist, Y. Reese, C.-Y. Shih, and H. Wiesnmann (1999), Asteroidal water within fluid inclusions-bearing halite in an H5 chondrite, Monahans, *Science, 285*, 1377–1379.

9

Cosmogenic Nuclides in Antarctic Meteorites

Gregory F. Herzog[1], Marc W. Caffee[2], and A. J. Timothy Jull[3]

9.1. INTRODUCTION

We discuss applications to Antarctic meteorites of cosmogenic nuclides. Cosmogenic nuclides are nuclei produced by cosmic rays (Table 9.1), mostly before the meteorites collided with Earth. They help us estimate how long the meteorites were exposed to cosmic rays, a length of time called the cosmic ray exposure age, and the duration of a meteorite's stay on Earth, a length of time called the terrestrial age. Cosmic ray exposure ages have been reviewed by *Eugster et al.* [2006] and *Herzog and Caffee* [2014]; *Jull* [2006] has summarized data for terrestrial ages. A comprehensive review of all relevant material would occupy more space than is available and so we limit the discussion to articles that demonstrate special features of the Antarctic collection or that exemplify broader trends seen (or not) in the non-Antarctic collection.

In the 1960s and 1970s, people who studied meteorites used what was at that time an indispensable tool, the Catalogue of the Natural History Museum, London [*Hey*, 1966]. That catalogue included only four Antarctic finds: the L chondrite Adelie Land, the IAB iron Neptune Mountains, the pallasite Thiel Mountains, and the ungrouped iron Lazarev; and so the notion of studying cosmic-ray exposure histories of Antarctic meteorites would have seemed constricted. The early harvests from the Yamato Mountains and Allan Hills instantly changed

matters for the better. The first Yamato meteorites were collected by a Japanese team in 1969 [see *Kojima*, 2006]. By 1977, U.S. expeditions had commenced and have continued ever since. These annual expeditions regularly retrieved hundreds of meteorites from the Antarctic ice. As the number of interesting samples rapidly increased, the limits on laboratory instrument time and staffing soon rivaled sample availability as a key factor in experimental design.

The new flood of meteorite arrivals sparked interest in both old and new questions related to cosmogenic nuclides. The technical task for the Antarctic meteorites remained the same: to measure the concentrations (atom g^{-1}) of stable cosmogenic nuclides and either the concentrations or the activities (usually in dpm/kg = decays minute^{-1} kg^{-1}) of radioactive ones. The conversion from concentration to activity or vice versa is given by radioactive decay law (equation 9.1),

$$\text{Activity}\left(\text{dpm}/\text{kg}\right)=-\frac{d[\text{A}]}{dt}=\lambda[\text{A}]$$
$$=\text{Concentration}\left(\text{atom g}^{-1}\right)/t_{1/2}\left(\text{Ma}\right)\times 1.3179\times 10^{-9},$$

$$(9.1)$$

which incorporates the current IUPAC recommendation that $1\ a = 31556925.445\ s$. In this expression, $t_{1/2}$ is the half-life of the nuclide of interest.

About the same time that the acquisition of large numbers of Antarctic meteorites began, accelerator mass spectrometry (AMS) became important for the measurement of cosmogenic radionuclides. With AMS, samples of 0.01–0.5 g could be analyzed, a mass range hundreds to thousands of times smaller than accessible with older methods. The increase in sensitivity revolutionized the

[1] *Department of Chemistry and Chemical Biology, Rutgers University, Piscataway, NJ*
[2] *Department of Physics, Purdue University, W. Lafayette, IN*
[3] *NSF–Arizona Accelerator Mass Spectrometry Laboratory, Departments of Geosciences and Physics, University of Arizona, Tucson*

35 Seasons of U.S. Antarctic Meteorites (1976–2010): A Pictorial Guide to the Collection, Special Publication 68,
First Edition. Edited by Kevin Righter, Catherine M. Corrigan, Timothy J. McCoy and Ralph P. Harvey.

Table 9.1. Useful[a] cosmogenic nuclides.

Nuclide	Half-life (Ma)
Radionuclides	
^{14}C	0.005730
^{59}Ni	0.076
^{41}Ca	0.1034
^{81}Kr	0.229[b]
^{36}Cl	0.301
^{26}Al	0.717
^{10}Be	1.387[c]
^{53}Mn	3.74
^{129}I	15.7
Stable nuclides	
^{3}He	
$^{6,7}Li$	
^{21}Ne	
^{38}Ar	
^{83}Kr	
^{126}Xe	
Gd	
Sm	
W	
Pt	

[a]Many shorter-lived radionuclides such as ^{22}Na and ^{54}Mn are also useful, but in most Antarctic meteorites they have decayed to levels below current detection limits and are not considered here. For a listing see, e.g., *Herzog and Caffee* [2014].

[b]*Baglin*, 2008.

[c]*Chmeleff et al.*, 2010; *Korschinek et al.*, 2010.

field. Some early examples of AMS studies of Antarctic meteorites are *Nishiizumi et al.* [1979a, b; 1981; 1983].

Many, but not all, of the scientific goals remained unchanged. The most important one is to devise a history of exposure to cosmic rays for each specimen, a history that explains the measured concentrations of cosmogenic nuclides. In the simplest but nonetheless fairly common cases these cosmic-ray exposure (CRE) histories are defined by a small number of parameters.

• **A terrestrial age, T_{Terr}**: how long after the time of fall (T_{Fall}) the meteorite spent in the Antarctic after it landed.

• **A CRE age, T_{Exp}**: the time the meteorite took to travel to Earth as a relatively small body (a meteoroid). In some cases the meteoroid may undergo collisions and in others the meteoroid may retain some memory of earlier cosmic-ray irradiation on the parent body.

• **An ejection age, $T_{Ej} = T_{Terr} + T_{Exp}$**: the total time interval from launch from the parent body to the present, which is important for establishing source or launch pairing, discussed further below.

• **Geometric conditions of irradiation or "shielding conditions"**: (1) how big the body (meteoroid) hosting the protometeorite was just before it passed through the Earth's atmosphere; and (2) the position of the protome-

teorite in relation to the preatmospheric surface of the meteoroid. The meteoroid is usually approximated as a sphere of radius R, and the position of the protometeorite by a depth d from the surface.

More complicated irradiation histories occur. Some meteorites, lunar ones especially, retain a record of cosmic-ray irradiation that preceded the launch from the parent body. Occasionally, meteoroids may undergo collisions while in transit, and the resulting changes in size and shape can affect the concentrations of the radionuclides to varying degrees.

The Antarctic meteorites also opened some new lines of investigation. By virtue of their large numbers, they were expected to include unusual or previously unknown types of objects. Moreover, the Antarctic, a long-term cold-storage locker, seemed likely to preserve meteorites with very old terrestrial ages. If so, then a study of the meteorite census as a function of terrestrial age might reveal changes in the fluxes of meteorites to Earth over time [*Zolensky et al.*, 2006]. In this connection, pairing information is important in sorting out the rates and incoming mass distributions of different meteorite types. We distinguish two kinds of pairing. Two meteorites are said to be launch or source paired if a single event ejected them from a parent body. Source-paired meteorites may follow different paths to Earth and hence have different CRE ages, but as noted above, they should share an ejection age. Two meteorites are said to be fall paired if they derive from a meteoroid that broke up in the Earth's atmosphere. Fall-paired meteorites should have the same terrestrial age and a common CRE age as well.

9.2. TERRESTRIAL AGES

9.2.1. Simplest Calculation of Terrestrial Ages

The terrestrial age, T_{Terr}, of a meteorite refers to the time elapsed between a meteorite's fall (T_{Fall}) and its recovery (or more rigorously, the present). We can calculate terrestrial ages in several ways. The most common, generally applicable, and dependable methods rely on measurements of cosmogenic radionuclides. The simplest though least accurate among them makes use of the integrated form of the radioactive decay law (equation 9.1):

$$A(T_{Terr}) = A_{Fall}e^{-\lambda T_{Terr}} ; \quad T_{Terr} = -\frac{1}{\lambda}Ln\left(\frac{A(T_{Terr})}{A_{Fall}}\right), \quad (9.2)$$

where $A(T_{Terr})$ denotes the activity of the radionuclide measured in the laboratory (on a recorded date), and A_{Fall} the activity at the time of fall. By convention, the present is taken as $T_{present} = 0$ and the terrestrial age as positive so that we have $T_{Terr} = -T_{Fall}$.

9.2.2. Choice of Radionuclide

As a rule of thumb, the radionuclide chosen for calculating a terrestrial age should have a half-life, $t_{1/2}$, comparable to the terrestrial age of the sample; this rule follows from standard propagation of error considerations. Specifically, we have for the relative uncertainty of T_{Terr}

$$\left|\frac{\Delta T_{Terr}}{T_{Terr}}\right| = +\sqrt{\left|\frac{\Delta A_{Fall}}{\lambda T_{Terr} A_{Fall}}\right|^2 + \left|\frac{\Delta \lambda}{\lambda}\right|^2 + \left|\frac{\Delta A(T_{Terr})}{\lambda T_{Terr} A(T_{Terr})}\right|^2}, \quad (9.3)$$

where Δ denotes the uncertainty of an experimental quantity. All three terms on the right may contribute significantly to the overall uncertainty. At large values of λT_{Terr}, measurement difficulties usually lead to increased values of $\frac{\Delta A(T_{Terr})}{A(T_{Terr})}$. At small values of the terrestrial age, the terms λT_{Terr} in the denominator approach zero and increase the relative uncertainty. Consequently, in the determination of terrestrial age it is usually best to choose a radionuclide with a half-life comparable to the terrestrial age.

For a meteorite that fell within the last 15 ka or so, ^{14}C would be optimal, while for one that fell 100 ka ago, ^{41}Ca would be the better choice. With AMS, the relative uncertainty of the measured ^{14}C activity, $\frac{\Delta A(T_{Terr})}{A(T_{Terr})}$, is usually 3%–10%, mainly because of blank corrections.

9.2.3. Activity Measurements

Historically, the radionuclide activities $A(T_{Terr})$ were measured by low-level decay counting. Today, decay counting is mostly limited to short-lived ($t_{1/2} < 1000$ a) species such as ^{7}Be, ^{60}Co, and ^{54}Mn. Some γ counting of longer-lived ^{26}Al continues and ^{81}Kr concentrations are measured by conventional (low-energy) mass spectrometry. For the radionuclides ^{14}C, ^{41}Ca, ^{36}Cl, ^{41}Ca, ^{53}Mn, ^{60}Fe, and ^{129}I, however, AMS has become the method of choice, although even with the greater sensitivity of AMS relative to decay counting, activities may be below detection limits. When the activities are too low to measure, terrestrial ages are presented as lower bounds. The practical upper limits for quantitative determinations of terrestrial age are about 30–40 ka and 400–1500 ka for ^{14}C and ^{36}Cl, respectively.

9.2.4. Activity at the Time of Fall (A_{Fall}) and the Uncertainty of the Terrestrial Age

The use of equation 9.2 to calculate a terrestrial age requires knowledge of the radionuclide activity at the time of fall, A_{Fall}. Absent independent historical information about the fall, this value cannot be measured for a find and therefore must be estimated. In the simplest approximation for A_{Fall}, we use average or selected activities measured for fresh meteorite falls comparable to the meteorite whose terrestrial age, T_{Terr}, we want to know. These data define *ranges* of possible values for A_{Fall}, and the resulting imprecision contributes appreciably to the overall uncertainty of the terrestrial age (equation 9.3).

For example, the minimum activity for ^{14}C in ordinary chondrite falls is about 20 dpm/kg and the maximum about 60 dpm/kg, depending on the size of the precursor meteoroid, its composition, and the position of the analyzed sample within it. Lower values are unusual and are most often found in objects that were atypically small or large in space. In the likely worst case then, A_{Fall} for ^{14}C may differ by 50% from the midpoint of the total range. Inspection of equation 9.2 shows that a 50% uncertainty ^{14}C$_{Fall}$ for a meteorite with $T_{Terr} = 1/\lambda$ translates to a minimum $\Delta T_{terr}/T_{terr}$ of 50%.

Most fresh falls, however, contain between 40 and 60 dpm/kg ^{14}C. It follows that for $T_{Terr} = 1/\lambda$, the typical (relative) uncertainty attributable to ΔA_{Fall} would be approximately 15%–20%. Relative uncertainties of the measured ^{14}C activities after corrections for blank normally range from <1% to 5% but may exceed 50% for samples that give signals close to blank levels. Blanks typically correspond to a terrestrial age of about 30,000 years; a 20% error in the ^{14}C determination corresponds to an absolute error in the terrestrial age of about 1650 years, which in many contexts is negligible (see discussion below).

9.2.5. First Terrestrial Ages for Antarctic Meteorites from ^{14}C and ^{26}Al

Fireman et al. [1979] and *Fireman and Norris* [1981] used ^{14}C measurements to calculate the terrestrial ages of 12 Allan Hills (ALH) meteorites, which included a eucrite, a diogenite, and several L and H chondrites. For A_{Fall}, they adopted the value of 57 ± 3 dpm/kg, a value *Fireman* [1978] determined for the L6 chondrite Bruderheim; the stated uncertainty made no allowance for the influence of shielding. Their results established a range of terrestrial ages beginning at 10 ka and extending upward (Table 9.2). Later observations agreed with this early important result, verifying Fireman's techniques.

At about the same time, the late 1970s, John Evans and John Wacker at Battelle Pacific Northwest Laboratories began a program to measure cosmic-ray effects, and particularly ^{26}Al ($t_{1/2} = 0.7$ Ma) activities in Antarctic meteorites. Measured activities in ordinary chondrites at the time of fall typically lie between 30 dpm/kg and 120 dpm/kg. The substitution of reasonable values for the uncertainty of a measured ^{26}Al activity in an Antarctic find (5%–10% in most instances) into equation 9.3 shows

Table 9.2. First terrestrial ages, T_{Terr} (ka), of Antarctic meteorites based on $^{14}C^a$.

Meteorite	T_{Terr}
ALH A76005	>34
ALH A76006	>32
ALH A76007	>34
ALH A76008	>32
ALH A77003	21^{+4}_{-3}
ALH A77004	>33
ALH A77214	>25
ALH A77256	11.1
ALH A77272	>38
ALH A77282	~30
ALH A77294	30
ALH A77297	>35

Sources: Fireman et al., 1979; Fireman and Norris, 1981.

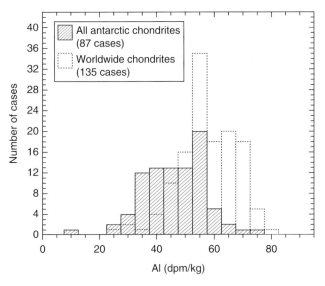

Figure 9.1. Distribution of ^{26}Al activities in Antarctic meteorites. Adapted from *Evans et al.* [1982].

that 100 ka is about the minimum terrestrial age one can determine with ^{26}Al by using equation 9.2; shorter terrestrial ages have relative uncertainties of 50% or more.

Evans et al. [1982] measured the ^{26}Al activities of about 100 meteorites, mostly ordinary chondrites, including most of the stones listed in Table 9.2. In their calculations they adopted $A_{Fall} = 59 \pm 9$ dpm/kg for L chondrites and 55 ± 8 dpm/kg for H chondrites. With a few exceptions, the terrestrial ages were too small to calculate with much confidence as discussed in the previous paragraph. Looking for more information, Evans and coworkers pooled the data for many meteorites and treated them statistically. Figure 9.1 reproduces their histograms comparing the ^{26}Al activities measured for Antarctic and non-Antarctic chondrites. These authors noted that the Antarctic meteorites are distributed toward lower ^{26}Al contents, i.e., from the average of about 60 for (non-Antarctic) falls to one of 49 dpm/kg (Figure 9.1). If attributed only to older terrestrial age, the 20% decrease in ^{26}Al activity corresponds to an average residence time of ~200 ka in the Antarctic. A later publication [*Evans and Reeves*, 1987] included data for 220 samples but did not change these conclusions.

Aware of the dangers of relying on one rather long-lived isotope in the calculation of terrestrial ages, Evans and colleagues did not themselves make a calculation of average residence time (~200 ka). Instead, they cautiously observed that a few stones *probably* had very old terrestrial ages. Ambiguity arises because either a small or a large meteoroid size in space could have depressed the rate of ^{26}Al production in space, so that the actual value of A_{Fall} was smaller than the adopted value of A_{Fall}, ~60 dpm/kg. In this case the calculated terrestrial ages would be too old. Put another way, meteorites that arrived very recently on Earth with lower ^{26}Al activities would have the same measured activities as meteorites

that arrived earlier but with higher ^{26}Al activities. It is also possible that in a few cases short exposure times for meteoroids in space might have prevented ^{26}Al from building up to its maximum possible value of ~60 dpm/kg in ordinary chondrites [see also *Wacker*, 1993], again leading to inflated terrestrial ages. Irradiation by solar cosmic rays could have raised the ^{26}Al activity at the time of fall, although very few meteorites retain these effects. Finally, weathering can affect terrestrial ages; we discuss this point in section 2.13 below.

Uncertainties in A_{Fall} notwithstanding, taken together the early results available from ^{14}C and ^{26}Al indicated that most Antarctic meteorites from Victoria Land, at least, had terrestrial ages between 10 ka and 200 ka. These early works also showed that ^{26}Al measurements could flag unusual objects (those in the wings of the activity distributions) for further study. On the strength of these arguments, John Annexstad, editor of the *Antarctic Meteorite Newsletter*, wrote in the newsletter of August 1978, "Plans are underway to have several specimens gamma counted and residence times published in the catalogue (of Antarctic meteorites). We are also exploring the use of thermoluminescence for measuring residence terrestrial ages." Ten years later NASA funding helped realize these plans.

9.2.6. Natural Thermoluminescence and Terrestrial Ages

When cosmic rays slow and stop in meteoroids, they create defects or higher-energy lattice sites called "traps" in the molecular structure. During the slowing of cosmic rays, electrons are dislodged and may occupy these traps, which are stable in quartz and in K-feldspars [*Lian*, 2007].

In space, the concentration of the trapped electrons, n, increases with the duration of the irradiation up to a limiting or saturation value. As soon as the electrons are trapped, they can decay to lower energy states through a process of thermal de-excitation that emits light termed natural thermoluminescence (NTL). Thus, NTL can be loosely regarded as a decay product of the (unstable) population of trapped electrons. The rate of decay increases with temperature and other factors.

As with production of cosmogenic radionuclides, production of the trapped electrons mostly ends with a meteorite's fall because the Earth's atmosphere attenuates the cosmic-ray flux by a factor of ~1000. Decays, however, continue and so the concentration of trapped electrons decreases as time passes. In the laboratory, heating of a meteorite sample to high temperature (from ~100 °C to 450 °C) greatly hastens this decay or draining (in the jargon of the trade) or annealing of the trapped electrons, which is accompanied by the emission of NTL. Measured as a glow curve and integrated over suitable temperature ranges, the NTL data give signals related to the trapped electron concentration at the time of measurement. To the extent that prior annealing in the Antarctic occurs at a constant or known rate, NTL may serve as a basis for calculating terrestrial ages [see, e.g., *Sears and Hasan*, 1986].

9.2.7. Quantitative Relation Between NTL Data and Terrestrial Age

The calculation of terrestrial ages from NTL measurements is model dependent [*McKeever et al.*, 1982], and we have never seen an equation in closed form from which we could make a calculation ourselves. Nevertheless and although over the years Sears and coworkers have emphasized different methods for calculating terrestrial ages, the general pattern of results has changed little.

9.2.8. Terrestrial Ages Based on NTL Measurements

Sears and Hasan [1986] did not calculate terrestrial ages explicitly but examined the distribution of a quantity, $\log\left(\dfrac{NTL_{low-T}}{NTL_{high-T}}\right)$, for 20 Antarctic meteorites (Figure 9.2). This quantity is expected to decrease with increasing terrestrial age. Indeeed, as observed for ^{26}Al, the distributions for both Antarctic and non-Antarctic finds show shifts to lower values by a factor of ~2 relative to the one for falls. The respective averages for falls and for Antarctic finds are 2.8 and 1.3. (The data set of *Hasan et al.* [1987] includes three additional meteorites but is otherwise similar.) Assuming that the selections of meteorites for ^{26}Al and NTL are similar and that the shift toward lower NTL values in Antarctic stones reflects primarily terrestrial

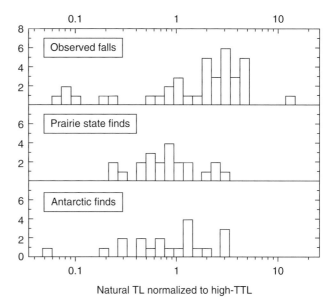

Figure 9.2. Thermoluminescence (TL) measurements for Antarctic and non-Antarctic meteorites. Adapted from *Sears and Hasan* [1986].

age, an assumption not universally accepted, a comparison of the NTL and ^{26}Al results implies an effective half-life of ~0.2 Ma for the decay of NTL, about the same as that of ^{81}Kr or ^{36}Cl.

9.2.9. The ^{26}Al and NTL Surveys of the U.S. ANSMET Collection of Antarctic Meteorites

In 1987, the Meteorite Working Group, which oversees the U.S. collection, curation, and distribution of Antarctic meteorites, announced that it had approved "two types of surveys of Antarctic meteorites which will identify meteorites with particularly short or long terrestrial ages or unusual thermal or radiation histories" (*Antarctic Meteorite Newsletter*, vol. 10, of February 1987), as John Annexstad had suggested eight years earlier. About one year after the announcement, the *Antarctic Meteorite Newsletter* of February 1988, Table 4, presented NTL measurements from Derek Sears' laboratory for 199 meteorites from Lewis Cliffs. And in the following issue of August, 1988, 110 more measurements appeared along with 94 measurements of ^{26}Al activities from John Evans, John Wacker, and James Reeves of Battelle Northwest Laboratories. The determination of NTL in Antarctic meteorites was to continue for fourteen years [*Sears et al.*, 2011].

In the active days of the NTL survey of Antarctic meteorites, the curators at NASA's Johnson Space Center took considerable pains to sample material at least 1 cm from fusion crusts. They did so in order to avoid the possibility that atmospheric heating had bleached the samples, that is, had erased some or the entire NTL signal. Ironically, the proposal of *Miono and Nakanishi* [1994] stands this practice on its head. These authors examined material from just under meteorite fusion crusts, where atmospheric heating reset the NTL clock to count up from Earth arrival, rather than down from irradiation in space. As it turns out, the temperature dependence of NTL decay limits the accuracy of this method [*Akridge et al.*, 2000].

To summarize, from the basic physics, we know that some portion of the variability of NTL measurements reflects terrestrial age. The precise portion may be difficult to establish in general because of the possibilities of varying rates of decay, anomalous fading, and solar heating. Given the existing state of the science, we believe cosmogenic nuclides are the most robust means by which to determine terrestrial age. If one is interested, however, in a statistical profile of a large group of meteorites or in investigating pairing relations [e.g., *Ninagawa et al.*, 2011], then a cautious interpretation of NTL measurements may provide useful guidance. NTL has a role to play in constructing exposure histories of individual meteorites. Figure 11 of the recent review by *Sears et al.* [2013] presents a compilation of terrestrial ages for Antarctic and other meteorites inferred from NTL measurements. *Mokos et al.* [2000] present a cautionary tale of Antarctic H chondrites meteorites for which cosmogenic radionuclide measurements forced a reassessment of groupings based on NTL.

9.2.10. Terrestrial Ages Based on ³⁶Cl for Meteorites from the Yamato Mountains

To improve the estimation of terrestrial ages obtained from ¹⁴C and ²⁶Al, whose half-lives were at the respective short and long ends of the terrestrial age distribution, a radionuclide with an intermediate half-life was needed. In the late 1970s, a group at the University of Rochester led by David Elmore developed the AMS techniques needed to measure ³⁶Cl (half-life, 0.3 Ma) with exquisite sensitivity. And while the early measurements of ¹⁴C and ²⁶Al were taking place in the United States, in Tokyo, a team that included Kuni Nishiizumi and Masatake Honda had begun measuring several radionuclide activities in the Antarctic meteorites collected by Japan's National Institute of Polar Research [*Nishiizumi et al.*, 1978; *Nishiizumi et al.*, 1979a, b]. Shortly thereafter, *Nishiizumi et al.* [1981] presented ³⁶Cl activities (in meteorite metal) along with ²⁶Al activities (in bulk samples) obtained by counting. Figure 9.3 shows the results of calculations that model the production of ³⁶Cl in space and hence the activity expected at the time of fall. For a large majority of meteorites (those with exposure ages >1 Ma and derived from meteoroids with radii less than 65 cm),

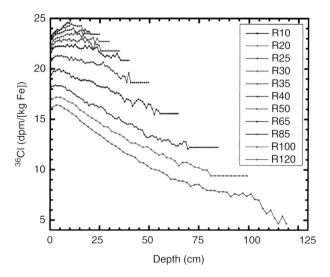

Figure 9.3. Model calculations [*Ammon et al.*, 2009] for ³⁶Cl activities in meteoritic metal. The numbers following the letter *R* denote the meteoroid radius in cm.

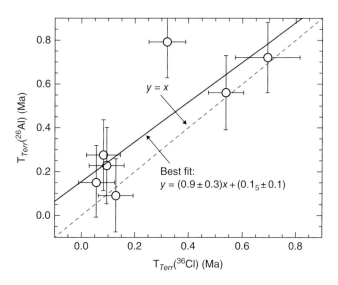

Figure 9.4. Comparison of terrestrial ages calculated from the activities of ²⁶Al and of ³⁶Cl [in Antarctic meteorites *Nishiizumi et al.*, 1981].

the range of variation of A_{Fall} of ³⁶Cl in the metal of stony meteorites is relatively small, from 19 to 24.5 dpm/(kg Fe). The narrowness of the range helps improve the accuracy of terrestrial ages based on ³⁶Cl.

Figure 9.4 compares the ³⁶Cl with the ²⁶Al terrestrial ages of seven ALH meteorites [*Nishiizumi et al.*, 1981]. Overall, the two sets of results agree reasonably well. The range is from 50 ka to 800 ka; typical uncertainties are 70 ka and 160 ka for the terrestrial ages based on ³⁶Cl and ²⁶Al, respectively. *Nishiizumi et al.* [1981] concluded that "the terrestrial ages of Antarctic meteorites are less than ~3 × 10⁵ y, although longer terrestrial ages cannot be excluded for unexamined meteorites," in agreement with the results of *Evans et al.* [1982].

9.2.11. Terrestrial Ages of Yamato and Other Icefields

From the 1980s onward, evidence accumulated that meteorites collected from different icefields and within different parts of the same icefield have different terrestrial ages. *Kigoshi and Matsuda* [1986] presented ^{14}C terrestrial ages from Yamato meteorites; *Nishiizumi et al.* [1989a] presented 135 terrestrial ages based on ^{36}Cl analyses made for meteorites collected in the Allan Hills, the Yamato Mountains, and other locations (Figure 9.5). *Nishiizumi and coworkers* concluded that "the terrestrial ages of Allan Hills meteorites vary from 0.2 to 1 Ma and are clearly longer than those of Yamato meteorites and other Antarctic meteorites" [1989a]. The apparent reason for this difference in terrestrial ages is that meteorites fall in various locations and then travel (in the ice) to the recovery location over considerable distances, along different paths, and at varying velocities [*Whillans and Cassidy*, 1983]. A similar dispersion of terrestrial ages is reported for the Frontier Mountains site [*Welten et al.*, 2006]. *Folco et al.* [2006] noted that the terrestrial ages of meteorites could also be used to constrain the age of the ice itself.

Nishiizumi et al. [1989b] singled out for special attention the H5 chondrite Allan Hills 82102, which was found *in* rather than *on* the surface of the ice at the Far Western Icefield. Its terrestrial age is ~11 ka. Given the slow rates

of Antarctic glacier movement and assuming that the ice surrounding the meteorite did not thaw and refreeze, this result implies that Allan Hills 82102 was transported for only a short distance. By extension, the young terrestrial ages of Yamato meteorites, which *Jull et al.* [1993] confirmed for many additional stones, and the older terrestrial ages of Allan Hills Main Icefield meteorites may imply transport over shorter and longer distances, respectively, if the incoming ice moves at a constant rate.

Using NTL data, *Benoit et al.* [1993] reached a qualitatively similar conclusion about the relative terrestrial ages of Yamato and Allan Hills meteorites. They also suggested that meteorites from the Far Western Icefield of Allan Hills tended to be younger than those of other Allan Hills areas. In a more detailed study based mostly on ^{14}C, *Jull et al.* [1998a] found that many meteorites from the Allan Hills Near, Middle, and Far Western icefields had terrestrial ages of <50 ka, values well within the range of ^{14}C dating. In contrast, a much lower percentage of meteorites from the main icefield had low terrestrial ages (Table 9.3, adapted from *Jull* [2000]). The large span of the results (Figure 9.5) emphasizes the need to choose the radionuclide with a task-appropriate half-life.

9.2.12. Terrestrial Ages from Two or More Radionuclide Activities

The determination of terrestrial ages using cosmogenic nuclides is limited in part by our confidence in the radionuclide activity at the time of fall, A_{Fall}. While this parameter often can be constrained well enough to allow the determination meaningful terrestrial ages, it would be

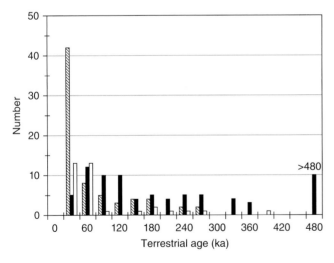

Figure 9.5. Terrestrial age distributions of meteorites. Yamato site (slanted bars) taken from *Jull et al.* [1993, 1998a, unpublished data] and *Nishiizumi et al.* [1989a, 1999]. The Allan Hills Main Icefield (black bars) and Elephant Moraine (white bars) taken from *Nishiizumi et al.* [1999], *Michlovich et al.* [1995], *Jull et al.*, [1998b], and *Jull* [unpublished data, 2003]. Figure reprinted from *Jull* [2006] by permission of the University of Arizona Press.

Table 9.3. Percentage of meteorites recovered from different Antarctic locations and with terrestrial ages falling in different time periods (adapted from *Jull* [2000]).

	% Meteorites per Radionuclide and T_{Terr}		
Recovery Location	^{14}C <25 ka	^{14}C or ^{36}Cl 25–70 ka	^{36}Cl or ^{81}Kr >70 ka
Far Western Ice field, Allan Hills	65%	35%	~0%
Yamato Mountains	27%	54%	19%
Middle Western Ice field, Allan Hills	30%	70%	~0%
Elephant Moraine	14%	37%	49%
Lewis Cliff	10%	25%	65%
Frontier Mountains	8%	31%	61%[a]
Allan Hills Main Ice field	7%[c]	39%	54%[b]

[a] *Welten et al.* [2000].
[b] Estimate of *Nishiizumi et al.* [1999] on 102 samples.
[c] *Jull et al.* [1998a] and unpublished data.

preferable to have a method with greater precision and natural to suppose that combinations of radionuclide measurements might provide the basis of such a method. Equation 9.4 is obtained by forming a ratio of activities measured for two nuclides in one meteorite.

$$\frac{A_1(T_{Terr})}{A_2(T_{Terr})} = \frac{A_{1,Fall}}{A_{2,Fall}} e^{-(\lambda_1 - \lambda_2)T_{Terr}}; \lambda_1 > \lambda_2; T_{Exp} > 3.5/\lambda_2. \quad (9.4)$$

We assume that nuclide 1 has the shorter half-life and that the CRE lasted long enough for nuclide 2 to reach >98% of its maximum (or saturation) value. A first question concerns the range of terrestrial ages for which this equation is best applied. Comparison of equation 9.4 with equation 9.2

$$A(T_{Terr}) = A_{Fall} e^{-\lambda T_{Terr}} \quad (9.2)$$

suggests a formal analogy in which $\frac{A_1(T_{Terr})}{A_2(T_{Terr})}$, $\frac{A_{1,Fall}}{A_{2,Fall}}$, and $(\lambda_1 - \lambda_2)$ play the roles of $A(T_{Terr})$, A_{Fall}, and λ, respectively. Thus, the ratio $\frac{A_1(T_{Terr})}{A_2(T_{Terr})}$ behaves like a radionuclide that decays with half-life equal to $\frac{\ln(2)}{(\lambda_1 - \lambda_2)}$. Just as in the single-nuclide approach, we expect equation 9.4 to yield the best results for terrestrial ages comparable to this value. For example, if we take as our two nuclides ^{14}C ($\lambda = 0.1210$ ka^{-1}) and ^{10}Be ($\lambda = 0.0005$ ka^{-1}), we see that $\lambda(^{14}$C$) - \lambda(^{10}$Be$) \approx \lambda(^{14}$C$)$ and conclude that equation 9.4 gives optimal results over a time scale equal to a few half-lives of ^{14}C. With typical H chondrite values for $A_{1,Fall}$ (^{14}C) = 50 dpm/kg and $A_{2,Fall}$ (^{10}Be) = 20 dpm/kg, equation 9.4 becomes

$$\frac{^{14}C}{^{10}Be} = \frac{50}{20} e^{-0.1205 T_{Terr}} \sim 2.5 e^{-\lambda_{14} T_{Terr}}, \quad (9.5)$$

where T_{Terr} is in ka. In cases where the cosmic-ray irradiation lasted for fewer than five half-lives of ^{10}Be, a correction based on the CRE age must be applied (for details, see *Herzog and Caffee* [2014]).

One might wonder whether this method offers any great advantage, inasmuch as most of the uncertainties propagated in equation 9.2 carry over into equation 9.4. Indeed, the precision of the terrestrial age obtained using the two-nuclide equation 9.5 is generally poorer than that obtained using the one-nuclide equation 9.2. For one key term, however, the relative uncertainty associated with equation 9.4 is smaller and for this reason the overall accuracy of T_{Terr} is better. In particular, in equation 9.2,

the factor that ultimately limits the accuracy of a terrestrial age is the activity of that nuclide at the time of fall. As suggested above in section 2.5, A_{Fall} depends on the size of the meteoroid, the location of our specimen within that meteoroid, and the duration of cosmic-ray exposure, all of which may be unknown. The corresponding term that appears in equation 9.4 is the *ratio* of activities, $A_{1,Fall}/A_{2,Fall}$. Both modeling calculations and direct experimental measurements confirm that when the two nuclides are chosen appropriately, $A_{1,Fall}/A_{2,Fall}$ is much less sensitive to size and depth effects than is the value of A_{Fall} for one nuclide, and that where $A_{1,Fall}/A_{2,Fall}$ varies, it does so in predictable ways [*Neupert et al.*, 1997; *Jull et al.* 1994 (Knyahinya); *Kring et al.*, 2001 (Gold Basin); *Nishiizumi and Caffee*, 1998; *Lavielle et al.*, 1999 (small iron meteorites)]. This ratio is also nearly independent of the meteorite's elemental composition for ^{14}C and ^{10}Be and certain other pairs of nuclides [*Jull et al.*, 1994].

Nishiizumi et al. [1997] used ^{36}Cl and ^{10}Be activities in equation 9.4 to calculate terrestrial ages for iron meteorites. The ^{36}Cl/^{10}Be ratio has an effective half-life of 382 ka, and with high-precision measurements of activities, ΔT_{Terr} can be reduced to about ±40 ka. *Welten et al.* [2000] calculated terrestrial ages from equation 9.4 using ^{41}Ca and ^{36}Cl activities measured in the metal phase of chondrites and small iron meteorites. The ^{41}Ca/^{36}Cl ratio has an effective half-life of 157 ka, and the ^{36}Cl/^{10}Be ratio has one of 383 ka. For further discussion, see *Welten et al.* [2006] and references therein.

Although it is neither practical nor necessary to obtain a terrestrial age for every Antarctic meteorite, a sizable database is useful. *Welten et al.* [2011a] have been separating metal phases from chondrites and using the ^{36}Cl/^{10}Be method to obtain terrestrial ages. With results in hand for ~490 chondrites, these authors find that all but a few Antarctic meteorites arrived within the last 650 ka and that a majority have terrestrial residence times of less than 50 ka. A small but significant fraction of the meteorites have radionuclide activities so low that reliable terrestrial ages cannot be determined.

More detailed studies on Antarctic meteorites may use three or more nuclides in order to calculate terrestrial ages and to correct for undersaturation. *Welten et al.* [2006] studied 36 meteorites from the Frontier Mountains blue icefield, which contains a classic "trap" for meteorites. The terrestrial ages ranged from 6 to 525 ka.

9.2.13. Terrestrial Age and Terrestrial Alteration

Weathering was documented early in the history of the Antarctic meteorite collection [*Biswas et al.*, 1980; *Gooding*, 1981; *Lipschutz*, 1982], and to this day weathering remains one of the classification parameters for Antarctic meteorites. To first order, one would expect alteration to increase with terrestrial age and to be slower in Antarctic than in similar non-Antarctic finds. Ultimately, weathering reduces a meteorite to dust and makes recovery impractical. Unsurprisingly then, in the Antarctic and elsewhere the probability of recovering a meteorite tends to decrease with increasing terrestrial age (Figure 9.5) [*Gibson and Bogard*, 1978; *Bland et al.*, 1996, 1998; *Lee and Bland*, 2004; *Al-Kathiri et al.*, 2005; *Zurfluh et al.*, in press]. *Lee and Bland* [2004] compared the degrees of weathering observed in meteorites found in the hot deserts of midlatitudes or in the cold desert of Antarctica. From a consideration of terrestrial ages, they too concluded that the warmer areas weather much more rapidly, but they recognized that terrestrial age alone could not account for all differences in degree of weathering. *Lee and Bland* [2004] argued that weathering was slower in Antarctic meteorites because ice would have protected them from both high temperatures and from the action of liquid water. *Tyra et al.* [2007] noted that a series of CM meteorites from Elephant Moraine displayed evidence of terrestrial alteration, including uptake of terrestrial oxygen and carbon.

The details of the relationship between weathering and terrestrial age are complex, however, because the measured concentrations of cosmogenic nuclides can change through sporadic processes other than radioactive decay, processes whose rates may vary and whose net effects are not uniquely determined by terrestrial age [*Welten*, 1999]. First, meteorites can gain mass, mainly in the form of oxygen, water, and to a lesser extent carbon dioxide; such gains lead to lower radionuclide concentrations. They can also lose mass through the selective physical removal of certain phases; such losses may either raise or lower concentrations in the remaining sample depending on which phases are removed. In general, the gain of mass seems to be more probable [*Schultz et al.*, 1991]. Corrections based on complete elemental analyses are possible in principle, although the corrections introduce new assumptions and uncertainties. Second, any terrestrial matter incorporated may bring with it cosmogenic radionuclides produced here on Earth. A prime example is ^{10}Be. Cosmic rays produce it at relatively high rates in the atmosphere. This so-called meteoric ^{10}Be falls with precipitation to the Earth's surface, where it may be adsorbed by or combine chemically with meteoritic matter. Third, prolonged residence within the topmost decimeter of the Antarctic surface directly exposes meteorites to cosmic rays. If the meteorite remains at a constant depth, the activity of a radionuclide, say ^{36}Cl, due to new production on Earth grows on a time scale set by the half life. As the cosmic-ray fluxes in the Antarctic are about a thousand times lower than in space, the newly produced activity becomes comparable to the residual extraterrestrial activity after about 10 half-lives, the time required for a 1000 decrease

due to decay. The Lazarev iron meteorite is a rare example in which terrestrial production has become significant [*Nishiizumi et al.*, 1987]. In most Antarctic meteorites, and especially those with shorter terrestrial ages, the net effects of weathering or terrestrial and exposure ages are thought to be negligible.

Welten et al. [2008] determined the oldest currently known terrestrial age for a stony meteorite, 3.0 ± 0.3 Ma for the chondrite FRO 01149, based on ^{26}Al and ^{10}Be. The meteorite had a mass of only 1.5 g. In this sample ^{41}Ca has decayed below current detection limits. They noted that the ^{10}Be and ^{36}Cl values were probably affected by terrestrial weathering over this time, and in particular by the incorporation of meteoric ^{10}Be.

A recurring question in studies of Antarctic meteorites is how much alteration may have preceded arrival on Earth. The discovery of thin, patchy deposits of salts on martian meteorites from the Antarctic [see *Jull et al.*, 1988] stimulated speculation that the salts formed on Mars when brine evaporated. Two related observations indicated that this hypothesis should be regarded cautiously. First, the salt-bearing martian meteorite Elephant Moraine 79001 had a terrestrial age of about 12 ± 2 ka, providing ample time for formation of the salts on Earth. Second and more directly, similar salt deposits grew on the meteorites while in terrestrial storage [*Grady et al.*, 1989]! Similar issues arose later when it was claimed that the martian meteorite Allan Hills 84001 (Plate 69) contained biogenic material [e.g., *McKay et al.* 1996]. *Jull et al.* [1998b] measured a terrestrial age of 12 ± 1 ka. They also investigated ^{14}C in organic phases from ALH A84001 and noted that much of this organic material could be identified from ^{14}C concentrations as having a terrestrial origin. Through ongoing attention to ^{14}C along with ^{12,13}C, recent work has continued the task of assigning meteoritic carbonates to terrestrial or extraterrestrial sources [e.g., *Losiak and Velbel*, 2011; *Velbel*, 2012].

9.2.14. Pairing

A meteoroid undergoes tremendous mechanical and thermal stress as it passes through the Earth's atmosphere and when it lands. These stresses usually result in a catastrophic breakup. The fragments that survive on the Earth's surface create what is called a strewn field. When a large fall is witnessed, it is not unusual for meteorite collectors, scientists, and interested rock hounds to search out and recover dozens or even hundreds of fragments. The meteorite fragments that belong to the same strewn field are said to be paired or, more strictly, fall paired. The meteorites of Antarctica were subjected to these same entry stresses and fragmentation, so we expect many Antarctic meteorites to be fall paired. The observation that the average recovered mass of Antarctic meteorites is smaller than that of non-Antarctic meteorites implies that smaller fragments stand a better chance of survival in the Antarctic and that pairing is common there.

The identification of pairing groups has both practical and scientific importance. An especially interesting specimen may be so small that the mass available does not meet the scientific demand. Under these circumstances a researcher may willingly substitute a fall-paired specimen. Beyond practical considerations, the identification of pairs is essential for identifying ancient strewn fields and reconstructing their total masses, sorting meteorite fluxes by type, and associating individual lunar and martian meteorites with specific ejection events.

How do we recognize pairing relationships? The first clues are chemical and/or petrologic similarity and geographical proximity. Based on these considerations, Antarctic team members note tentative or preliminary pairings as they collect the samples. Such pairing assignments are not rigorous, however, and to raise confidence in them it is desirable to show that two meteorites assigned to a likely pair have the same terrestrial age and the same CRE age. Unfortunately, with hundreds if not thousands of potentially paired meteorites on hand, it is not possible to obtain these data in every case. In practice then, when a few tentative pairings in a larger group are confirmed by terrestrial and CRE ages, it is then usually assumed that the entire suite is paired.

Pairing studies of the meteorites harvested at the Frontier Mountains (FRO) in Northern Victoria Land show how terrestrial exposure ages can be used to sort out distinct falls present in a single area. Expeditions spanning several decades collected over 450 meteorites in this classic "meteorite trap." By interpreting the measured activities of several radionuclides in 23 samples, *Welten et al.* [2006] were able to identify two large pairing groups. One pairing group, associated with FRO 90174, consists of H3-6 chondrites, and is identified as an H chondrite breccia. Its terrestrial age is ~100 ka and the group is estimated to include over 50% of all the chondrites in this stranding area. A second pairing group suggested by physical characteristics comprises samples associated with FRO 90001, which has a terrestrial age of ~40 ka. This group consists of seven members. None of them has an extensive fusion crust, so it is likely that the breakup occurred on or in the ice, rather than in the atmosphere. A few of the FRO meteorites belong to neither of the larger pairing groups. The paired stones FRO 93009 and FRO 01172 are among the oldest at the site, having terrestrial ages of ~500 ka. Interestingly, they were found on opposite sides of the Frontier Mountains, suggesting an origin in a large meteoroid that fragmented in the atmosphere and fell at a location "upflow" (i.e., upstream as the glaciers flow) from the two recovery locations. After landing, the two fragments were transported to opposite sides of the stranding area by separate ice flows. Taken together, the pairings at FRO suggest that the basic dynamics of ice flow in the region have not changed much in the last 500 ka. As it happens, the Frontier Mountains area is also the location of the oldest terrestrial find, FRO 01149. This sample was found on a bedrock surface, where *Welten et al.* [2008] argue that it fell. It is safe to say that only in Antarctica could one find a meteorite having a terrestrial age of ~3 Ma on a bedrock surface! When viewed in light of independent evidence [*Höfle*, 1989] showing that the bedrock was overridden by glacier at some point in its history, the presence of this terrestrially ancient meteorite indicates that the glacial event(s) occurred before the meteorite landed >3 Ma ago.

Meteorites recovered from the stranding area at the Queen Alexandria Range provide a second example illustrating the significance of pairing studies [*Welten et al.*, 2011b] (Plate 9). This area has yielded about 3500 fragments, including 660 L5 chondrites and nearly 1500 LL5 chondrites. It was initially assumed these meteorites came from two large showers. The terrestrial ages of 11 representative fragments (iron fractions) all have the same terrestrial age, consistent with the inference of a single L/LL5 chondrite fall.

As implied above, the terrestrial ages of Antarctic meteorites have implications for regional glaciology [e.g., *Spaulding et al.*, 2012]. For further details, we refer the reader to *Cassidy et al.* [1992] and *Harvey et al.* [2014 (this volume)].

9.3. EXPOSURE AGES

There is no reason to doubt that solar and galactic cosmic rays have permeated the solar system for 4.56 Ga. Attenuated one thousand fold by a few meters of solid material and traveling at near-light speeds, most of these high-energy particles leave few traces that we can detect. The exceptions occur when the irradiated object is small, ≤1 m in size (as are most meteoroids and all micrometeoroids), and when the matter is located near an exposed surface. If a collision excavates a meteoroid from deep below the surface of an asteroidal or planetary parent body and the enforcers of Murphy's Law are distracted, we may expect the meteoroid to record a single, uniform period of irradiation that ended when the object landed on Earth after a time equal to its exposure age T_{Exp}. This is the so-called one-stage model for an exposure history and it works surprisingly well for many meteorites. For example, it explains to first order clusters in the CRE age distributions of L and H chondrites [*Graf and Marti*, 1995, and references therein].

Herzog and Caffee [2014] discuss the calculation of CRE ages in detail. The calculations depend on use of the relation relating the activity of a cosmogenic radionuclide to the duration of exposure

$$A_{Fall} = P_A \left(1 - e^{-\lambda_A T_{Exp}}\right) \qquad (9.6)$$

or a variant derived from it such as

$$\frac{S}{A_{Fall}} = \frac{P_S T_{exp}}{P_A \left(1 - e^{-\lambda_A T_{exp}}\right)}. \qquad (9.7)$$

In equations 9.5 and 9.6, P denotes the production rate of a cosmogenic nuclide production rate (in space) and S the concentration of a stable cosmogenic radionuclide such as ^{21}Ne. When the product $\lambda_A T_{Exp}$ is large, equation 9.6 simplifies to

$$\frac{S}{A_{Fall}} = \frac{P_S T_{Exp}}{P_A} \qquad (9.8)$$

The nuclides A and S are chosen to minimize the variability of P_S/P_A, whose value must be known independently. A full discussion is beyond our scope, but two constraints on this approach are directly relevant to the discussion below. As noted in connection with terrestrial ages, the value of A_{Fall} is not known for finds. In practice, to solve equation 9.5 or 9.6, we need a value for the terrestrial age, which in turn means that we usually need to measure not one but two or more cosmogenic radionuclides (see equation 9.4) to obtain T_{Exp}. Second, equation 9.7 presupposes that the meteorite retains no memory of any earlier irradiation. This assumption seems to hold for many meteorites from Mars, but not from the Moon. For simplicity, therefore, we begin our brief discussion of the CRE ages of Antarctic meteorites with the simpler martian meteorites.

9.3.1. Martian Meteorites

In 1962, the unique "SNC" group of meteorites comprised six stones distinguished by their mineralogy and composition. By 1999 six more like them had been discovered, five in the Antarctic and one in Los Angeles, California, USA. During the same period, circumstantial evidence for the idea that these objects came from Mars gained acceptance, which stimulated intense interest and helped maintain funding for the Antarctic collection programs. As of October 2013, the count of martian meteorites stood at 125, 28 of them from the Antarctic and 65 from Northwest Africa. Just as the first terrestrial ages

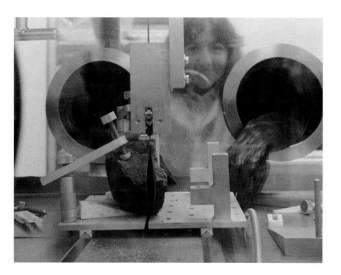

Figure 9.6a. Robbie Score, finder of EET A79001, cutting a slice from the meteorite.

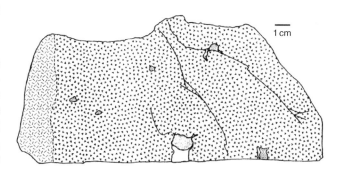

Figure 9.6b. Schematic drawing of a slice of EET A79001 showing different lithologies.

Lithology A   Lithology C

Lithology B   White druse

measured for the Victoria Land meteorites established the broad outlines of the measurements to follow, so too the CRE histories of the first recovered Antarctic martian meteorites were consistent with and representative of later findings for other martian meteorites. We discuss as representative examples two of those meteorites, EET A79001 (olivine-phyric) and ALH A77005 (lherzolitic), both of which were acquired by ANSMET (Plates 70 and 72, respectively).

EET A79001 (7.9 kg) was discovered in 1979 and was only recently displaced by Tissint (recovered in Morocco) as the second most massive meteorite from Mars. EET A79001 has three different lithologies (Figure 9.6), whose descriptions we take from *Martinez and Gooding* [1986]. Lithology A, the most abundant, consists mainly of "medium-grained, feldspathic pyroxenite with abundant megacrysts of olivine and some megacrysts of pyroxene." Lithology B is a "coarse-grained equivalent of Lithology B, but without the megacrysts." And Lithology C is a complicated mixture of glass and microcrystalline materials. A white druse consists of salts that apparently formed in place, a portion of them on Earth.

Using revised production rates for noble gases, *Eugster et al.* [1997] recalculated CRE ages for EET A79001 based on earlier analyses of ^{3}He, ^{21}Ne, and ^{38}Ar. The ^{21}Ne CRE ages, T_{21}, which are probably the most reliable, were approximately 0.7 Ma. *Jull and Donahue* [1988] measured the ^{14}C concentration and obtained a terrestrial age of 12 ± 2 ka based on $A_{Fall} = 65 \pm 11$ dpm/ kg for a body with the average composition of shergottites. CRE ages can also be calculated from the activities at the time of fall of ^{26}Al, ^{10}Be, and ^{53}Mn, all of which have half-lives comparable to or longer than T_{21}. *Pal et al.* [1986] reported ^{10}Be ages of 0.7 ± 0.2 Ma and 0.9 ± 0.2 Ma for EET A79001A and B, respectively based on composition-corrected production rates of 22 and 26 dpm/kg. *Sarafin et al.* [1985] measured both the ^{26}Al and the ^{10}Be activities of EET A79001. For each radionuclide we have $A = P_A \left(1 - e^{-\lambda_A T_{exp}}\right) e^{-\lambda_A T_{Terr}}$.

Adopting the values $P(^{26}$Al$) = 84$ dpm/kg and $P(^{10}$Be$) = 21.2$ dpm/kg, Sarafin and coworkers solved these two equations iteratively and found $T_{Exp} = 0.78 \pm 0.14$ Ma and $T_{Terr} = 0.32 \pm 0.17$ Ma. Both *Pal et al.* [1986] and *Sarafin et al.* [1985] assumed a ^{10}Be half-life of 1.6 Ma, rather than the currently accepted value of 1.387 Ma. This lower half-life decreases the CRE ages estimated from ^{10}Be by about 10%. More important than the details, however, and as noted by several early workers, EET A79001 has a CRE age on the order of 1 Ma. By comparison, the CRE ages of basaltic and other shergottites are older. Recent work by *Berezhnoy et al.* [2010] and especially by *Nishiizumi et al.* [2011] show the existence of a prominent 1-Ma peak in the CRE age distribution of phyric shergottites, strongly suggesting a single launching event and neighboring locations of the stones on Mars.

One of the minor mysteries concerning martian meteorites has been the apparent absence of material that lay close enough to the martian surface to record cosmic-ray irradiation prior to launch. *Hidaka et al.* [2009] searched for evidence of such pre-exposure by measuring Sm isotope abundances in a group of 12 martian meteorites that included EET A79001. These authors concluded that EET A79001, among others, shows shifts in Sm isotope abundances that require irradiation on Mars. Ironically, however, neither a simple nor even a multistage exposure history of the intact meteoroid and/or its precursor in the parent body can simultaneously explain the Sm data and

Table 9.4. Terrestrial and exposure ages (T_{Terr} and T_{Exp}) of martian meteorite ALH A77005.

Nuclide	T_{Terr} (ka)	T_{Exp} (Ma)	Source
^{81}Kr, ^{38}Ar	0.19 ± 0.07	3.6 ± 0.4	a
^{81}Kr, ^{36}Cl, ^{10}Be	0.21 ± 0.07	2.5 ± 0.3	b, c
^{10}Be		2.8 ± 0.6	d
^{3}He, ^{21}Ne, ^{38}Ar		3.3 ± 0.6	e
^{3}He, ^{21}Ne, ^{38}Ar		2.9 ± 0.7	f

a *Schultz and Freundel*, [1984]
b *Nishiizumi et al.*, [1994]
c *Nishiizumi et al.*, [1986]
d *Pal et al.*, [1986]
e *Bogard et al.*, [1984] recalculated by *Eugster et al.*, [1997]
f *Miura et al.*, [1995]

satisfy constraints from short-lived cosmogenic radionuclides, which imply a prelaunch burial depth of at least 150 cm. The authors conclude, therefore, that EET 79001 assimilated martian regolithic material. The assimilated material evidently carries a Sm but not a ^{21}Ne signature altered by martian preirradiation.

ALH A77005, 482 g, was the first martian meteorite discovered in in the Antarctic. The Martian Meteorite Compendium [*Meyer*, 2012; http://curator.jsc.nasa. gov/antmet/mmc/ALH77005.pdf] summarizes the main experimental results. ALH A77005 consists of olivine (~55%) and pyroxene (~30%), with a balance mostly of maskelynite and melt. All the major phases show signs of heavy shock, which is also suggested by the presence of melt pockets and glassy (pseudotachylite) veins.

Table 9.4 presents published terrestrial and CRE ages for ALH A77005. The weighted mean CRE age is 2.93 ± 0.20, very close to the value of 2.87 ± 0.20 recommended by *Nyquist et al.* [2001]. We base the ages given in Table 9.4 on older half-lives: ^{10}Be (1.5 Ma) and ^{81}Kr (0.213 Ma) rather than the currently accepted values, in order to facilitate comparisons with ages in the literature, which assume the older half-lives. The data hint that the ALH A77005 lherzolite is somewhat older than most basaltic shergottites.

In Table 9.5 we have compiled ejection and terrestrial ages of martian meteorites collected in the Antarctic. The CRE age distributions for Antarctic and for non-Antarctic martian meteorites (Figure 9.7) are indistinguishable. Most martian meteorites fall into one of a small number, five to seven, of groups. This clustering of ages implies a correspondingly small number of launches from a few locations on Mars. Radiometric dating (mostly Sm/Nd with some Rb/Sr) yields a clumpy distribution of much older ages clustering about 180 Ma or 600 Ma for shergottites and about 1.2 Ga for nakhlites. The outlier ALH 84001 and two recent martian finds from North Africa have a much older age of 4.1 Ga [see, e.g., *Nyquist*

Table 9.5. Ejection and terrestrial ages (T_{Ej} and T_{Terr}) of selected martian meteorites found in the Antarctic.[a]

Meteorite	Classification	T_{Ej} (Ma)	T_{Terr} (ka)
ALH A77005	Shergottite	3.32 ± 0.55	190 ± 70
ALH 84001	Orthopyroxenite (unique)	14.4 ± 0.7	13 ± 1
EET A79001	Shergottite	0.65 ± 0.20	12 ± 1
LAR 06319	Shergottite	~3.3	—
LEW 88516	Shergottite	4.1 ± 0.6	21.5 ± 1.5
MIL 03346	Nakhlite	9.5 ± 1.0	—
QUE 94201	Shergottite	2.6 ± 0.5	290 ± 50
RBT 04261	Shergottite	3.0 ± 0.6	<60
RBT 04262	Shergottite	2.0 ± 0.5	710 ± 60[b]
Y 793604	Shergottite	4.4 ± 1.0	—
Y 980459	Shergottite	1.1 ± 0.2	—
Y 000027	Shergottite	4.9	—
Y 000593, 749, 802	Nakhlite	12.1 ± 0.7	55 ± 20

[a]References to the original data for these and other martian meteorites may be found in *Jull* [2006], *Meyer* [2012], and *Herzog and Caffee* [2014].
[b]*Jull*, unpublished data.

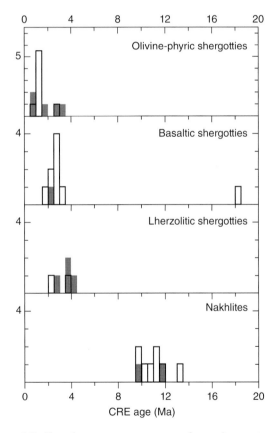

Figure 9.7. Cosmic-ray exposure ages of martian meteorites. Open and filled bars are for non-Antarctic and Antarctic meteorites, respectively.

et al., 2001; *Park et al.*, 2014]. That the radiometric ages greatly exceed the CRE ages is no surprise as launch must postdate the formation of the rocks. More puzzling is that the radiometric ages are much younger than estimates of the average age of the martian surface based on crater counting. This discrepancy constitutes what is often called the martian age paradox. Any of several explanations could help resolve it: (1) The material in younger martian terrain may be better suited to survive launch into space by virtue of its mechanical properties. (2) The crater counting dates may be wrong. (3) The ages reported for martian meteorites may date a local resetting event that is, somehow, not reflected the cratering record, rather than recent crystallization from a magma.

None of the martian meteorites has a CRE age older than 20 Ma (Figure 9.7). Taken as a group and with only a few exceptions (lunar, CI, and CM meteorites), the CRE ages of martian meteorites are younger than those of most other stony meteorites. The difference can be explained in terms of the energetics of the collisions that produced the meteorites and the orbital dynamics of the meteorites' subsequent transport. *Gladman* [1997] modeled the orbital evolution of a chaotic swarm of fragments blasted by a collision from the surface of Mars. In accord with the distribution shown in Figure 9.7, they found that the orbits of the fragments evolved rapidly, within 1 Ma, into Earth-crossing orbits and in such a way that most of the fragments fell into the Sun with 10 Ma. The older CRE ages of many asteroidal meteoroids reflect the lower energies of asteroidal collisions and the longer times they take to reach the resonances that propel them into Earth-crossing orbits.

9.3.2. Lunar Meteorites

The identification of ALH A81005 (Plate 64) as the first meteorite from the Moon was a major event in meteoritics. As noted in the *Antarctic Meteorite Newsletter*, vol. 5, No. 4, November 1982, the meteorite "has been characterized as

Table 9.6. Pre-ejection lunar depths ($D_{2\pi}$) and durations ($T_{2\pi}$), and ejection, CRE exposure, and terrestrial ages (T_{Ej}, T_{Exp}, and T_{Terr}) of selected lunar meteorites found in the Antarctic.[a]

Meteorite	$D_{2\pi}$ (g/cm^2)	$T_{2\pi}$ (Ma)	T_{Ej} (Ma)	T_{Exp} (Ma)	T_{Terr} (ka)
ALH A81005	150–180	520	0.04 ± 0.02	<0.05	18 ± 1
Asuka 881757	>1000		0.9 ± 0.1	0.9 ± 0.1	<50
EET 87521, EET 96008	540–600	26 >10 ± 1[c]	0.08 ± 0.04	<0.01	80 ± 30
LAP 02205[b]		>43[c]	0.055 ± 0.005	<0.01	20 ± 5
MAC 88104, MAC 88105	360–400	630 ± 200	0.28 ± 0.02	0.04–0.05	230 ± 20
MET 01210			0.95	<0.01	~20
MIL 05035		>1–3[c]			
PCA 02007	—		0.95	<0.01	~30
QUE 93069, QUE 94269	65–80	1000 ± 400	0.16 ± 0.02	0.15 ± 0.02	10 ± 2
QUE 94281	270–320	400 ± 60	0.05 ± 0.03	<0.05	23.5 ± 1.8
Y 791197	4–8	450	<0.1	<0.1	30–90
Y 793169	500	50 ± 10	1.1 ± 0.2	1.1 ± 0.2	<50
Y 793274, Y 981031	140–180	600 ± 150	0.032 ± 0.003	<0.01	20–35
Y 793885	~90		—	—	>36
Y 82192, 3, 86032[d]	>1000	<10 Ma	11	80 ± 8	80 ± 8

[a]References to the original data for these and other lunar meteorites may be found in *Lorenzetti et al.* [2005], *Jull* [2006], *Righter and Gruener* [2013], and *Herzog and Caffee* [2014].

[b]Paired with LAP 02224/02226/02436/03632.

[c]*Fernandes et al.* [2009]. These are lower limits because they assume maximum ^{38}Ar production rates that apply only at the very surface of the Moon.

[d]Lorenzetti et al., 2005.

an anorthositic breccia and appears to be a very rare bird." Although lunar meteorites from "hot" deserts now outnumber Antarctic ones, the latter are more "pristine," that is, they show less weathering, which may simplify the interpretation of experimental results. The first measurements of ^{26}Al and ^{10}Be in ALH A81005 gave low activities, which limited the transit time to Earth to 1 Ma [*Tuniz et al.*, 1983]; measurements of NTL suggested an even shorter one, 2500 y, a value also compatible with extremely low densities of cosmic-ray tracks [*Sutton and Crozaz*, 1983]. Later reports of ^{14}C, ^{41}Ca, ^{36}Cl, ^{26}Al, ^{10}Be, and ^{53}Mn activities enabled the construction of a much more detailed exposure history [*Nishiizumi et al.*, 1991; *Jull and Donahue*, 1992]. With five activities available for modeling, these authors were able to assign to ALH A81005 a terrestrial age of 9 ka and a transit time of only 2 ka.

Consistent with these observations, theoretical studies carried out in the 1960s give short transit times from the Moon [e.g., *Arnold*, 1965; *Wetherill*, 1968], as does recent work [e.g., *Gladman et al.*, 1995]. More specifically, *Gladman et al.* [1995] concluded that much lunar orbital debris ejected with low velocities would reach the Earth in under 10 ka and that the spatial distribution of the meteorites, even from single launch events, would cover the globe. Thus, the relatively short CRE age of ALH A81005 confirms expectations and foreshadowed similar experimental results in years to come.

By comparing values of T_{Ej} for meteorites such as ALH A81005 we can identify meteorites that are launch paired.

With supplementary information from remote sensing and the mineralogy and petrology of the objects, it even may be possible to identify the source location on the Moon [*Gnos et al.*, 2004]. Table 9.6, adapted from *Jull* [2006] with additional material from *Righter and Gruener* [2013], summarizes terrestrial and ejection ages of selected lunar meteorites recovered in the Antarctic.

So far omitted from this discussion is irradiation on the Moon itself. To push the time horizon to the period predating exposure as a meteoroid, one needs to measure cosmogenic nuclides with half-lives longer than the ejection age. In this respect, the stable rare gas isotope ^{21}Ne and to a lesser degree the isotopes ^{38}Ar, ^{83}Kr, and ^{126,128}Xe have proved useful. *Eugster* [1989] and *Lorenzetti et al.* [2005] have reviewed the exposure ages of lunar meteorites. *Herzog and Caffee* [2014] discuss the general process of modeling a detailed cosmic-ray exposure history and refer to many papers in the primary literature. In short, in the simplest case, one assumes a single period of cosmic-ray exposure on the Moon characterized by a duration $T_{2\pi}$ at a depth d followed by irradiation in space in a body of radius R at a depth d for a time $T_{4\pi}$, followed by near quiescence on Earth for a time T_{Terr}. The modeler seeks values of these six parameters that lead to a calculated match to the observed activities or concentrations of cosmogenic nuclides.

The fitting of experimental results indicates that Antarctic lunar meteorites often record 10 Ma or more of cosmic-ray exposure in the topmost two or three meters of the Moon, that is, in the lunar regolith. These estimates of $T_{2\pi}$, and particularly the values greater than 100 Ma, should be regarded with some reserve. Ample evidence from studies of samples returned by the *Apollo* missions suggests that meteoroid bombardment "gardens" the lunar regolith. The bombardment can move material either horizontally or vertically. Vertical motion increases or decreases shielding or, equivalently, the nuclear particle fluxes and production rates of cosmogenic nuclides; horizontal motion can bring the protometeorite into or out of the cosmic-ray shadows cast by the surrounding topography, which also affects shielding. In a complex irradiation scenario of this kind, $T_{2\pi}$ and $D_{2\pi}$ represent averages over all the conditions of lunar irradiation.

Why should lunar meteorites, unlike both asteroidal and martian meteorites, tend to record preirradiation on the parent body? Launches from the Moon take less energy. Lower energy means that impactors may be smaller (and more numerous) and implies a lower probability of excavating deeper-lying material [*Warren*, 1994; *Mileikowsky et al.* [2000]; *Wieler*, 2002; *Nyquist et al.*, 2001; *Herzog and Caffee*, 2014].

9.3.3. Meteor Streams

During the 1990s, M. E. Lipschutz, S. P. Wolf, and their coworkers analyzed and interpreted the trace element concentrations of numerous H and other chondrites from the Antarctic. They inferred that non-Antarctic meteorites had thermal histories different from those of meteorites from Victoria Land, but not from Queen Maud Land [*Dennison and Lipschutz*, 1987; *Wolf and Lipschutz*, 1995a, b; *Michlovich et al.*, 1995, and references therein]. In discussing the difference, *Denison and Lipschutz* [1987] argued that "As the mean terrestrial ages of the two groups (Victoria Land and non-Antarctic meteorites) differ by about 10^5 yr, a change in meteoroid flux with time could be invoked." More broadly, the authors suggested that the Earth samples "streams" of meteoroids (objects with similar orbital elements) whose fluxes may ebb or flow.

While astronomical evidence establishes firmly the existence of cometary and asteroidal meteoroid streams [*Lipschutz et al.*, 1997; *Wiegert et al.*, 2009; *Williams and Jones*, 2007], the stream lifetimes have been controversial. Studies of L chondrites especially show that the terrestrial fluxes of meteorites from a single collisional event may continue for up to 500 Ma [*Schmitz and Häggström*, 2003; *Heck et al.*, 2004], but to our knowledge no one has suggested that streams can preserve their orbital elements for so long. *Wetherill* [1986] asserted streams may not persist for more than 10 ka. Modeling by *Wiegert et al.* [2009] for particles with radii up to 10 cm follows the evolution of the terrestrial meteoroid flux from a single event for over 50 ka, that is, from the creation of a stream to its

dispersion into sporadic meteoroids. The maximum particle radius of 10 cm considered in these calculations is not so different from preatmospheric radii estimated for many Antarctic meteorites, which suggests that the influence of the original stream may persist for up to 50 ka. *Pauls and Gladman* [2005] carried out detailed simulations following the orbital parameters of stream fragments in Earth-intersecting orbits. They concluded that dissipation occurs in time periods of ~10^4 to 10^5 years.

The association of meteorites with meteor streams based on trace element evidence has met with limited acceptance. Although of high quality, the trace element data are discussed in a way that is difficult for non-statisticians to comprehend. Other debated criticisms have focused on the possibilities that (1) differences in the trace-element profiles may reflect weathering rather than orbital effects; and (2) unrecognized pairs were erroneously counted separately, thereby muddling the find statistics [see *Koeberl and Cassidy*, 1991].

In our view, however, the most significant challenges to the stream hypothesis have come from the implications of cosmogenic nuclide measurements. *Schultz et al.* [1991] and *Schultz and Weber* [1996] measured cosmogenic noble gases in H chondrites from Victoria Land. The range of CRE ages, from 4 to 70 Ma, requires a major stretch of estimates of stream life times cited above. In addition the concentration patterns of cosmogenic ^{3}He, ^{21}Ne, and ^{38}Ar and radiogenic ^{4}He and ^{40}Ar show that some of the meteoroids in this group of H chondrites lost ^{3}He and ^{4}He in the last few Ma, probably while at or near perihelion, while others did not. It is difficult to see how a single astronomical stream could accommodate objects with such different orbital elements.

9.4. CONCLUSIONS

From the 1970s onwards, the return of Antarctic meteorites gave an enormous boost to the study of meteoritics in general and of cosmogenic nuclides and cosmic-ray exposure histories in particular. With the investigation of Antarctic lunar meteorites has come improved understanding of meteoroid ejection process and the transit to Earth. Through their greater numbers and clustering, the exposure ages of Antarctic martian meteorites have helped to identify major launch events from Mars. Newly documented and surprisingly short exposure ages for many CI and CM chondrites have pointed strongly to a source in a nearby orbit [*Nishiizumi and Caffee*, 2012]. The distribution of terrestrial ages of Antarctic finds tells us that to first order, the Earth has been sampling a similar population of meteorites for up to 2 Ma and serves as a warning or perhaps an invitation to consider the influence of terrestrial weathering on meteoroid arrival rates or fluxes as a function of time. At present, the hope of seeing changes in the fluxes of meteorites to the Antarctic ice sheet goes unfulfilled, but with meteorite pairing relations better understood and more terrestrial ages known, progress on that front can be expected. For the practicing meteoriticist looking for a sample of, say, Semarkona, neither the *Antarctic Meteorite Newsletter* nor the *NIPR Meteorite Newsletter* may ever supplant the *Catalogue of the Natural History Museum, London*, or its electronic successors. But those newsletters describe finds with a pedigree, abundance, and availability unrivaled elsewhere.

Acknowledgments. We thank Rainer Wieler, an anonymous reviewer, and Kevin Righter for many helpful comments. This work was supported in part by the NASA Cosmochemistry program.

REFERENCES

Akridge, J. M. C., P. H. Benoit, and D. W. G. Sears (2000), Terrestrial age measurements using natural thermoluminescence of a drained zone under the fusion crust of Antarctic ordinary chondrites, *Meteoritics Planet. Sci., 35*, 869–874.

Al-Kathiri, A., B. A. Hofmann, A. J. T. Jull, and E. Gnos (2005), Weathering of meteorites from Oman: Correlation of chemical and mineralogical weathering proxies with ^{14}C terrestrial ages and the influence of soil chemistry, *Meteoritics Planet. Sci., 40*, 1215–1239.

Ammon, K., J. Masarik, and I. Leya (2009), New model calculations for the production rates of cosmogenic nuclides in iron meteorites, *Meteoritics Planet. Sci., 44*, 485–503.

Arnold, J. R. (1965), The origin of meteorites as small bodies: II. The model, *Astrophys. J., 141*, 1536–1547.

Baglin, C. M. (2008), Nuclear data sheets for A = 81, *Nuclear Data Sheets, 109*, 2257–2437.

Benoit, P. H., A. J. T. Jull, S. W. S. McKeever, and D. W. G. Sears (1991), The natural thermoluminescence of meteorites: VI. Carbon-14, thermoluminescence and the terrestrial ages of meteorites, *Meteoritics, 28*, 196–203.

Berezhnoy, A. A., T. E. Bunch, P. Ma, G. F. Herzog, K. Knie, G. Rugel, T. Faestermann, and G. Korschinek (2010), Al-26, Be-10, and Mn-53 in martian meteorites (Abstract #5306), *Meteoritics Planet. Sci., 45*, A13.

Biswas, S., H. T. Ngo, and M. E. Lipschutz (1980), Trace element contents of selected Antarctic meteorites: I. Weathering effects and ALHA A77005, A77257, A77278 and A77299, *Zeitschrift f. Naturforsch., 35a*, 191–196.

Bland, P. A., F. J. Berry, T. B. Smith, S. J. Skinner, and C. T. Pillinger (1996), The flux of meteorites to the Earth and weathering in hot desert ordinary chondrite finds, *Geochim. Cosmochim. Acta, 60*, 2053–2059.

Bland, P. A., A. S. Sexton, A. J. T. Jull, A. W. R. Bevan, F. J. Berry, D. M. Thornley, T. R. Astin, D. T. Britt, and C. T. Pillinger (1998), Climate and rock weathering: A study of

terrestrial age dated ordinary chondritic meteorites from hot desert regions, *Geochim. Cosmochim. Acta, 62*, 3169–3184.

Bogard, D. D., L. E. Nyquist, and P. Johnson (1984), Noble gas contents of shergottites and implications for the martian origin of SNC meteorites, *Geochim. Cosmochim. Acta, 48*, 1723–1739.

Cassidy, W., R. Harvey, J. Schutt, G. Delisle, and K. Yanai (1992), The meteorite collection sites of Antarctica, *Meteoritics Planet. Sci., 27*, 490–525.

Chmeleff, J., F. von Blanckenburg K., Kossert, and D. Jakob (2010), Determination of the ^{10}Be half-life by multicollector ICP-MS and liquid scintillation counting, *Nuclear Instru. Meth. Phys. Res., B 268*, 192–199.

Dennison, J. E., and M. E. Lipschutz (1987), Chemical studies of H chondrites: II. Weathering effects in the Victoria Land, Antarctic population and comparison of two Antarctic populations with non-Antarctic falls, *Geochim. Cosmochim. Acta, 51*, 741–754.

Eugster, O. (1989), History of meteorites from the Moon collected in Antarctica, *Science, 245*, 1197–1202.

Eugster, O., A. Weigel, and E. Polnau (1997), Ejection times of martian meteorites, *Geochim. Cosmochim. Acta, 61*, 2749–2757.

Eugster, O., G. F. Herzog, K. Marti, and M. W. Caffee (2006), Recent irradiation and cosmic-ray exposure ages, in *Meteorites and the Early Solar System II*, edited by D. S. Lauretta and H. Y. McSween Jr., U. Arizona Press, Tucson, Arizona (pp. 829–851).

Evans, J. C., and J. H. Reeves (1987), ^{26}Al survey of Antarctic meteorites, *Earth Planet. Sci. Lett., 82*, 223–230.

Evans, J. C., J. H. Reeves, and L. A. Rancitelli (1982), Aluminum-26: Survey of Victoria Land Meteorites, in Catalog of Meteorites from Victoria Land, Antarctica, 1978–1980, edited by U. B. Marvin and B. Mason, *Smithsonian Contrib. Earth Sci., 24*, 70–74.

Fernandes, V. A., R. Burgess, and A. Morris (2009), ^{40}Ar-^{39}Ar age determinations of lunar basalt meteorites Asuka 881757, Yamato 793169, Miller Range 05035, La Paz Icefield 02205, Northwest Africa 479, and basaltic breccia Elephant Moraine 96008, *Meteoritics Planet. Sci., 44*, 805–821.

Fireman, E. L. (1978), Carbon-14 in lunar soil and in meteorites, *Lunar Planet. Sci. Conf., 9*(2), New York, Pergamon Press (pp. 1647–1654).

Fireman, E. L., and T. L. Norris (1981), Ages and composition of gas trapped in Allan Hills and Byrd core ice, *Earth Planet. Sci. Lett., 60*, 339–350.

Fireman, E. L., L. A. Rancitelli, and T. Kirsten (1979), Terrestrial ages of four Allan Hills meteorites: Consequences for Antarctic ice, *Science, 203*, 453–455.

Folco, L., K. C. Welten, A. J. T. Jull, K. Nishiizumi, and A. Zeoli (2006), Meteorites constrain the age of Antarctic ice at the Frontier Mountain blue ice field (northern Victoria Land), *Earth Planet. Sci. Lett., 248*, 209–216.

Gibson, E. K., Jr., and D. D. Bogard 1978. Chemical alterations of the Holbrook chondrite resulting from terrestrial weathering. *Meteoritics, 13*, 277–289.

Gladman, B. (1997), Destination Earth: Martian meteorite delivery, *Icarus, 130*, 228–246.

Gladman, B. J., J. A. Burns, M. J. Duncan, and H. F. Levison (1995), The dynamical evolution of lunar impact ejecta, *Icarus, 118,* 302–321.

Gnos, E., B. A. Hofmann, A. Al-Kathiri, S. Lorenzetti, O. Eugster, M. J. Whitehouse, I. M. Villa, A. J. T. Jull, J. Eikenberg, B. Spettel, U. Krähenbühl, I. A. Franchi, and R. C. Greenwood (2004), Pinpointing the source of a lunar meteorite: Implications for the evolution of the Moon, *Science, 305,* 657–660.

Gooding, J. L. (1981), Mineralogical aspects of terrestrial weathering effects in chondrites from Allan Hills, Antarctica. *Proc.12th Lunar Planet. Sci. Conf.,* 1105–1122.

Grady, M. M., E. K. Gibson Jr., I. P. Wright, and C. T. Pillinger (1989), The formation of weathering products on the LEW 85320 ordinary chondrite: Evidence from carbon and oxygen stable isotope compositions and implications for carbonates in SNC meteorites, *Meteoritics, 24,* 1–7.

Graf, T., and K. Marti (1995), Collisional history of H chondrites, *J. Geophys. Res. (Planets), 100,* 21,247–21,263.

Harvey, R. P., J. Schutt, and J. Karner (2014), Fieldwork methods of the U.S. Antarctic Search for Meteorites program, in *35 Seasons of U. S. Antarctic Meteorites (1976–2010): A Pictorial Guide to the Collection, Special Publication 68,* edited by K. Righter, C. M. Corrigan, R. P. Harvey, and T. J. McCoy, American Geophysical Union/John Wiley & Sons, Washington, D. C.

Hasan, F. A., M. Haq, and D. W. G. Sears (1987), The natural thermoluminescence of meteorites: I. Twenty-three Antarctic meteorites of known [26]Al content, *Proc. 17th Lunar Planet. Sci. Conf., Part 2, J. Geophys. Res., 92,* E703–E709.

Heck, P. R., B. Schmitz, H. Baur, A. N. Halliday, and R. Wieler (2004), Fast delivery of meteorites to Earth after a major asteroid collision, *Nature, 430,* 323–325.

Herzog, G. F., and M. W. Caffee (2014), 1.13 Cosmic-ray exposure ages of meteorites, in *Treatise on Geochemistry,* 2nd ed, vol. 1, edited by H. D. Holland and K. K. Turekian, Elsevier, Oxford, UK (pp. 419–453).

Hey, M. H. (1966), *Catalogue of Meteorites.* Trustees of the British Museum (Natural History), London.

Hidaka, H., S. Yoneda, and K. Nishiizumi (2009), Cosmic-ray exposure histories of martian meteorites studied from neutron capture reactions of Sm and Gd isotopes, *Earth Planet. Sci. Lett., 288,* 564–571.

Höfle, H. C. 1989. The glacial history of the Outback Nunataks Area in western North Victoria Land, *Geol. Jahrb, E38,* 335–355.

Jull, A. J. T. (2000), Terrestrial ages of meteorites, in *Accretion of Extraterrestrial Matter throughout Earth's History,* edited by B. Peuker-Ehrenbrink and B. Schmitz, Kluwer Academic/ Plenum, New York (pp. 241–266).

Jull, A. J. T. (2006), Terrestrial ages of meteorites, in *Meteorites and the Early Solar System II,* edited by D. S. Lauretta and H. Y. McSween Jr., University of Arizona Press, Tucson (pp. 889–905).

Jull, A. J. T., and D. J. Donahue (1988), Terrestrial ^{14}C age of the Antarctic shergottite, EETA 79001, *Science, 242,* 417–419.

Jull, A. J. T., and D. J. Donahue (1992), [14]C Terrestrial ages of two lunar meteorites, ALHA81005 and EET 87521, *Lunar Planet. Sci., 23,* 637–638.

Jull, A. J. T., S. Cheng, J. L. Gooding, and M. H. Velbel (1988), Isotopic composition and mineralogy of weathering product carbonates from the Antarctic meteorite LEW 85320, *Science, 242,* 417–419.

Jull, A. J. T., Y. Miura, E. Cielaszyk, D. J. Donahue, and K. Yanai (1993), AMS ^{14}C ages of Yamato achondritic meteorites, *Proc. NIPR Symp. Antarctic Meteorites, 6,* 374–380.

Jull, A. J. T., D. J. Donahue, R. C. Reedy, and J. Masarik (1994), A carbon-14 depth profile in the L5 chondrite Knyahinya, *Meteoritics, 29,* 649–651.

Jull, A. J. T., S. Cloudt, and E. Cielaszyk (1998a), ^{14}C terrestrial ages of meteorites from Victoria Land, Antarctica and the infall rate of meteorites, in *Meteorites: Flux with time and impact effects,* edited by G. J. McCall, R. Hutchison, M. M. Grady, and D. Rothery, Geological Society of London Special Publication 140 (pp. 75–91).

Jull, A. J. T., C. Courtney, D. A. Jeffrey, and J. W. Beck (1998b), Isotopic evidence for a terrestrial source of organic compounds found in martian meteorites ALH 84001 and EETA 79001, *Science, 279,* 366–369.

Kigoshi, K., and E. Matsuda (1986), Radiocarbon datings of Yamato meteorites. *Lunar and Planetary Inst. International Workshop on Antarctic Meteorites,* 58–60.

Koeberl, C., and W. A. Cassidy (1991), Differences between Antarctic and non-Antarctic meteorites: An assessment, *Geochim. Cosmochim. Acta, 55,* 3–18.

Kojima, H. (2006), The history of meteoritics and key meteorite collections: Fireballs, falls and finds. *Geological Society Special Publications No. 256,* Geological Society, Brassmill Lane, Bath, UK, 291–304.

Korschinek, G., Bergmaier A., Faestermann T., Gerstmann U. C., Knie K., Rugel G., Wallner A., Dillmann I., Dollinger G., Lierse von Gostomski Ch., Kossert K., Maiti M., Poutivtsev M., and Remmert A. (2010), A new value for the half-life of [10]Be by heavy-ion elastic recoil detection and liquid scintillation counting, *Nuclear Instruments & Methods in Physical Research, B268,* 187–191.

Kring, D. A., A. J. T. Jull, L. R. McHargue, P. A. Bland, D. H. Hill, and F. J. Berry (2001), Gold Basin meteorite strewn field, Mojave Desert, northwestern Arizona: Relic of a small late Pleistocene impact event, *Meteoritics Planet. Sci., 36,* 1057–1066.

Lavielle, B., K. Marti, J.-P. Jeannot, K. Nishiizumi, and M. Caffee (1999), The [36]Cl-[36]Ar-[40]K-[41]K records and cosmic ray production rates in iron meteorites, *Earth Planet. Sci. Lett., 170,* 93–104.

Lee, M. R., and P. A. Bland (2004), Mechanisms of weathering of meteorites recovered from hot and cold deserts and the formation of phyllosilicates, *Geochim. Cosmochim. Acta, 68,* 893–916.

Lian, O. B. (2007), Luminescence Dating: Thermoluminescence, in *Encyclopedia of Quaternary Science,* edited by S. Elias, Elsevier, Amsterdam (pp. 1480–1491).

Lipschutz, M. E. (1982), Weathering effects in Antarctic meteorites, in *Catalog of Meteorites from Victoria Land, Antarctica, 1978–1980*, edited by U. B. Marvin and B. Mason, *Smithsonian Contrib. Earth Sci., 24*, 67–69.

Lipschutz, M. E., S. E. Wolf, and R. T. Dodd (1997), Meteoroid streams as sources for meteorite falls: A status report, *Planet. Space Sci., 45*, 517–523.

Lorenzetti, S., H. Busemann, and O. Eugster (2005), Regolith history of lunar meteorites, *Meteoritics Planet. Sci., 40*, 315–327.

Losiak, A. I., and M. A. Velbel (2011), Evaporite formation during the weathering of Antarctic meteorites: A weathering census analysis based on the ANSMET database, *Meteoritics Planet. Sci., 46*, 443–458.

Martinez, R., and J. L. Gooding (1986), New saw-cut surfaces of EETA79001. *Antarctic Meteorite Newsletter, 9*(1), 23–29.

McKay, D. S., E. K. Gibson Jr., K. L. Thomas-Keprta, H. Vali, C. S. Romanek, S. J. Clemett, X. D. F. Chillier, C. R. Maechling, and R. N. Zare (1996), Search for past life on Mars: Possible relic biogenic activity in martian meteorite ALH84001, *Science, 273*, 924–930.

McKeever, S. W. S. (1982), Dating of meteorite falls using thermoluminescence: Application to Antarctic meteorites, *Earth Planet. Sci. Lett., 58*, 419–429.

Meyer, C. (2012), Martian meteorite compendium. Available at http://curator.jsc.nasa. gov/antmet/mmc. Last accessed 8 Sept 2013.

Michlovich, E. S., S. F. Wolf, M.-S. Wang, S. Vogt, D. Elmore, and M. E. Lipschutz (1995), Chemical studies of H chondrites: 5. Temporal variations of sources. *J. Geophys. Res., 100*, 3317–3333.

Mileikowsky, C., F. A. Cucinotta, J. W. Wilson, B. Gladman, G. Horneck, L. Lindegren, J. Melosh, H. Rickman, M. Valtonen, and J. Q. Zheng (2000), Natural transfer of viable microbes in space, *Icarus, 145*, 391–427.

Miono, S., and A. Nakanishi (1994), Terrestrial ages of the Antarctic meteorites measured by thermoluminescence of the fusion crust: II. *Proceedings of the NIPR Symposium on Antarctic Meteorites (1994), 7*, 225–229.

Miura, Y. N., K. Nagao, N. Sugiura, H. Sagawa, and K. Matsubara (1995), Orthopyroxenite ALH84001 and Shergottite ALHA77005: Additional evidence for a martian origin from noble gases, *Geochim. Cosmochim. Acta, 59*, 2105–2113.

Mokos, J. L., L. Franke, P. Scherer, L. Schultz, and M. E. Lipschutz (2000), Cosmogenic radionuclides and noble gases in Antarctic H chondrites with high and normal natural thermoluminescence levels, *Meteoritics. Planet. Sci., 35*, 713–721.

Neupert, U., R. Michel, I. Leya, S. Neumann, L. Schultz, P. Scherer, G. Bonani, I. Hajdas, S. Ivy-Ochs, P. W. Kubik, and M. Suter (1997), Ordinary chondrites from the Açfer region: A study of exposure histories, *Meteoritics Planet. Sci. (abstract), 32*, A98–99.

Ninagawa, K., A. Inoue, N. Imae, and H. Kojima (2011), Thermoluminescence study of Japanese Antarctic meteorites XIII, *Antarctic Meteorites, 35*, 60–61.

Nishiizumi, K., and M. W. Caffee (1998), Measurements of cosmogenic calcium-41 and calcium-41/chlorine-36 terrestrial ages, *Meteoritics Planet. Sci., 33*, A117.

Nishiizumi, K., and M. W. Caffee (2012), Exposure histories of CI1 and CM1 carbonaceous chondrites, *Lunar Planet. Sci., 43*, 2758.

Nishiizumi, K., M. Imamura, and M. Honda (1978), Cosmic-ray induced ^{53}Mn in Yamato-7301 (j), -7305 (k), and -7304 (m), meteorites, *Mem. Natl. Inst. Polar Res., 8*(Special issue), 209–219, National Institute of Polar Research.

Nishiizumi, K., M. Imamura, and M. Honda (1979a), Cosmic ray produced radionuclides in Antarctic meteorites. *Mem. Natl. Inst. Polar Res., 12*(Special issue), 209–219, 161–177, National Institute of Polar Research.

Nishiizumi, K., J. R. Arnold, D. Elmore, R. D. Ferraro, H. E. Gove, R. C. Finkel, R. P. Beukens, K. H. Chang, and L. R. Kilius (1979b), Measurements of ^{36}Cl in Antarctic meteorites and Antarctic ice using a van de Graaff accelerator, *Earth Planet. Sci. Lett., 45*, 285–292.

Nishiizumi, K., M. T. Murrell, J. R. Arnold, D. Elmore, R. D. Ferraro, H. E. Gove, and R. C. Finkel (1981), Cosmic ray produced ^{36}Cl and ^{53}Mn in Allan Hills: 77 meteorites, *Earth Planet. Sci. Lett., 52*, 31–38.

Nishiizumi, K., J. R. Arnold, D. Elmore, X. Ma, D. Newman, and H. E. Gove (1983), ^{36}Cl and ^{53}Mn in Antarctic meteorites and ^{10}Be-^{36}Cl dating of Antarctic ice, *Earth Planet. Sci. Lett., 62*, 407–417.

Nishiizumi, K., J. Klein, R. Middleton, D. Elmore, P. W. Kubik, and J. R. Arnold (1986), Exposure history of shergottites, *Geochim. Cosmochim. Acta, 50*, 1017–1021.

Nishiizumi, K., J. Klein, R. Middleton, and J. R. Arnold (1987), Long-lived cosmogenic nuclides in the Derrick Peak and Lazarev iron meteorites, *Lunar Planet. Sci., 18*, 724–725.

Nishiizumi, K., D. Elmore, and P. W. Kubik (1989a), Update on terrestrial ages of Antarctic meteorites, *Earth Planet. Sci. Lett., 93*, 299–313.

Nishiizumi, K., A. J. T. Jull, G. Bonani, M. Suter, W. Wolfli, D. Elmore, P. W. Kubik, and J. R. Arnold (1989b), Age of Allan Hills 82102, a meteorite found inside the ice, *Nature, 340*, 550–552.

Nishiizumi, K., J. R. Arnold, J. Klein, D. Fink, R. Middleton, P. W. Kubik, P. Sharma, D. Elmore, and R. C. Reedy (1991), Exposure histories of lunar meteorites: ALH81005, MAC88104, and Y791197, *Geochim. Cosmochim. Acta, 55*, 3149–3155.

Nishiizumi, K., M. W. Caffee, and R. C. Finkel (1994), Exposure histories of ALH 84001 and ALHA 77005, *Meteoritics, 29*, 511.

Nishiizumi, K., M. W. Caffee, J. P. Jeannot, B. Lavielle, and M. A. Honda (1997), systematic study of the cosmic-ray-exposure history of iron meteorites: Beryllium-10/Chlorine-36/Beryllium-10 terrestrial ages (abstract), *Meteoritics Planet. Sci., 32*, A100.

Nishiizumi, K., M. W. Caffee, and K. C. Welten (1999), Terrestrial ages of Antarctic meteorites: Update to 1999, in *Workshop on Extraterrestrial Materials from Cold and Hot Deserts*, LPI Contribution No. 997, Lunar and Planetary Institute, Houston, Texas (p. 64).

Nishiizumi, K., K. Nagao, M. W. Caffee, A. J. T. Jull, and A. J. Irving (2011), Cosmic-ray exposure chronologies of depleted olivine-phyric shergottites, *Lunar Planet. Sci., 42*, 2371.

Nyquist, L. E., D. D. Bogard, C.-Y. Shih, A. Greshake, D. Stöffler, and O. Eugster (2001), Ages and geologic histories of martian meteorites, *Space Sci. Rev., 96*, 105–164.

Pal, D. K., C. Tuniz, R. K. Moniot, W. Savin, T. H. Kruse, and G. F. Herzog (1986), Beryllium-10 contents of Shergottites, Nakhlites, and Chassigny, *Geochim. Cosmochim. Acta, 50*, 2405–2409.

Park, J., D. D. Bogard, L. E. Nyquist, and G. F. Herzog (2014), Issues in dating young rocks from another planet: Martian shergottites, in $^{40}Ar/^{39}Ar$ *Dating: from Geochronology to Thermochronology, from Archaeology to Planetary Sciences*, edited by F. Jourdan, D. F. Mark, and C. Verati, The Geological Society, GSL Special Publication SP378 pp. 297–316.

Pauls, A., and B. Gladman (2005), Decoherence time scales for "meteoroid streams," *Meteoritics Planet. Sci., 40*, 1241–1256.

Righter, K., and J. Gruener (2013), Lunar meteorite compendium. Available at http://curator. jsc. nasa. gov/antmet/lmc. Accessed 8 Sept 2013.

Sarafin, R., U. Herpers, P. Signer, R. Wieler, G. Bonani, H. J. Hofmann, E. Morenzoni, M. Nessi, M. Suter, W. Wolfli (1985), Be-10, Al-26, Mn-53, and light noble gases in the Antarctic shergottite EETA 79001 (A), *Earth Planet. Sci. Lett., 75*, 72–76.

Schmitz, B., and T. Häggström (2003), Sediment-dispersed extraterrestrial chromite traces a major asteroid disruption event, *Science, 300*, 961–964.

Schultz, L., and M. Freundel (1984), Terrestrial ages of Antarctic meteorites (abs), *Meteoritics, 19*, 310.

Schultz, L., and H. Weber (1996), Noble gases and H chondrite meteoroid streams: No confirmation, *J. Geophys. Res., 101*, 21177–21181.

Schultz, L., H. W. Weber, and F. Begemann (1991), Noble gases in H-chondrites and potential differences between Antarctic and non-Antarctic meteorites, *Geochim. Cosmochim. Acta, 55*, 59–66.

Sears, D. W. G., and F. A. Hasan (1986), Thermoluminescence and Antarctic meteorites, in *International Workshop on Antarctic meteorites*, edited by J. O. Annexstad, L. Schultz, and H. Wänke, LPI Technical Report 86-01, Lunar and Planetary Institute, Houston, TX (pp. 83–100).

Sears, D. W. G., J. Yozzo, and C. Ragland (2011), The natural thermoluminescence of Antarctic meteorites and their terrestrial ages and orbits: A 2010 update, *Meteoritics Planet. Sci., 46*, 79–91.

Sears, D. W. G., K. Ninakawa, and A. K. Singhvi (2013), Luminescence studies of extraterrestrial materials: Insights into their recent radiation and thermal histories and their metamorphic history, *Chem. Erde, 73*, 1–37.

Spaulding, N. E., V. B. Spikes, G. S. Hamilton, P. A. Mayewski, N. W. Dunbar, R. P. Harvey, J. Schutt, and A. V. Kurbatov (2012), Ice motion and mass balance at the Allan Hills blueice area, Antarctica, with implications for paleoclimate reconstructions, *J. Glaciology, 58*, 399–406.

Sutton, S. R., and G. Crozaz (1983), Thermoluminescence and nuclear particle tracks in ALHA-81005: Evidence for a brief transit time, *Geophys. Res. Lett., 10*, 809–812.

Tuniz, C., D. K. Pal, R. K. Moniot, W. Savin, T. H. Kruse, G. F. Herzog, and J. C. Evans, (1983), Recent cosmic ray exposure history of ALHA81005, *Geophys. Res. Lett., 10*, 804–806.

Tyra, M. A., J. Farquhar, B. A. Wing, G. K. Benedix, A. J. T. Jull, T. Jackson, and M. H. Thiemens (2007), Terrestrial alteration of carbonate in a suite of Antarctic CM chondrites: Evidence from oxygen and carbon isotopes, *Geochim. Cosmochim. Acta, 71*, 782–795.

Velbel, M. A. (2012), Terrestrial weathering of ordinary chondrites in nature and continuing during laboratory storage and processing: Review and implications for Hayabusa sample integrity (abstract), *Meteoritics Planet. Sci., 48*, doi:10.1111/j.1945-5100.2012.01405.x.

Wacker, J. F. (1993), Aluminum-26 activities in meteorites, *Lunar Planet. Sci., 24*, 1471–1472.

Warren, P. (1994), Lunar and martian meteorite delivery services, *Icarus, 111*, 338–353

Welten, K. C. 1999. Concentrations of siderophile elements in nonmagnetic fractions for Antarctic H- and L-chondrites: A quantitative approach on weathering effects, *Meteoritics Planet. Sci., 34*, 259–270.

Welten, K. C., K. Nishiizumi, and M. W. Caffee (2000), Update on terrestrial ages of Antarctic meteorites, *Lunar Planet. Sci., 31*, 2077.

Welten, K. C., K. Nishiizumi, M. W. Caffee, D. J. Hillegonds, J. A. Johnson, A. J. T. Jull, R. Wieler, and L. Folco (2006), Terrestrial ages, pairing and concentration mechanism of Antarctic chondrites from Frontier Mountains, Antarctica, *Meteoritics Planet. Sci., 41*, 1081–1094.

Welten, K. C., L. Folco, K. Nishiizumi, M. W. Caffee, A. Grimberg, M. M. M. Meier, and F. Kober (2008), Meteoritic and bedrock constraints on the glacial history of Frontier Mountain in northern Victoria Land, Antarctica, *Earth Planet. Sci. Lett., 270*, 308–315.

Welten, K. C., Caffee M. W., and Nishiizumi K. (2011a), Update on terrestrial ages and pairing studies of Antarctic meteorites, *Antarctic Meteorites, 35*, 87–88.

Welten, K. C., M. W. Caffee, D. J. Hillgeonds, T. J. McCoy, J. Masarik, and K. Nishiizumi (2011b), Cosmogenic radionuclides in L5 and LL5 chondrites from Queen Alexandra Range, Antarctica: Identification of a large L/LL5 chondrite shower with a preatmospheric mass of approximately 50,000 kg, *Meteoritics Planet. Sci., 46*, 177–196.

Wetherill, G. W. (1968), Dynamical studies of asteroidal and cometary orbits and their relation to the origin of meteorites, in *Origin and Distribution of the Elements*, edited by L. H. Ahrens and E. Ingerson, Pergamon Press, New York (pp. 423–443).

Wetherill, G. W. (1986), Meteorites: Unexpected Antarctic chemistry, *Nature, 319*, 357–358.

Whillans, I. M., and W. A. Cassidy (1983), Catch a falling star: Meteorites and old ice, *Science, 222*, 55–57.

Wiegert, P., J. Vaubaillon, and M. Campbell-Brown (2009), A dynamical model of the sporadic meteoroid complex, *Icarus, 201*, 295–310.

Wieler, R. (2002), Cosmic-ray-produced noble gases in meteorites, in *Noble Gases in Geochemistry and Cosmochemistry*, Reviews in Mineralogy and Geochemistry. *Min. Rev. Geochem., 47*, 125–170.

Williams, I. P., and D. C. Jones (2007), How useful is the 'mean stream' in discussing meteoroid stream evolution? *Mon. Not. R. Astron. Soc., 375*, 595–603.

Wolf, S. E., and M. E. Lipschutz (1995a), Chemical studies of H chondrites: 4. New data and comparison of Antarctic suites, *J. Geophys. Res., 100*, 3297–3316.

Wolf, S. E., and M. E. Lipschutz (1995b), Chemical studies of H chondrites: 6. Antarctic/non-Antarctic compositional differences revisited, *J. Geophys. Res., 100*, 3335–3349.

Zolensky, M., P. Bland, P. Brown, and I. Halliday (2006), Flux of extraterrestrial materials, in *Meteorites and the Early Solar System II,* edited by D. S. Lauretta and H. Y. McSween Jr., University of Arizona Press, Tucson (pp. 869–888).

Zurfluh, F. J., B. A. Hofmann, E. Gnos, U. Eggenberger, and A. J. T. Jull (in press), Terrestrial age estimation of ordinary chondrites from Oman based on a refined weathering scale and other physical and chemical weathering parameters. *Meteoritics Planet. Sci.*

10

A Statistical Look at the U.S. Antarctic Meteorite Collection

Catherine M. Corrigan[1], Linda C. Welzenbach[1], Kevin Righter[2], Kathleen M. McBride[2],
Timothy J. McCoy[1], Ralph P. Harvey[3], and Cecilia E. Satterwhite[2]

10.1. INTRODUCTION

Over the last century, the Antarctic ice has been an amazing source for the collection of meteoritic material. Meteorite recoveries in Antarctica prior to 1969 were largely incidental and accidental to exploration or investigation of the continent. As early as 1912, an Australian expedition led by Douglas Mawson found an L5 chondritic meteorite [*Mawson*, 1915]. In 1961, an iron was recovered from the Humboldt Mountains in Adelie Land by Russian geologists and a pallasite in the Thiel Mountains by the United States Geological Survey (USGS). A fourth (another iron meteorite) was found in 1964 in the Neptune Mountains, also by the USGS [*Cassidy*, 2003].

The earliest systematic collection of meteorites by a U.S. team in Antarctica began in 1976, when Dr. William Cassidy (University of Pittsburgh) established the ANSMET program. Initially done in collaboration with Japanese colleagues [*Yoshida*, 2010], the U.S. program quickly evolved into a sophisticated field and curatorial program with the Smithsonian Institution, NASA, and NSF (National Science Foundation) as partners. The average number of meteorites collected during ANSMET expeditions has steadily increased for a number of reasons that include longer field seasons, better identification of blue icefields, multiple teams (including those dedicated to reconnaissance), and the use of a variety of aerial vehicles to access more remote field sites [*Harvey et al.*, 2014 (this volume)]. The field teams, composed of volunteer scientists, teachers, writers, and even a few astronauts, have returned more meteorites in ~35 years than were collected by scientific institutions around the world in the past few hundred years.

In this chapter, we examine the U.S. Antarctic meteorite collection from a statistical perspective, comparing data from individual field sites, seasons, and other meteorites collected in terms of types of meteorites and the collection as a whole. One of the prime early drivers for consistent and methodical characterization of the entire U.S. Antarctic collection was to allow such statistical comparisons. Early statistical assessments of the U.S. Antarctic collection include *Harvey and Cassidy* [1989], *Score and Lindstrom* [1990], and *Cassidy and Harvey* [1991], which together examined mass distributions and the relative frequency of meteorite types for the collection as well as comparisons to a defined set of modern falls.

More recent studies have looked at these statistics in more detail. Principal component analysis was used by Lipschutz and colleagues [*Wolf and Lipschutz*, 1995a, 1995b, *Michlovich et al.*, 1995] to analyze labile trace element data for H chondrites and argue that the flux of H chondrites changed with time between the Antarctic meteorites and modern falls. *Harvey* [1995] used model size distributions to deconstruct the contribution of wind movement, meteorite supply, and search losses to the Antarctic collection. The mass-based statistics of the collection were updated by *Cassidy* [2003], while size distribution comparisons were similarly reexamined by *Harvey* [2003] and *Cassidy* [2003]. More recently, a series of abstracts by *McBride and Righter* [2010] and *Welzenbach and McCoy* [2006] have examined various aspects of statistics from the Antarctic collection, including comparison with modern falls and Saharan meteorites.

[1] *Department of Mineral Sciences, National Museum of Natural History, Smithsonian Institution*
[2] *Jacobs Technology, NASA Johnson Space Center*
[3] *Case Western Reserve University, Department of Earth Environmental and Planetary Sciences*

35 Seasons of U.S. Antarctic Meteorites (1976–2010): A Pictorial Guide to the Collection, Special Publication 68,
First Edition. Edited by Kevin Righter, Catherine M. Corrigan, Timothy J. McCoy and Ralph P. Harvey.
© 2015 American Geophysical Union. Published 2015 by John Wiley & Sons, Inc.

Table 10.1. Number of Antarctic Meteorites by Type (Expanded Classification)

Type	Number
Acapulcoite	15
Acapulcoite/Lodranite	5
Achondrite ungrouped	12
Angrite	2
Aubrite	40
Brachinite	4
C2 ungrouped	6
C3 ungrouped	1
CB	12
CH	6
Chondrite ungrouped	3
CH3	3
CK4	23
CK5	65
CK6	8
CM1 and CM1/2	33
CM2	270
CO3	209
CR1	2
CR2	72
CR3	1
CV3	87
Diogenite	67
En chondrite ungrouped	8
EH impact melt	4
EH3	53
EH4	16
EH5	4
EH6	1
EL3	20
EL4	4
EL5	1
EL6	17
Eucrite	3
Eucrite (brecciated)	117
Eucrite (unbrecciated)	34
FCr (chondrite)	4
H impact melt	12
H3	60
H4	324
H5	3249
H6	2301
Howardite	93
Irons	111
Kakangari-like	1
L impact melt	13
L metal	4
L3	246
L4	322
L5	2986
L6	3748
L7	4
LL impact melt	6
LL3	36
LL4	57

Table 10.1. (Continued)

Type	Number
LL5	2616
LL6	1305
LL7	4
Lodranite	6
Lunar	24
Mesosiderite	37
Nakhlite	4
Pallasite	26
R chondrite	27
Shergottite	10
ALH 84001	1
Ureilite	65
Winonaite	1

The U.S. Antarctic meteorite program has, as of this publication, retrieved 20,327 individual specimens collected from 50 formally named field sites (Tables 10.1, 10.2, and 10.3) on the East Antarctic Plateau adjacent to the Transantarctic Mountains. A few glaciological and geographically distinct field sites were grouped together during the early days of naming field areas that might today have been split due to the large numbers of meteorites found in these areas. The Grosvenor Mountains (GRO) meteorites, for example, actually represent meteorites from individual icefields such as Otway Massif, Mt. Cecily, Mt. Bumstead, Mt. Raymond, Mt. Block, Block Peak, and Mauger Nunataks. The Japanese, Chinese, South Koreans and a number of European countries also have programs that send meteorite hunters to the ice. The EUROMET program recovered more than 250 meteorites from the Frontier Mountains in 1991 and ~200 meteorites from the Allan Hills region in 1988 [e.g., *Delisle and Sievers*, 1991; *Delisle et al.*, 1993]. The Japanese Antarctic collection, housed at the National Institute of Polar Research, consists of ~17,000 meteorites (as of 2006; a 2012 expedition added more), though only 25% have been classified [*Kojima*, 2006; *Yamaguchi et al.*, 2012]. The Chinese Antarctic program (CHINARE) has about 10,000 meteorites [*Liu et al.*, 2004].

Of the 20,327 meteorites retrieved from the Antarctic by the U.S. program to date, 18,926 have been classified as of this publication (~93%). The U.S. efforts have focused on systematic characterization and classification of all meteorites, while other programs concentrate their efforts on the most scientifically interesting meteorites. Both approaches yield similar insights into early solar system processes. We focus on the U.S. collection, which has the unique distinction of being a statistically significant, well-characterized population, allowing comparison with similarly sized populations of meteorites observed to fall and non-Antarctic finds.

We classify the U.S. Antarctic meteorite collection into ~40 general types (i.e., all L chondrites are counted as the

Table 10.2. Number of objects recovered by year.

Year	Stones	Meteorites	Terrestrial*
1976	11	11	0
1977	302	300	2
1978	308	305	3
1979	75	73	2
1980	102	100	2
1981	317	315	2
1982	108	108	0
1983	289	281	8
1984	274	272	2
1985	366	360	6
1986	590	558	32
1987	685	670	15
1988	910	876	34
1989	0	0	0
1990	1,100	1,098	2
1991	601	600	1
1992	255	254	1
1993	853	840	13
1994	608	596	12
1995	238	235	3
1996	394	388	6
1997	1,091	1,086	5
1998	192	190	2
1999	1,041	1,031	10
2000	759	756	3*
2001	336	335	1
2002	924	885	39*
2003	1,355	1,353	2*
2004	1,231	1,228	3*
2005	236	233	5
2006	854	842	13
2007	710	700	10
2008	521	519	2
2009	1,021	1,014	8
2010§	1,234	1,237	4
2011§	302	302	0
2012§	394	393	1
Total recovered	20,580	20,327	254
Total classified		18,926	
% Terrestrial		1.23	

*Terrestrial number also includes tiny pieces of fusion crust.
§Samples from these seasons are still being characterized as of November 2013,
so additional terrestrial material may be identified among the collection.

same type, for example, as are all martian meteorites; the number of meteorites of each type can be seen in the expanded classification in Table 10.1). While the vast majority (~90%) of Antarctic meteorites collected each year are ordinary chondrites, each year a few meteorites are found that are enigmatic, either exhibiting properties outside the range previously defined for a given group of meteorites or necessitating the formation of an entirely new group of meteorites. These meteorites represent pieces of the cosmochemical puzzle that challenge our

understanding of how our solar system formed and evolved. As examples, the first of the newly defined CH chondrites (ALH 85085; Plate 24) was recovered in Antarctica [*Weisberg and Righter*, 2014 (this volume)] and the acapulcoite-lodranite clan [*Mittlefehldt and McCoy*, 2014 (this volume)] was only represented by the type meteorites prior to the Antarctic program. Many anomalous, poorly understood meteorites remain in the collection awaiting future recoveries, technological innovations, or paradigm shifts to fully understand their

Table 10.3. Number of samples recovered at all field sites (as of November 2013).

Abbrev.	Locality	Specimens	Terrestrial	Field Season Year (# Meteorites)
ALH	Allan Hills	1,628	13	1976(9), 1977(299), 1978(259), 1979(53), 1980(32), 1981(315), 1982(45), 1983(83), 1984(263), 1985(158), 1986(13), 1987(7), 1990(15), 1994(20), 1995(10), 1997(12), 1999(7), 2003(2), 2009(10), 2010(3)
BEC	Beckett Nunatak	2	0	1992(2)
BOW	Bowden Neve	1	0	1985(1)
BTN	Bates Nunataks	15	0	1978(4), 2000(11)
BUC	Buckley Island	30	1	2010(29)
CMS	Cumulus Hills	79	0	2004(79)
CRA	Mt. Cranfield	1	0	2003(1)
CRE	Mt. Crean	1	0	2001(1)
DAV	David Glacier	9	0	1992(9)
DEW	Mt. DeWitt	2	0	1996(2)
DOM	Dominion Range	1,574	6	1985(11), 2003(140), 2008(519), 2010(898)
DNG	D'Angelo Bluff	4	0	2006(4)
DRP	Derrick Peak	11	0	1978(9), 2000(2)
EET	Elephant Moraine	2,204	17	1979(10), 1982(17), 1983(197), 1984(9), 1986(3), 1987(360), 1990(992), 1992(192), 1996(348), 1999(59)
FIN	Finger Ridge	9	0	2000(3), 2001(6)
GDR	Gardner Ridge	4	0	1998(2), 2012(2)
GEO	Geologists Range	32	0	1985(2), 1999(30)
GRA	Graves Nunataks	354	2	1995(33), 1998(184), 2006(135), 2012(12)
GRO	Grosvenor Mountains	412	4	1985(19), 1995(166), 2003(170), 2006(48), 2012(5)
HOW	Mt. Howe	5	0	1988(4), 2010(1)
ILD	Inland Forts	1	0	1983(1)
KLE	Klein Ice Field	1	0	1998(1)
LAP	LaPaz Ice Field	1,676	1	1991(3), 2002(237), 2003(848), 2004(416), 2010(171)
LAR	Larkman Nunatak	1,037	13	2004(79), 2006(621), 2012(324)
LEW	Lewis Cliff	1,960	90	1985(168), 1986(535), 1987(280), 1988(767), 1990(1), 1993(89), 1997(28), 1999(2)
LON	Lonewolf Nunataks	10	0	1994(10)
MBR	Mount Baldr	2	0	1976(2)
MAC	MacAlpine Hills	1,065	38	1987(21), 1988(105), 2002(479), 2004(422)
MCY	MacKay Glacier	69	0	1992(4), 2005(65)
MET	Meteorite Hills	1,133	4	1978(28), 1996(38), 2000(738), 2001(325)
MIL	Miller Range	2,339	23	1985(1), 1999(30), 2003(113), 2005(166), 2007(700), 2009(1004), 2011(302)
ODE	Odell Glacier	3	0	2001(3)
OTT	Outpost Nunatak	1	0	1980(1)

PAT	Patuxent Range	183	1	1991(52), 2010(131)
PCA	Pecora Escarpment	599	0	1982(29), 1991(490), 2002(80)
PGP	Purgatory Peak	1	0	1977(1)
PRA	Mt. Pratt	23	1	2004(22)
PRE	Mt. Prestrud	17	0	1995(17)
QUE	Queen Alexandra Range	3,481	36	1986(1), 1987(2), 1990(86), 1993(751), 1994(566), 1997(1046), 1999(903), 2002(90)
RBT	Roberts Massif	229	0	2003(19), 2004(210)
RKP	Reckling Peak	138	3	1978(5), 1979(10), 1980(67), 1986(6), 1992(47)
SAN	Sandford Cliffs	60	0	2003(60)
SCO	Scott Glaciers	42	0	1998(3), 2006(34), 2012(5)
STE	Stewart Hills	1	0	1991(1)
SZA	Szabo Bluff	44	0	2012(44)
TEN	Tentacle Ridge	2	0	2000(2)
TIL	Thiel Mountains	41	0	1982(16), 1991(25)
TYR	Taylor Glacier	3	0	1982(1), 2005(2)
WIS	Wisconsin Range	33	0	1990(4), 1991(29)
WSG	Mt. Wisting	9	0	1995(9)
Total		20,580	254	20,327

Note: The 1977–78 and 1978–79 field seasons were joint ventures with the ANSMET and NIPR (Japan) field parties. Many of the samples were split in half and reside in both collections. However, some smaller samples were allocated to either one or the other collection. As a result, the U.S. collection contains only 248 meteorites from the 1977–78 season (52 are at NIPR) and only 228 meteorites from 1978–79 (77 are at NIPR). The numbers in this table include all of the samples collected by the field parties.

Table 10.4. Examples of large pairing groups in the U.S. Antarctic Meteorite Collection.

Specimen	Mass (g)	Type	Number in Pairing Group	Total Mass (g)
ALH 83009	1.70	aubrite	20	3,587.09
CMS 04061	8,465.0	pallasite	19	170,545.9
LAP 02205	1,226.3	lunar—basalt	6	1,930.43
MIL 03346	715.2	nakhlite	4	1,871.0
ALH 76005	1,425.0	polymict eucrite	14	4,292.3
EET 87503	1,734.5	howardite	9	4,071.8
DRP 78001	15,200.0	IIAB iron	11	263,644.5
LAP 91900	786.87	diogenite	9	2,867.4
EET 87507	36.192	CK chondrite	48	2,315.4
ALH 83100	3,019.0	CM chondrite	21	5,380.5
EET 96005	1.34	CM chondrite	15	1,124.9
EET 87711	5.69	CR chondrite	50	1,341.07
MIL 07099	13.08	CO chondrite	169	8,946.04
DOM 08004	294.5	CO chondrite	7	1,251.7
MIL 07590	3.85	CV chondrite	34	539.7
QUE 90201	1,282.5	L/LL chondrite	>123	>24,874.6
ALH 77 011	291.5	L3.3 to L3.5	78	11,824.5
PCA 82518	21.878	EH3 chondrite	21	408.12

origin [*McCoy*, 2014 (this volume)]. Some of these unusual meteorites, like the first-ever recognized lunar meteorite ALH A81005 (Plate 64) and the ancient hydrous phase–bearing martian meteorite ALH 84001 (Plate 69) (argued to contain evidence of past microbial life) are even paradigm shifting. The chapters in this volume on chondrites, achondrites, and lunar and martian meteorites are all testament to the strength that this pivotal program has added to our understanding of the origin and evolution of our solar system.

While inevitably a few terrestrial rocks are collected along with the meteorites (the policy of bringing back questionable rocks ensures that even potential meteorites are at least examined in the lab) the number of "false positives" is remarkably low. There have been only 254 terrestrial rocks collected in the 37 seasons since 1976 (the 1989 season was cancelled; Table 10.2). This number is just 1.23% of the total meteorites collected during those seasons, a remarkable testament to the ease of recognition of a meteorite relative to a terrestrial rock, and the power of the human senses to be able to distinguish them. Notably, almost 40% of these terrestrial rocks were gathered at the Lewis Cliffs Ice Tongue, a site that is full of small, dark-colored glacial moraine material scraped from the walls of the surrounding Lewis Cliffs.

In a statistical evaluation of the U.S. Antarctic meteorite collection and, in particular, in comparison to other collections (e.g., falls, non-Antarctic), pairing relationships need to be considered. Two or more meteorites are considered paired if they were originally part of the same meteorite that broke apart upon entering the Earth's atmosphere or hitting the ground; and statistical analysis of the modern falls show that 1 in 10 falls consists of 10 individuals or more [*Harvey and Cassidy*, 1989]. As an example, thousands of individual stones from the Allende, Mexico, shower [*Clarke et al.*, 1970] or the Chelyabinsk, Russia, shower [*Popova et al.*, 2013] are paired as single meteorites. The U.S. Antarctic meteorite collection contains a large number of pairing groups of a broad range of meteorite types, including achondrites, chondrites, and irons (Table 10.4). The classification process has identified a number of unequilibrated chondrite pairing groups (e.g., ALH A77011 and PCA 82518; Table 10.4), but many equilibrated chondrite groups have no doubt gone unrecognized due to limited resources for classification and low levels of interest among the scientific community. One exception is the relatively well characterized pairing group of QUE 90201 (Table 10.4; Plate 9), with thousands of pairs of L/LL5 material [*Welten et al.*, 2011]. An interesting observation is that many of these paired specimens are nearly whole fusion crusted stones, strongly suggesting that showers produced during atmospheric breakups are well preserved in Antarctica, while fragmentation during ice movement appears to be a far less common phenomenon. All of the numbers presented in this chapter are without correction for pairing groups.

10.2. OUTPUT BY SEASON AND SITE

ANSMET field seasons have yielded anywhere from 11 meteorites to a whopping 1353 meteorites, collected during the 2003–2004 field season. Table 10.2 shows the number of meteorites that have been collected during

Figure 10.1. Where numerous samples are found in an area less than a square meter, sample numbers are given one GPS location (left-hand image). Flags at MacAlpine Hills, Antarctica, where a large number of (up to 20 very small) meteorites were recovered in a few square meters (right-hand image).

each ANSMET season since 1976. Much of the reason for consistently high numbers in the seasons since the year 2000 was the introduction of a reconnaissance team. This team (usually consisting of four members) spends the season "hopping" between a number of meteorite recovery field sites in order to either complete the systematic search of a site or, equally important, to scout out unexplored sites with concentrations worthy of a full season by the main, larger (usually eight-person) team. There have been a few seasons where the recon team actually collected more meteorites than the main team. During the 2004 season, for example, the recon team collected at a number of new, meteorite-rich sites, including the Cumulus Hills site and the Larkman Nunataks site, and did a return trip to the MacAlpine Hills where they collected a large number of (very small) meteorites (Figure 10.1). The main team stayed at the LaPaz Icefields during the 2004 season, a meteorite-rich area where persistent windy conditions and long distances can limit specimen recovery. The number of field team members (with the exception of splitting them into a main team and recon team) does not necessarily correlate with the number of meteorites collected. For example, the 1982 team only collected 108 meteorites with 10 meteorite hunters, while the six-member 1986 team collected 558 meteorites (plus 32 terrestrial rocks).

The main factor in determination of the number of meteorites discovered in a field site (apart from the weather!) is the sites themselves. Concentration mechanisms are not fully understood [see *Harvey et al.*, 2014 (this volume); *Cassidy*, 2003; *Harvey*, 2003], but some field sites have distinctly more recoverable meteorites stranded on the surface than others. Table 10.3 and Figure 10.2 show the number of meteorites collected from each field site (terrestrial rocks excluded). While return visits to a field site are certainly a factor, this may be slightly misleading, as the team will not return to sites that are determined not to have many meteorites. Mt. Pratt, for example, was visited by the recon team during the 2004–2005 season and will likely not be revisited. Twenty-three meteorites were found there in three days of searching. Larkman Nunataks, however, was also visited during that season (79 meteorites were collected during a few days of searching and 8 days of sitting in the tent through a windstorm, though many more were knowingly left behind). The 2006 and 2012 field teams returned there for a full season, recovering over 1000 more meteorites (many not yet classified).

It can be seen from Figure 10.2 and Table 10.3 that the most productive field sites for meteorite collection have been the Queen Alexandra Range (QUE), the Miller Range (MIL), and Elephant Moraine (EET), each with more than 2200 meteorites collected, and QUE with nearly 3500. However, it isn't necessarily the number of meteorites that makes a place great for collection. For example, the QUE meteorites come from about a half dozen distinct

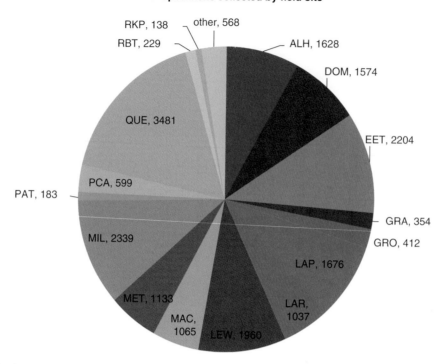

Number of specimens collected by field site

RKP, 138
other, 568
RBT, 229
ALH, 1628
DOM, 1574
EET, 2204
QUE, 3481
GRA, 354
PAT, 183
PCA, 599
GRO, 412
MIL, 2339
LAP, 1676
MET, 1133
LAR, 1037
MAC, 1065
LEW, 1960

Figure 10.2. This pie chart shows the number of meteorites collected from each field site (terrestrial rocks excluded) where 100+ meteorites have been collected. In over 30 seasons, samples have been collected from 50 named field sites, but only 16 have produced more than 100 meteorites, and only 9 sites have produced more than 1000.

Figure 10.3. Approaching storm shows blowing snow obscuring the ground at the 2004 La Paz Icefield campsite.

icefields and moraines, some very highly localized and others including shower falls of more than 1000 likely specimens [*Harvey*, 2003]. Similarly, the Allan Hills (ALH) meteorites come from several distinct icefields. While "only" yielding 1628 meteorites to date, a wide variety of interesting meteorites have been recovered, including the first meteorites collected by ANSMET, the first lunar meteorite found on Earth (ALH A81005; Plate 64), the type CH chondrite (ALH 85085; Plate 24), and the only known martian orthopyroxenite (ALH 84001; Plate 69). The Allan Hills continues to be an important source of meteorites, having been visited by organized meteorite collection efforts, or other Antarctic science programs, during 20 of the 35+ seasons of ANSMET history (most recently in 2010), many of which yielded samples.

Weather can also limit how many meteorites are collected in a season. ANSMET teams do not work when blowing or newly fallen snow limits visibility (Figure 10.3), and the resulting number of "tent days" directly reduces the number of meteorites recovered during any season. For example, one of the lowest totals recorded by an ANSMET field team was the 2012 reconnaissance team working in the Graves Nunatak area where 63 meteorites were recovered in the nine workable days Mother Nature provided.

10.3. STATISTICS OF ANTARCTIC METEORITE COLLECTION SITES

One of the most important questions surrounding the collection of meteorites in Antarctica is whether or not we are collecting a representative sampling of what is actually there. In addition, it is perfectly reasonable to wonder if there are mechanisms at play that are resulting in our selectively collecting certain types of meteorites as opposed to others.

Prior to considering the Antarctic meteorite collection as a whole, it is worth considering the variability in the collection between field sites. Table 10.3 and Figure 10.2 show the number of meteorites collected from each field site the U.S. teams have visited in Antarctica. When an individual field site is coarsely subdivided into meteorite types (e.g., carbonaceous chondrites, achondrites, H, L, LL) (Figure 10.4), ordinary chondrites comprise the vast majority in every case, with carbonaceous and achondrites sampling the largest remaining percentages. However, collection sites with fewer meteorites exhibit wider variation in abundance. For example, 10% of the 352 total Graves Nunataks meteorites are achondrites, a percentage far higher than other field areas.

Interestingly, field sites with at least 1000 total meteorites converge at ~90% ordinary chondrites with the remaining ~10% including other types of chondrites, achondrites, stony-iron and irons. These ratios exhibit minimal changes between 1000 meteorites (MAC, MET) and 3400 meteorites (Queen Alexandra Range). Essentially, field sites with 1000+ meteorites exhibit a robust ratio of ordinary chondrites/other meteorite types, suggesting that this ratio is likely ~90/10 for the entire population of Antarctic meteorites. A corollary of this convergence is that the assertion that one field site is preferable to another for collection of unusual meteorites likely reflects some combination of the statistics of small numbers (e.g., the GRA meteorites), dominance by a pairing group (e.g., the CMS pallasites; Table 10.4; Plate 77), or the selection of an individual field site based on the appearance that it would be more likely to yield a certain type of meteorite (e.g., martian).

Looking at the ordinary chondrite abundances across the field sites represented in Figure 10.4, it can be seen that the proportions of each type (H, L, or LL) vary widely across these sites. As noted earlier, the statistics presented here have not been corrected for pairing. It is likely that the varying proportions of H, L, and LL group ordinary chondrites between field sites reflect the contributions of large, unpaired meteorite showers.

The assertion that large, unpaired showers skew the ordinary chondrite statistics is based on three main observations. First, as noted earlier, shower falls are much more common than most realize; most meteorite specimens are part of a shower, not solo falls [*Harvey and Cassidy*, 1989]. A second key observation is made by those involved in the recovery and classification of the Antarctic meteorites. As noted earlier, most meteorites tend to be whole, fusion-crusted stones, suggesting that breakup during atmospheric passage dominates over breakup in the ice in producing pairings. While proximal, fragmental stones are certainly found having broken on the Antarctic ice, they are rare and do not dominate pairing groups. Indeed, obvious proximal fragments are typically collected and labeled as a single meteorite on the ice.

The third key observation is the lack of pairing of equilibrated ordinary chondrites during classification. Although pairing of equilibrated ordinary chondrites was typically suggested in the first decade of the program, the limited classification approach typically assignment of chemical group from oil immersion techniques and petrologic type from visual examination of a chip [*Lunning et al.*, 2012], is insufficient for robust pairing. Pairing of the enormous number of ordinary chondrites that come through the classification pipeline each year is far beyond the capabilities of the existing curatorial system, requiring far more time than this simple statistic is probably worth.

A few in-depth efforts have been made to assess the relative contributions of pairing to the Antarctic population. *Scott* [1989] conducted a study of pairing of the meteorites found in Victoria Land and the Thiel Mountains (TIL) Regions in Antarctica, concluding that most equilibrated ordinary chondrites are paired with 1–3 other meteorites. *Lindstrom and Score* [1995] examined

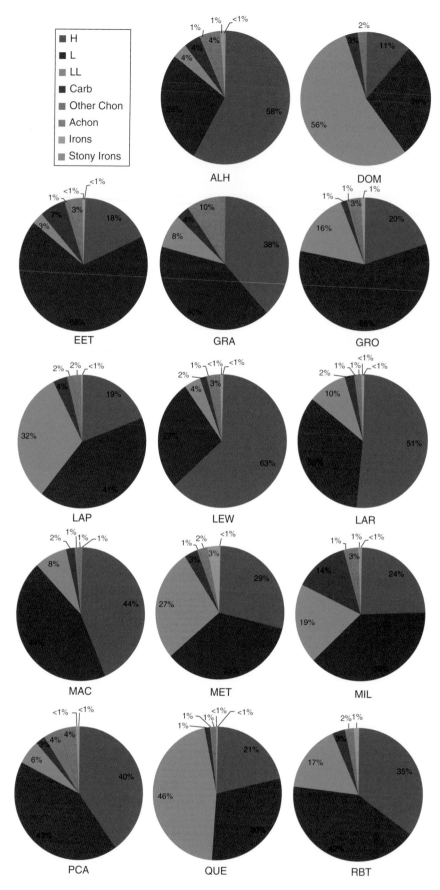

Figure 10.4. Meteorite types collected by field site.

the sizes of documented pairing groups in the Antarctic collection to that point, suggesting a somewhat higher number of ~5. *Welten et al.* [2011] used petrologic, mineral chemical, and cosmogenic nuclide data to suggest that nearly 2000 ordinary chondrites from Queen Alexandra Range, roughly 60% of the total from that area, are paired (Plate 9). This furthers the argument put forth by Lindstrom and Score [1995] that pairing is distinctly nonlinear, with most meteorites paired with few or any other meteorites, while a small number of pairing groups dominate pairing statistics.

10.3.1. Mass Distribution

An alternative way to approach the examination of meteorite statistics is to look at the cumulative mass distribution of a population relative to the number of actual meteorites represented. *Harvey and Cassidy* [1989] and *Huss* [1991] both point out that the mass of meteorites found in Antarctic field sites (Allan Hills and Yamato) peaks at ~10 g, while those of modern witnessed stone falls peak at ~5 kg. This conclusion is logical in light of the fact that small (<2 cm) meteorites are much easier to spot on Antarctic ice than they are in non-Antarctic locations. Additionally, systematic collection of meteorites in Antarctica and elsewhere (Roosevelt County, New Mexico, for example, as discussed in *Huss* [1991]) recovers more small meteorites than random searches for potential meteorite falls. Figure 10.5 shows a diagram from *Huss* [1991, Figure 1] updated by *Righter et al.* [2006] that shows the number of meteorites collected in these icefields, Roosevelt County, and modern falls versus total mass. If these meteorites were thoroughly examined and put into pairing groups, the number of meteorites after pairing would certainly decrease, though the mass of many individual meteorites would rise. This would shift the curves from each of the field sites shown in Figure 10.5 to the right (increased mass), making it more similar to the Modern Stone Falls measured by *Hughes* [1981], while also dropping the "number of meteorites" on the y axis. In addition, if more of the small modern falls were actually recovered [*Huss*, 1991], the "modern falls" curve would move up and to the left, as the number of meteorites increased and the mass decreased.

10.3.2. Does the Antarctic Population Represent a Complete Sampling of Astromaterials?

An important aspect of the Antarctic population is the fact that, as a whole, it contains 91% ordinary chondrites, as shown by individual field sites that have produced over 1000 meteorites (Figure 10.6 and Table 10.3). This is further evidence that the number of meteorites recovered in a field area is the critical parameter in sampling a representative assemblage of meteorites, far more so than the

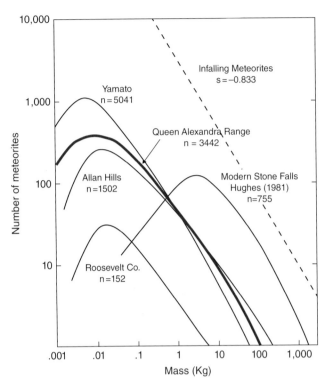

Figure 10.5. Mass frequency diagram of icefields, Roosevelt County, and modern falls vs. total mass (from *Huss* [1991], modified by *Righter et al.* [2006] to include the QUE meteorites). If the Antarctic meteorites were put into pairing groups, the number of meteorites would decrease, with a corresponding increase in individual mass. This would shift the curves from each of the field sites to the right (increased mass), closer to Modern Stone Falls, also reducing the number of meteorites.

specific field site where they are collected. The other 10% of the population is dominated by carbonaceous chondrites, which make up 4% of all U.S. Antarctic meteorites, and achondrites, with HED meteorites composing 2%.

To understand whether the U.S. Antarctic collection is representative of material that fell to Earth over recent history (terrestrial ages of Antarctic meteorites average about 30 Ka, with the oldest around 2 Ma), it is reasonable to compare broad statistics of the collection to other similar collections. Here we choose modern observed meteorite falls [*Grady et al.*, 2000] and hot desert meteorites from Northern Africa [*Welzenbach and McCoy*, 2006, *McBride and Righter*, 2010]. Both the modern falls population and the hot desert meteorite population sample more than 1000 meteorites, perhaps indicative that they should also be a representative sampling.

At a broad level, it is noteworthy that the percentage of ordinary chondrites differs significantly between the Antarctic population (~90%) and the falls and hot desert meteorites (~80%) (Figure 10.6). It is also interesting that while individual Antarctic sites demonstrate dramatic variations in the relative abundances of H, L, and LL chondrites, the cumulative populations of modern falls,

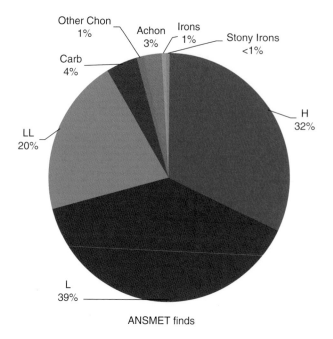

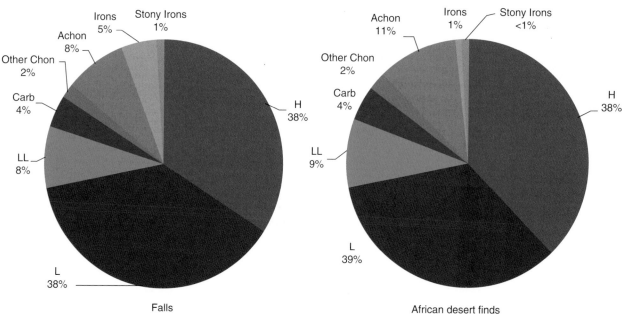

Figure 10.6. A comparison of modern falls, African and Omani desert finds, and Antarctic meteorites (labeled ANSMET finds). Note the similarity between these locations of the abundances of H and L chondrites. Also interesting is the total number of carbonaceous chondrites compared with the variation in numbers of achondrites and iron meteorites.

the U.S. Antarctic meteorites, and hot desert meteorites exhibit relatively constant ratios of these three types of meteorites. In contrast, the hot desert meteorites exhibit significantly higher abundances of rare R chondrites, primitive achondrites, martian meteorites, and lunar meteorites. This almost certainly reflects economic pressures that place high monetary value on rarer meteorites and low value on ordinary chondrites. Thus, the number of ordinary chondrites either collected or, more likely,

classified and described is deflated in the case of hot desert meteorites. If the 90% of ordinary chondrites in the U.S. Antarctic population serves as a guide (Figure 10.6), at least twice as many ordinary chondrites should have been recovered from the hot deserts. Conversely, if the modern falls serve as a guide, the Antarctic population contains roughly twice as many ordinary chondrites as expected, probably reflecting a lack of pairing of ordinary chondrites in the Antarctic

Table 10.5. Antarctic meteorites compared to falls (global) and the hot deserts of Africa.

Meteorite Class	U.S. Antarctic Program	Falls—Global	Oman/North African Deserts
Ordinary chondrites (OC)	17,293	852	6,790
Carbonaceous chondrites (CC)	798	44	381
CM	303	15	23
Enstatite chondrites	128	17	106
R chondrites	27	1	81
Primitive achondrites (PA)	96	9	199
How/Euc/Dio (HED)	314	61	432
Other achondrites	54	11	75
Lunar	24	0	112
Martian	15	5	73
Irons	111	49	67
Total	18,931	1,049	8,316
Total non-OC	1,638	197	1,526
% OC/All Else	91.3	81.2	81.6
% HED/All Else	1.7	5.8	5.2
% CC/non-OC	48.7	22.3	25.0
% CM/CC	38.0	34.1	6.0
% R chon/non-OC	1.6	0.5	5.3
% PA/non-OC	5.9	4.6	13.0
% Martian/non-OC	0.9	2.5	4.8
% Lunar/non-OC	1.5	0.0	7.3
% Irons/non-OC	6.8	24.9	4.4

Note: Primitive achondrites = acapulcoites, lodranites, brachinites, ureilites and winonaites.
Sources: Grady, 2000; Welzenbach and McCoy, 2006; McBride and Righter, 2010.

population coupled with the ease of relatively complete recovery of small stones from the ice.

Perhaps the most striking mismatch between modern falls and the U.S. Antarctic population is in the abundance of iron meteorites. Iron meteorites, with only 111 samples, only represent 0.6% of the entire U.S. Antarctic population (Table 10.5; again, without pairing). This number is surprisingly low when compared with the number of iron meteorites that are collected as falls or finds around the world. Approximately 5% of meteorite falls worldwide are iron meteorites, and 40% of the non-Antarctic finds worldwide are irons [*Grady*, 2000]. Interestingly, the number of irons collected in the hot desert meteorite collections of the world is also low: only 0.8% of the meteorites collected from the hot deserts are irons. These percentages would rise somewhat if the dominant ordinary chondrites were subjected to rigorous pairing, but a factor of 2–3 decrease in the total number of Antarctic meteorites would not increase the percentage of irons at levels comparable to either modern falls or non-Antarctic finds. Also of note are the average masses of irons from both hot and cold deserts. No mass over 25 kg has been returned from the Antarctic ice (though 15 Derrick Peak iron meteorites, the largest of which is 138 kg, were collected from a nunatak, which is solid ground that rises above the surface of the ice; Plate 78). So, are Antarctic irons anomalously low in abundance?

Within the ice flows of Antarctica, the relatively high density of the iron meteorites coupled with their high heat capacity might prevent them from reaching the surface of the moving ice flows and being recovered. But *Nagata* [1978] argues that a specimen would have to be ~0.8 m for this mechanism to apply, which seems inconsistent with the lack of sizes smaller than this as well. Instead, perhaps iron meteorite recoveries are enhanced among the finds and falls in the populated parts of the world (where metallic specimens are distinct from the background terrestrial rock) rather than depleted in Antarctica (where meteorites of any kind are distinctive).

Despite the discrepancies in the numbers and masses of iron meteorites, it is interesting to note that the Antarctic seems to be producing more interesting irons than hot deserts and falls. 40% of the U.S. Antarctic iron meteorites are ungrouped. Ungrouped irons make up 15% of all nondesert irons (and less than 5% of North African irons collected.) The small size and high percentage of ungrouped irons likely go hand in hand, as small irons might be more readily perturbed from their source asteroids in zones of the asteroid belt far from meteorite-delivering resonances [see *Mittlefehldt and McCoy*, 2014 (this volume)]. Aside from the high abundance of ordinary chondrites, Antarctica seems to sample an underabundance of HED achondrites relative to falls and hot deserts.

The apparent underrepresentation of HEDs in Antarctica might simply reflect a statistical overabundance of ordinary chondrites owing to a lack of pairing among the latter. A decrease in unique ordinary chondrites by a factor of 2–3 would increase the percentage of HEDs at levels similar to falls and hot deserts.

At the same time, this decrease would serve to increase the abundance of carbonaceous chondrites in Antarctica to levels that exceed the modern falls. In this sense, the difference between hot and cold deserts is striking. Hot deserts have a paucity of CM chondrites relative to other types of carbonaceous chondrites or the population as a whole. Most of the Antarctic CM chondrites are relatively pristine, with intact fusion crust, and show a young terrestrial age of ~10,000 years, both of which suggest that the Antarctic is a relatively gentle environment that preserves smaller samples. Further, it is likely that the cold storage of Antarctica would preserve hydrated carbonaceous chondrites better than would the hot desert environment, in which both chemical (e.g., rain) and physical (e.g., freeze-thaw) weathering would aggressively destroy such meteorites. Interestingly, type 1 carbonaceous chondrites, in which virtually all mafic silicates have altered to hydrated phyllosilicates, attest to the relative importance of chemical and physical preservation. CM1 chondrites were first discovered in Antarctica and are almost exclusively known from that continent (Plate 14). Their survival suggests that chemical weathering has produced minimal alteration of these indurated samples. In addition, there are two CR1 Antarctic meteorites, and these may be all we have in the world's collections (Plate 27). In contrast, CI1 chondrites are absent, despite the collection of 20,000+ samples. Unlike CM1 chondrites, CI1 chondrites are both heavily aqueously altered and poorly lithified. The absence of chondrules prevents them from having cohesive strength. While there are only five known CI meteorites in the world's collections (Alais, Ivuna, Orgueil, Tonk, and Revelstoke), none have been found in the Antarctic or in the world's hot deserts (with the exception of some possible meta-CI's in the Japanese collection [*Tonui et al.*, 2003; *Zolensky et al.*, 2005]. Given the extremely fragile nature of these meteorites, it could be that even modest physical weathering in the extreme physical conditions in Antarctica effectively destroys these rare samples.

REFERENCES

Cassidy, W. A. (2003), *Meteorites, Ice and Antarctica: A Personal Account*, Cambridge University Press, Cambridge, UK.

Cassidy, W. A., and R. P. Harvey (1991), Are there real differences between the modern falls and the Antarctic finds? *Geochimica et Cosmochimica Acta, 55*, 99–104.

Clarke, R. S., Jr., E. Jarosewich, B. Mason, J. Nelen, M. Gomez, and J. R. Hyde (1970), The Allende, Mexico, meteorite shower, *Smithsonian Contributions to the Earth Sciences, 5*, 53 pp.

Delisle, G., and J. Sievers (1991), Sub-ice topography and meteorite finds near the Allan Hills and the Near Western Ice Field, Victoria Land, Antarctica, *J Geophys. Res., 96, 15*, 577–15.

Delisle, G., I. Franchi, A. Rossi, and R. Wieler (1993), Meteorite finds by EUROMET near Frontier Mountain, North Victoria Land, Antarctica, *Meteoritics, 28*, 126–129.

Grady, M. M. (2000), *Catalogue of Meteorites*. Cambridge University Press, Cambridge, UK.

Harvey, R. P. (1995), Moving targets: The effect of supply, wind movement, and search losses on Antarctic meteorite size distributions, in *Workshop on Meteorites from Cold and Hot Deserts*, edited by L. Schultz, J. O. Annexstad, and M. E. Zolensky, Lunar and Planetary Institute Technical Report, 95-02, 34–36.

Harvey, R. P. (2003), The origin and significance of Antarctic meteorites, *Chemie der Erde, 63*, 93–147.

Harvey, R. P., and W. A. Cassidy (1989), A statistical comparison of Antarctic finds and modern falls: Mass frequency distributions and relative abundance by type, *Meteoritics, 24*, 9–14.

Harvey, R. P., J. Schutt, and J. M. Karner (2014), Fieldwork methods of the U.S. Antarctic Search for Meteorites program, in *35 Seasons of U.S. Antarctic Meteorites (1976–2010): A Pictorial Guide to the Collection, Special Publication 68*, edited by K. Righter, C. M. Corrigan, R. P. Harvey, and T. J. McCoy, American Geophysical Union/John Wiley & Sons, Washington, D. C.

Hughes, D. W. (1981), Meteorite falls and finds: Some statistics, *Meteoritics, 25*, 41–56.

Huss, G. R. (1991), Meteorite mass distributions and differences between Antarctic and non-Antarctic meteorites, *Geochim. Cosmochim. Acta, 55*, 105–111.

Kojima, H. (2006), The history of Japanese Antarctic meteorites, *Geological Society, London, Special Publications, 256*, 291–303.

Lindstrom, M., and R. Score (1995), Populations, pairings and rare meteorites in the US Antarctic meteorite collection (abstract), in *Workshop on Meteorites from Cold and Hot Deserts*, edited by L. Schultz, J. O. Annexstad, and M. E. Zolensky, *LPI Technical Report* 95-02, 43–44, Lunar and Planetary Institute, Houston, Texas.

Liu, J.-Z., Y.-L, Zou, C.-L. Li, L. Xu, and Z.-Y. Ouyang (2004), A study on the recovery and classification of meteorites from the Mt. Grove region of Antarctica, *Chin. J. Astron. Astrophys, 4*, 166–175.

Lunning, N. G., C. M. Corrigan, L. C. Welzenbach, and T. J. McCoy (2012), Using immersion oils to classify equilibrated ordinary chondrites from Antarctica, *Lunar and Planetary Science Conference, XLIII*, #1566.

Mawson, D. (1915), *The Home of the Blizzard* (Vol. *II*, p. 11), William Heinemann, London.

McBride, K. M., and K. Righter (2010), Comparison of U.S. Antarctic meteorite collection to other cold and hot desert and modern falls. *Meteoritics and Planetary Science, 45*, suppl. #5343.

McCoy, T. J. (2014), Meteorite misfits: Fuzzy clues to solar system processes, in *35 Seasons of U.S. Antarctic Meteorites (1976–2010): A Pictorial Guide to the Collection, Special Publication 68,* edited by K. Righter, C. M. Corrigan, R. P. Harvey, and T. J. McCoy, American Geophysical Union/ John Wiley & Sons, Washington, D.C.

Michlovich, E. S., S. F. Wolf, M. Wang, S. Vogt, D. Elmore, and M. E. Lipschutz (1995), Chemical studies of H chondrites: 5. Temporal variations of sources, *J. Geophys. Res., 100,* 3317–3333.

Mittlefehldt, D. W., and T. J. McCoy (2014), Achondrites and Irons: Products of Magmatism on Strongly Heated Asteroids, in *35 Seasons of U S. Antarctic Meteorites (1976–2010): A Pictorial Guide to the Collection, Special Publication 68,* edited by K. Righter, C. M. Corrigan, R. P. Harvey, and T. J. McCoy, American Geophysical Union/John Wiley & Sons, Washington, D. C.Nagata, T. (1978), A possible mechanism of concentration of meteorites within the meteorite ice field in Antarctica, *Memoirs of National Institute of Polar Research,* 8(Special issue), 70–92,

Popova, O. P., and 58 others including the Chelyabinsk Airburst Consortium (2013), Chelyabinsk airburst, damage assessment, meteorite recovery, and characterization, *Science Express,* doi: 10.1126/science.1242642.

Righter K., R. P. Harvey, J. Schutt, C. Satterwhite, K. M. McBride, T. J. McCoy, and L. Welzenbach (2006), The great diversity of planetary materials returned from the Queen Alexandra Range, Antarctica, *Meteoritics and Planetary Science, 41*(Suppl.), #5363.

Score, R., and M. M. Lindstrom (1990), Guide to the U.S. collection of Antarctic meteorites 1976–1988 (Everything you wanted to know about the meteorite collection), *Antarctic Meteorite Newsletter, 13, 1.*

Scott, E. R. D. (1989), Pairing of meteorites from Victoria Land and the Thiel Mountains, Antarctica, *Smithsonian Contribution to the Earth Sciences, 28,* 103–111.

Tonui E. K., M. E. Zolensky, M. E. Lipschutz, M-S. Wang, and T. Nakamura (2003), Yamato 86029: Aqueously altered and thermally metamorphosed CI-like chondrite with unusual textures. *Meteoritics and Planetary Science 38,* 269–292.

Weisberg, M., and K. Righter (2014), Primitive asteroids: Expanding the range of known primitive materials, in *35 Seasons of U.S. Antarctic Meteorites (1976–2010): A Pictorial Guide to the Collection, Special Publication 68,* edited by K. Righter, C. M. Corrigan, R. P. Harvey, and T. J. McCoy, American Geophysical Union/John Wiley & Sons, Washington, D. C.

Welten, K. C., M. W. Caffee, D. J. Hillegonds, T. J. McCoy, J. Masarik, and K. Nishiizumi (2011), Cosmogenic radionuclides in L5 and LL5 chondrites from Queen Alexandra Range, Antarctica: Identification of a large L/LL5 chondrite shower with a pre-atmospheric mass of approximately 50,000 kg, *Meteoritics and Planetary Science, 46,* 177–196.

Welzenbach, L. C., and T. J. McCoy (2006), Meteorites from hot and cold deserts: What's there, what's missing, and why we should care, *Meteoritics and Planetary Science, 41*(Suppl.), A215.

Wolf, S. F., and M. E. Lipschutz (1995a), Chemical studies of H chondrites: 4. New data and comparison of Antarctic suites, *J. Geophys. Res., 100,* 3297–3316.

Wolf, S. F., and M. E. Lipschutz (1995b), Chemical studies of H chondrites: 6. Antarctic/non-Antarctic compositional differences revisited, *J. Geophys. Res., 100,* 3335–3349.

Yamaguchi, A., N. Imae, H. Kojima, S. Ozawa, and M. Kimura (2012), Curation of Antarctic meteorites at the National Institute of Polar Research, *Antarctic Meteorites, XXXV, 64.*

Yoshida, M. (2010), Discovery of the Yamato meteorites in 1969, *Polar Science, 3,* 272–284.

Zolensky, M., P. Abell, and E. Tonui (2005), Metamorphosed CM and CI carbonaceous chondrites could be from the breakup of the same Earth-crossing asteroid, *Lunar and Planetary Science Conference, XXXVI,* #2084.

INDEX

Acapulco (meteorite), 81
Acapulcoite-lodranite clan, 81–82
Acapulcoites, 81–82
Accelerator mass spectrometry (AMS), 153–54
Achondrites, 8, 9f, 79–95
  chondrite links with, 148
  interest in, 48, 49f
  primitive, acapulcoite-lodranite clan of, 81–82
Advanced Very High Resolution Radiometer (AVHRR), 30
AI-rich chondrules, 145
^{26}Al, 155–56, 156f
  Antarctic Meteorite Collection, U.S. and, 157–58
  T_{Terr} and, 158
ALH A77307, 74
ALH A81189, 73
  presolar grains in, 74
ALH 84001, 132, 134, 137–39, 137f, 138f, 161, 178
ALH 84025, 87, 88
ALH 85151, 65
ALH A77005, 134–36, 135, 136, 139, 164, 164t
  T_{Terr} ages, 164, 164t
ALH A77283, 10, 10f
ALH A78019, 84, 85f
ALH A78262, 84
ALH A81001, 92
ALH A81004, 112f
ALH A81005, 21, 21f, 53–4, 53f, 54f, 108, 131, 165–66, 178
  composition of, 109
  CRE age of, 166
Allan Hills, 6, 7f. *See also specific ALH meteorites*
  ANSMET season III, 14–15, 14f
  ANSMET season VI, 18–20
  Battlements Nunatak Icefield, 19
  Expedition A, 9
  Far Western Icefield, 19
  geodetic network at, 11, 12f
    second remeasurement of, 16–17, 17f
  icefields, 19–20, 19f
  Main Icefield, 8, 19
  Near Western Icefield, 19, 20f
  productivity, 181
AMS. *See* Accelerator mass spectrometry
Anderson, Duwayne, 5
Angra dos Reis (meteorite), 92, 94–95
Angrites, 92–95
  compositions, 93–94
  LEW 86010, 94

LEW 87051, 94
  parent body, 94
  plutonic, 93
  volcanic, 93
Annexstad, John O., 7, 11, 15–17
  curation and, 43
ANSMET. *See* Antarctic Search for Meteorites Program, U.S.
Antarctica
  expeditions
    Japanese, 4
    U.S., 4–5
  HEDs, 90, 185–86
  meteorites, 2f, 23–24, 79
    cosmogenic nuclides in, 153–68
    history of, 1–2
    hot desert meteorites compared with, 183–84, 184f, 185t
    iron, 82
    Japanese interest in, 3–4
    martian, 131–40
    mass distribution of, 183, 183f
    modern recovery efforts, 23–30
    number of, by field site, 176t
    number of, by type, 174t
    number of, by year, 175t
    pairings of, 178, 178t
    searching for, 31–35
    T_{Terr} for, 155–56, 156t
Antarctic Meteorite Collection, U.S., 66f
  ^{26}Al surveys of, 157–58
  astromaterials represented by, 183–86, 184f, 185t
  chondrites, 65
  classification, 174–75
  collection site statistics and, 181–86
  curation and allocation of samples in, 43–61
  curators, 44t
  HEDs, 80
  meteorites from strongly heated asteroids, 80–81
  NTL surveys of, 157–58
  number of, by year, 175t
  output by season and site, 178–81
  pairings in, 178, 178t
  statistical assessments, 173
  statistical look at, 173–86
  ureilites, 80
Antarctic meteorite collections, 174
Antarctic meteorite lab, 51
Antarctic Meteorite Newsletter, 8, 46
Antarctic Program, U.S. (USAP), 24

35 Seasons of U.S. Antarctic Meteorites (1976–2010): A Pictorial Guide to the Collection, Special Publication 68,
First Edition. Edited by Kevin Righter, Catherine M. Corrigan, Timothy J. McCoy and Ralph P. Harvey.
© 2015 American Geophysical Union. Published 2015 by John Wiley & Sons, Inc.